# Studies in Computational Intelligence

Volume 533

*Series editor*

Janusz Kacprzyk, Polish Academy of Sciences, Warsaw, Poland
e-mail: kacprzyk@ibspan.waw.pl

For further volumes:
http://www.springer.com/series/7092

*About this Series*

The series "Studies in Computational Intelligence" (SCI) publishes new developments and advances in the various areas of computational intelligence—quickly and with a high quality. The intent is to cover the theory, applications, and design methods of computational intelligence, as embedded in the fields of engineering, computer science, physics and life sciences, as well as the methodologies behind them. The series contains monographs, lecture notes and edited volumes in computational intelligence spanning the areas of neural networks, connectionist systems, genetic algorithms, evolutionary computation, artificial intelligence, cellular automata, self-organizing systems, soft computing, fuzzy systems, and hybrid intelligent systems. Of particular value to both the contributors and the readership are the short publication timeframe and the world-wide distribution, which enable both wide and rapid dissemination of research output.

Jiuping Xu · Ziqiang Zeng

# Fuzzy-Like Multiple Objective Multistage Decision Making

Springer

Prof. Dr. Dr. Jiuping Xu
Uncertainty Decision-Making Laboratory
Sichuan University
Chengdu
People's Republic of China

Asiss. Prof. Dr. Ziqiang Zeng
Uncertainty Decision-Making Laboratory
Sichuan University
Chengdu
People's Republic of China

ISSN 1860-949X ISSN 1860-9503 (electronic)
ISBN 978-3-319-34314-3 ISBN 978-3-319-03398-3 (eBook)
DOI 10.1007/978-3-319-03398-3
Springer Cham Heidelberg New York Dordrecht London



Printed on acid-free paper

Springer is part of Springer Science+Business Media (www.springer.com)

# Preface

Since Richard Bellman introduced the concept of dynamic programming in the 1940s to describe the process of solving problems where it was necessary to find the best decisions in sequence, multistage decision making has been found in many important practical decision making problem applications, such as problems in management science, operations research, engineering management, and construction planning. With the rapid development of science and technology, the application of appropriate multistage decision making has been playing an increasingly important role. Particular examples can be found in manufacturing, construction, transportation, and engineering management. In most real-world problems, decision makers face multiple and conflicting objectives in a hybrid uncertain environment where fuzziness and randomness co-exist, so they usually need to make decisions sequentially at different points in time and space and at different levels for components or systems. This complex decision making environment has motivated the authors to write a book examining fuzzy-like multiple objective multistage decision making.

The need to address uncertainty in multiple objective multistage decision-making is widely recognized, as uncertainties exist in many system components. As a result, the inherent complexity and uncertainty existing in real-world multiple objective multistage decision making problems has essentially placed them beyond conventional deterministic optimization methods. Fuzzy set theory was developed to solve problems where the descriptions of the activities and observations are imprecise, vague, and uncertain. The term "fuzzy" refers to a situation in which there are not well-defined boundaries to the set of activities or observations to which the descriptions apply. This fuzziness exists in many practical applications, such as in equipment failure rate, effective monthly working days, transportation and stay time, arrival rate of vehicles, and facilities servicing rates. This has led to the development of the concept of fuzzy multiple objective multistage decision making. In some other cases, especially in large scale construction projects, the practical problems in the decision making process often need to be considered in a hybrid uncertain environment where fuzziness and randomness co-exist. For example, over the years, the advancements in engineering technology have provided more alternatives for construction operations and, subsequently, decision makers have been faced with increasingly more imprecise information. Therefore, fuzzy random variables that can take into account both fuzziness and randomness are becoming more favored by decision makers to describe the uncertainty and vague information encountered in reality. The complexity of these decision making problems has led to the development of fuzzy random multiple objective multistage decision making. Generally, for convenience, the fuzzy multiple objective multistage decision making and the fuzzy random multiple objective multistage decision making are called by the combined name fuzzy-like multiple objective multistage decision making.

This book uses real-life dynamic decision making problems to develop the general framework for the fuzzy-like multiple objective multistage decision making, and includes the basic theories behind the framework, the modeling process, the model analysis and the algorithm design. In addition, applications of this methodology to realistic world problems are given. The issues selected are focused on engineering applications, including dynamic machine allocation, closed multiclass queuing network optimization, inventory management, facilities planning, and transportation assignments.

The book has 7 chapters which explain the different aspects of the fuzzy-like multiple objective multistage decision making principles. Chapter 1 focuses on the description of the nature of multiple objective multistage decision-making, and presents state-of-the-art surveys of its theoretical and methodological advancements. The first section discusses the physical structure of the general model and subsequently, the mathematical foundation is explained in detail including the principle of optimality and maximum principle. The three typical problems in this field; multiple objective multistage decision making under a specified termination time, multiple objective multistage decision making under an implicit termination time, and multiple objective multistage decision making under an infinite termination time, are respectively presented.

Based on the concepts introduced in Chapter 1, Chapter 2 extends the multiple objective multistage decision making into fuzzy-like multiple objective multistage decision making. The basic theories for fuzzy sets, fuzzy variables, and fuzzy random variables are reviewed. Then, the methodology for fuzzy-like multiple objective multistage decision making including the development of the research paradigm, the nature of the space-time trade-off, functional space, and future prospects are discussed.

In Chapter 3, a statement for a dynamic machine allocation problem is presented as a practical application of fuzzy multiple objective multistage decision making. The modeling process for the dynamic machine allocation problem is explained and then a model analysis for a dynamic machine allocation problem is discussed, for which a theoretical algorithm is designed for small scale problems. To solve large scale problems, a dynamic programming based particle swarm optimization algorithm is developed. Finally, a case study for the Shuibuya Hydropower Project in China is given as an application example.

Chapter 4, which has a similar structure to Chapter 3, mainly focuses on multistage decision making for closed multiclass queuing network problems under a fuzzy environment. In this chapter, the multistage queuing decision problem in concrete transportation systems is discussed. Specifically, the transportation queuing network in a hydropower construction project concrete pouring system is a closed multiclass system which has limited vehicles and unloading equipment. Decision makers need to dynamically allocate the vehicles and unloading equipment to the queuing networks to minimize total operational costs and construction duration. The properties of the closed multiclass queuing networks are analyzed and a multistage optimal control model is developed for the scheduling by considering the multiple objectives, the allocation constraints, and the uncertainties in the system. To deal with the uncertainties, a fuzzy expected value concept is introduced and its equivalent crisp model derived. Using particle swarm optimization, which utilizes both local and global experiences during the search process, the particular nature of the model motivates the development of an antithetic method-based particle swarm optimization algorithm. This antithetic particle-updating mechanism is used rather than the traditional updating method to automatically control the particle-updating in the feasible solution space. The results and a sensitivity analysis for the Jinping-I Hydropower Project, China, are presented to highlight the performance of the optimization method, which is proved to be very effective and efficient compared to the actual data from the project and the standard PSO algorithm.

The main objective of Chapter 5 is to apply fuzzy random multiple objective multistage decision making to inventory management. In this chapter, an inventory management problem for a large-scale construction project is presented. The modeling process for the large-scale construction project inventory management problem is explained and then a model analysis for this problem is discussed in which a theoretical algorithm is designed for small scale problems. To solve larger scale problems, a dynamic programming based particle swarm optimization algorithm is developed. Finally, a case study for the China Yangtze Three Gorges Project is given as an application example.

Chapter 6 presents an application of fuzzy random multiple objective multistage decision making for facilities planning. The facilities planning problem has two important parts; site layout planning and facilities location. This chapter focuses on the current frontiers of these two problems; dynamic construction site layout planning and dynamic temporary facilities location problems.

Chapter 7 primarily focuses on a discussion of the multistage decision making for a two stage-based dynamic transportation assignment problem under a fuzzy random environment. Dynamic transportation assignment problems are often encountered in many practical systems, such as urban planning traffic networks, closed traffic network systems, field service support systems, container transportation and flow-shop-type production systems. Because of the quantity limitations and other constraints, decision makers need to determine a suitable allocation by dynamically allocating traffic to ensure that the total operational cost, transportation duration and total cost are minimized. Specifically, for large scale hydropower projects, several transportation carrier types and multiple stages are proposed, which indicates that the two stage-based dynamic transportation assignment problem is a multi-objective dynamic programming process.

Fuzzy-like multiple objective multistage decision making is an effective methodology for handling decision making problems. This book attempts to integrate multiple objective multistage decision making and fuzzy-like theory and explain their practical applications. While practical applications are the most important content in this book, the theoretical discussion, especially for the functional space for fuzzy-like multiple objective multistage decision making, still needs further research. Future discussions could focus on compact support, fixed compact support, properties, locality, semi-locality, positive powers, duals, and the normalizing of the functional space for fuzzy-like multiple objective multistage decision making. From the application aspect, in the modern world there are more practical problems which have increasing complexity. The description of these problems using multiple objective multistage decision making may also be an interesting focus in the future.

With its emphasis on problem solving and practical application, this book is ideal for researchers, practitioners, engineers, graduate students, and upper-level undergraduates with civil engineering backgrounds in applied mathematics, management science, operations research, engineering management, construction planning, and transportation optimization.

This monograph has been supported by the Key Program of the National Natural Science Foundation of China (Grant No. 70831005), and the Research Foundation of the Ministry of Education for the Doctoral Program of Higher Education of China (Grant No. 20130181110063). Authors take this opportunity to thank Prof. Janusz Kacprzyk for his kind support and valuable insights and information, from whom the authors have received significant enlightenment in the development of the multistage fuzzy control theory. For discussion and advice, the authors also would like to thank researchers from the Uncertainty Decision Making Laboratory of Sichuan University, particularly, L.M. Yao, Z.M. Tao, Z.M. Li, Q.R. Liu, J. Meng, C.Y. Feng, who have done intensive work in this field and have made a number of corrections. Finally, the authors express their deep gratitude to all of Springer's editorial staff.

Sichuan University, Jiuping Xu
September 2013 Ziqiang Zeng

# Contents

# List of Tables

# List of Figures

# Acronym

| | |
|---|---|
| AM | antithetic method |
| CAD | computer aided design |
| CFRD | concrete faced rockfill dam |
| CMQNP | closed multiclass queueing networks problem |
| CPSO | contraction particle swarm optimization |
| DCSLP | dynamic construction site layout planning |
| DMAP | dynamic machine allocation problem |
| DP | dynamic programming |
| DTAP | dynamic transportation assignment problem |
| DTFLP | dynamic temporary facilities location problem |
| EETT | endurable earliness termination time |
| EF | equipment flows |
| EHDC | Ertan Hydropower Development Company, Ltd. |
| ELTT | endurable lateness termination time |
| EVM | expected value model |
| F-MOMSDM | fuzzy multiple objective multistage decision making |
| FEVM | fuzzy expected value model |
| FPP | facilities planning problem |
| FPWT | failure probability-work time |
| FR-MOMSDM | fuzzy random multiple objective multistage decision making |
| FRV | fuzzy random variable |
| FST | fuzzy set theory |
| GA | genetic algorithm |
| IF | information flows |
| IMP-LCP | inventory management problem for large-scale construction projects |
| MF | material flows |
| MODP | multi-objective dynamic programming |
| MOGA | multiple objective genetic algorithm |
| MOMSDM | multiple objective multistage decision making |
| MOPSO | multi-objective particle swarm optimization algorithm |

| | |
|---|---|
| MTTW | mean time to work |
| PDF | probability density function |
| PF | personnel flows |
| PMX | partially mapped crossover |
| PSO | particle swarm optimization |
| SA | simulated annealing |
| SE | safety/environment concerns |
| TGP | Three Gorges Project |
| TS | two stage |
| UP | users' preference |
| YRWHDC | Yellow River Water and Hydroelectric Power Development Corporation |

# 1
# Multiple Objective Multistage Decision Making

Decision making has been a focus of reflection by many thinkers since ancient times. The great philosophers, such as Aristotle, Plato, and Thomas Aquinas, discussed the capacity of human decision making claiming that it is this capacity that distinguishes us from animals (Figueira et al. 2005 [163]). With the modernization of both scientific methods and society, appropriate decision making has become increasingly important in many areas of human activity, including engineering, management, and economic development. In most practical problems, decisions have to be made sequentially at different points in time and space, and at different system levels. These problems where decisions are made sequentially are called sequential decision problems (Bellman 1957 [164]; Bellman and Dreyfus 1962 [165]). Since these decisions are made in a number of stages, they are also referred to as multistage decision problems (Rao 2009 [166]). At the same time there is an increasing awareness in decision makers so that it is necessary to incorporate multiple and conflicting objectives in the multistage decision processes. This has led to the development of the concept of multiple objective multistage decision making (MOMSDM). This chapter focuses on a description of the nature of the MOMSDM, and presents state-of-the-art surveys on its theoretical and methodological advancements. The first section discusses the physical structure of a general MOMSDM model. Then, the MOMSDM mathematical foundation is explained in detail including the principle of optimality and the maximum principle. Three typical problems, i.e., a MOMSDM under a specified termination time, a MOMSDM under an implicit termination time, and a MOMSDM under an infinite termination time, are respectively presented.

## 1.1 Elements of General Model

Multistage decision making processes can be separated into a number of sequential steps, or stages, which are completed in one or more ways. The options for completing these stages are known as decisions. The condition of the process at a given

J. Xu and Z. Zeng, *Fuzzy-Like Multiple Objective Multistage Decision Making*, Studies in Computational Intelligence 533,
DOI: 10.1007/978-3-319-03398-3_1, © Springer International Publishing Switzerland 2014

stage is known as the state at that stage; each decision affects the transition from the current state to a state associated with the next stage.

In modern life, especially in engineering, economic, industrial, scientific and even political spheres, there are many MOMSDM processes. MOMSDM modeling has long been applied to many areas in the fields of mathematics, science, engineering, business, medicine, information systems, biomathematics, artificial intelligence, and others. MOMSDM applications have increased as recent advances have been made in areas such as scheduling problems in large scale construction projects, water resource management, reservoir operations, forest resource management, power system optimization, and some areas where computational intelligence is required.

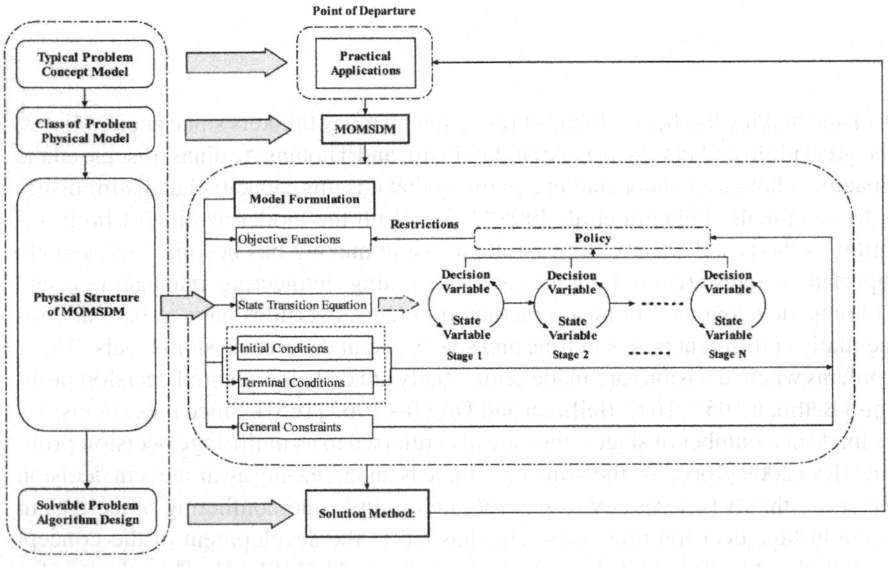


**Fig. 1.1** Physical structure of MOMSDM

For all the above applications, there are commonly 9 elements in the models, all of which characterize the MOMSDM's physical structure. These elements are, respectively, stage, termination time, state, decisions, policies, initial and terminal conditions, general constraints, state transition equations and objective functions. When applied to practical problems, a multiple objective multistage decision making process is one in which a number of single-stage processes are connected in a series so that the output of one stage is the input of the succeeding stage. Strictly speaking, this type of process should be called a serial multistage decision process since the individual stages are connected head to tail with no recycling. Serial multiple objective multistage decision problems arise in many types of practical problems. Figure 1.1 illustrates the physical structure of the MOMSDM. Details of the

MOMSDM elements are discussed in the following subsections. Here, we define the symbols for the MOMSDM as follows:

*Subscripts*

$i$= objective functions index, where $i = 1,2,\cdots,m$;

$j$= general constraint functions index, where $j = 1,2,\cdots,p$;

$k$= stages index, where $k = 1,2,\cdots,T$, $T$ is the termination time;

*Variables*

$\mathbf{x}(k)$= state vector in Stage $k$;

$\mathbf{u}(k)$= decision vector in Stage $k$;

*Functions*

$J_i(\mathbf{x}(0),\mathbf{x}(T),\pi)$= objective function of MOMSDM;

$g_j(\mathbf{x}(\cdot),\mathbf{u}(\cdot))$= general constraint functions of MOMSDM;

$f_k(\mathbf{x}(k),\mathbf{u}(k))$= state transition function of MOMSDM;

*Spaces*

$I$= initial state space;

$X$= state space;

$U$= decision space;

$\pi = (\mathbf{u}(0),\mathbf{u}(1),\cdots,\mathbf{u}(T))$ is a policy for the complete decision making process;

$U(\pi)$= termination state space.

There are three cases for the multiple objective multistage decision making model:

(1) MOMSDM under a specified termination time

If the termination time $T$ is fixed in advance (i.e., $T = a$, where $a$ is the stage amount), this means that the number of stages in the decision making process has been previously determined. In this case, the decision making is called MOMSDM under a specified termination time, and its general model is expressed as:

$$\begin{cases} \max \{J_1(\mathbf{x}(0),\mathbf{x}(T),\pi), J_2(\mathbf{x}(0),\mathbf{x}(T),\pi),\cdots,J_n(\mathbf{x}(0),\mathbf{x}(T),\pi)\} \\ \text{s.t.} \begin{cases} \mathbf{x}(k+1) = f_k(\mathbf{x}(k),\mathbf{u}(k)), \\ \mathbf{x}(0) \in I, \\ T = a, \\ g_j(\mathbf{x}(\cdot),\mathbf{u}(\cdot)) \le 0,\ j = 1,2,\cdots,p, \\ \mathbf{x}(k) \in X,\ k = 1,2,\cdots,T, \\ \mathbf{u}(k) \in U,\ k = 1,2,\cdots,T. \end{cases} \end{cases} \tag{1.1}$$

A typical example problem of a MOMSDM under a specified termination time can be found in construction management, production operations, and transportation planning. Figure 1.2 (Xu and Wei 2012 [167]) shows an application example of a MOMSDM for the concrete production and distribution planning in a construction supply chain under a specified termination time. The state of the construction supply chain is transferred stage by stage, and ends at a fixed termination time. Objectives for the producer, distributor, and customer are considered for the optimization of the construction materials supply in multiple stages. Constraints for cement transport to the distributor and the customer are also taken into account in each stage.

(2) MOMSDM under an implicit termination time

In a more general case, the termination time $T$ can be assumed to be determined implicitly by a subsidiary condition of the form $\mathbf{x}(T) \in U(\pi)$, where $U(\pi)$ is the

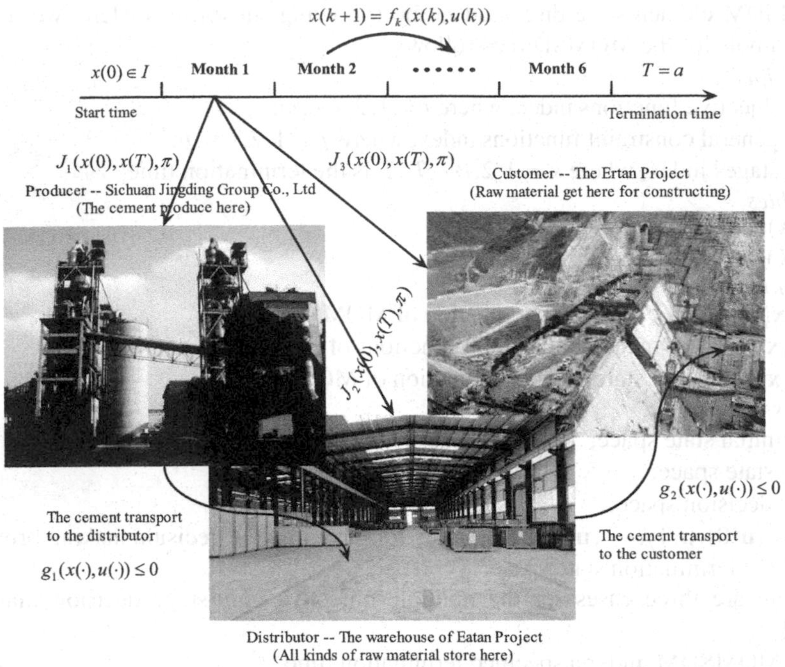


**Fig. 1.2** Application example of MOMSDM under a specified termination time

termination state space. In this case, the process terminates when the state of the system under control enters, for the first time, a specified subset of the state space (i.e., $U(\pi)$). Thus, its general model is expressed as:

$$\begin{cases} \max \{J_1(\mathbf{x}(0),\mathbf{x}(T),\pi), J_2(\mathbf{x}(0),\mathbf{x}(T),\pi), \cdots, J_n(\mathbf{x}(0),\mathbf{x}(T),\pi)\} \\ \text{s.t.} \begin{cases} \mathbf{x}(k+1) = f_k(\mathbf{x}(k),\mathbf{u}(k)), \\ \mathbf{x}(0) \in I, \\ \mathbf{x}(T) \in U(\pi), \\ g_j(\mathbf{x}(\cdot),\mathbf{u}(\cdot)) \le 0, \ j = 1,2,\cdots,p, \\ \mathbf{x}(k) \in X, \ k = 1,2,\cdots,T, \\ \mathbf{u}(k) \in U, \ k = 1,2,\cdots,T. \end{cases} \end{cases} \tag{1.2}$$

First discussed by Bellman and Zadeh (1970) [285], practical applications for this model have been widely employed in infrastructure management, equipment allocation and project scheduling. Take a hydropower construction project as an example. Usually, in a hydropower construction project, the concrete production system has several groups of concrete mixing buildings which are located at different sites on the dam area. To construct these concrete mixing buildings efficiently, several construction teams are dispatched in multiple stages. The state of the system is the

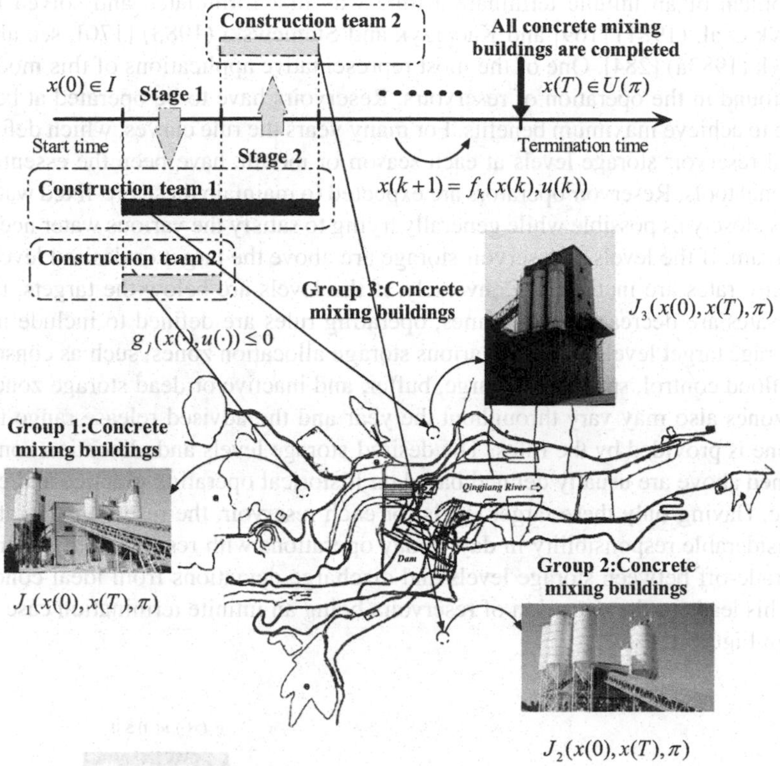


**Fig. 1.3** Application example of MOMSDM under an implicit termination time

number of finished concrete mixing buildings. Thus the implicit termination occurs when all concrete mixing buildings are completed as shown in Figure 1.3.

(3) MOMSDM under an infinite termination time

Formulations with an infinite termination time are mostly used for cases with long planning horizons, such as water resource management (Nandalal and Bogardi 2007 [168]), urban planning and development and outer space exploration. The number of stages in this case is assumed to be $T = \infty$, therefore, its general model is expressed as:

$$\begin{cases} \max \{J_1(\mathbf{x}(0),\mathbf{x}(T),\pi), J_2(\mathbf{x}(0),\mathbf{x}(T),\pi), \cdots, J_n(\mathbf{x}(0),\mathbf{x}(T),\pi)\} \\ \text{s.t.} \begin{cases} \mathbf{x}(k+1) = f_k(\mathbf{x}(k),\mathbf{u}(k)), \\ \mathbf{x}(0) \in I, \\ T = \infty, \\ g_j(\mathbf{x}(\cdot),\mathbf{u}(\cdot)) \leq 0, \ j = 1,2,\cdots,p, \\ \mathbf{x}(k) \in X, \ k = 1,2,\cdots, \\ \mathbf{u}(k) \in U, \ k = 1,2,\cdots. \end{cases} \end{cases} \tag{1.3}$$

The problem of an infinite termination time was first formulated and solved by Kacprzyk et al. (1981) [169] and Kacprzyk and Staniewski (1983) [170]; see also Kacprzyk (1983a) [284]. One of the most representative applications of this model can be found in the operation of reservoirs. Reservoirs have to be operated at best practice to achieve maximum benefits. For many years the rule curves, which define the ideal reservoir storage levels at each season or month, have been the essential operational tools. Reservoir operators are expected to maintain these pre-fixed water levels as closely as possible while generally trying to satisfy the various water needs downstream. If the levels of reservoir storage are above the target or desired levels, the release rates are increased. Conversely, if the levels are below the targets, the release rates are decreased. Sometimes, operating rules are defined to include not only storage target levels, but also various storage allocation zones, such as conservation, flood control, spill or surcharge, buffer, and inactive or dead storage zones. Those zones also may vary throughout the year and the advised release range for each zone is provided by the rules. The desired storage levels and allocation zones mentioned above are usually defined based on historical operating practice and experience. Having only these target levels for each reservoir, the reservoir operator has considerable responsibility in day-to-day operations with respect to the appropriate trade-off between storage levels and discharge deviations from ideal conditions. This leads to the operation of reservoirs being an infinite termination case as shown in Figure 1.4.

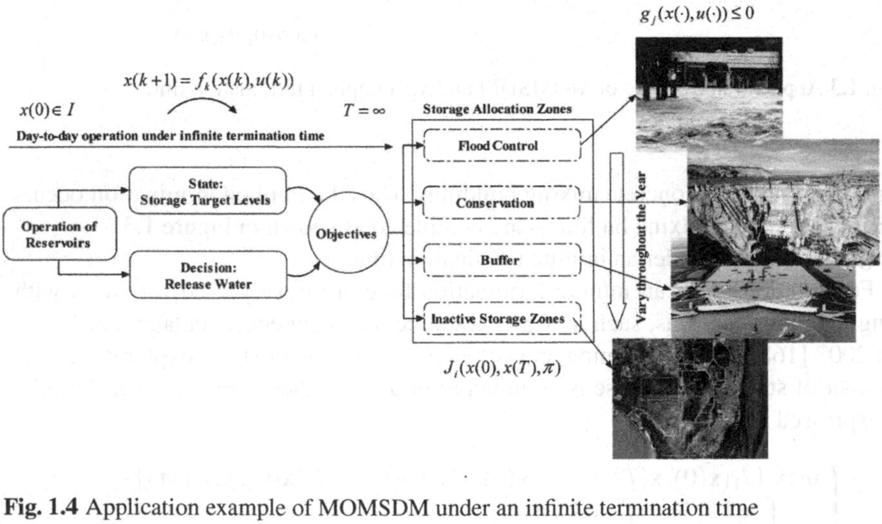


**Fig. 1.4** Application example of MOMSDM under an infinite termination time

In the following sections some basic knowledge is given about the elements which characterize the features of MOMSDM, including the stage, termination time, state, decision, policy, initial and terminal conditions, general constraints, state transition equations and objective functions.

### 1.1.1 Stage and Termination Time

When a decision maker makes multistage decisions, it is important to know how many stages the decision making process has, and when the decision making process ends. This part will discuss the relative elements of the MOMSDM, i.e, stage and termination time.

#### Stage

Generally, the decision making process in a system can be divided into several sequential time periods. Every time period is said to be a stage. To emphasize the sequence of decisions, the concept of the stage is important. The definition of a stage in MOMSDM is as follows,

**Definition 1.1.** *(Kacprzyk 1983 [284]) A stage in multiple objective multistage decision making is said to be a sequential time period in the decision making process of a system.*

As applied to dynamic programming, a multistage decision process is one in which a number of single-stage processes are connected in a series so that the output of one stage is the input of the succeeding stage. Strictly speaking, this type of process should be called a serial multistage decision process since the individual stages are connected head to tail with no recycling. Serial multistage decision problems arise in many types of practical problems. A few examples are given below and many others can be found in the literature.

*Example 1.1.* In large scale construction projects, the contractors usually use the following policy to improve construction throughput within limited construction durations. The complete construction duration is divided into several periods. Preventive maintenance is done during the interval between every two adjacent periods. Therefore, equipment that breaks down in a certain period cannot resume work until preventive maintenance has been conducted. Moreover, every period is divided into several stages. Since the span of a period may range from one week to several months depending on the type of equipment and construction project, the span of a stage may accordingly range from one day to several weeks. This policy is widely used in concrete faced rockfill dam (CFRD) construction projects. For the dump trucks, the span of a period is usually about one month, and the span of a stage is around one week. The construction policy for CFRD construction projects is illustrated in Figure 1.5.

In real world problems, the stage variables are usually discrete. Thus, it is possible to use integers, i.e., 1, 2, $\cdots$, $T$. In fact, a MOMSDM is said to be discrete if the decision making process rather than the state variables is discrete. There are also cases where the stage variables are continuous, so decisions can be made any time. For example, some Markov Decision Processes can be characterized as continuous.

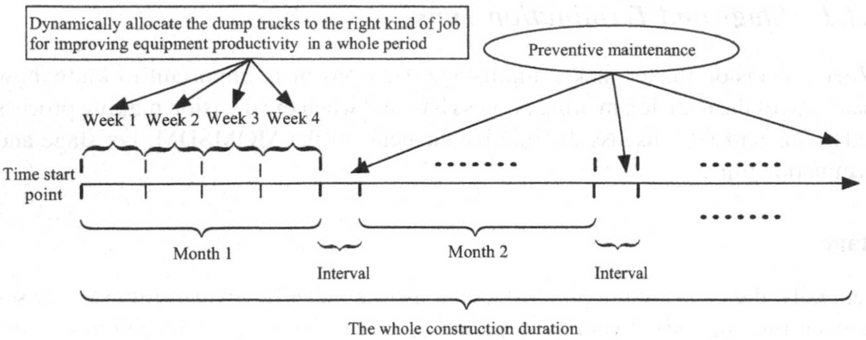


**Fig. 1.5** Construction strategy in CFRD construction projects

### Termination Time

The termination time $T$ is the end time period for the complete decision making process of a system. There are three termination time conditions: (i) a fixed and specified termination time; (ii) an implicit termination time; (iii) an infinite termination time.

If the system under control is a time-invariant finite-state deterministic system, then the termination time $T$ is fixed a priori. In this case, $\{1,2,\cdots,N\}$ are usually used to number the decision process stages

In a more general case, the termination time $T$ is often assumed to be determined implicitly by a subsidiary condition of the form $\mathbf{x}_T \in \Omega$, where $\Omega$ is a specified subset of the state space termed the termination set. Thus, the process terminates when the state of the system under control enters, for the first time, a specified subset of the state space. This case plays an important role in optimal control theory and in Markovian decision processes [285]. Some of the more relevant research on this subject are cited in the list of references.

The problem of an infinite termination time was first formulated and solved by Kacprzyk et al. (1987) [286] and Kacprzyk and Staniewski (1980, 1983) [287, 291]; see also Kacprzyk (1983b) [289]. Formulations with an infinite termination time are mostly used for cases with long planning horizons, and low varying goals and constraints. The problem with an infinite termination time and a deterministic system was first formulated and solved by Kacprzyk and Staniewski (1983) [291].

*Example 1.2.* Suppose the initial state and/or the terminal state are given, but the terminal time $T$ is subject to optimization. Let $\{(x^*(t),u^*(t))|t \in (0,TJ)\}$ be an optimal state-control trajectory pair and let $T^*$ be the optimal terminal time. Then if the terminal time were fixed at $T^*$, the pair $\{(x^*(t),u^*(t))|t \in (0,T^*J)\}$ would satisfy the conditions of the Minimum Principle. In particular,

$$u^*(t) = \arg\min_{u\in U} H(x^*(t),u,p(t)), \quad \forall t \in [0,T^*], \tag{1.4}$$

where $p(t)$ is the solution to the adjoining equation. What is lost because the terminal time is free, is regained through the derivation of an extra condition as follows. It can be argued that if the terminal time was fixed at $T^*$ and the initial state was fixed at the given $x(0)$, but the initial time was also subject to optimization, it would be optimal to start at $t = 0$. This means that the first order variation of the optimal cost with respect to the initial time must be zero:

$$\nabla_t J^*(t, x^*(t))|_{t=0} = 0. \tag{1.5}$$

The above equation can be written along the optimal trajectory as,

$$\nabla_t J^*(t, x^*(t)) = -H(x^*(t), u^*(t), p(t)), \quad \forall t \in [0, T^*], \tag{1.6}$$

so the preceding two equations yield

$$H(x^*(0), u^*(0), p(0)) = 0. \tag{1.7}$$

Since the Hamiltonian was shown earlier to be constant along the optimal trajectory, we obtain for the case of a free terminal time

$$H(x^*(t), u^*(t), p(t)) = 0, \forall t \in [0, T^*]. \tag{1.8}$$

### 1.1.2 State

A state $\mathbf{x}(k)$ is used to describe the characteristic of a stage. The state variable can be discrete or continuous. However, in real world problems, there are actually no continuous state variables, since only discrete entities are met, and the inherent precision in physical measurement is also discrete. Nevertheless, when dealing with some problems, mathematical continuity is a helpful assumption. The state can also be a vector, which can describe the different aspects of a system through its components.

**Definition 1.2.** *(Rao 2009 [166]) A state of multiple objective multistage decision making is an information transfer and incorporates elements that characterize the features of its corresponding stage.*

The state $\mathbf{x}(k)$, in general, incorporates information about the sequence of decisions made so far. In some cases, the state may be the complete sequence, but in other cases only partial information is sufficient; for example, in a case where the set of all states can be partitioned into equivalence classes, with each represented by the last decision. In some simpler problems, the length of the sequence, also called the stage at which the next decision is to be made, suffices. The initial state, which reflects the situation in which no decision has yet been made, is called the goal state and denoted $\mathbf{x}_0$. State space is defined as,

**Definition 1.3.** *(Kacprzyk 1997 [162]) A state space for multiple objective multistage decision making is said to be a nonempty set, in which every element is a state that possibly appears in the decision making process.*

**Fixed Terminal State**

Suppose that in addition to the initial state $x(0)$, the final state $x(T)$ is given. Then the preceding informal derivations still hold except that the terminal condition $C^*(T,x) = h(x)$ is not true. In effect, here we have

$$C^*(T,x) = \begin{cases} 0 \text{ if } x = x(T), \\ \infty \text{ otherwise.} \end{cases} \tag{1.9}$$

Thus, $C^*(T,x)$ cannot be differentiated with respect to $x$, and the terminal boundary condition $p(T) = \nabla h(x^*(T))$ for the adjoining equation does not hold. However, for compensation, there is an extra condition

$$x(T) : \text{given}, \tag{1.10}$$

thus maintaining the balance between the boundary conditions and the unknowns. If only some of the terminal states are fixed, that is

$$x_i(T) : \text{given}, \forall i \in I, \tag{1.11}$$

where $I$ is some index set, we have partial boundary conditions

$$P_i(T) = \frac{\partial h(x^*(T))}{\partial x_j}, \forall j \notin I, \tag{1.12}$$

for the adjoining equation.

*Example 1.3.* Consider the problem of finding a curve of minimum length connecting two points $(0,\alpha)$ and $(T,\beta)$. This is a fixed endpoint variation. We have

$$\dot{x}(t) = u(t), x(0) = \alpha, \quad x(T) = \beta, \tag{1.13}$$

and the cost is

$$\int_0^T \sqrt{1 + U(t)^2 dt}. \tag{1.14}$$

The adjoining equation is

$$J(t) = 0, \tag{1.15}$$

implying that

$$p(t) = \text{constant}, \quad \forall t \in [0,T]. \tag{1.16}$$

Minimization of the Hamiltonian,

$$\min_{u \in R}[\sqrt{1 + u^2 + p(t)u}], \tag{1.17}$$

yields

$$u^*(t) = \text{constant}, \quad \forall t \in [0,T]. \tag{1.18}$$

thus the optimal trajectory $\{x^*(t) | t \in [0,T]\}$ is a straight line. Since this trajectory must pass through $(0,\alpha)$ and $(T,\beta)$, an (a priori obvious) optimal solution is obtained as shown in Figure 1.6.

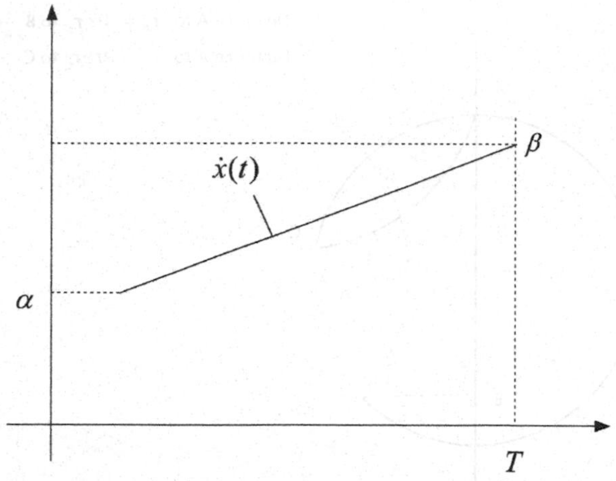


**Fig. 1.6** Optimal solution of the problem of connection the two points $(0,\alpha)$ and $(T,\beta)$ with a minimum length curve

*Example 1.4.* (The Brachistochrone Problem)

In 1696, Johann Bernoulli [17] challenged the mathematical world of his time with a problem that played an instrumental role in the development of the calculus of variations: Given two points $A$ and $B$, find a curve connecting $A$ and $B$ such that a body moving along the curve under the force of gravity reaches $B$ in minimum time (see Figure 1.7). Let $A$ be $(0,0)$ and $B$ be $(T,-b)$ with $b>0$. Then it can be seen that the problem is to find $\{x(t)|t\in[0,T]\}$ with $x(0)=0$ and $x(T)=b$, which minimizes

$$\int_0^T \frac{\sqrt{1+(\dot{x}(t))^2}}{\sqrt{2\gamma x(t)}}dt, \tag{1.19}$$

where $\gamma$ is the acceleration due to gravity. Here $\{(t,-x(t))|t\in[0,TJ]\}$, is the desired curve, the term $1+(\dot{x}(t))^2dt$ is the length of the curve from $x(t)$ to $x(t+dt)$, and the term $\sqrt{2\gamma x(t)}$ is the velocity of the body upon reaching level $x(t)$ [if $m$ and $v$ denote the mass and the velocity of the body, the kinetic energy is $mv^2/2$, at which level $x(t)$ must be equal to the change in potential energy, which is $m\gamma x(t)$; this yields $v=\sqrt{2\gamma x(t)}$].

$\dot{x}=u$ is introduced into the system , and a fixed terminal state problem $[x(0)=0$ and $x(T)=b]$ is obtained. Letting

$$g(x,u)=\frac{\sqrt{1+u^2}}{\sqrt{2\gamma x}}, \tag{1.20}$$

the Hamiltonian is

$$H(x,u,p)=g(x,u)+pu. \tag{1.21}$$

The Hamiltonian is minimized by setting to zero its derivative with respect to $u$,

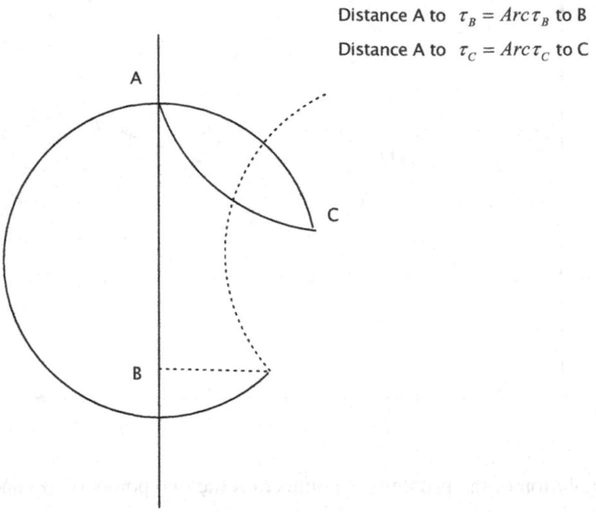


**Fig. 1.7** Formulation and optimal solution of the brachistochrone problem

$$p = -\nabla_u g(x^*(t), u^*(t)). \tag{1.22}$$

The Minimum Principle states that the Hamiltonian is constant along an optimal trajectory, i.e.,

$$g(x^*(t), u^*(t)) - \nabla_u g(x^*(t), u^*(t)) u^*(t) = constant, \forall t \in [0, T]. \tag{1.23}$$

Using the expression for g, this can be written as

$$\frac{\sqrt{1+(u^*(t))^2}}{\sqrt{2\gamma x^*(t)}} - \frac{(u^*(t))^2}{\sqrt{1+(u^*(t))^2}\sqrt{2\gamma x^*(t)}} = constant, \forall t \in [0, T]. \tag{1.24}$$

or equivalently

$$\frac{1}{\sqrt{1+(u^*(t))^2}\sqrt{2\gamma x^*(t)}} = constant, \forall t \in [0, T]. \tag{1.25}$$

Using the relation $x^*(t) = u^*(t)$, this yields

$$x^*(t)(1 + x^*(t)^2) = C, \forall t \in [0, T]. \tag{1.26}$$

for some constant $C$. Thus an optimal trajectory satisfies the differential equation

$$x^*(t) = \sqrt{\frac{C - x^*(t)}{x^*(t)}}, \forall t \in [0, T]. \tag{1.27}$$

The solution to this differential equation was known in Bernoulli's time to be a cycloid; see Figure 1.7. The unknown parameters of the cycloid are determined by the boundary conditions $x^*(0) = 0$ and $x^*(T) = b$.

**Free Initial State**

If the initial state $x(0)$ is not fixed but is subject to optimization, there is

$$J^*(0,x^*(0)) \leq J^*(0,x), \forall x \in \Re^n, \tag{1.28}$$

yielding

$$\nabla_x J^*(0,x^*(0)) = 0, \tag{1.29}$$

and the extra boundary condition for the adjoining equation

$$p(0) = 0. \tag{1.30}$$

Also if there is a cost $\ell(x(0))$ on the initial state, i.e., the cost is

$$\ell(x(0)) + \int_0^T g(x(t),u(t))dt + h(x(T)), \tag{1.31}$$

the boundary condition becomes

$$p(0) = -\nabla \ell(x^*(0)). \tag{1.32}$$

This is followed by setting the gradient to zero with respect to $x$ of $\ell(x) + J(0,x)$, i.e.,

$$\nabla_x \{\ell(x) + J(0,x)\}|_{x=x^*(0)} = 0. \tag{1.33}$$

### *1.1.3 Decision and Policy*

Making a decision without a policy is fairly common in practice, but does not often end well. In a MOMSDM process, policies allow for decisions to be made comfortably, intelligently, and simply. Policies give decision making four benefits: (1) Independent goals can be established as policies provide a conscious and directed series of choices. (2) A measurement standard for goal achievement can be established through policies. (3) Policies convert values to action as they allow for the documentation of the most efficient and practical planning methods. (4) Policies allow limited resources to be committed and agreed upon in an orderly way while being mindful of cost and materials constraints. This part will discuss the relevant decision and policy elements of the MOMSDM.

**Decision**

In every stage, the decision-maker needs to make different choices, so the state in the next stage can be determined. The choices the decision-maker selects are is said to be a decision.

**Definition 1.4.** *The decision in multiple objective multistage decision making is the choice that the decision-maker selects in every stage.*

In this book, $\mathbf{u}(k)$ is used to represent the decision variable in Stage $k$.

The decision space $D(\mathbf{x})$ is the set of possible or "eligible" choices for the next decision $\mathbf{u}$. It is a function of the state $\mathbf{x}$ in which the decision $\mathbf{u}$ is to be made. The constraints on a possible next-state transformation from a state $\mathbf{x}$ can be imposed by suitably restricting $D(\mathbf{x})$. If $D(\mathbf{x}) = \emptyset$, so that there are no eligible decisions in state $\mathbf{x}$, then $\mathbf{x}$ is a terminal state.

**Definition 1.5.** *The decision space in multiple objective multistage decision making is the set of possible or "eligible" choices for the next decision.*

Decisions are made at points of time referred to as decision epochs. Let $T$ denote the set of decision epochs. This subset of the non-negative real line may be classified in two ways: as either a discrete set or a continuum, and as either a finite or an infinite set. When discrete, decisions are made at all decision epochs. When it is a continuum, decisions may be made at

1. all decision epochs (continuously),
2. random points in time when certain events occur, such as arrivals to a queuing system, or
3. opportune times chosen by the decision maker.

When decisions are made continuously, the sequential decision problems are best analyzed using control theory methods based on dynamic system equations.

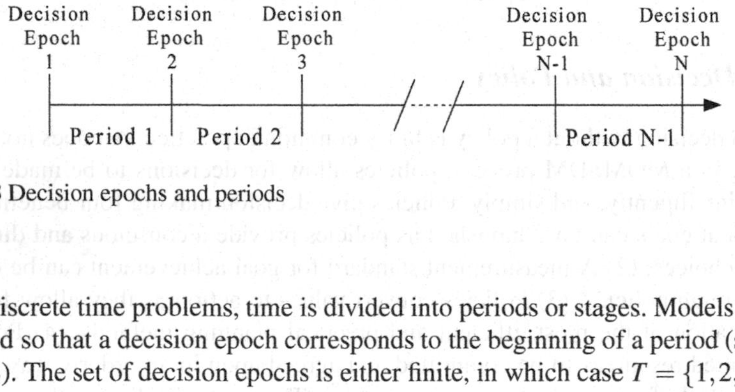


**Fig. 1.8** Decision epochs and periods

In discrete time problems, time is divided into periods or stages. Models are formulated so that a decision epoch corresponds to the beginning of a period (see Figure 1.8). The set of decision epochs is either finite, in which case $T = \{1, 2, \cdots, N\}$ for some integer $N < \infty$, or infinite, in which case $T = \{1, 2, \cdots\}$. This is written $T = \{1, 2, \cdots, N\}, N \le \infty$ to include both cases. When $T$ is an interval, it is denoted by either $T = [0, N]$ or $T = [0, \infty]$. Elements of $T$ (decision epochs) are denoted by $t$ and usually referred to as "time $t$". When $N$ is finite, the decision problem is called a finite horizon problem; otherwise it is called an infinite horizon problem. Most of this book focuses on infinite horizon models and a convention adopted whereby in finite horizon problems, decisions are not made at decision epoch $N$ and is included for final system state evaluation. Consequently, the last decision is made at decision epoch $N-1$, frequently referred to as an $N-1$ period problem.

**Policy**

A policy is any rule for making decisions which yield an allowable sequence of decisions; and an optimal policy is a policy which maximizes a pre-assigned function for the final state variables. A more precise definition of a policy is not as readily obtained as might be thought. Although not too difficult for deterministic processes, stochastic processes require more care. For any particular process, it is not difficult to render an exact concept. The key word is, of course, "allowable" [164]. Here the definition of $(k,l)$-subpolicy, $k$-subpolicy and policy are given.

**Definition 1.6.** *(Bellman 1957 [164]) Let $u_k(x_k)$ be the decision under state $x_k$ in stage $k$, then $\forall 0 \le k \le l \le T$, the function sequence $\{u_k(x_k), u_{k+1}(x_{k+1}), \cdots, u_{l-1}(x_{l-1})\}$ is said to be a $(k,l)$-subpolicy, i.e., $\pi_{k,l}$. When $l = T$, the function sequence $\{u_k(x_k), u_{k+1}(x_{k+1}), \cdots, u_{T-1}(x_{T-1})\}$ is said to be a $k$-subpolicy, i.e., $\pi_k$. When $k = 0$, $l = T$, the function sequence $\{u_0(x_0), u_1(x_1), \cdots, u_{T-1}(x_{T-1})\}$ is said to be a policy, i.e., $\pi$.*

### *1.1.4 Initial and Terminal Conditions*

In a multiple objective multistage decision making process, the initial and terminal conditions can be divided into two cases, fixed and free.

Let $I$ be the initial state space, and $U(\pi)$ be the terminal state space, then the initial and terminal conditions can be expressed as follows,

$$\mathbf{x}(0) \in I, \quad \mathbf{x}(T) \in U(\pi). \tag{1.34}$$

According to the types of initial and terminal conditions, the multiple objective serial multistage decision problems can be classified into three categories:

1. *Initial value problem.* If the value of the initial state variable, $\mathbf{x}_{n+1}$, is prescribed, the problem is called an initial value problem.

2. *Final value problem.* If the value of the final state variable, $\mathbf{x}_1$ is prescribed, the problem is called a final value problem. Note that a final value problem can be transformed into an initial value problem by reversing the directions of $\mathbf{s}_i$ , $i = 1, 2, \cdots, n+1$.

3. *Boundary value problem.* If the values of both the input and output variables are specified, the problem is called a boundary value problem. These three types of problems are shown schematically in Figure 1.9, where the symbol $\mapsto$ is used to indicate a prescribed state variable.

### *1.1.5 Constraints*

In many practical problems, the design variables cannot be chosen arbitrarily; rather, they have to satisfy certain specified functions and other requirements. The restrictions that must be satisfied to produce an acceptable design are collectively called design constraints. Constraints that represent limitations on the behavior or performance of the system are termed behavior or functional constraints. Constraints that

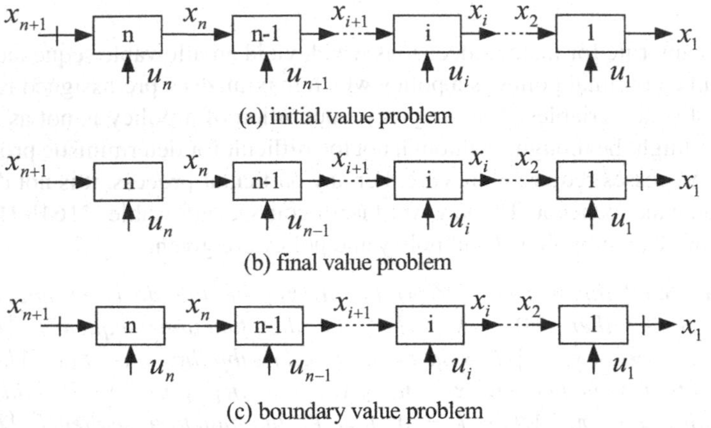


**Fig. 1.9** Types of multiple objective multistage problems

represent physical limitations on the design variables, such as availability, fabricability, and transportability, are known as geometric or side constraints.

The general constraints describe the restrictions on the state and decision variables. Here, $g_j(x,\xi) \leq 0,\ j = 1,2,\cdots,p$ is used to express the MOMSDM's general constraints.

### *1.1.6 State Transition Equations*

The state transition function $f$, a function of $\mathbf{x}$ and $\mathbf{u}$, specifies the next-state that results from making a decision $\mathbf{u}$ in state $\mathbf{x}$.

In a multiple objective multistage decision making process, the next state can be determined by the state and decision in the current stage. Here $f_k(\mathbf{x}(k),\mathbf{u}(k))$ are used to express the state transition function, thus

$$\mathbf{x}(k+1) = f_k(\mathbf{x}(k),\mathbf{u}(k)). \tag{1.35}$$

The above equation is known as the state transition equation.

**Time Lags**

In many applications the system state $\mathbf{x}(k+l)$ depends not only on the preceding state $\mathbf{x}(k)$ and control $\mathbf{u}(k)$ but also on earlier states and controls. In other words, states and controls influence future states with some time lag. Such situations can be handled by state augmentation, where the state is expanded to include an appropriate number of earlier states and controls.

For simplicity, assume that there is at most a single period time lag in the state and control; that is, the system equation has the form

$$\mathbf{x}(k+1) = f_k(\mathbf{x}(k), \mathbf{x}(k-1), \mathbf{u}(k), \mathbf{u}(k-1)), \quad k = 1, 2, \cdots, T-1. \tag{1.36}$$

Time lags of more than one period can be handled similarly.

### 1.1.7 Objective Functions

Conventional design procedures aim at finding an acceptable or adequate design that merely satisfies the functional and other requirements of the problem. In general, there will be more than one acceptable design, and the purpose of optimization is to choose the best of the many acceptable designs available. Thus a criterion has to be chosen for comparing the different alternative acceptable designs and for selecting the best. The criterion with respect to which design is optimized, when expressed as a function of the design variables, is known as the criterion or merit or objective function. The choice of objective functions is governed by the nature of the problem. The objective function for minimization is generally taken as the weight in aircraft and aerospace structural design problems. In civil engineering structural designs, the objective is usually taken as the minimization of cost. The maximization of mechanical efficiency is the obvious choice of an objective in mechanical engineering systems design. Thus the choice of the objective function appears to be straightforward in most design problems. However, there may be cases where the optimization with respect to a particular criterion may lead to results that may not be satisfactory with respect to another criterion. For example, in mechanical design, a gearbox transmitting the maximum power may not have the minimum weight. Similarly, in structural design, the minimum weight design may not correspond to the minimum stress design, and the minimum stress design, again, may not correspond to the maximum frequency design. Thus the selection of the objective function can be one of the most important decisions in the optimum design process.

The objective functions for the MOMSDM must have the following form

$$J_k(\mathbf{x}(k), \pi_k) = \Gamma_k[\mathbf{x}(k), \mathbf{u}(k), J_{k+1}(\mathbf{x}(k+1), \pi_{k+1})]. \tag{1.37}$$

Separability and monotonicity of the functions defined in the above equation have been used to derive the recursive formula for the MOMSDM. Here, definitions are given for those properties.

**Definition 1.7.** *(Abo-Sinna 2004 [195]) (Separability and monotonicity) The objective function $F_l$ is said to be separable if there exist functions $F_l^n$, $n = 1, \cdots, N$, defined on $R^n$ and functions $\phi_l^n$, $n = 2, \cdots, N$, defined on $R^2$ satisfying, for $n = 2, \cdots, N$*

$$F_l^n(f_{l1}(x_1), \cdots, f_{ln}(x_n)) = \phi_l^n[F_l^{n-1}(f_{l1}(x_1), \cdots, f_{ln-1}(x_{n-1})), f_{ln}(x_n)] \tag{1.38}$$

$$F_l^N(f_{l1}(x_1), \cdots, f_{lN}(x_N)) = F_l(f_{l1}(x_1), \cdots, f_{lN}(x_N)). \tag{1.39}$$

*Similarly, the constraint function $G_m$ is separable, if there exist functions $G_m^n$, $n = 1, \cdots, N$, defined on $R^n$ and $\phi_m^n$, $n = 2, \cdots, N$, defined on $R^n$ satisfying, for $n = 2, \cdots, N$*

$$G_n^m(g_{m1}(x_1),\cdots,g_{mn}(x_n)) = \psi_m^n[G_m^{n-1}(g_{m1}(x_1),\cdots,g_{mn-1}(x_{n-1})),g_{mn}(x_n)] \quad (1.40)$$

$$G_m^N(g_{m1}(x_1),\cdots,g_{mN}(x_N)) = G_m(g_{m1}(x_1),\cdots,g_{mN}(x_N)). \quad (1.41)$$

The objective function has three different forms.

### Sum Form

It is common to see the objective functions of practical problems organized in sum form. Typical application examples can be widely found in production planning, supply chain management and transportation optimization. The sum form of the objective function is:

$$J_k(\mathbf{x}(k),\pi_k) = \sum_{j=k}^{T-1} L_j(\mathbf{x}(j),\mathbf{u}(j)) + K(\mathbf{x}(T)). \quad (1.42)$$

### Product Form

In some specific problems, such as the reservoir operation and information management, the objective functions may be organized in product form as below:

$$J_k(\mathbf{x}(k),\pi_k) = \prod_{j=k}^{T-1} L_j(\mathbf{x}(j),\mathbf{u}(j)) \cdot K(\mathbf{x}(T)). \quad (1.43)$$

### Max-Min Form

For the max-min problem often encountered in networks optimization, the max-min form of the objective functions are usually employed as below:

$$J_k(\mathbf{x}(k),\pi_k) = \max(\min)_{k\le j\le T} L_j. \quad (1.44)$$

If all objective and constraint functions are separable, the MOMSDM problem is considered to be separable and monotone if all functions $\phi_l^n$ and $\psi_m^n$ strictly increase with respect to the first argument for each fixed second argument. Specifically, for each $y \in R$,

$$\phi_l^n(s,y) > \phi_l^n(s',y) \quad \text{iff } s > s' \quad \textit{and} \quad \varphi_m^n(s,y) > \varphi_m^n(s',y) \quad \text{iff } s > s' \quad (1.45)$$

for every $l = 1,\cdots,L$, $m = 1,\cdots,M$, and $n = 2,\cdots,N$.

When no confusion occurs, Eq. (1.38) and (1.39) can be written as

$$F = f_1 \cdot f_2 \cdot \cdots \cdot f_n. \quad (1.46)$$

*Example 1.5.* (i) Let $F^1(x_1,x_2,x_3) = \Sigma_{j=1}^3 f_j(x_j)$ and $F^2(x_1,x_2,x_3) = \max_j\{f_j^2(x_j)\}$. Then $F = (F^1,F^2)$ is separable. Note,

$$F^2(x_1,x_2,x_3) = \max\{f_1(x_1), \max\{f_2(x_2), f_3(x_3)\}\}. \quad (1.47)$$

The vector operator "." represents addition in the first component and "*max*" in the second component, i.e.,

$$\phi_1(f_1, \phi_2(f_2, f_3)) = \begin{bmatrix} f_1^1 + \phi_2^1 \\ \max\{f_1^2, \phi_2^2\} \end{bmatrix}. \quad (1.48)$$

(ii) Let $F = (F^1, F^2)$ with $F^1 = f_1^1(x_1)f_2^2(x_2) + f_3^1(x_3)$ and $F^2 = f_1^2(x_1) + f_2^2(x_2)f_3^2(x_3)$. Then $F$ is not separable (since $F^1$ cannot be represented as a function $\phi_1(f_1, \phi_2(f_2, f_3))$.

**Definition 1.8.** *(Abo-Sinna 2004 [195]) Assuming the separability of a MOMSDM problem, an nth subproblem $P_n(s)$ can be defined for each $n = 1, \cdots, N$ and each $s = (s_1, \cdots, s_m) \in R^m$ as follows:*

$$\begin{cases} \min\{F_1^n(f_{11}(x_1), \cdots, f_{1n}(x_n)), \cdots, F_L^n(f_{L1}(x_1), \cdots, f_{Ln}(x_n))\} \\ \text{s.t.} \begin{cases} G_m^n(g_{m1}(x_1), \cdots, g_{mn}(x_n)) \le s_m, \quad m = 1, \cdots, M, \\ x_1 \in X_1, \cdots, x_n \in X_n. \end{cases} \end{cases} \quad (1.49)$$

*It is clear that when $n = N$ and $s = 0$, the above problem coincides with a MOMSDM problem. The nth subproblem $P_n(s)$ for each $s \in R^m$ is again a multiple objective programming problem.*

**Definition 1.9.** *(Abo-Sinna 2004 [195]) Let $(x_1^0, \cdots, x_n^0)$ be a feasible solution to problem $P_n(s)$. It is called efficient, if there exists no feasible solution $(x_1, \cdots, x_2)$ such that*

$$F_l^n(f_{l1}(x_1), \cdots, f_{ln}(x_n)) \le F_l^n(f_{l1}(x_1^0), \cdots, f_{ln}(x_n^0)) \quad (1.50)$$

*for all l and*

$$F_k^n(f_{k1}(x_1), \cdots, f_{kn}(x_n)) < F_k^n(f_{k1}(x_1^0), \cdots, f_{kn}(x_n^0)) \quad (1.51)$$

*for at least one index $k \in \{1, \cdots, L\}$.*

**Lemma 1.1.** *(Abo-Sinna 2004 [195]) Let the MOMSDM problem be separable and the separation be monotone, and $F_l^n(f_{l1}(x_1), \cdots, f_{ln}(x_n))$, $l = 1, \cdots, L$, and $G_m^n(g_{m1}(x_1), \cdots, g_{mn}(x_n))$, $m = 1, \cdots, M$ be continuous functions of $(x_1, \cdots, x_n)$. If $(x_1^0, \cdots, x_n^0)$ is any efficient solution for problem $P_n(s)$, then $(x_1^0, \cdots, x_{n-1}^0)$ is an efficient solution for problem $P_{n-1}\lfloor s^{n-1}(x_n^0, s)\rfloor$. Where*

$$s^{n-1}(x_n, s) = \lfloor s_1^{n-1}(x_n, s), \cdots s_M^{n-1}(x_n, s)\rfloor, \quad (1.52)$$

*and*

$$s_m^{n-1}(x_n, s) = \sup\{\xi \in R | \varphi_m^n(\xi, g_{mn}(x_n)) \le s_m\}, \quad m = 1, \cdots, M. \quad (1.53)$$

Note that the converse of the result of Lemma 1.1 does not necessarily hold. However, the following theorem can be stated.

**Lemma 1.2.** *(Abo-Sinna 2004 [195]) Let the MOMSDM problem be separable and the separation be monotone, and* $F_l^n(f_{l1}(x_1),\cdots,f_{ln}(x_n))$, $l=1,\cdots,L$, *and* $G_m^n(g_{m1}(x_1),\cdots,g_{mn}(x_n))$, $m=1,\cdots,M$ *be continuous functions of* $(x_1,\cdots,x_n)$. *Let* $(x_1^0,\cdots,x_n^0)$ *be any efficient solution of the following multiple objective programming problem:*

$$\begin{cases} \min\{F_1^n(f_{11}(x_1),\cdots,f_{1n}(x_n)),\cdots,F_L^n(f_{L1}(x_1),\cdots,f_{Ln}(x_n))\} \\ s.t.\ \{\ (x_1,\cdots,x_n)\in\bigcup_{x_n}E_{n-1}\lfloor s^{n-1}(x_n,s)\rfloor X\{x_n\}. \end{cases} \tag{1.54}$$

*Then* $(x_1^0,\cdots,x_n^0)$ *constitutes an efficient solution to problem* $P_n(s)$. *Here* $E_{n-1}\lfloor s^{n-1}(x_n,s)\rfloor$ *is the set of all efficient solutions to problem* $P_{n-1}\lfloor s^{n-1}(x_n,s)\rfloor$.

Consequently, from Lemmas 1.1 and 1.2, the following recursive relations of MOMSDM problems, for $n=2,\cdots,N$ can be obtained

$$Q^n(s)=\min Q^n(Q^{n-1}\lfloor s^{n-1}(x_n,s)\rfloor,f_n(x_n)), \tag{1.55}$$

where $f_n(x_n)=[f_{1n}(x_n),\cdots,f_{2n}(x_n)]^T$, $Q^n(s)$ is the image of the set $E_n(s)$ under the mapping $F^2(f_1,\cdots,f_n)=[F_1^n,\cdots,F_L^n]^T$, and the right-hand side is understood to be the set of minimal points of the set

$$\{y\in R^L; y=Q^n(v,f_n(x_n)), v\in Q^{n-1}\lfloor s^{n-1}(x_n,s)\rfloor\}. \tag{1.56}$$

Conventional design procedures aim at finding an acceptable or adequate design that merely satisfies the functional and other requirements of the problem. In general, there is more than one acceptable design, and the purpose of the optimization is to choose the best of the many acceptable designs available. Thus a criterion has to be chosen to compare the different alternative acceptable designs and for selecting the best. The criterion with respect to which design is optimized, when expressed as a function of the design variables, is known as the criterion or merit or objective function. The choice of objective function is governed by the nature of problem. The objective function for minimization is generally taken as the weight in aircraft and aerospace structural design problems. In civil engineering structural designs, the objective is usually taken as the minimization of cost. The maximization of mechanical efficiency is the obvious choice for an objective in a mechanical engineering systems design. Thus the choice of the objective function appears to be straightforward in most design problems. However, there may be cases where the optimization with respect to a particular criterion may lead to results that may not be satisfactory with respect to another criterion. For example, in mechanical design, a gearbox transmitting the maximum power may not have the minimum weight. Similarly, in structural design, the minimum weight design may not correspond to the minimum stress design, and the minimum stress design, again, may not correspond to the maximum frequency design. Thus the selection of the objective function can be one of the most important decisions in the whole optimum design process. In some situations, there may be more than one criterion to be satisfied simultaneously. For example, a gear pair may have to be designed for minimum weight and maximum efficiency while transmitting a specified horsepower. An optimization problem involving multiple objective functions is known as a multi-objective programming problem. With multiple objectives there arises a possibility of conflict, and one simple way to handle the problem is to construct an overall objective function as a linear combination of the conflicting multiple objective functions.

## 1.2 Mathematical Foundation of MOMSDM

To harmonize application practices with design theories, this section presents the mathematical foundation of the MOMSDM. Systems can be represented by mathematical models of many different forms, such as algebraic equations, differential or integral equations, finite state machines, Petri nets, and rules. They are used particularly in the natural sciences and engineering disciplines such as physics, biology, electrical and computer engineering, and in the social sciences such as economics or sociology. Engineers, computer scientists, physicists and economists use mathematical models most extensively. A mathematical model should be a representation of the essential aspects of an existing system (or a system to be constructed) (Eykhoff 1974 [197]). This model should express the knowledge of that system in usable form.

### *1.2.1 The Principle of Optimality*

In this section, the discrete time deterministic decision-making process under a fixed and specified termination time is discussed. Here we suppose the termination time to be $N$, the state space, $R^n$, and the decision space, $R^m$. The state transition equation is then

$$x_{k+1} = f_k(x_k, u_k), \tag{1.57}$$

where $k = 0, 1, \cdots, N-1$. The allowable decision-making set is said to be $U_k(x_k) \subset R^m$. The objective functions meet the following recurrence relation:

$$J_k(x_k, \pi_k) = \Gamma_k[x_k, u_k, J_{k+1}(x_{k+1}, \pi_{k+1})] \tag{1.58}$$

where $k = 0, 1, \cdots, N-1$. When $k = N$, $J_N$ is only a function of $x_N$.

*Remark 1.1.* In the above decision-making process, if there are constraints for the states in every stage, then these constraints can be added into the allowable decision-making set. In fact, let $\hat{X}_{k+1}$ be the constraint set of $x_{k+1}$, $0 \le k \le N-1$, the original allowable decision-making set is $U_k^0(x_k)$, then the new allowable decision-making set can be defined as:

$$U_k(x_k) \triangleq \{u | u \in U_k^0(x_k), f_k(x_k, u) \in \hat{X}_{k+1}\},$$

in which $u_k \in U_k(x_k)$ meets not only the decision-making constraints, but also the state constraints.

Let $X_0$ be the initial state set, and $x_0$, the initial state. Then the attainable state set $X_k$, $1 \le k \le N$, is defined as follows:

**Definition 1.10.** *$X_k$ is said to be the attainable state set for stage k, if and only if*

$$X_k \triangleq \{x|x = f_{k-1}(x_{k-1}, u_{k-1}), x_{k-1} \in X_{k-1}, u_{k-1} \in U_{k-1}(x_{k-1})\} \tag{1.59}$$

*where $k = 1, 2, \cdots, N$.*

**Definition 1.11.** *The policy $\pi_k^*$ is said to be the optimal k-subpolicy, if and only if*

$$\forall x_k \in X_k, \quad J_k(x_k, \pi_k^*) = \min_{\pi_k \in \Pi_{k,a}} J_k(x_k, \pi_k), \tag{1.60}$$

*where $\Pi_{k,a}$ is the allowable k-subpolicy set, $k = 0, 1, \cdots, N-1$.*

In this section, it is supposed that there exists an optimal $k$-subpolicy $\pi_k^*$ for all $k \in \{0, 1, \cdots, N-1\}$.

**Lemma 1.3.** *Let $F(x,z)$: $X \times Z \to R$, $z(\omega)$: $W \to Z$, $Z \subset R$. If $F$ is a monotonically non-decreasing function of z, and $z(\omega)$ has a minimal solution $\omega^*$ in $W$, then*

$$F(x, \min_{\omega \in W} z(\omega)) = \min_{\omega \in W} F(x, z(\omega)). \tag{1.61}$$

*Proof.* Since $\min_{\omega \in W} z(\omega) = z(\omega^*)$, $\omega^* \in W$, then

$$F(x, \min_{\omega \in W} z(\omega)) = F(x, z(\omega^*)) \geq \inf_{\omega \in W} F(x, z(\omega)).$$

On the other hand, since $\forall \omega \in W$, $z(\omega^*) \leq z(\omega)$, according to the monotonic and non-decreasing of $F$, thus,

$$F(x, \min_{\omega \in W} z(\omega)) = F(x, z(\omega^*)) \leq F(x, z(\omega)), \quad \forall \omega \in W,$$

therefore,

$$F(x, \min_{\omega \in W} z(\omega)) \leq \inf_{\omega \in W} F(x, z(\omega)).$$

Hence $F(x, \min_{\omega \in W} z(\omega)) = \min_{\omega \in W} F(x, z(\omega))$ holds. □

Based on (1.58), we can derive

$$J_0(x_0, \pi_0) = \Gamma_{0,k}[x_0, \pi_{0,k}, J_k(x_k, \pi_k)], \tag{1.62}$$

where $k = 1, 2, \cdots, N-1$. Here, it should be noted that

$$\begin{aligned} \Gamma_{0,k} = \Gamma_0(x_0, u_0(x_0), \Gamma_1(f_0[x_0, u_0(x_0)], u_1(x_1), \\ \Gamma_2(\cdots, \Gamma_{k-1}(x_{k-1}, u_{k-1}(x_{k-1}), J_k(x_k, \pi_k))))), \end{aligned} \tag{1.63}$$

and the $x_k$ in (1.62) and (1.63) is defined as

$$x_k = f_{0,k}(x_0, \pi_{0,k}) \triangleq f_{k-1}(u_{k-1}(x_{k-1}), f_{k-2}(u_{k-2}(x_{k-2}), \cdots, u_1(x_1), f_0(u_0(x_0), x_0))). \tag{1.64}$$

**Proposition 1.1.** *If $\Gamma_k$ is a monotonically non-decreasing function of $J_{k+1}$, then the allowable policy $\pi_0^*$ is the optimal policy, if and only if*

$$\forall x_0 \in X_0, \quad J_0(x_0,\pi_0^*) = \min_{\pi_{0,k}\in\Pi_{0,k,a}} [\Gamma_{0,k}(x_0,\pi_{0,k}, \min_{\pi_k\in\Pi_{k,a}} J_k(x_k,\pi_k))], \tag{1.65}$$

*where $1 \le k \le N-1$, $x_k$ is defined as (1.63).*

*Proof.* First the necessity is proven. Let $\pi_0^*$ be the optimal policy, then $\forall k$, $1 \le k \le N-1$, based on (1.62), so it can be derived that $\forall x_0 \in X_0$,

$$J_0(x_0,\pi_0^*) = \min_{\pi_0\in\Pi_{0,a}} J_0(x_0,\pi_0) = \min_{\pi_{0,k}\in\Pi_{0,k,a}} \{ \min_{\pi_k\in\Pi_{k,a}} \Gamma_{0,k}[x_0,\pi_{0,k},J_k(x_k,\pi_k)]\}.$$

Since $\Gamma_k$ is a monotonically non-decreasing function of $J_{k+1}$, according to (1.63), It follows from Lemma 1.3 that

$$\min_{\pi_k\in\Pi_{k,a}} \Gamma_{0,k}[x_0,\pi_{0,k},J_k(x_k,\pi_k)] = \Gamma_{0,k}[x_0,\pi_{0,k}, \min_{\pi_k\in\Pi_{k,a}} J_k(x_k,\pi_k)],$$

which implies (1.65).

Then the sufficiency is proved. If (1.65) holds, $\forall \pi_0 \in \Pi_{0,a}$, $\pi_k \in \Pi_{k,a}$ is a subpolicy of $\pi_0$, it is obvious that

$$J_k(x_k,\pi_k) \ge \min_{\pi_k\in\Pi_{k,a}} J_k(x_k,\pi_k),$$

since $\Gamma_{0,k}$ is a monotonically non-decreasing function of $J_k$, it follows from (1.62) that

$$\begin{aligned} J_0(x_0,\pi_0) = \Gamma_{0,k}[x_0,\pi_{0,k},J_k(x_k,\pi_k)] &\ge \Gamma_{0,k}[x_0,\pi_{0,k},\min\nolimits_{\pi_k\in\Pi_{k,a}} J_k(x_k,\pi_k)] \\ &\ge \min\nolimits_{\pi_{0,k}\in\Pi_{0,k,a}}\{\Gamma_{0,k}[x_0,\pi_{0,k},\min\nolimits_{\pi_k\in\Pi_{k,a}} J_k(x_k,\pi_k)]\}. \end{aligned}$$

According to (1.65), the right side of the above inequality is equal to $J_0(x_0,\pi_0^*)$, thus

$$\forall \pi_0 \in \Pi_{0,a}, \quad J_0(x_0,\pi_0) \ge J_0(x_0,\pi_0^*).$$

This means $\pi_0^*$ is the optimal policy. □

**Corollary 1.1.** *$\pi_k^*$ is the optimal k-subpolicy, if and only if*

$$J_k(x_k,\pi_k^*) = \min_{u_k\in U_k(x_k)} \{\Gamma_k[x_k,u_k, \min_{\pi_{k+1}\in\Pi_{k+1,a}} J_{k+1}(x_{k+1},\pi_{k+1})]\} \tag{1.66}$$

*where $k = 0,1,\cdots,N-1$, $\forall x_k \in X_k$, note that $J_N$ is only a function of $x_N$ and*

$$x_{k+1} = f_k(x_k,u_k(x_k)). \tag{1.67}$$

*Proof.* Since the proposition 1.1 is just a special case for the process $[0,N]$, now the cases for processes $[N-1,N]$, $[N-2,N]$, $\cdots$, are considered, note that

$$\left.\begin{aligned} \pi_{k,k+1} &= u_k(x_k), \\ f_{k,k+1}(x_k,\pi_{k,k+1}) &= f_k(x_k,u_k(x_k)). \end{aligned}\right\} \tag{1.68}$$

It follows from (1.65) and (1.68) that (1.66) and (1.67) hold. □

In order to simplify (1.66), the optimal value function $V_k(x_k)$ $(k = 0, 1, \cdots, N-1)$ is defined as follows:

$$V_k(x_k) \triangleq \min_{\pi_k \in \Pi_{k,a}} J_k(x_k, \pi_k), \quad \forall x_k \in X_k. \tag{1.69}$$

where $k = 0, 1, \cdots, N-1$, and

$$V_N(x_N) = J_N(x_N). \tag{1.70}$$

Thus, (1.66) can be transformed as

$$\begin{cases} V_k(x_k) = \min_{u_k \in U_k(x_k)} \Gamma_k(x_k, u_k, V_{k+1}(x_{k+1})), \quad \forall x_k \in X_k, \\ V_N(x_N) = J_N(x_N), \end{cases} \tag{1.71}$$

where $k = 0, 1, \cdots, N-1$. The equations (1.71) are called the basic equations for the dynamic programming for the MOMSDM.

**Theorem 1.1.** *(The principle of optimality) Let $\Gamma_k$ be a monotonic increasing function of $J_{k+1}$, $\pi_k^*$ be any k-subpolicy of the optimal policy $\pi_0^*$, $1 \leq k \leq N-1$, if $x_k = f_{0,k}(x_0, \pi_{0,k}^*)$, then $\pi_k^*$ is the optimal k-subpolicy.*

*Proof.* Since $\Gamma_k$ is a monotonic increasing function of $J_{k+1}$, according to (1.63), it is obvious that $\Gamma_{0,k}$ is a monotonic increasing function of $J_k$. Supposing the theorem is not true, then there exists a $k$, $1 \leq k \leq N-1$, $\pi_k \in \Pi_{k,a}$, such that

$$J_k(x_k, \pi_k) < J_k(x_k, \pi_k^*).$$

Thus

$$J_0(x_0, \hat{\pi}_0) = \Gamma_{0,k}(x_0, \pi_{0,k}^*, J_k(x_k, \pi_k)) < \Gamma_{0,k}(x_0, \pi_{0,k}^*, J_k(x_k, \pi_k^*)) = J_0(x_0, \pi_0^*),$$

in which $\hat{\pi}_0 \in \Pi_a$ is a combination policy of $\pi_{0,k}^*$ and $\pi_k$, so therefore the above inequality is inconsistent with the assumption that $\pi_0^*$ is the optimal policy. □

The conclusion to the above theorem is that an optimal policy has the property that whatever the initial state and initial decision are, the remaining decisions must constitute an optimal policy with regard to the state resulting from the first decision.

### *1.2.2 Expanded State Attainable Set*

Let $J_k$ be a function defined by

$$J_k = \sum_{j=0}^{k-1} L_j(\mathbf{x}(j), \mathbf{u}(j)). \tag{1.72}$$

where $k = 1, 2, \cdots, n$.

Then, it is necessary to define the expanded state variable and the expanded state vector, so it is much more convenient to discuss the problem in Euclidean space.

**Definition 1.12.** $x^0(k)$ *is an expanded state variable if and only if*

$$x^0(0) = 0, \quad x^0(k) = J_k, \quad k = 1, 2, \cdots, n. \tag{1.73}$$

**Definition 1.13.** $y(k)$ *is an expanded state vector if and only if*

$$y(k) = (x^0(k), \boldsymbol{x}(k)), \quad k = 0, 1, \cdots, n. \tag{1.74}$$

*Remark 1.2.* Let $\mathbf{Y}$ be the space of all $\mathbf{y}(k)$ $(k = 0, 1, \cdots, n)$, then $\mathbf{Y} \subset \mathbf{R^2}$.

To simplify the expression, the state transition equation is described as follow. Let $\mathbf{f^k}$ be a function defined by

$$\mathbf{x}(k+1) = \mathbf{f^k}(\mathbf{x}(k), \mathbf{u}(k)), \quad k = 0, 1, \cdots, n-1. \tag{1.75}$$

Then, let $\mathbf{F^k}$ be a function defined by

$$\mathbf{F^k} = (J_{k+1}, \mathbf{f^k}), \quad k = 0, 1, \cdots, n-1. \tag{1.76}$$

Therefore, $\mathbf{y}(0) = (0, \mathbf{x}(0)) = (0, \mathbf{a}), \mathbf{y}(k+1) = \mathbf{F^k}(\mathbf{y}(k), \mathbf{u}(k)), k = 0, 1, \cdots, n-1$.

Based on the above, the state attainable set and the expanded state attainable set can be given as follows.

The state attainable set of the $2^{nd}$ stage $\mathbf{X^1}(\mathbf{x}(0))$ which is derived from the initial condition $\mathbf{x}(0) = \mathbf{a}$ is defined by

$$\mathbf{X^1}(\mathbf{x}(0)) = \{\mathbf{x}(1) | \mathbf{x}(1) = \mathbf{f^0}(\mathbf{x}(0), \mathbf{u}(0)), \mathbf{u}(0) \in \mathbf{U}_0\},$$

where $\mathbf{U}_0$ is the constraint set of the decision variable $\mathbf{u}(0)$ for the $1^{st}$ stage.

**Definition 1.14.** $\boldsymbol{X^k}(\boldsymbol{x}(k-1))$ *is the state attainable set of the* $(k+1)^{th}$ *stage which is derived from* $\mathbf{x}(k-1)$ *if and only if*

$$\boldsymbol{X^k}(\boldsymbol{x}(k-1)) = \{\boldsymbol{x}(k) | \boldsymbol{x}(k) = \boldsymbol{f^{k\text{-}1}}(\boldsymbol{x}(k-1), \boldsymbol{u}(k-1)), \boldsymbol{u}(k-1) \in \boldsymbol{U}_{k-1}(\boldsymbol{x}(k-1))\}, \tag{1.77}$$

$\forall \boldsymbol{x}(k-1) \in \boldsymbol{X^{k\text{-}1}}, k = 2, 3, \cdots, n.$

The expanded state attainable set of the $2^{nd}$ stage $\mathbf{Y^1}(\mathbf{y}(0))$ which is in the corresponding expanded space $\mathbf{Y}$ is defined by

$$\mathbf{Y^1}(\mathbf{y}(0)) = \{\mathbf{y}(1) | \mathbf{y}(1) = \mathbf{F^0}(\mathbf{y}(0), \mathbf{u}(0)), \mathbf{u}(0) \in \mathbf{U}_0\}.$$

**Definition 1.15.** $\boldsymbol{Y^k}(\boldsymbol{y}(k-1))$ *is the expanded state attainable set of the* $(k+1)^{th}$ *stage which is derived from* $\mathbf{y}(k-1)$ *if and only if*

$$\boldsymbol{Y^k}(\boldsymbol{y}(k-1)) = \{\boldsymbol{y}(k) | \boldsymbol{y}(k) = \boldsymbol{F^{k\text{-}1}}(\boldsymbol{y}(k-1), \boldsymbol{u}(k-1)), \boldsymbol{u}(k-1) \in \boldsymbol{U}_{k-1}(\boldsymbol{x}(k-1))\}, \tag{1.78}$$

$\forall \boldsymbol{y}(k-1) \in \boldsymbol{Y^{k\text{-}1}}, k = 2, 3, \cdots, n.$

All of the state attainable sets of the $(k+1)^{th}$ stage $\mathbf{X^k}$ are defined by

$$\mathbf{X^k} = \bigcup\{\mathbf{X^k}(\mathbf{x}(k-1))|\mathbf{x}(k-1) \in \mathbf{X^{k\text{-}1}}\}, \quad k = 2,3,\cdots,n.$$

All of the expanded state attainable sets of the $(k+1)^{th}$ stage $\mathbf{Y^k}$ are defined by

$$\mathbf{Y^k} = \bigcup\{\mathbf{Y^k}(\mathbf{y}(k-1))|\mathbf{y}(k-1) \in \mathbf{Y^{k\text{-}1}}\}, \quad k = 2,3,\cdots,n.$$

Up to now, a series of concepts about the solution space have been listed. Here, the concepts for the conditional function of the optimal value and the function of the optimal value are defined for further preparation of the proofs.

### 1.2.3 Conditional Function of Optimal Value

The conditional function of the optimal value of the $2^{nd}$ stage $fov^1(\mathbf{x}(1)|\mathbf{y}(0))$ is defined by

$$fov^1(\mathbf{x}(1)|\mathbf{y}(0)) = \min\{J_1(\mathbf{u}(0))|\mathbf{u}(0) \in \mathbf{U}_0, \mathbf{f^0}(\mathbf{x}(0),\mathbf{u}(0)) = \mathbf{x}(1)\}.$$

The function of optimal value of the $2^{nd}$ stage $FOV^1(\mathbf{x}(1))$ is defined by

$$FOV^1(\mathbf{x}(1)) = \min\{J_1(\mathbf{u}(0))|\mathbf{u}(0) \in \mathbf{U}_0, \mathbf{f^0}(\mathbf{x}(0),\mathbf{u}(0)) = \mathbf{x}(1)\}.$$

Here, the value of the conditional function of the optimal value $fov^1(\mathbf{x}(1)|\mathbf{y}(0))$ is calculated under the condition that the value of the expanded state vector $\mathbf{y}(0)$ is fixed. Since $\mathbf{y}(0) = (0,\mathbf{x}(0)) = (0,\mathbf{a})$ is the only fixed value in the $2^{nd}$ stage, the conditional function of the optimal value of the $2^{nd}$ stage $fov^1(\mathbf{x}(1)|\mathbf{y}(0))$ is the same as the function of the optimal value of the $2^{nd}$ stage $FOV^1(\mathbf{x}(1))$ in this special case, i.e. $fov^1(\mathbf{x}(1)|\mathbf{y}(0)) = FOV^1(\mathbf{x}(1))$. But there are differences between the two functions when $k = 2,3,\cdots,n$.

**Definition 1.16.** *$fov^k(\boldsymbol{x}(k)|\boldsymbol{y}(k-1))$ is the conditional function of the optimal value of the $(k+1)^{th}$ stage if and only if*

$$fov^k(\boldsymbol{x}(k)|\boldsymbol{y}(k-1)) = \min\left\{ J_k(\boldsymbol{u}(k-1)) \left| \begin{array}{l} \boldsymbol{u}(k-1) \in \boldsymbol{U}_{k-1}(\boldsymbol{x}(k-1)), \\ J_k = x^0(k-1) + \\ L_{k-1}(\boldsymbol{x}(k-1),\boldsymbol{u}(k-1)), \\ \boldsymbol{f^{k\text{-}1}}(\boldsymbol{x}(k-1),\boldsymbol{u}(k-1)) = \boldsymbol{x}(k) \end{array} \right. \right\},$$

$k = 2,3,\cdots,n.$

**Definition 1.17.** *$FOV^k(\boldsymbol{x}(k))$ is the function of the optimal value of the $(k+1)^{th}$ stage if and only if*

$$FOV^k(\boldsymbol{x}(k)) = \min\left\{ J_k(\cdot) \left| \begin{array}{l} \boldsymbol{u}(0) \in \boldsymbol{U}_0, \boldsymbol{u}(1) \in \boldsymbol{U}_1(\boldsymbol{x}(1)), \cdots, \\ \boldsymbol{u}(k-1) \in \boldsymbol{U}_{k-1}(\boldsymbol{x}(k-1)), \\ \boldsymbol{f^0}(\boldsymbol{x}(0),\boldsymbol{u}(0)) = \boldsymbol{x}(1), \boldsymbol{f^1}(\boldsymbol{x}(1),\boldsymbol{u}(1)) = \boldsymbol{x}(2), \\ \cdots, \boldsymbol{f}^{k-1}(\boldsymbol{x}(k-1),\boldsymbol{u}(k-1)) = \boldsymbol{x}(k) \end{array} \right. \right\},$$

$k = 2,3,\cdots,n.$

Based on the above, the conditional lower bound surface and the lower bound surface can be defined.

The conditional lower bound surface of the $2^{nd}$ stage is defined as follows:

Let $\Pi^1(\mathbf{y}(0))$ be the conditional lower bound surface of the $2^{nd}$ stage which is defined by

$$\Pi^1(\mathbf{y}(0)) = \{\mathbf{y}(1)|x^0(1) = fov^1(\mathbf{x}(1)|\mathbf{y}(0)), \mathbf{x}(1) \in \mathbf{X^1}(\mathbf{x}(0))\}.$$

Let $\Pi^1$ be the lower bound surface of the $2^{nd}$ stage which is defined by

$$\Pi^1 = \{\mathbf{y}(1)|x^0(1) = FOV^1(\mathbf{x}(1)), \mathbf{x}(1) \in \mathbf{X^1}\}.$$

Since it is known $fov^1(\mathbf{x}(1)|\mathbf{y}(0)) = FOV^1(\mathbf{x}(1))$, it is obvious that $\Pi^1(\mathbf{y}(0)) = \Pi^2$ in this special case. However, the situations are different when $k = 2, 3, \cdots, n$, as in the following definitions.

**Definition 1.18.** *$\Pi^k(y(k-1))$ is the conditional lower bound surface of the $(k+1)^{th}$ stage if and only if*

$$\Pi^k(y(k-1)) = \{y(k)|x^0(k) = fov^k(x(k)|y(k-1)), x(k) \in \boldsymbol{X^k}(x(k-1))\},$$

$\forall y(k-1) \in \boldsymbol{Y^{k\text{-}1}}, k = 2, 3, \cdots, n.$

**Definition 1.19.** *$\Pi^k$ is the lower bound surface of the $(k+1)^{th}$ stage if and only if*

$$\Pi^k = \{y(k)|x^0(k) = FOV^k(x(k)), x(k) \in \boldsymbol{X^k}\}, \quad k = 2, 3, \cdots, n.$$

According to Def. 1.16, 1.17 and the above statement for $fov^1(\mathbf{x}(1)|\mathbf{y}(0))$ and $FOV^1(\mathbf{x}(1))$, it is obvious that

$$FOV^k(\mathbf{x}(k)) \leq fov^k(\mathbf{x}(k)|\mathbf{y}(k-1)), \quad \forall \mathbf{y}(k-1) \in \mathbf{Y^{k\text{-}1}}, \quad k = 1, 2, \cdots, n. \tag{1.79}$$

Assume that $\{\mathbf{u}^*(0), \cdots, \mathbf{u}^*(n-1)\}$ is the optimal solution sequence of the amount of the $i^{th}$ kind of construction machines which are allocated to a designated job; $\{\mathbf{x}^*(0) = \mathbf{a}, \mathbf{x}^*(1), \cdots, \mathbf{x}^*(n)\}$ is the optimal solution sequence of the amount of the $i^{th}$ kind of available construction machines; $\{\mathbf{y}^*(0), \cdots, \mathbf{y}^*(n)\}$ is the optimal solution sequence of the corresponding expanded vectors, then we have the following lemmas.

**Lemma 1.4.** *The point $\mathbf{y}^*(k)$ is on the lower bound surface of the $(k+1)^{th}$ stage $\Pi^k$, $k = 1, 2, \cdots, n$.*

*Proof.* When $k = n$, assume that $\mathbf{y}^*(n) \notin \Pi^n$, then there exists a sequence $\{\bar{\mathbf{u}}(0), \cdots, \bar{\mathbf{u}}(n-1)\}$ such that the corresponding expanded vector $\bar{\mathbf{y}}(n) = (\bar{J}_n, \bar{\mathbf{x}}(n))$ satisfies

$$\bar{J}_n < FOV^n(\bar{\mathbf{x}}(n)), \quad \bar{\mathbf{x}}(n) = \mathbf{x}^*(n).$$

Thus,

$$\bar{J}_n < FOV^n(\mathbf{x}^*(n)) = J^*,$$

where $J^*$ is the optimal value of the corresponding optimal control indicator function. That means the sequence $\{\bar{\mathbf{u}}(0),\cdots,\bar{\mathbf{u}}(n-1)\}$ is better than the optimal solution sequence $\{\mathbf{u}^*(0),\cdots,\mathbf{u}^*(n-1)\}$, but is impossible.

When $1 \leq k \leq n-1$, assume that $\mathbf{y}^*(k) \notin \Pi^k$, then there exists a sequence $\{\bar{\mathbf{u}}(0),\cdots,\bar{\mathbf{u}}(k-1)\}$ such that the corresponding expanded vector $\bar{\mathbf{y}}(k) = (\bar{J}_k, \bar{\mathbf{x}}(k))$ satisfies

$$\bar{\mathbf{x}}(k) = \mathbf{x}^*(k),$$

$$\bar{J}_k < J_k^* = \sum_{j=0}^{k-1} L_j(\mathbf{x}^*(j), \mathbf{u}^*(j)), \tag{1.80}$$

Now consider the sequence $\{\bar{\mathbf{u}}(0),\cdots,\bar{\mathbf{u}}(k-1),\mathbf{u}^*(k),\cdots,\mathbf{u}^*(n-1)\}$, let

$$\hat{J}_n = \bar{J}_k + \sum_{j=k}^{n-1} L_j(x^*(j), u^*(j)). \tag{1.81}$$

From Eq. (1.80) and (1.81), there is

$$\hat{J}_n < \sum_{j=0}^{n-1} L_j(\mathbf{x}^*(j), \mathbf{u}^*(j)) = J^*.$$

This means that the sequence $\{\bar{\mathbf{u}}(0),\cdots,\bar{\mathbf{u}}(k-1),\mathbf{u}^*(k),\cdots,\mathbf{u}^*(n-1)\}$ is better than the optimal solution sequence $\{\mathbf{u}^*(0),\cdots,\mathbf{u}^*(n-1)\}$, but is impossible. The proof is completed. □

From Lemma 1.4, it is not difficult to prove the following lemma.

**Lemma 1.5.** *The point $\mathbf{y}^*(k)$ is on the conditional lower bound surface of the $(k+1)^{th}$ stage $\Pi^k(\mathbf{y}(k-1))$, $k = 1,\cdots,n$.*

*Proof.* Let $\mathbf{y}^*(k) = (J_k^*, \mathbf{x}^*(k))$, where $k = 1,2,\cdots,n$. According to Def. 1.16 and the statement for $fov^1(\mathbf{x}(1)|\mathbf{y}(0))$, there is

$$J_k^* \geq fov^k(\mathbf{x}^*(k)|\mathbf{y}^*(k-1)),$$

but according to Lemma 1.4, it is

$$J_k^* = FOV^k(\mathbf{x}^*(k)). \tag{1.82}$$

On the other hand, from Eq. (1.79), it is

$$fov^k(\mathbf{x}^*(k)|\mathbf{y}^*(k-1)) \geq FOV^k(\mathbf{x}^*(k)).$$

Hence,

$$fov^k(\mathbf{x}^*(k)|\mathbf{y}^*(k-1)) = J_k^*. \tag{1.83}$$

The above equation means that point $\mathbf{y}^*(k)$ is on the conditional lower bound surface of the $(k+1)^{th}$ stage $\Pi^k(\mathbf{y}(k-1))$, $k = 1,2,\cdots,n$. The proof is completed. □

### 1.2.4 Support Function of Surface

Now the support function of the surface concept is introduced in the following definition.

**Definition 1.20.** *Let $S^k(\boldsymbol{x}(k)|\boldsymbol{y}(k-1))$ be a finite function which is continuous for all real values $\boldsymbol{x}(k)$ and is defined on the state attainable set of the $(k+1)^{th}$ stage $\boldsymbol{X^k}(\boldsymbol{x}(k-1))$ which is derived from $\boldsymbol{x}(k-1)$, then $S^k(\boldsymbol{x}(k)|\boldsymbol{y}(k-1))$ is called the support function of the conditional lower bound surface of the $(k+1)^{th}$ stage $\Pi^k(\boldsymbol{y}(k-1))$ on the point $\hat{\boldsymbol{x}}(k)$, if and only if there exists a real number $\alpha^k$ such that:*

$$S^k(\boldsymbol{x}(k)|\boldsymbol{y}(k-1)) - \alpha^k \le fov^k(\boldsymbol{x}(k)|\boldsymbol{y}(k-1)), \quad \forall \boldsymbol{x}(k) \in \boldsymbol{X^k}(\boldsymbol{x}(k-1)), \tag{1.84}$$

$$S^k(\hat{\boldsymbol{x}}(k)|\boldsymbol{y}(k-1)) - \alpha^k = fov^k(\hat{\boldsymbol{x}}(k)|\boldsymbol{y}(k-1)), \tag{1.85}$$

$k = 1, 2, \cdots, n.$

**Definition 1.21.** *(Johnsonbaugh and Pfaffenberger 2002 [38]) A solution $\boldsymbol{x}_g^*$ is called a global optimal solution, if $\boldsymbol{x}_g^*$ satisfies: $\forall \boldsymbol{x} \in R^n$, $f(\boldsymbol{x}) \ge f(\boldsymbol{x}_g^*)$. A solution $\boldsymbol{x}_l^*$ is called a locally optimal solution, if $\boldsymbol{x}_l^*$ satisfies: $\forall \boldsymbol{x} \in N_\delta(\boldsymbol{x}_l^*)$, $f(\boldsymbol{x}) \ge f(\boldsymbol{x}_l^*)$, where $N_\delta(\boldsymbol{x}_l^*)$ is a $\delta$-neighborhood of the solution $\boldsymbol{x}_l^*$.*

**Proposition 1.2.** *(Johnsonbaugh and Pfaffenberger 2002 [38]) Let $f(\boldsymbol{x})$ be a function that is differentiable at $\boldsymbol{x}^*$. If $\boldsymbol{x}^*$ is a locally optimal solution of $f(\boldsymbol{x})$, then $\nabla f(\boldsymbol{x}^*) = 0$.*

**Lemma 1.6.** *Let $S^k(\boldsymbol{x}(k)|\boldsymbol{y}^*(k-1))$ be the support function of the conditional lower bound surface of the $(k+1)^{th}$ stage $\Pi^k(\boldsymbol{y}^*(k-1))$ on the point $\boldsymbol{x}^*(k)$, if $S^k(\boldsymbol{x}(k)|\boldsymbol{y}^*(k-1))$, $fov^k(\boldsymbol{x}^*(k)|\boldsymbol{y}^*(k-1))$, and $FOV^k(\boldsymbol{x}^*(k))$ are differentiable to its argument $\boldsymbol{x}(k)$, and $\boldsymbol{x}^*(k)$ is the interior point of $\boldsymbol{X^k}(\boldsymbol{x}^*(k-1))$ then*

$$\nabla S^k(\boldsymbol{x}^*(k)|\boldsymbol{y}^*(k-1)) = \nabla fov^k(\boldsymbol{x}^*(k)|\boldsymbol{y}^*(k-1)) = \nabla FOV^k(\boldsymbol{x}^*(k)), \quad k = 1, 2, \cdots, n.$$

*Proof.* From Eq. (1.84) and (1.85), it is

$$S^k(\mathbf{x}(k)|\mathbf{y}^*(k-1)) - fov^k(\mathbf{x}(k)|\mathbf{y}^*(k-1)) \le \alpha^k,$$

$$S^k(\mathbf{x}^*(k)|\mathbf{y}^*(k-1)) - fov^k(\mathbf{x}^*(k)|\mathbf{y}^*(k-1)) = \alpha^k,$$

i.e. $S^k(\mathbf{x}(k)|\mathbf{y}^*(k-1)) - fov^k(\mathbf{x}(k)|\mathbf{y}^*(k-1))$ arrives at the maximum value $\alpha^k$ at point $\mathbf{x}^*(k)$, $k = 1, 2, \cdots, n$. Since $S^k(\mathbf{x}(k)|\mathbf{y}^*(k-1))$ and $fov^k(\mathbf{x}^*(k)|\mathbf{y}^*(k-1))$ are differentiable of its argument $\mathbf{x}(k)$, according to Proposition 1.2, so it is known that

$$\nabla(S^k(\mathbf{x}^*(k)|\mathbf{y}^*(k-1)) - fov^k(\mathbf{x}^*(k)|\mathbf{y}^*(k-1))) = 0, \quad k = 1, 2, \cdots, n,$$

i.e.

$$\nabla S^k(\mathbf{x}^*(k)|\mathbf{y}^*(k-1)) = \nabla fov^k(\mathbf{x}^*(k)|\mathbf{y}^*(k-1)), \quad k = 1, 2, \cdots, n.$$

From Eq. (1.82) and (1.83), it is

$$FOV^k(\mathbf{x}^*(k)) = fov^k(\mathbf{x}^*(k)|\mathbf{y}^*(k-1)),$$

From Eq. (1.79), it is

$$FOV^k(\mathbf{x}(k)) \le fov^k(\mathbf{x}(k)|\mathbf{y}^*(k-1)),$$

i.e. $FOV^k(\mathbf{x}(k)) - fov^k(\mathbf{x}(k)|\mathbf{y}^*(k-1))$ arrives at the maximum value 0 at point $\mathbf{x}^*(k)$, $k = 1,2,\cdots,n$. Since $FOV^k(\mathbf{x}(k))$ and $fov^k(\mathbf{x}^*(k)|\mathbf{y}^*(k-1))$ are differentiable in their argument $\mathbf{x}(k)$, according to Proposition 1.2, it is

$$\nabla fov^k(\mathbf{x}^*(k)|\mathbf{y}^*(k-1)) = \nabla FOV^k(\mathbf{x}^*(k)), \quad k = 1,2,\cdots,n.$$

The proof is completed. □

Then, the generalized Hamilton function concept is introduced in the following definition.

**Definition 1.22.** *Let $S^k(\boldsymbol{x}(k)|\boldsymbol{y}(k-1))$ be the support function for the conditional lower bound surface of the $(k+1)^{th}$ stage $\Pi^k(\boldsymbol{y}(k-1))$ at $\boldsymbol{x}(k)$, then $H^k(\boldsymbol{x}(k),\boldsymbol{u}(k),S^{k+1})$ is the generalized Hamilton function if and only if*

$$H^k(\boldsymbol{x}(k),\boldsymbol{u}(k),S^{k+1}) = -L_k(\boldsymbol{x}(k),\boldsymbol{u}(k)) + S^{k+1}(\boldsymbol{x}(k+1)|\boldsymbol{y}(k)), \tag{1.86}$$

$k = 0,1,\cdots,n-1.$

Based on Lemma 1.5, and the above statements and definitions, the most important lemma can be proved as at below.

**Lemma 1.7.** *It is said that $S^{k+1}(\boldsymbol{x}(k+1)|\boldsymbol{y}^*(k))$ is the support function for the conditional lower bound surface of the $(k+1)^{th}$ stage $\Pi^{k+1}(\boldsymbol{y}^*(k))$ at point $\boldsymbol{x}^*(k+1)$, if and only if $H^k(\boldsymbol{x}^*(k),\boldsymbol{u}(k),S^{k+1})$ arrives at the maximum value at point $\boldsymbol{u}^*(k)$, where $\boldsymbol{u}^*(k) \in U_k(\boldsymbol{x}^*(k))$, $k = 0,1,\cdots,n-1$.*

*Proof.* Let $S^{k+1}(\mathbf{x}(k+1)|\mathbf{y}^*(k))$ be the support function for the conditional lower bound surface of the $(k+1)^{th}$ stage $\Pi^{k+1}(\mathbf{y}^*(k))$ at point $x^*(k+1)$, then

$$-fov^{k+1}(\mathbf{x}(k+1)|\mathbf{y}^*(k)) + S^{k+1}(\mathbf{x}(k+1)|\mathbf{y}^*(k)) \le \alpha^{k+1}, \quad \forall \mathbf{x}(k+1) \in \mathbf{X}^{\mathbf{k+1}}(\mathbf{x}^*(k)). \tag{1.87}$$

Let $\bar{\mathbf{u}}(k) \in \mathbf{U}_k(\mathbf{x}(k))$, and $\bar{\mathbf{x}}(k+1) = \mathbf{f}^{\mathbf{k}}(\mathbf{x}^*(k),\bar{\mathbf{u}}(k))$, by Def. 1.16 and the statement of $fov^1(\mathbf{x}(1)|\mathbf{y}(0))$, then

$$-x^{0*}(k) - L_k(\mathbf{x}^*(k),\bar{\mathbf{u}}(k)) \le -fov^{k+1}(\bar{\mathbf{x}}(k+1)|\mathbf{y}^*(k)).$$

With Eq. (1.87), and by adding $S^{k+1}(\bar{\mathbf{x}}(k+1)|\mathbf{y}^*(k))$ to both sides of the above inequality, then

$$-x^{0*}(k) - L_k(\mathbf{x}^*(k),\bar{\mathbf{u}}(k)) + S^{k+1}(\bar{\mathbf{x}}(k+1)|\mathbf{y}^*(k)) \le \alpha^{k+1},$$

i.e.

$$H^k(\mathbf{x}^*(k),\mathbf{u}(k),S^{k+1}(\mathbf{x}(k+1)|\mathbf{y}^*(k))) \leq \alpha^{k+1} + x^{0*}(k), \quad \mathbf{u}(k) \in \mathbf{U}_k(\mathbf{x}^*(k)).$$

On the other hand, from Eq. (1.83) in Lemma 1.5, then

$$fov^{k+1}(\mathbf{x}^*(k+1)|\mathbf{y}^*(k)) = x^{0*}(k) + L_k(\mathbf{x}^*(k),\mathbf{u}^*(k)). \tag{1.88}$$

By Def. 1.20, we have

$$-fov^{k+1}(\mathbf{x}^*(k+1)|\mathbf{y}^*(k)) + S^{k+1}(\mathbf{x}^*(k+1)|\mathbf{y}^*(k)) \leq \alpha^{k+1},$$

i.e.

$$H^k(\mathbf{x}^*(k),\mathbf{u}^*(k),S^{k+1}(\mathbf{x}^*(k+1)|\mathbf{y}^*(k))) = \alpha^{k+1} + x^{0*}(k).$$

This means $H^k(\mathbf{x}^*(k),\mathbf{u}(k),S^{k+1}(\mathbf{x}(k+1)|\mathbf{y}^*(k)))$ arrives at the maximum value $\alpha^{k+1} + x^{0*}(k)$ at point $\mathbf{u}^*(k)$, $k = 0,1,\cdots,n-1$.

On the contrary, if $H^k(\mathbf{x}^*(k),\mathbf{u}(k),S^{k+1}(\mathbf{x}(k+1)|\mathbf{y}^*(k)))$ arrives at the maximum value at point $\mathbf{u}^*(k)$, then there exists $\beta^k$ such that

$$-L_k(\mathbf{x}^*(k),\mathbf{u}^*(k)) + S^{k+1}(\mathbf{x}(k+1)|\mathbf{y}^*(k)) \leq \beta^k, \tag{1.89}$$

where $\mathbf{u}(k) \in \mathbf{U}_k(\mathbf{x}^*(k))$,

$$-L_k(\mathbf{x}^*(k),\mathbf{u}^*(k)) + S^{k+1}(\mathbf{x}^*(k+1)|\mathbf{y}^*(k)) = \beta^k. \tag{1.90}$$

Let $\alpha^{k+1} = \beta^k - x^{0*}(k)$, by Eq. (1.88) and (1.90) then

$$-fov^{k+1}(\mathbf{x}^*(k+1)|\mathbf{y}^*(k)) + S^{k+1}(\mathbf{x}^*(k+1)|\mathbf{y}^*(k)) = \beta^k - x^{0*}(k) = \alpha^{k+1}.$$

On the other hand, with Eq. (1.89), then

$$-x^{0*}(k) - L_k(\mathbf{x}^*(k),\mathbf{u}(k)) + S^{k+1}(\mathbf{x}(k+1)|\mathbf{y}^*(k)) \leq \alpha^{k+1},$$

where $\mathbf{u}(k) \in \mathbf{U}_k(\mathbf{x}^*(k))$.

From Def. 1.16 and the statement for $fov^1(\mathbf{x}(1)|\mathbf{y}(0))$ and Eq. (1.83), then

$$-fov^{k+1}(\mathbf{x}(k+1)|\mathbf{y}^*(k)) + S^{k+1}(\mathbf{x}(k+1)|\mathbf{y}^*(k)) \leq \alpha^{k+1}, \quad \forall \mathbf{x}(k+1) \in \mathbf{X}^{\mathbf{k+1}}(\mathbf{x}^*(k)).$$

According to Def 1.20, it is known that $S^{k+1}(\mathbf{x}(k+1)|\mathbf{y}^*(k))$ is the support function for the conditional lower bound surface of the $(k+1)^{th}$ stage $\Pi^{k+1}(\mathbf{y}^*(k))$ on the point $\mathbf{x}^*(k+1)$, $k = 0,1,\cdots,n-1$. The proof is completed. □

**Theorem 1.2.** *Let $\{\boldsymbol{u}^*(0),\cdots,\boldsymbol{u}^*(n-1)\}$ be the optimal solution sequence of decision variables, and $\{\boldsymbol{x}^*(0) = \boldsymbol{a},\boldsymbol{x}^*(1),\cdots,\boldsymbol{x}^*(n)\}$ be the optimal solution sequence of the state variables. Assuming that the partial derivative of $L_k(\boldsymbol{x}(k),\boldsymbol{u}(k))$ and $f^k(x(k),u(k))$ for the argument $x_i(k)$ exist in the definitional domain, and $\boldsymbol{x}^*(k)$ is an interior point of $\boldsymbol{X}^{\boldsymbol{k}}(\boldsymbol{x}^*(k-1))$, where $k = 0,1,\cdots,n$. Then there exists a function*

*$S^k(\boldsymbol{x}(k)|\boldsymbol{y}^*(k-1))$ which is the support function for the conditional lower bound surface of the $(k+1)^{th}$ stage $\Pi^k(\boldsymbol{y}^*(k-1))$ at point $\boldsymbol{x}^*(k)$, $k=1,2,\cdots,n$, such that,*
*(i)*

$$\frac{\partial S^k(\boldsymbol{x}^*(k)|\boldsymbol{y}^*(k-1))}{\partial x_i(k)} = \frac{\partial H^k(\boldsymbol{x}^*(k),\boldsymbol{u}^*(k),S^{k+1}(\boldsymbol{x}^*(k+1)|\boldsymbol{y}^*(k)))}{\partial x_i(k)}, \tag{1.91}$$

*where $k=1,2,\cdots,n-1$;*
*(ii)*

$$H^k(\boldsymbol{x}^*(k),\boldsymbol{u}^*(k),S^{k+1*}) = \max[H^k(\boldsymbol{x}^*(k),\boldsymbol{u}(k),S^{k+1})|\boldsymbol{u}(k)\in U_k(\boldsymbol{x}^*(k))], \tag{1.92}$$

*where $S^{k+1*}=S^{k+1}(\boldsymbol{x}^*(k+1)|\boldsymbol{y}^*(k))$, $S^{k+1}=S^{k+1}(\boldsymbol{x}(k+1)|\boldsymbol{y}^*(k))$, $k=0,1,\cdots,n-1$;*
*(iii)*

$$\frac{\partial S^n(\boldsymbol{x}^*(n)|\boldsymbol{y}^*(n-1))}{\partial x_i(n)} = 0. \tag{1.93}$$

*Proof.* From Eq. (1.79), (1.82) and (1.83), then

$$FOV^k(\mathbf{x}(k)) \le fov^k(\mathbf{x}(k)|\mathbf{y}^*(k-1)), \quad k=1,2,\cdots,n,$$

$$FOV^k(\mathbf{x}^*(k)) = fov^k(\mathbf{x}^*(k)|\mathbf{y}^*(k-1)), \quad k=1,2,\cdots,n.$$

From Def. 1.20, it is known that $FOV^k(\mathbf{x}(k))$ is the support function for the conditional lower bound surface of the $(k+1)^{th}$ stage $\Pi^k(\mathbf{y}^*(k-1))$ at point $\mathbf{x}^*(k)$ (here $\alpha^k=0$). Now it is necessary to only prove that function $FOV^k(\mathbf{x}^*(k))$ satisfies (i), (ii), (iii), so it is only the support function $S^k(\mathbf{x}(k)|\mathbf{y}^*(k-1))$ which needs to be found.

Let $\hat{\mathbf{X}}^{\mathbf{k}}$ be a subset of $\mathbf{X}^{\mathbf{k}}$, $\hat{\mathbf{X}}^{\mathbf{k}}$ is defined as follows,

$$\hat{\mathbf{X}}^{\mathbf{k}} = \{\mathbf{x}(k)|\mathbf{x}(k) = f^{k-1}(\mathbf{x}(k-1),\mathbf{u}^*(k-1)),\mathbf{x}(k-1)\in\mathbf{X}^{\mathbf{k-1}}\}.$$

Let $\hat{F}^k(\mathbf{x}(k))$ be a function defined on the subset $\hat{\mathbf{X}}^{\mathbf{k}}$ as follows,

$$\begin{aligned}&\hat{F}^k(\mathbf{x}(k))\\ &=\min\{FOV^{k-1}(\mathbf{x}(k-1))+L_{k-1}(\mathbf{x}(k-1),\mathbf{u}^*(k-1))\\ &\quad|\mathbf{f}^{\mathbf{k-1}}(\mathbf{x}(k-1),\mathbf{u}^*(k-1))=\mathbf{x}(k),\mathbf{x}(k-1)\in\mathbf{X}^{\mathbf{k-1}}\}.\end{aligned}$$

From the above definition, then

$$FOV^k(\mathbf{x}(k)) \le \hat{F}^k(\mathbf{x}(k)), \quad \forall \mathbf{x}(k)\in\hat{\mathbf{X}}^{\mathbf{k}}, \tag{1.94}$$

$$FOV^k(\mathbf{x}^*(k)) = \hat{F}^k(\mathbf{x}^*(k)).$$

From Eq. (1.82), then

$$FOV^{k+1}(\mathbf{x}^*(k+1)) = FOV^k(\mathbf{x}^*(k)) + L_k(\mathbf{x}^*(k),\mathbf{u}^*(k)). \tag{1.95}$$

According to Eq. (1.94) and the definition for $\hat{F}^k(\mathbf{x}(k))$, then

$$FOV^{k+1}(\mathbf{x}(k+1)) \le \hat{F}^{k+1}(\mathbf{x}(k+1)) \le FOV^k(\mathbf{x}(k)) + L_k(\mathbf{x}(k),\mathbf{u}^*(k)), \quad (1.96)$$

$\forall \mathbf{x}(k) \in \mathbf{X^k}, \mathbf{x}(k+1) \in \hat{\mathbf{X}}^{\mathbf{k+1}}$.

Since, $\mathbf{x}(k+1) = \mathbf{f^k}(\mathbf{x}(k),\mathbf{u}(k))$, Eq. (1.95) and (1.96) as transformed as follows

$$-L_k(\mathbf{x}^*(k),\mathbf{u}^*(k)) + FOV^{k+1}(\mathbf{f^k}(\mathbf{x}^*(k),\mathbf{u}^*(k))) - FOV^k(\mathbf{x}^*(k)) = 0, \quad (1.97)$$

$$-L_k(\mathbf{x}(k),\mathbf{u}^*(k)) + FOV^{k+1}(\mathbf{f^k}(\mathbf{x}(k),\mathbf{u}^*(k))) - FOV^k(\mathbf{x}(k)) \le 0, \quad (1.98)$$

$\forall \mathbf{x}(k) \in \mathbf{X^k}$.

Let

$$\Upsilon(\mathbf{x}(k)) = -L_k(\mathbf{x}(k),\mathbf{u}^*(k)) + FOV^{k+1}(\mathbf{f^k}(\mathbf{x}(k),\mathbf{u}^*(k))) - FOV^k(\mathbf{x}(k)).$$

From Eq. (1.97) and (1.98), it is known that $\Upsilon(\mathbf{x}(k))$ arrives at the maximum value 0 at point $\mathbf{x}^*(k)$. According to the assumptions of this theorem and Proposition 1.2, then

$$\nabla\Upsilon(\mathbf{x}^*(k)) = 0. \quad (1.99)$$

From Def. 1.21, then

$$\Upsilon(\mathbf{x}(k)) = H^k(\mathbf{x}(k),\mathbf{u}^*(k),FOV^{k+1}(\mathbf{f^k}(\mathbf{x}(k),\mathbf{u}^*(k)))) - FOV^k(\mathbf{x}(k)).$$

From Eq. (1.99), then

$$\frac{\partial FOV^k(\mathbf{x}^*(k))}{\partial x_i(k)} = \frac{\partial H^k(\mathbf{x}^*(k),\mathbf{u}^*(k),FOV^{k+1}(\mathbf{x}^*(k+1)))}{\partial x_i(k)}, \quad k = 1,2,\cdots,n-1. \quad (1.100)$$

This means that $FOV^k(\mathbf{x}(k))$ satisfies (i), $k = 1,2,\cdots,n-1$.

Since it has been previously proved that $FOV^k(\mathbf{x}(k))$ is the support function for the conditional lower bound surface of the $(k+1)^{th}$ stage $\Pi^k(\mathbf{y}^*(k-1))$ at point $\mathbf{x}^*(k)$, according to Lemma 1.7, then, it can be seen that $FOV^{k+1}(\mathbf{x}(k+1))$ satisfies (ii), $k = 0,1,\cdots,n-1$.

From Def. 1.17, Lemma 1.4 and Eq. (1.82), then

$$FOV^n(\mathbf{x}(n)) \ge \min[FOV^n(\mathbf{x}(n))|\mathbf{x}(n) \in \mathbf{X^n}] = J^* = FOV^n(\mathbf{x}^*(n)),$$

i.e. $FOV^n(\mathbf{x}(n))$ arrives at the minimum value at point $\mathbf{x}^*(n)$, so then

$$\frac{\partial FOV^n(\mathbf{x}^*(n))}{\partial x_i(n)} = 0.$$

Hence, $FOV^n(\mathbf{x}(n))$ satisfies (iii). The proof is completed. □

### 1.2.5 Existence of Linear Support Function

However, the Hamilton function in the Theorem 1.2 is in a generalized form, and may not be linear. The generalized form of the Hamilton function is transformed into a linear function as follows:

$$H^k(\mathbf{x}(k),\mathbf{u}(k),\lambda(k+1)) = -L_k(\mathbf{x}(k),\mathbf{u}(k)) + \lambda^T(k+1)\mathbf{f^k}(\mathbf{x}(k),\mathbf{u}(k)), \quad (1.101)$$

$k = 0,1,\cdots,n-1$. The most important thing is to prove the existence of a linear support function as follows:

$$S^{k+1}(\mathbf{f^k}(\mathbf{x}(k),\mathbf{u}(k))) = \lambda^T(k+1)\mathbf{f^k}(\mathbf{x}(k),\mathbf{u}(k)), \quad \forall i \in \Psi, \quad k=0,1,\cdots,n-1.$$

Now, the problem is whether there is a linear support function $S^{k+1}(\mathbf{f^k}(\mathbf{x}(k),\mathbf{u}(k))) = \lambda^T(k+1)\mathbf{f^k}(\mathbf{x}(k),\mathbf{u}(k))$ or not. That means the existence of the linear support function needs to be proved. As a result, the following lemmas are needed.

**Lemma 1.8.** *Assume that $f^k(\mathbf{x}(k),\mathbf{u}(k))$ is a linear function of its argument $\mathbf{u}(k)$, and $\mathbf{U}_k(\mathbf{x}^*(k))$ is a convex set. The state attainable set of the $(k+1)^{th}$ stage $\mathbf{X^k}(\mathbf{x}(k-1))$ which is derived from $\mathbf{x}(k-1)$ is a convex set, $k = 1,2,\cdots,n$.*

*Proof.* Let $\delta \in [0,1]$, $\forall \mathbf{x}^1(k),\mathbf{x}^2(k) \in \mathbf{X^k}(\mathbf{x}(k-1))$, so it is necessary to prove that $\delta\mathbf{x}^1(k) + (1-\delta)\mathbf{x}^2(k) \in \mathbf{X^k}(\mathbf{x}(k-1))$, $k = 1,2,\cdots,n$. In fact, from Def. 1.14 and the statement of $\mathbf{X^2}(\mathbf{x}(1))$, then

$$\begin{aligned}
&\delta\mathbf{x}^1(k) + (1-\delta)\mathbf{x}^2(k) \\
&= \delta\mathbf{f^{k\text{-}1}}(\mathbf{x}(k-1),\mathbf{u}^1(k-1)) + (1-\delta)\mathbf{f^{k\text{-}1}}(\mathbf{x}(k-1),\mathbf{u}^2(k-1)) \quad (1.102)\\
&= \mathbf{f}^{k-1}(\mathbf{x}(k-1),\delta\mathbf{u}^1(k-1) + (1-\delta)\mathbf{u}^2(k-1)),
\end{aligned}$$

and

$$\delta\mathbf{u}(k) \in \mathbf{U}_k(\mathbf{x}^*(k)), \quad (1-\delta)\mathbf{u}(k) \in \mathbf{U}_k(\mathbf{x}^*(k)).$$

According to assumption,

$$\delta\mathbf{u}(k) + (1-\delta)\mathbf{u}(k) \in \mathbf{U}_k(\mathbf{x}^*(k)). \quad (1.103)$$

From Def. 1.14 and the statement of $\mathbf{X^2}(\mathbf{x}(1))$, Eq. (1.102) and (1.103), it is known that

$$\delta\mathbf{x}^1(k) + (1-\delta)\mathbf{x}^2(k) \in \mathbf{X^k}(\mathbf{x}(k-1)), \quad k = 1,2,\cdots,n.$$

This means $\mathbf{X^k}(\mathbf{x}(k-1))$ is a convex set, $k = 1,2,\cdots,n$. The proof is completed. □

**Definition 1.23.** *Let $z \in \boldsymbol{R}^2$, if $\forall\delta \in [0,1], \forall \boldsymbol{y}^1(k),\boldsymbol{y}^2(k) \in \boldsymbol{Y}^k(\boldsymbol{y}(k-1))$, then the set $\boldsymbol{Y}^k(\boldsymbol{y}(k-1)) \subset \boldsymbol{R}^2$ is called convex in direction $z$, if and only if there exists $\gamma \geq 0$, such that $\delta\boldsymbol{y}^1(k) + (1-\delta)\boldsymbol{y}^2(k) - \gamma z \in \boldsymbol{Y}^k(\boldsymbol{y}(k-1))$.*

**Lemma 1.9.** *The expanded state attainable set of the $(k+1)^{th}$ stage $\mathbf{Y^k}(\mathbf{y}(k-1))$, which is derived from $\mathbf{y}(k-1)$, is convex in direction $\boldsymbol{e} = (1,0)$, $k = 1,2,\cdots,n$.*

*Proof.* Let $\delta \in [0,1]$, $\forall \mathbf{y^1}(k), \mathbf{y^2}(k) \in \mathbf{Y^k}(\mathbf{y}(k-1))$, from Def. 1.23, it is necessary to prove that $\gamma \geq 0$ exists, such that $\delta \mathbf{y^1}(k) + (1-\delta)\mathbf{y^2}(k) - \gamma \mathbf{e} \in \mathbf{Y^k}(\mathbf{y}(k-1))$, $k = 1,2,\cdots,n$. In fact, from Def. 1.15 and the statement of $\mathbf{Y^2}(\mathbf{y}(1))$, then

$$\begin{aligned}
&\delta \mathbf{y^1}(k) + (1-\delta)\mathbf{y^2}(k) - \gamma \mathbf{e} \\
&= \delta \mathbf{F^{k\text{-}1}}(\mathbf{y}(k-1), \mathbf{u}^1(k-1)) + (1-\delta)\mathbf{F^{k\text{-}1}}(\mathbf{y}(k-1), \mathbf{u}^2(k-1)) - (\gamma, 0) \\
&= \delta(J_k(\mathbf{u}^1(k-1)), \mathbf{f^{k\text{-}1}}(\mathbf{x}(k-1), \mathbf{u}^1(k-1))) + (1-\delta) \\
&\quad (J_k(\mathbf{u}^2(k-1)), \mathbf{f^{k\text{-}1}}(\mathbf{x}(k-1), \mathbf{u}^2(k-1))) - (\gamma, 0) \\
&= (J_k(\delta \mathbf{u}^1(k-1) + (1-\delta)\mathbf{u}^2(k-1)) - \gamma, \mathbf{f^{k\text{-}1}}(\mathbf{x}(k-1), \delta \mathbf{u}^1(k-1) + (1-\delta)\mathbf{u}^2(k-1))).
\end{aligned} \tag{1.104}$$

Let $\gamma = 0$, then

$$\begin{aligned}
&(J_k(\delta \mathbf{u}^1(k-1) + (1-\delta)\mathbf{u}^2(k-1)) - \gamma, \mathbf{f^{k\text{-}1}}(\mathbf{x}(k-1), \delta \mathbf{u}^1(k-1) + (1-\delta)\mathbf{u}^2(k-1))) \\
&= (J_k(\delta \mathbf{u}^1(k-1) + (1-\delta)\mathbf{u}^2(k-1)), \mathbf{f^{k\text{-}1}}(\mathbf{x}(k-1), \delta \mathbf{u}^1(k-1) + (1-\delta)\mathbf{u}^2(k-1))) \\
&= \mathbf{F^{k\text{-}1}}(\mathbf{y}(k-1), \delta \mathbf{u}^1(k-1) + (1-\delta)\mathbf{u}^2(k-1)).
\end{aligned}$$

From Def. 1.15 and the statement for $\mathbf{Y^2}(\mathbf{y}(1))$ and Eq. (1.103), then

$$\delta \mathbf{y^1}(k) + (1-\delta)\mathbf{y^2}(k) - \gamma \mathbf{e} = \mathbf{F^{k\text{-}1}}(\mathbf{y}(k-1), \delta \mathbf{u}^1(k-1) + (1-\delta)\mathbf{u}^2(k-1)) \in \mathbf{Y^k}(\mathbf{y}(k-1)).$$

This means $\mathbf{Y^k}(\mathbf{y}(k-1))$ $\mathbf{y}(k-1)$ is convex in direction $\mathbf{e} = (1,0)$, $k = 1,2,\cdots,n$. The proof is completed. □

**Lemma 1.10.** *If the expanded state attainable set of the $(k+1)^{th}$ stage $\mathbf{Y^k}(\mathbf{y}(k-1))$, which is derived from $\mathbf{y}(k-1)$ is convex in direction $\mathbf{e} = (1,0)$, then the conditional function of the optimal value of the $(k+1)^{th}$ stage $fov^k(\mathbf{x}(k)|\mathbf{y}(k-1))$ is a convex function, $k = 1,2,\cdots,n$.*

*Proof.* Assume that $\mathbf{Y^k}(\mathbf{y}(k-1))$ is convex in direction $\mathbf{e} = (1,0)$. For $k = 1,2,\cdots,n$, let $\mathbf{x}^1(k), \mathbf{x}^2(k) \in \mathbf{X^k}(\mathbf{x}(k-1))$, choose $x^{0,1}(k), x^{0,2}(k)$, such that $\mathbf{y}^1(k) = (x^{0,1}(k), \mathbf{x}^1(k))$, $\mathbf{y}^2(k) = (x^{0,2}(k), \mathbf{x}^2(k))$, $\mathbf{y}^1(k)$, $\mathbf{y}^2(k) \in \mathbf{Y^k}(\mathbf{y}(k-1))$. From Def. 1.23, $\forall \delta \in [0,1]$, $\gamma \geq 0$ exists, such that

$$\delta \mathbf{y^1}(k) + (1-\delta)\mathbf{y^2}(k) - \gamma \mathbf{e} \in \mathbf{Y^k}(\mathbf{y}(k-1)),$$

i.e.

$$\delta \mathbf{x}^1(k) + (1-\delta)\mathbf{x}^2(k) \in \mathbf{X^k}(\mathbf{x}(k-1)).$$

Hence, $\mathbf{X^k}(\mathbf{x}(k-1))$ is a convex set. Now let $\overline{\mathbf{x}^1(k)}, \overline{\mathbf{x}^2(k)} \in \mathbf{X^k}(\mathbf{x}(k-1))$, let

$$\overline{x^{0,1}(k)} = fov^k(\overline{\mathbf{x}^1(k)}|\mathbf{y}(k-1)), \quad \overline{x^{0,2}(k)} = fov^k(\overline{\mathbf{x}^2(k)}|\mathbf{y}(k-1)),$$

$$\overline{\mathbf{y}^1(k)} = (\overline{x^{0,1}(k)}, \overline{\mathbf{x}^1(k)}), \quad \overline{\mathbf{y}^2(k)} = (\overline{x^{0,2}(k)}, \overline{\mathbf{x}^2(k)}),$$

then we have $\overline{\mathbf{y}^1(k)}, \overline{\mathbf{y}^2(k)} \in \mathbf{Y^k}(\mathbf{y}(k-1))$. Since $\mathbf{Y^k}(\mathbf{y}(k-1))$ is convex in direction $\mathbf{e} = (1,0)$, $\forall \delta \in [0,1]$, $\bar{\gamma} \geq 0$ exists, such that

$$\delta \overline{\mathbf{y}^1(k)} + (1-\delta)\overline{\mathbf{y}^2(k)} - \bar{\gamma}\mathbf{e} \in \mathbf{Y^k}(\mathbf{y}(k-1)),$$

i.e.

$$\delta fov^k(\overline{\mathbf{x}^1(k)}|\mathbf{y}(k-1)) + (1-\delta) fov^k(\overline{\mathbf{x}^2(k)}|\mathbf{y}(k-1)) - \bar{\gamma} \geq fov^k(\delta\overline{\mathbf{x}^1(k)} + (1-\delta)\overline{\mathbf{x}^2(k)}|\mathbf{y}(k-1)).$$

Since $\bar{\gamma} \geq 0$, then

$$\delta fov^k(\overline{\mathbf{x}^1(k)}|\mathbf{y}(k-1)) + (1-\delta) fov^k(\overline{\mathbf{x}^2(k)}|\mathbf{y}(k-1)) \geq fov^k(\delta\overline{\mathbf{x}^1(k)} + (1-\delta)\overline{\mathbf{x}^2(k)}|\mathbf{y}(k-1)).$$

This means $fov^k(\mathbf{x}(k)|\mathbf{y}(k-1))$ is a convex function, $k = 1, 2, \cdots, n$. The proof is completed. □

Based on Lemma 1.10, the following lemma can be proved.

**Lemma 1.11.** *If the conditional function for the optimal value of the* $(k+1)^{th}$ *stage* $fov^k(\boldsymbol{x}(k)|\boldsymbol{y}(k-1))$ *is convex on* $\boldsymbol{X}^k(\boldsymbol{x}(k-1))$*, and* $\hat{\boldsymbol{x}}(k)$ *is any interior point of* $\boldsymbol{X}^k(\boldsymbol{x}(k-1))$*, then there exists a linear function* $S^k(\boldsymbol{x}(k)) = \lambda^T(k)\boldsymbol{x}(k)$*, which is the support function for the conditional lower bound surface of the* $(k+1)^{th}$ *stage* $\Pi^k(\boldsymbol{y}(k-1))$ *at point* $\hat{\boldsymbol{x}}(k)$ $k = 1, 2, \cdots, n$.

*Proof.* Let $\overline{\mathbf{Y}}^{\mathbf{k}}$ be a subset of $\mathbf{Y}^{\mathbf{k}}(\mathbf{y}(k-1))$, $k = 1, 2, \cdots, n$, we define it as below

$$\overline{\mathbf{Y}}^{\mathbf{k}} = \left\{ \mathbf{y}(k) \left| \begin{array}{l} fov^k(\mathbf{x}(k)|\mathbf{y}(k-1)) \leq x^0(k) \leq \sup[fov^k(\mathbf{x}(k)|\mathbf{y}(k-1))| \\ \mathbf{x}(k) \in \mathbf{X}^{\mathbf{k}}(\mathbf{x}(k-1))], \mathbf{x}(k) \in \mathbf{X}^{\mathbf{k}}(\mathbf{x}(k-1)) \end{array} \right. \right\}.$$

Firstly, it is necessary to prove that $\overline{\mathbf{Y}}^{\mathbf{k}}$ is a convex set. Let $\mathbf{y}^1(k) = (x^{0,1}(k), \mathbf{x}^1(k))$, $\mathbf{y}^2(k) = (x^{0,2}(k), \mathbf{x}^2(k))$, $\mathbf{y}^1(k), \mathbf{y}^2(k) \in \overline{\mathbf{Y}}^{\mathbf{k}}$, so it is defined

$$\mathbf{y}^k(\delta) = \delta\mathbf{y}^1(k) + (1-\delta)\mathbf{y}^2(k), \quad 0 \leq \delta \leq 1.$$

According to Lemma 1.10, $\mathbf{X}^{\mathbf{k}}(\mathbf{x}(k-1))$ is a convex set. Let $\mathbf{x}^1(k), \mathbf{x}^2(k) \in \mathbf{X}^{\mathbf{k}}(\mathbf{x}(k-1))$, then

$$\delta\mathbf{x}^1(k) + (1-\delta)\mathbf{x}^2(k) \in \mathbf{X}^{\mathbf{k}}(\mathbf{x}(k-1)).$$

On the other hand, it is obvious that

$$x^{0,1}(k) \geq fov^k(\mathbf{x}^1(k)|\mathbf{y}(k-1)), \quad x^{0,2}(k) \geq fov^k(\mathbf{x}^2(k)|\mathbf{y}(k-1)),$$

since $fov^k(\mathbf{x}(k)|\mathbf{y}(k-1))$ is convex, then we can get

$$\delta x^{0,1}(k) + (1-\delta)x^{0,2} \geq fov^k(\delta\mathbf{x}^1(k) + (1-\delta)\mathbf{x}^2(k)|\mathbf{y}(k-1)).$$

Since

$$x^{0,1}(k) \leq \sup[fov^k(\mathbf{x}(k)|\mathbf{y}(k-1))|\mathbf{x}(k) \in \mathbf{X}^{\mathbf{k}}(\mathbf{x}(k-1))],$$
$$x^{0,2}(k) \leq \sup[fov^k(\mathbf{x}(k)|\mathbf{y}(k-1))|\mathbf{x}(k) \in \mathbf{X}^{\mathbf{k}}(\mathbf{x}(k-1))],$$

it is obvious that

$$\delta x^{0,1}(k) + (1-\delta)x^{0,2} \le \sup[fov^k(\mathbf{x}(k)|\mathbf{y}(k-1))|\mathbf{x}(k) \in \mathbf{X^k}(\mathbf{x}(k-1))].$$

This means $\mathbf{y}^k(\delta) \in \overline{\mathbf{Y}}^{\mathbf{k}}$, so $\overline{\mathbf{Y}}^{\mathbf{k}}$ is a convex set.

According to the theorems for convex sets, $\forall \hat{\mathbf{y}}(k) = (\hat{x}^0(k), \hat{\mathbf{x}}(k))$ which is a boundary point of $\overline{\mathbf{Y}}^{\mathbf{k}}$, is a support plane $\mathbf{P^k}$ which is defined below

$$\mathbf{P^k} = \{\mathbf{y}(k)|\mathbf{y}(k) = (x^0(k), \mathbf{x}(k)), a^0(k)x^0(k) + \mathbf{a}^T(k)\mathbf{x}(k) = b(k)\},$$

where $a^0(k), \mathbf{a}(k)$ and $b(k)$ are all nonnegative real values. Then

$$a^0(k)\hat{x}^0(k) + \mathbf{a}^T(k)\hat{\mathbf{x}}(k) = b(k),$$

$$a^0(k)x^0(k) + \mathbf{a}^T(k)\mathbf{x}(k) \ge b(k), \quad \forall \mathbf{y}(k) \in \overline{\mathbf{Y}}^{\mathbf{k}}.$$

Now let $\hat{\mathbf{x}}(k)$ be an interior point of $\mathbf{X^k}(\mathbf{x}(k-1))$, let $\hat{\mathbf{y}}(k) = (fov^k(\hat{\mathbf{x}}(k)|\mathbf{y}(k-1)), \hat{\mathbf{x}}(k))$, then from the definition for $\overline{\mathbf{Y}}^{\mathbf{k}}$, it can be seen that $\hat{\mathbf{y}}(k)$ is a boundary point of $\overline{\mathbf{Y}}^{\mathbf{k}}$, so then

$$a^0(k)fov^k(\hat{\mathbf{x}}(k)|\mathbf{y}(k-1)) + \mathbf{a}^T(k)\hat{\mathbf{x}}(k) = b(k),$$

$$a^0(k)fov^k(\mathbf{x}(k)|\mathbf{y}(k-1)) + \mathbf{a}^T(k)\mathbf{x}(k) \ge b(k), \quad \forall \mathbf{x}(k) \in \mathbf{X^k}(\mathbf{x}(k-1)).$$

Here, it must be that $a^0(k) \neq 0$, otherwise we have

$$\mathbf{a}^T(k)\hat{\mathbf{x}}(k) = b(k),$$

$$\mathbf{a}^T(k)\mathbf{x}(k) \ge b(k), \quad \forall \mathbf{x}(k) \in \mathbf{X^k}(\mathbf{x}(k-1)).$$

This means that $\mathbf{a}^T(k)\mathbf{x}(k) - b(k) = 0$ is just the support plane for $\mathbf{X^k}(\mathbf{x}(k-1))$ at point $\hat{\mathbf{x}}(k)$, i.e. $\hat{\mathbf{x}}(k)$ is a boundary point of $\mathbf{X^k}(\mathbf{x}(k-1))$, but it is impossible as it is assumed that $\hat{\mathbf{x}}(k)$ is an interior point of $\mathbf{X^k}(\mathbf{x}(k-1))$.

As $a^0(k) \neq 0$, let $\alpha^k = -\frac{b(k)}{a^0(k)}$, $\lambda(k) = -\frac{\mathbf{a}(k)}{a^0(k)}$, then

$$\lambda(k)\hat{\mathbf{x}}(k) - \alpha^k = fov^k(\hat{\mathbf{x}}(k)|\mathbf{y}(k-1)),$$

$$\lambda(k)\mathbf{x}(k) - \alpha^k \le fov^k(\mathbf{x}(k)|\mathbf{y}(k-1)), \quad \forall \mathbf{x}(k) \in \mathbf{X^k}(\mathbf{x}(k-1)).$$

From Def. 1.20, then

$$S^k(\mathbf{x}(k)) = \lambda^T(k)\mathbf{x}(k), \quad k = 1, 2, \cdots, n.$$

This is just the linear support function for the conditional lower bound surface of the $(k+1)^{th}$ stage $\Pi^k(\mathbf{y}(k-1))$ at point $\hat{\mathbf{x}}(k)$. The proof is completed. □

Now the theorem for the existence of the linear support function is proved.

**Theorem 1.3.** *If $\hat{\mathbf{x}}(k)$ is an interior point of $\mathbf{X^k}(\mathbf{x}(k-1))$, then there exists a $\lambda(k)$ such that $S^k(\mathbf{x}(k)) = \lambda^T(k)\mathbf{x}(k)$ is a linear support function for the conditional lower bound surface of the $(k+1)^{th}$ stage $\Pi^k(\mathbf{y}(k-1))$ at point $\hat{\mathbf{x}}(k)$ $k = 1, 2, \cdots, n$.*

*Proof.* According to Lemma 1.9, 1.10 and 1.11, the theorem of the existence of the linear support function is proven as below.

From Lemma 1.9, it is known that the expanded state attainable set for the $(k+1)^{th}$ stage $\mathbf{Y^k}(\mathbf{y}(k-1))$ which is derived from $\mathbf{y}(k-1)$ is convex in direction $\mathbf{e} = (1,0)$, $k = 1,2,\cdots,n$. Hence, the conditional function for the optimal value of the $(k+1)^{th}$ stage $fov^k(\mathbf{x}(k)|\mathbf{y}(k-1))$ is a convex function according to Lemma 1.10. From Lemma 1.11, it is known that there exists a $\lambda(k)$ such that $S^k(\mathbf{x}(k)) = \lambda^T(k)\mathbf{x}(k)$ is a linear support function for the conditional lower bound surface of the $(k+1)^{th}$ stage $\Pi^k(\mathbf{y}(k-1))$ at point $\hat{\mathbf{x}}(k)$ $k = 1,2,\cdots,n$. The proof is completed. □

Since the the existence of the linear support function has been proved, now it is possible to define the linear form for the Hamilton function.

**Definition 1.24.** *Let $S^{k+1}(\boldsymbol{x}(k+1)) = \lambda^T(k+1)\boldsymbol{x}(k+1)$ be the linear support function for the conditional lower bound surface of the $(k+1)^{th}$ stage $\Pi^{k+1}(\boldsymbol{y}(k))$ at point $\boldsymbol{x}(k+1)$, then $H^k(\boldsymbol{x}(k),\boldsymbol{u}(k),\lambda(k+1))$ is the linear Hamilton function if and only if*

$$H^k(\boldsymbol{x}(k),\boldsymbol{u}(k),\lambda(k+1)) = -L_k(\boldsymbol{x}(k),\boldsymbol{u}(k)) + \lambda^T(k+1)\boldsymbol{f}^k(\boldsymbol{x}(k),\boldsymbol{u}(k)), \quad (1.105)$$

*where*

$$\boldsymbol{f}^k(\boldsymbol{x}(k),\boldsymbol{u}(k)) = \boldsymbol{x}(k+1),$$

$k = 0,1,\cdots,n-1$.

### 1.2.6 The Maximal Theory

Now, it is possible to prove the theorem crucial for finding the optimal solution to the problem based on Theorem 1.2 and 1.3.

**Theorem 1.4.** *(The maximal theory) Let $\{\boldsymbol{u}^*(0),\cdots,\boldsymbol{u}^*(n-1)\}$ be the optimal solution sequence for the decision variables, and $\{\boldsymbol{x}^*(0) = \boldsymbol{a},\boldsymbol{x}^*(1),\cdots,\boldsymbol{x}^*(n)\}$ be the optimal solution sequence for the state variables. Assume that $\boldsymbol{x}^*(k)$ is an interior point of $\mathbf{X^k}(\boldsymbol{x}^*(k-1))$, where $k = 1,2,\cdots,n$. Then there exists a $\lambda^*(k)$, $k = 1,2,\cdots,n$, such that,*

*(i)*

$$\lambda_i^*(k) = \frac{\partial H^k(\boldsymbol{x}^*(k),\boldsymbol{u}^*(k),\lambda^*(k+1))}{\partial x_i(k)}, \quad k = 1,2,\cdots,n-1; \quad (1.106)$$

*(ii)*

$$H^k(\boldsymbol{x}^*(k),\boldsymbol{u}^*(k),\lambda^*(k+1)) = \max[H^k(\boldsymbol{x}^*(k),\boldsymbol{u}(k),\lambda^*(k+1))|\boldsymbol{u}(k) \in U_k(\boldsymbol{x}^*(k)] \quad (1.107)$$

$k = 0,1,\cdots,n-1$;

*(iii)*

$$\lambda_i^*(n) = 0. \quad (1.108)$$

*Proof.* From Theorem 1.3, since $\mathbf{x}^*(k)$ is an interior point of $\mathbf{X^k}(\mathbf{x}^*(k-1))$, there exists a $\lambda^*(k)$ such that $\hat{S}^k(\mathbf{x}^*(k)) = \lambda^*(k)\mathbf{x}^*(k)$ is a linear support function for the conditional lower bound surface of the $(k+1)^{th}$ stage $\Pi^k(\mathbf{y}^*(k-1))$ at point $\mathbf{x}^*(k)$, $k = 1, 2, \cdots, n$. From Def. 1.24, then

$$\frac{\partial H^k(\mathbf{x}^*(k),\mathbf{u}^*(k),\lambda^*(k+1))}{\partial x_i(k)} = -\frac{\partial L_k(\mathbf{x}(k),\mathbf{u}(k))}{\partial x_i(k)} + \frac{\partial \hat{S}^{k+1}(\mathbf{x}^*(k+1))}{\partial x_i(k+1)} \times \frac{\partial \mathbf{f^k}(\mathbf{x}^*(k),\mathbf{u}^*(k))}{\partial x_i(k)}.$$

According to Lemma 1.6, then

$$\frac{\partial \hat{S}^{k+1}(\mathbf{x}^*(k+1))}{\partial x_i(k+1)} = \frac{\partial FOV^{k+1}(\mathbf{x}^*(k+1))}{\partial x_i(k+1)}. \tag{1.109}$$

This means

$$\frac{\partial H^k(\mathbf{x}^*(k),\mathbf{u}^*(k),\lambda^*(k+1))}{\partial x_i(k)} = \frac{\partial H^k(\mathbf{x}^*(k),\mathbf{u}^*(k),FOV^{k+1}(\mathbf{x}^*(k+1)))}{\partial x_i(k)}. \tag{1.110}$$

From Eq. (1.100) in Theorem 1.2, then

$$\frac{\partial FOV^k(\mathbf{x}^*(k))}{\partial x_i(k)} = \frac{\partial H^k(\mathbf{x}^*(k),\mathbf{u}^*(k),FOV^{k+1}(\mathbf{x}^*(k+1)))}{\partial x_i(k)}. \tag{1.111}$$

From Eq. (1.109), (1.110) and (1.111), then

$$\frac{\partial H^k(\mathbf{x}^*(k),\mathbf{u}^*(k),\lambda^*(k+1))}{\partial x_i(k)} = \frac{\partial \hat{S}^k(\mathbf{x}^*(k))}{\partial x_i(k)}. \tag{1.112}$$

Since

$$\frac{\partial \hat{S}^k(\mathbf{x}^*(k))}{\partial x_i(k)} = \lambda_i^*(k),$$

so

$$\lambda_i^*(k) = \frac{\partial H^k(\mathbf{x}^*(k),\mathbf{u}^*(k),\lambda^*(k+1))}{\partial x_i(k)}, \quad k = 1, 2, \cdots, n-1.$$

This means $\lambda_i^*(k)$ satisfies (i).

Since

$$\hat{S}^k(\mathbf{x}^*(k)) = \lambda^{T*}(k)\mathbf{x}^*(k)$$

is a linear support function for the conditional lower bound surface of the $(k+1)^{th}$ stage $\Pi^k(\mathbf{y}^*(k-1))$ at point $\mathbf{x}^*(k)$, according to Lemma 1.7, so it is known that

$$\hat{S}^{k+1}(\mathbf{x}^*(k+1)) = \lambda^{T*}(k+1)\mathbf{x}^*(k+1)$$

satisfies (ii), for $k = 0, 1, \cdots, n-1$.

According to Theorem 1.2, it is proved that

$$\frac{\partial FOV^n(\mathbf{x}^*(n))}{\partial x_i(n)} = 0.$$

From Eq. (1.109), then

$$\frac{\partial \hat{S}^n(\mathbf{x}^*(n))}{\partial x_i(n)} = \frac{\partial FOV^n(\mathbf{x}^*(n))}{\partial x_i(n)}.$$

Since

$$\frac{\partial \hat{S}^n(\mathbf{x}^*(n))}{\partial x_i(n)} = \lambda_i^*(n),$$

then

$$\lambda_i^*(n) = 0.$$

This means $\lambda_i^*(n)$ satisfies (iii). The proof is completed. □

The maximal theory proved above plays an important role in the design of a theoretical algorithm for solving MOMSDM problems, which is discussed and employed throughout the remainder of this book.

## 1.3 MOMSDM under a Specified Termination Time

The termination time is a very important element in the MOMSDM process. In the real world, most practical MOSMSDM problems have a specified termination time. In this section, we introduce a MOMSDM under a fixed and specified termination time, and discuss the characteristics of this case in more detail.

### *1.3.1 System with a Specified Termination Time*

A multistage decision making process is finite if there are only a finite number of stages in the process and a finite number of states associated with each stage. It is often convenient and sometimes necessary to incorporate stage numbers as a part of the definition of the state. For example, in a linear search problem there are N distinct decisions that must be made, and they are assumed to be made in a specified order. It is assumed that $N$, also called the horizon, is finite and known. The first decision, made at stage 1, is to decide which data item should be placed first in the array, the second decision, made at stage 2, is to decide which data item should be placed second in the array, and so on.

A typical formulation for a MOMSDM under a fixed and specified termination time is given as follows,

$$\begin{cases} \max\{J_1(\mathbf{x}(0),\mathbf{x}(T),\pi), J_2(\mathbf{x}(0),\mathbf{x}(T),\pi),\cdots,J_n(\mathbf{x}(0),\mathbf{x}(T),\pi)\} \\ \text{s.t.} \begin{cases} \mathbf{x}(k+1) = f_k(\mathbf{x}(k),\mathbf{u}(k)), \\ \mathbf{x}(0) \in I, \\ T = a, \\ g_j(\mathbf{x}(\cdot),\mathbf{u}(\cdot)) \le 0,\ j = 1,2,\cdots,p, \\ \mathbf{x}(k) \in X,\ k = 1,2,\cdots,T, \\ \mathbf{u}(k) \in U,\ k = 1,2,\cdots,T. \end{cases} \end{cases} \tag{1.113}$$

where $\mathbf{x}(k)$ and $\mathbf{u}(k)$ are state and decision variables; $I$ is the initial state space; $X$ is the state space; $U$ is the decision space.

### 1.3.2 The Structure of a Discrete Process

In this section the basic problem for the optimal control of a dynamic system over a finite horizon is formulated. The formulation is very general since the state space, control space, and uncertainty space are arbitrary and may vary from one state to the next. In particular, the system may be defined over a finite or infinite state space. The problem is characterized by the fact that the number of evolutionary stages in the system are finite and fixed (finite horizon), and by the fact that the control law is a function of the current state (perfect state information). However, the problems that arise when the termination time is not fixed or when the controller decides to terminate the process prior to the final time can be reduced to a fixed termination time case. Situations where the controller has imperfect state information can also be reduced to a problem with perfect state information. A variety of related problems can also be reduced into the basic problem form. Given is the discrete-time dynamic system

$$x_{k+1} = f_k(x_k, u_k, w_k), \quad k = 0, 1, \cdots, N-1,$$

where the state $x_k$ is an element of a space $s_k$, $k = 0, 1, \cdots, N$, the control $u_k$ is an element of a space $c_k$, $k = 0, 1, \cdots, N-1$, and the random "disturbance" $w_k$ is an element of a space $D_k$, $k = 0, 1, \cdots, N-1$. The control $u_k$ is constrained to take values from a given nonempty subset $U_k(X_k)$ of $C_k$, which depends on the current state $x_k$, $u_k$, and $U_k(X_k)$ for all $x_k \in S_k$, and $k = 0, 1, \cdots, N-1$. The random disturbance $w_k$ is characterized by a probability measure $P_k(x_k, u_k)$ defined from a collection of events in $D_k$. This probability measure may depend explicitly on $x_k$ and $u_k$ but not on the values of the prior disturbances $w_{k-1}, \cdots, w_0$. Here, the class of control laws (also called "policies") that consist of a finite sequence of function for all $x_k \in s_k$ are considered. Such control laws are termed admissible.

### 1.3.3 Linear Systems and Quadratic Returns

Many multistage decision processes have returns (cost or benefits) associated with each decision, and these returns may vary with both the stage and state of the process. The multistage decision process is deterministic if the outcome of each decision is exactly known.

In the real world, it is common that the MOMSDM state equation is a linear system. In this section the special case of a linear system

$$x_{k+1} = A_k x_k + B_k u_k + w_k, \quad k = 0, 1, \cdots, N-1,$$

and the quadratic return are considered

$$E\{x'_N Q_N x_N + \sum_{k=0}^{N-1} (x'_k Q_k x_k + u'_k R_k u_k)\}.$$

In these expressions, $x_k$ and $u_k$ are vectors of dimensions $n$ and $m$, respectively, and the matrices $A_k$, $B_k$, $Q_k$, $R_k$ are given and have an appropriate dimension. It is assumed that the matrices $Q_k$ are positive semi-definite and symmetrical, and the matrices $R_k$ are positive definite and symmetrical. The controls $U_k$ are unconstrained. The disturbances $w_k$ are independent random vectors with given probability distributions that do not depend on $x_k$ and $u_k$. Further, each $w_k$ has a zero mean and a finite second moment. The problem described above is a popular formulation for a regulation problem in which it is necessary to keep the system state close to the origin. Such problems are common in automatic control theory for a motion or a process. The quadratic cost function is often useful as it induces a high penalty for large deviations of the state from the origin, but a relatively small penalty for small deviations. Also, quadratic cost is frequently used, even when it is not entirely justified, because it leads to a practical analytical solution. A number of variations and generalizations have similar solutions. For example, the disturbances $w_k$ could have nonzero means and the quadratic cost could have the form

$$E\{(x_N-\bar{x}_N)'Q_N(x_N-\bar{x}_N)+\sum_{k=0}^{N-1}((x_k-\bar{x}_k)'Q_k(x_k-\bar{x}_k)+u_k'R_ku_k)\},$$

which expresses a desire to keep the state of the system close to a given trajectory $(\bar{x}_0, \bar{x}_1, \cdots, \bar{x}_N)$ rather than close to the origin. Another generalized version of the problem arises when $A_k$, $B_k$ are independent random matrices, rather than being known. This case is considered at the end of this section.

Applying the DP algorithm, there is

$$J_N(x_N)=x_N'Q_Nx_N,$$

$$J_k(x_k)=\min_{u_k} E\{x_k'Q_kx_k+u_k'R_ku_k+J_{k+1}(A_kx_k+B_ku_k+w_k)\}. \tag{1.114}$$

It turns out that the cost-to-go functions $J_k$ are quadratic and as a result the optimal control law is a linear function of the state. These facts can be verified by straightforward induction. Thus, Eq. (1.114) is written for $k=N-1$,

$$\begin{aligned}J_{N-1}(x_{N-1})=\min E\{&x_{N-1}'Q_{N-1}x_{N-1}+u_{N-1}'R_{N-1}u_{N-1}\\&+(A_{N-1}x_{N-1}+B_{N-1}u_{N-1}+w_{N-1})'Q_N\\&(A_{N-1}X_{N-1}+B_{N-1}u_{N-1}+w_{N-1})\},\end{aligned}$$

and the last quadratic form in the right-hand side is expanded. Then the fact $E\{w_{N-1}\}=0$ is used to eliminate the term $E\{w_{N-1}'Q_N(A_{N-1}x_{N-1}+B_{N-1}u_{N-1})\}$, and so,

$$\begin{aligned}J_{N-1}(x_{N-1})=&x_{N-1}'Q_{N-1}x_{N-1}+\min[u_{N-1}'R_{N-1}u_{N-1}\\&+u_{N-1}'B_{N-1}'Q_NB_{N-1}u_{N-1}+2x_{N-1}'A_{N-1}'Q_NB_{N-1}u_{N-1}]\\&+x_{N-1}'A_{N-1}'Q_NA_{N-1}x_{N-1}+E\{w_{N-1}'Q_Nw_{N-1}\}.\end{aligned}$$

By differentiating with respect to $U_{N-1}$ and by setting the derivative equal to zero, then

$$(R_{N-1}+B'_{N-1}Q_NB_{N-1})u_{N-1}=-B'_{N-1}Q_NA_{N-1}x_{N-1}.$$

The matrix multiplying $U_{N-1}$ on the left is a positive definite (and is hence invertible), since $R_{N-1}$ is a positive definite and $B'_{N-1}Q_NB_{N-1}$ is a positive semi-definite. As a result, the minimizing control vector is given by

$$u^*_{N-1}=-(R_{N-1}+B'_{N-1}Q_NB_{N-1})^{-1}B'_{N-1}Q_NA_{N-1}x_{N-1}.$$

By substitution into the expression for $J_{N-1}$, then

$$J_{N-1}(x_{N-1})=x'_{N-1}K_{N-1}x_{N-1}+E\{w'_{N-1}Q_Nw_{N-1}\},$$

where by straightforward calculation, the matrix $K_{N-1}$ is verified to be

$$K_{N-1}=A'_{N-1}(Q_N-Q_NB_{N-1}(B'_{N-1}Q_NB_{N-1}+R_{N-1})^{-1}B'_{N-1}Q_N)A_{N-1}+Q_{N-1}.$$

The matrix $K_{N-1}$ is clearly symmetric. It is also positive semidefinite. To see this, note that from the preceding calculation there is $x\in\Re^n$

$$\begin{aligned}x'K_{N-1}x=\min\nolimits_u[&x'Q_{N-1}x+u'R_{N-1}u\\&+(A_{N-1}x+B_{N-1}u)'Q_N(A_{N-1}x+B_{N-1}u)].\end{aligned}$$

Since $Q_{N-l}$, $R_{N-1}$, and $Q_N$ are positive semi-definite, the expression within the brackets is nonnegative. Minimization over $u$ preserves the non-negativity, so it follows that $x'K_{N-1}x\geq 0$ for all $x\in\Re^n$. Hence $K_{N-1}$ is a positive semi-definite.

Since $J_{N-1}$ is a positive semi-definite quadratic function (plus an inconsequential constant term), it is possible to proceed in a similar way and obtain from the DP equation (1.114) the optimal control law for stage $N-2$. As shown earlier, $J_{N-2}$ is a positive semi-definite quadratic function, and by proceeding sequentially, it is possible to obtain the optimal control law for every $k$. This has the form

$$u^*_k(x_k)=L_kx_k, \tag{1.115}$$

where the gain matrices $L_k$ are given by the equation

$$L_k=-(B'_kK_{k+1}B_k+R_k)^{-1}B'_kK_{k+1}A_k,$$

and where the symmetric positive semidefinite matrices $K_k$ are given recursively by the algorithm

$$K_N=Q_N, \tag{1.116}$$

$$K_k=A'_k(K_{k+1}-K_{k+1}B_k(B'_kK_{k+1}B_k+R_k)^{-1}B'_kK_{k+1})A_k+Q_k. \tag{1.117}$$

As with the DP, this algorithm starts at terminal time $N$ and proceeds backwards. The optimal cost is given by

$$J_0(x_0)=x'_0K_0x_0+\sum_{k=0}^{N-1}E\{w'_kK_{k+1}w_k\}.$$

The control law (1.115) is simple and attractive for implementation in engineering applications: the current state $x_k$ is being fed back as input through the linear feedback gain matrix $L_k$ as shown in Figure 1.10. This accounts in part for the popularity of the linear-quadratic formulation.

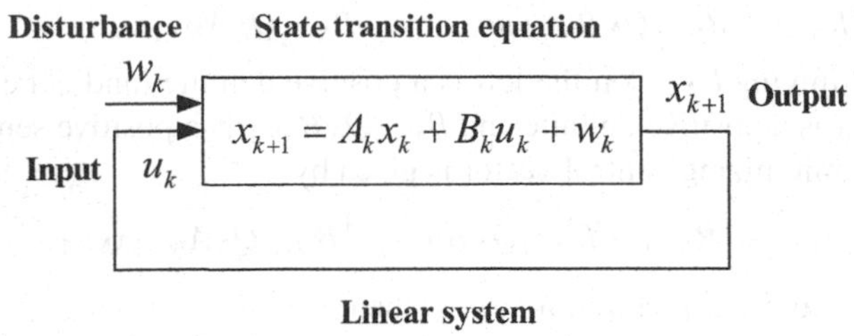


**Fig. 1.10** Linear feedback structure of the optimal controller for the linear-quadratic problem

## 1.3.4 Solution Method

The conventional algorithm for the basic problem is now outlined and its optimality shown by translating it into mathematical terms.

### Conventional Algorithm

Ideally, it would be best to use a DP algorithm to obtain closed-form expressions for $J_k$ or an optimal policy. In this book, however, a large number of models that allow for an analytical solution using the DP are discussed. Even if such models rely on oversimplified assumptions, they are often very useful as they may provide valuable insights into the structure of the optimal solution for more complex models, and they may form the basis for suboptimal control schemes. Further, the broad collection of analytically solvable models provides helpful guidelines for modeling when faced with a new problem as it is worth trying to pattern a model after one of the principal analytically tractable models.

$$x_k^i = \underline{x}_k + i \triangle x, \quad (i = 1, 2, \cdots, I(k)),$$

in which $I(k)$ is the maximal value of $i$ which makes $x_k^i \in X_k$. $\forall x_k^i$, $1 \leq i \leq I(k)$, for the decision variable $u_k$,

$$u_k^i(x_k^i) = \underline{u}_k(^i_k) + j \triangle u, \quad (j = 1, 2, \cdots, J(i,k))$$

Unfortunately, in many practical cases an analytical solution is not possible, and it is necessary to resort to a numerical execution of the DP algorithm. This may be quite time-consuming since the minimization in the DP must be carried out for each value of $X_k$. The state space must be discretized in some way if it is not already a finite set. The computational requirements are proportional to the number of possible values of $X_k$, so for complex problems the computational burden may be excessive. Nonetheless, the DP is the only general approach for sequential optimization under uncertainty, and even when it is computationally prohibitive, it can serve as the basis for more practical suboptimal approaches.

(1) The argument of the preceding proof provides an interpretation of $J_k(x_k)$ as the optimal cost for an $(N-k)$-stage problem starting at state $x_k$ and time $k$, and

ending at time $N$. $J_k(x_k)$ is known as the cost-to-go at state $x_k$ and time $k$, and refers to $J_k$ as the cost-to-go function at time $k$.

(2) In the second equation above, the minimum over $J(k+l)$ was moved inside the bracketed expression using a principle of optimality argument, that is the tail portion of an optimal policy is optimal for the tail sub-problem.

(3) In the third equation, the definition of $J(k+1)$ was used, and in the fourth equation the induction hypothesis was used. In the fifth equation, the minimization over $I(k)$ was converted to a minimization over $U_k$, using the fact that for any function $F$ of $x$ and $u$,

$$x_{k+1}^{ij} = f_k(x_k^i, u_k^{ij}),$$

in which $1 \leq i \leq I(k)$, $1 \leq j \leq J(i,k)$.

(4) The intuitive justification of the principle of optimality is very simple. If the truncated policy $\{u_i^*, u_{i+1}^*, \cdots, u_{N-1}^*\}$ were not optimal as stated, it would be possible to reduce the cost further by switching to an optimal policy for the sub-problem once $x_i$ is reached. Using an automobile travel analogy, suppose that the fastest route from Los Angeles to Boston passes through Chicago. The principle of optimality translates to the obvious fact that the Chicago to Boston portion of the route is also the fastest route for a trip that starts from Chicago and ends in Boston.

**Function Approximation Method**

The basic idea of this method is to take advantage of the known analytical properties of appropriate functions to approximate the optimal value function.The principle of optimality suggests that an optimal policy can be constructed in piecemeal fashion, first by constructing an optimal policy for the "tail sub-problem" involving the last stage, then by extending the optimal policy to the "tail sub-problem" involving the last two stages, and continuing in this manner until an optimal policy for the entire problem is constructed. The DP algorithm is based on this idea as it proceeds sequentially by solving all the tail sub-problems of a given time length, using the solution of the tail sub-problems of shorter time length. This algorithm is introduced through a deterministic example and a stochastic example.

$$\Phi_{n+1}(x) = \frac{2n+1}{n+1} x\Phi_n(x) - \frac{n}{n+1}\Phi_{n-1}(x),$$

$$\Phi_0(x) = 1, \quad \Phi_1(x) = x.$$

in the interval $[-1,+1]$, the orthogonality can be expressed as:

$$\int_{-1}^{+1} \Phi_l(x)\Phi_m(x)dx = \begin{cases} \frac{2}{2m+1}, & l = m, \\ 0, & l \neq m. \end{cases}$$

According to the principle of optimality, the "tail" portion of an optimal schedule must be optimal.

$$f(x_1, x_2, \cdots, x_n) = \sum_{i=0}^{R} a_i \prod_{s=1}^{n} \Phi_{t(i,s)}(x_s),$$

in which $\prod_{s=1}^{n} \Phi_{t(i,s)}(x_s)$ so that the optimal costs are computed as expected values. The coefficient $a_i$ can be determined as follows:

$$a_i = \alpha(i) \int_{-1}^{+1} dx_1 \cdots \int_{-1}^{+1} f(x_1, x_2, \cdots, x_n) \prod_{s=1}^{n} \Phi_{t(i,s)}(x_s) dx_n,$$

according to the properties of Legendre polynomial function, the above definite integral can be approximated using the following equation:

$$a_i = \alpha(i) \sum_{j_1=1}^{r} \sum_{j_2=1}^{r} \cdots \sum_{j_n=1}^{r} \lambda_{j_1} \lambda_{j_2} \cdots \lambda_{j_n} \cdot f(\mu_{j_1}, \mu_{j_2}, \cdots, \mu_{j_n}) \cdot \prod_{s=1}^{n} \Phi_{t(i,s)}(\mu_{j_s}) + \varepsilon,$$

assume that at the beginning of period $N-1$ the stock is $x_{N-1}$. Clearly, no matter what happened in the past, the inventory manager should order the amount of inventory that minimizes over $U_{N-1}$ the sum of the ordering cost and the expected holding/shortage cost. Thus, they should minimize over $U_{N-1}$ the sum $U_{N-1} + E\{R(x_N)\}$, which can be written as

$$\begin{aligned} &V_k(\mu_{j_1}, \mu_{j_2}, \cdots, \mu_{j_n}) \\ &= \min_{u \in U_k} \{L_k(\mu_{j_1}, \mu_{j_2}, \cdots, \mu_{j_n}, u) \\ &+ \textstyle\sum_{i=1}^{R} a_{k+1,i} \prod_{s=1}^{n} \Phi_{t(i,s)}[f_{k,s}(\mu_{j_1}, \mu_{j_2}, \cdots, \mu_{j_n}, u)]\}. \end{aligned}$$

This minimization is done by setting the derivative to zero with respect to $u_1$. This yields

$$R = \frac{(m+n)!}{m!n!},$$

Note that this is not a single control but rather a control function, a rule that tells us the optimal temperature $u_1 = u_1^*(x_1)$ for each possible state $x_1$.

**Discrete-Time Dynamic Programming Method**

Consider this chess match example [55]. There, a player can select a timid play (probabilities $P_d$ and $1-P_d$ for a draw or loss, respectively) or a bold play (probabilities $P_w$ and $1-P_w$ for a win or loss, respectively) in each game of the match. Thus it is necessary to seek to formulate a DP algorithm in such a way as to find the policy that maximizes the player's probability of winning the match. Note that here a maximization problem is being dealt with. The problem can be converted to a minimization problem by changing the sign of the cost function, but a simpler alternative, which is generally adopted, is to replace the minimization in the DP algorithm with a maximization.

$$x(k) = \bar{x}(k) + \delta x(k), \quad u(k) = \bar{u}(k) + \delta u(k).$$

Consider the general case of an N-game match, and let the state be the net score,

$$\begin{aligned} &\bar{J}_k(\bar{x}(k)) + a_k + V_x^k(\bar{x}(n))\delta x(k) + \tfrac{1}{2}\delta x(k)^T V_{xx}^k(\bar{x}(k))\delta x(k) \\ &= \min_{\delta u(k)} \{L_k(\bar{x}(k), \bar{u}(k)) + \bar{J}_{k+1}(\bar{x}(k+1)) + a_{k+1} + H_x^k \delta x(k) + H_u^k \delta u(k) \\ &+ \tfrac{1}{2}\delta x(k)^T A_k \delta x(k) + \delta u(k)^T B_k \delta x(k) + \tfrac{1}{2}\delta u(k)^T C_k \delta u(k) + e_k\}, \end{aligned}$$

in which $\bar{J}_k(\bar{x}(k))$ is the value of the objective function $J_k$ on the nominal trajectory, and the pseudo-Hamilton function $H$ is defined as

$$H_k(x(k),u(k),\psi) = L_k(x(k),u(k)) + \psi^T f_k(x(k),u(k)).$$

Let $V_x^k$ denote $\frac{\partial V_k}{\partial x(k)}$, $V_{xx}^k$ denote $\frac{\partial^2 V_k}{\partial x^2(k)}$,

$$\begin{aligned}
A_k &= H_{xx}^k + f_x^{kT} V_{xx}^{k+1}(\bar{x}(k+1)) f_x^k, \\
B_k &= H_{ux}^k + f_u^{kT} V_{xx}^{k+1}(\bar{x}(k+1)) f_x^k, \\
C_k &= H_{uu}^k + f_u^{kT} V_{xx}^{k+1}(\bar{x}(k+1)) f_u^k, \\
a_k &= V_k(\bar{x}(k)) - \bar{J}_k(\bar{x}(k)).
\end{aligned}$$

Using this system equation and denoting $g(i,u)$ as the expected cost per stage at state $i$ when control $U$ is applied, the DP algorithm can be rewritten as

$$C_k \delta u(k) + H_u^{kT} + B_k \delta x(k) = 0.$$

Set $C_k$ to be a positive definite matrix, then the minimal solution to $\delta u(k)$ should be

$$\delta u(k)^* = \alpha_k + \beta_k \delta x(k),$$

in which,

$$\begin{aligned}
\alpha_k &= -C_k^{-1} H_u^{kT}, \\
\beta_k &= -C_k^{-1} B_k.
\end{aligned}$$

Then we can get

$$\begin{aligned}
a_k &= a_{k+1} - \frac{1}{2} H_u^k C_k^{-1} H_u^{kT}, \\
V_x^k &= H_x^k - H_u^k C_k^{-1} B_k, \\
V_{xx}^k &= A_k - B_k^T C_k^{-1} B_k,
\end{aligned}$$

The two expressions in the above minimization correspond to the two available decisions (replace or not replace the machine). Then the termination condition is expressed as follows:

$$\begin{aligned}
a_N &= 0, \\
V_x^N(\bar{x}(N)) &= K_x(\bar{x}(N)), \\
V_{xx}^N(\bar{x}(N)) &= K_{xx}(\bar{x}(N)).
\end{aligned}$$

The above equations constitute the basic equations for differential dynamic programming.

The functions $J_k(C_k)$ denote the optimal expected cost for the tail sub-problem that starts at period $k$ with an initial inventory $x(k)$. These functions are computed recursively backward in time, starting at period $N-1$ and ending at period 0. The value $J_0(x(0))$ is the optimal expected cost when the initial stock at time 0 is $r_0$.

During the calculations, the optimal policy is simultaneously computed from the minimization in the right-hand side:

$$\alpha_k = -\varepsilon C_k^{-1} H_u^{kT} \quad (k = 0, 1, \cdots, N-1),$$

in which $0 \leq \varepsilon \leq 1$, then we can get

$$a_k = a_{k+1} - \varepsilon(1 - \frac{\varepsilon}{2}) H_u^k C_k^{-1} H_u^{kT}.$$

This nested set of optimization operations is performed from the inside-out (right-to-left), with the innermost optimization yielding the optimal choice for $d_n$:

$$a_0 = -\varepsilon(1 - \frac{\varepsilon}{2}) \sum_{j=0}^{N-1} H_u^j C_j^{-1} H_u^{jT}.$$

It should be noted that the sequence of decisions need not be limited to a fixed length $n$, but may be of indefinite length, terminating when a base case is reached. Different classes of DP problems may be characterized by how the states $S$, and hence the next-states $S'$ , are defined. It is often convenient to define the state $S$ not as a sequence of decisions made so far, with the next decision $d$ chosen from $D(S)$, but rather as the set from which the next decision can be chosen, so that $D(S) = ord \in S$.

Now the steps for the discrete-time dynamic programming algorithm are outlined:

**Step 1.** Given the initial trajectory $\{u^1(k)|k = 0, 1, \cdots, N-1\}$, which meets $\forall k$, $x^1(k) \in X_k$.

**Step 2.** Set the $lth$ iteration, $\xi^l \triangleq \{x^l(0), x^1(1), \cdots, x^l(N)\}$, let $j = l \cdot mod(N-1) - 1$, solve the following mathematical programming:

$$\begin{aligned}&\min\{L_j(x^l(j), u(j)) + L_{j+1}[f_j(x^l(j), u(j)), u(j+1)]|\\ &\quad u(j) \in U_j, u(j+1) \in U_{j+1}, f_{j+1}[f_j(x^1(j), u(j)), u(j+1)] = x^l_{j+2}k\}.\end{aligned}$$

**Step 3.** Let $\xi^{l*} \triangleq \{x^l_0, \cdots, x^l_j, x^{l*}_{j+1}, \cdots, x^l_N\}$, compute the objective function value $J_0(\xi^{l*})$.

**Step 4.** If $J_0(\xi^{l*}) < J_0(\xi^l)$, then $\xi^{l+1} = \xi^{l*}$, else $\xi^{l+1} = \xi^l$.

$$u^l_{k,j} = \varepsilon^l_j \alpha^l_k + \beta^l_k(x^l_{k,j} - x^l_k) + u^l_k,$$

$$x^l_{k+1,j} = f_k(x^l_{k,j}, u^l_{k,j}).$$

**Step 5.** Calculate the value of objective function $J^l_{0,j}$ corresponding to $\{u^1_{k,j}|k = 0, 1, \cdots, N-1\}$.

**Step 6.** If $J^l_{0,j} - J^l_0 < \frac{\varepsilon^l_j}{2} a^l_0$, then let $u^l_{k,j}$ take place of $u^l_k$ $(k = 0, 1, \cdots, N-1)$, and go back to step 2., otherwise let $\varepsilon^l_{j+1} = \frac{\varepsilon^l_j}{2}$, and use $\varepsilon^l_{j+1}$ to take the place of $\varepsilon^l_j$, and return to step 4.

The solution to a DP problem generally involves more than only computing the value of $f(S)$ for the goal state $S$. It is also often necessary to determine the initial optimal decision, and the optimal second decision that should be made in the next-state that results from the first decision, and so forth; that is, it may be necessary to determine the optimal sequence of decisions, also known as the optimal "policy", by what is known as a reconstruction process. To reconstruct these optimal decisions when evaluating $f(S) = opt_{d \in D(S)}\{C(d|S) \circ f(S')\}$ save the value of $d$, denoted $d^*$, that yields the optimal value of $f(S)$ at the time this value is computed by entering the value $d^*(S)$ in a table for each $S$. The main alternative to using such a policy table is to reevaluate $f(S)$ as needed, as the sequence of next-states are determined. This is an example of a space versus time tradeoff.

## 1.4 MOMSDM under an Implicit Termination Time

In systems with implicitly specified termination times, the process terminates when the outputs reach some pre-specified value. An iterative solution for deterministic systems was introduced by Bellman and Zadeh (1970) [285], and a graph-theoretic analysis has also been used to tackle the same problem, but a simpler solution can be derived using the branch-and-bound approach [162]. In this section, the MOMSDM under an implicit termination time is introduced.

### *1.4.1 System with an Implicit Termination Time*

Systems with implicitly defined termination times were first introduced by Bellman and Zadeh [285] in 1970. Instead of considering the termination time to be fixed (i.e., $N$), in this more general case, the termination time is assumed to be determined implicitly by a subsidiary condition of the form $x_N \in U$, where $U$ is a specified non-fuzzy subset of $X$ termed the termination set. Thus, the process terminates when the state of the system under control enters for the first time a specified subset of the state space. In a more general case, the termination time $T$ can be assumed to be determined implicitly by a subsidiary condition of the form

$$\mathbf{x}(T) \in U(\pi),$$

where $U(\pi)$ is the termination state space. In this case, the process terminates when the state of the system under control enters for the first time a specified subset of the state space (i.e., $U(\pi)$). Thus, its general model is expressed as:

$$\begin{cases} \max \{J_1(\mathbf{x}(0),\mathbf{x}(T),\pi), J_2(\mathbf{x}(0),\mathbf{x}(T),\pi), \cdots, J_n(\mathbf{x}(0),\mathbf{x}(T),\pi)\} \\ \text{s.t.} \begin{cases} \mathbf{x}(k+1) = f_k(\mathbf{x}(k),\mathbf{u}(k)), \\ \mathbf{x}(0) \in I, \\ \mathbf{x}(T) \in U(\pi), \\ g_j(\mathbf{x}(\cdot),\mathbf{u}(\cdot)) \leq 0, \ j = 1,2,\cdots,p, \\ \mathbf{x}(k) \in X, \ k = 1,2,\cdots,T, \\ \mathbf{u}(k) \in U, \ k = 1,2,\cdots,T. \end{cases} \end{cases} \tag{1.118}$$

where $\mathbf{x}(k)$ and $\mathbf{u}(k)$ are the state and decision variables, $I$ is the initial state space, $X$ is the state space, $U$ is the decision space, and $U(\pi)$ is the termination set of states.

### 1.4.2 Time-Delay Systems

Time delay is a typical phenomenon in a MOMSDM under an implicit termination time. Engineering systems often contain delayed elements such as recycling streams, transportation lags, and time delays associated with the measurement of output variables. These time delays can have a significant effect on the control of the system under consideration, so it is important to incorporate them into the mathematical model. This gives rise to differential equations with delayed arguments.

Due to the difficulty of solving the optimal control of nonlinear time-delay systems, Oh and Luus (1976) [173] suggested a suboptimal control approach based on the use of the Taylor series expansion of the delayed state to retain sufficient dynamic features of the original time-delay system, and then used a direct search to obtain the best feedback gain matrix to minimize the associated performance index. For small time delays this approach gave very good results.

In this section, however, instead of using only the piecewise constant control, piecewise linear continuous control is also used, so that better results can be obtained with a fewer number of time stages. Here, a single grid point ($N = 1$) is used. If a single grid point is used, then the problem of how to determine the initial profiles, which are required for the delay terms, does not arise. The only difficulty is in determining sufficiently good starting conditions for highly nonlinear systems to ensure convergence. Recent work by Luus et al. (1995) [174] showed that an efficient way of overcoming such difficulties is to use the Taylor series expansion for the delay terms to convert the given system into a non-delay system for which an optimal control policy can be readily established, which provides the starting control policy for the time-delay system. This approach is illustrated in the last example.

Consider a system described by a vector differential equation with a constant time delay $\tau$

$$\frac{d\mathbf{x}}{dt} = \mathbf{f}[\mathbf{x}(t), \mathbf{x}(t-\tau), \mathbf{u}(t)], \tag{1.119}$$

with the given initial state profile

$$\mathbf{x}(t) = \mathbf{g}(t), \quad -\tau \le t < 0, \tag{1.120}$$

and the initial condition

$$\mathbf{x}(0) = \mathbf{k}. \tag{1.121}$$

Here $\mathbf{x}$ is an $(n \times 1)$ state vector, $u$ is an $(m \times 1)$ control vector, $\tau$ is a constant delay factor and $\mathbf{g}$ is a given function of time that gives the initial profile for the state. Assume that each element in the control vector is bounded by

$$\alpha_j \le u_j \le \beta_j, \quad j = 1, 2, \cdots, m. \tag{1.122}$$

Associated with the system is a performance index which needs to be minimized:

$$I = \Phi(\mathbf{x}(t_f)), \tag{1.123}$$

where $\Phi$ is a function of the final state and the final time $t_f$ is specified. The optimal control problem is to find the control $\mathbf{u}(t)$ in the time interval $0 \leq t < t_f$ such that the performance index is minimized.

Divide the given time interval into $P$ subintervals $(0,t_1)$, $(t_1,t_2)$, $\cdots$, $(t_{P-1},t_P)$, each of length $L$, so that

$$L = \frac{t_f}{P}. \tag{1.124}$$

Instead of seeking only a piecewise constant control policy, iterative dynamic programming can be used to also find a piecewise linear control policy that minimizes the performance index. Conceptually there are no difficulties and the algorithm used earlier can be used. For integration of the differential equations with time delay the integration routine given by Hairer et al. (1987) [175] is used. To get an appreciation of iterative dynamic programming for the time delay problems, several examples are considered.

*Example 1.6.* Consider the linear time-delay system used for optimal control studies by Chan and Perkins (1973) [176], Oh and Luus (1976) [173], Dadebo and Luus (1992) [178], and Luus (1998b) [182]. The system is described using differential equations

$$\frac{dx_1}{dt} = x_2(t), \tag{1.125}$$

$$\frac{dx_2}{dt} = -10x_1(t) - 5x_2(t) - 2x_1(t-\tau) - x_2(t-\tau) + u(t), \tag{1.126}$$

$$\frac{dx_3}{dt} = 0.5(10x_1^2(t) + x_2^2(t) + u^2(t)), \tag{1.127}$$

with the initial state profile

$$x_1(t) = x_2(t) = 1.0, \quad -\tau \leq t \leq 0, \quad x_3(0) = 0. \tag{1.128}$$

The performance index to be minimized is

$$I = x_3(t_f). \tag{1.129}$$

where the final time is set at $t_f = 5$.

To solve this problem using iterative dynamic programming, a single grid point $N = 1$ is chosen, with $R = 3$ randomly chosen values. The initial control policy is zero, and the initial region size is 0.2. There are two passes, each consisting of 10 iterations, with a region reduction factor $\gamma = 0.80$ and a region restoration factor $\eta = 0.70$. To integrate the equations, the integration program from Hairer et al. (1987) [175] is used. Runs are made for different values of the delay factor $\tau$ and the number of stages for $P$ are 5, 10, and 20 for both piecewise constant and piecewise linear control.

Table 1.1 shows the results obtained using piecewise constant control. It can be seen that even with $P = 20$, the results are quite close to the results obtained using the control vector iteration procedure based on Pontryagin's maximum principle

**Table 1.1** Performance index for piecewise constant control policy using different values of delay factor and number of stages

| Time delay | Number of time steps | | | Control vector |
|---|---|---|---|---|
| $\tau$ | $P = 5$ | $P = 10$ | $P = 20$ | iteration |
| 0.10 | 2.5764 | 2.5709 | 2.5667 | 2.562 |
| 0.25 | 2.5374 | 2.5335 | 2.5283 | 2.526 |
| 0.50 | 2.6231 | 2.6189 | 2.6101 | 2.609 |
| 0.75 | 2.7918 | 2.7818 | 2.7704 | 2.769 |
| 1.00 | 2.9683 | 2.9471 | 2.9343 | 2.932 |

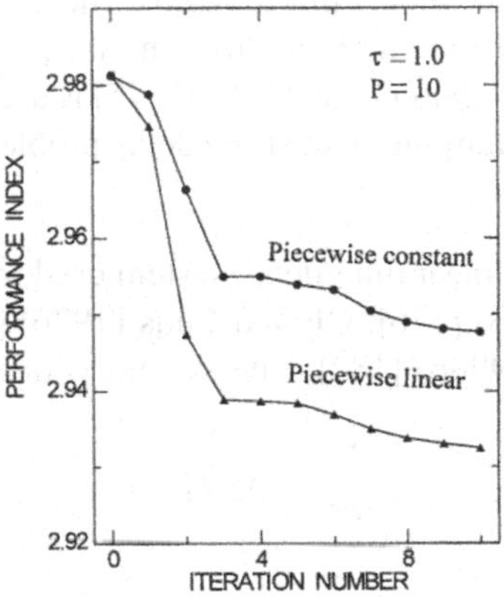


**Fig. 1.11** Convergence profile for Example 1.6 with the use of $N = 1$ and $R = 3$

(Oh and Luus 1976 [173]). The use of $P = 20$ already gives results very close to those reported by control vector iteration. The results for piecewise linear control are given in Table 1.2. It is obvious that the use of $P = 5$ gives an inadequate approximation of the optimal control. However, with only 10 stages, excellent results are obtained. However, the control vector iteration procedure based on Pontryagin's maximum principle was slow to converge, especially at higher values of $\tau$, and therefore the results obtained here for these cases are marginally better than those reported by Oh and Luus (1976) [173]. Here the convergence was very rapid, as shown in Figure 1.11 with $\tau = 1$, and $P = 10$, where after 10 iterations with $R = 3$, convergence to 2.9477 was obtained with piecewise constant control and 2.9325 was obtained with piecewise linear control. During the second pass, the convergence was obtained at the values reported in the Tables. The piecewise linear control policy with $P = 10$ with a delay factor $\tau = 1.0$, giving $I = 2.9302$, is given in Figure 1.12. A smoother control policy with $P = 20$, yielding $I = 2.9292$, was given by Luus (1998b) [182]. The computation time for the two passes with $P = 10$ was 0.7$s$, and with $P = 20$ was 1.5$s$ on a Core $i5$. Therefore, the computational effort here is negligible.

*Example 1.7.* Consider the linear time-delay system considered by Palanisamy et al. (1988) [177] for the optimal control of linear time-varying delay systems via a

**Table 1.2** Performance index for piecewise linear control policy using different values of delay factor $\tau$ and number of stages

| Time delay $\tau$ | Number of time steps $P=5$ | $P=10$ | $P=20$ | Control vector iteration |
|---|---|---|---|---|
| 0.10 | 2.5700 | 2.5653 | 2.5646 | 2.562 |
| 0.25 | 2.5332 | 2.5270 | 2.5258 | 2.526 |
| 0.50 | 2.6243 | 2.6076 | 2.6064 | 2.609 |
| 0.75 | 2.7996 | 2.7668 | 2.7658 | 2.769 |
| 1.00 | 2.9570 | 2.9302 | 2.9292 | 2.932 |

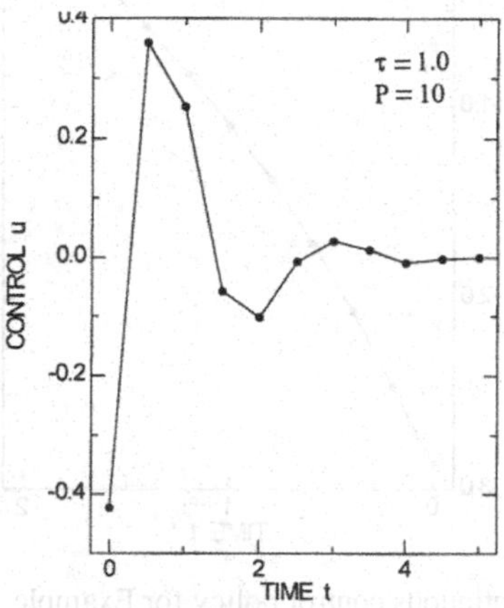


**Fig. 1.12** Piecewise linear continuous optimal control policy for Example 1.6 with $\tau = 1$

single term Walsh series, which was also used by Dadebo and Luus (1992) [178] to test iterative dynamic programming. The system is described using delay differential equations

$$\frac{dx_1(t)}{dt} = tx_1(t) + x_1(t-\tau) + u(t), \tag{1.130}$$

$$\frac{dx_2(t)}{dt} = x_1^2(t) + u^2(t), \tag{1.131}$$

with the initial state profile

$$x_1(t) = 1.0, \quad -1 \le t \le 0, x_2(0) = 0. \tag{1.132}$$

The optimal control problem is to find control $u$ in the time interval $0 \le t < 2$, so that the following performance index is minimized:

$$I = x_2(t_f), \quad t_f = 2. \tag{1.133}$$

The delay term $\tau = 1$.

To solve this problem using iterative dynamic programming, a single grid point $N = 1$ is again chosen, with $R = 3$ randomly chosen values, and $P = 10$ time stages of equal length. The initial control policy is zero, and the initial region size is 1.0. To establish piecewise linear control, ten passes, each consisting of 10 iterations, are used, with the region reduction factor $\gamma = 0.80$ and the region restoration factor $\eta = 0.70$. From the initial value of 80.3436, performance index $I$ is reduced very rapidly to 4.7968 in a computation time of 1.2$s$ on a Core $i$5.

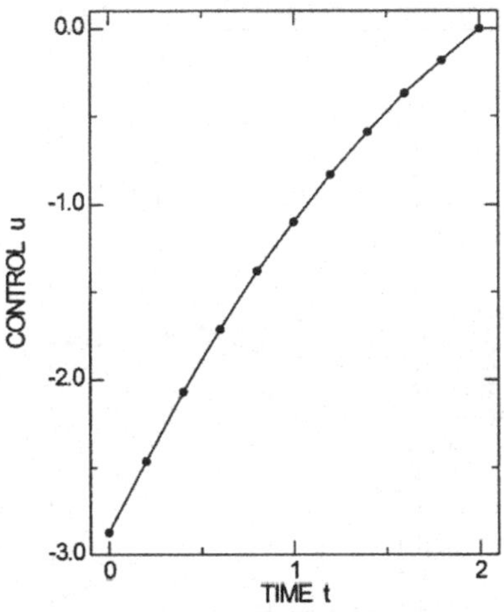


**Fig. 1.13** Piecewise linear continuous control policy for Example 1.7

This optimal control policy is given in Figure 1.13. It can be seen that the control policy is very smooth indeed and there is no need to have a larger number of stages to improve the accuracy.

## 1.4.3 Variable Stage Lengths

In the last section it was seen that when using piecewise constant control, increasing the number of time stages always increases the accuracy with which the control policy is approximated for a linear system with a quadratic performance index. However, there are nonlinear optimal control problems where an increase in the number of stages does not necessarily improve the performance index. For example, when the optimal control policy involves a sudden change from one extreme value to the other, the time of this switching is very important. In time optimal control problems this type of "bang-bang" control one can be anticipated. However, there are other occasions such a change in the optimal control policy is not expected. Therefore, a good approximation for the optimal control policy is determined not so much by the number of time stages, but how well the switching time can be approximated. Looking at a simple example as an illustration, consider the optimal control problem of Ko and Stevens (1971) [179] involving a first-order, reversible, exothermic chemical reaction $A \Leftrightarrow B$ in a chemical reactor. The material and energy balances of the reactor are given by:

$$\frac{dx_1}{dt} = (1 - x_1)k_1 - x_1 k_2, \tag{1.134}$$

$$\frac{dx_2}{dt} = 300[(1 - x_1)k_1 - x_1 k_2] - u(x_2 - 290), \tag{1.135}$$

with the initial condition

$$\mathbf{x}^T(0) = [0, 410]. \tag{1.136}$$

The state variable $x_1$ is the concentration of the desired component $B$, and $x_2$ denotes the absolute temperature. The control $u$ is a normalized heat transfer term dependent on the coolant flow rate, and is bounded by

$$0 \leq u \leq 0.5. \tag{1.137}$$

The reaction rate constants are dependent on the temperature in the following manner:

$$k_1 = 1.7536 \times 10^5 \exp(\frac{-1.1374 \times 10^4}{1.9872 x_2}) \tag{1.138}$$

and

$$k_2 = 2.4885 \times 10^{10} \exp(\frac{-2.2748 \times 10^4}{1.9872 x_2}). \tag{1.139}$$

The problem is to find $u$ in the time interval $0 \leq t < t_f$ such that the performance index

$$I = x_1(t_f), \quad t_f = 5 \tag{1.140}$$

is maximized. This problem corresponds to Case II of Reddy and Husain (1981) [180], who examined the difficulties in solving this optimal control problem using methods based on the gradient. Case I has been considered in some detail by Luus (1994) [181]. To obtain a piecewise constant optimal control policy by iterative dynamic programming presents no difficulties. However, what causes difficulties is choosing the number of time stages to solve this problem, since the improvement in the performance index is not very regular when the number of time stages increases, as is shown in Figure 1.14. However, there is a definite pattern, as three peaks occur at $P = 17$, $P = 26$, and $P = 35$. It can be observed that the use of 17 time stages gives a better value for the performance index than the use of 40 time stages. To understand the underlying reason, the optimal control profiles obtained for $P = 16$, 17, and 18 are examined. In Figure 1.15 with $P = 16$ it can be seen that after two time intervals with $t = 0.625$, it is a little late to switch to the maximum allowed value. Therefore, there is a smaller control value during the second time step.

In Figure 1.16 with $P = 17$, after the second time step at $t = 0.588$, the switching is approximately correct and therefore in the third time step the control is at the maximum allowed value. This enables the performance index $I = 0.72387$ to be obtained.

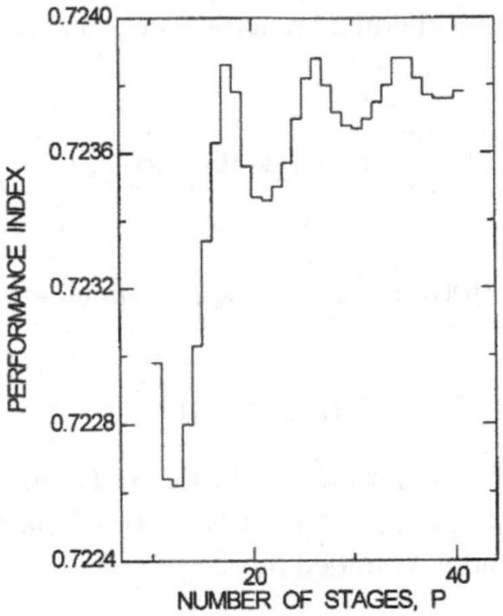


**Fig. 1.14** Effect of the number of time stages $P$ on the maximum value of the performance index

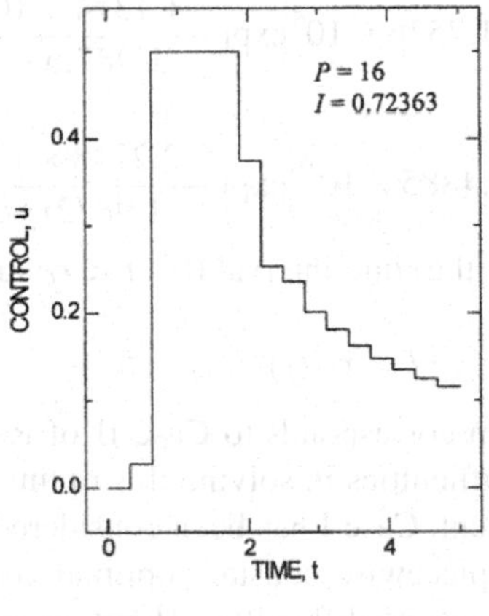


**Fig. 1.15** Optimal control policy obtained with $P = 16$ time stages

In Figure 1.17 with $P = 18$, after two time steps at $t = 0.555$ the switch is too early, and is made at a value less than the maximum, so the performance index of $I = 0.72378$ is not quite as good as the value obtained with $P = 17$.

Therefore, it can be seen that there is a need to use flexible stage lengths, rather than stages of constant length. The goal in this chapter was to show how flexible stage lengths can be used in iterative dynamic programming and the improvements that can be obtained. To finish off this example, the type of control policies that are obtained with flexible stage lengths are examined. Using 7 stages of varying length, a 6-stage optimal control policy is obtained as shown in Figure 1.18. Here both stages 2 and 3 gave the same upper bound for the control. The performance index

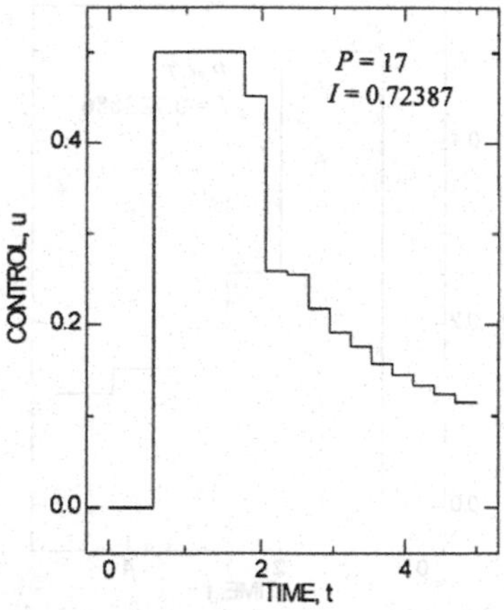


**Fig. 1.16** Optimal control policy obtained with $P = 17$ time stages

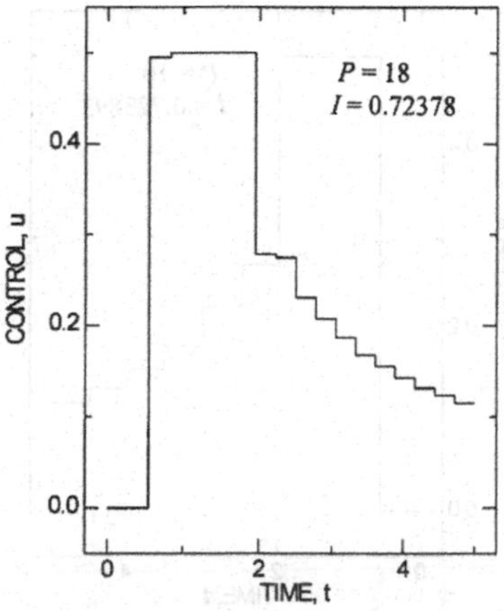


**Fig. 1.17** Optimal control policy obtained with $P = 18$ time stages

$I = 0.723886$ is marginally better than the value obtained with 17 stages of equal length.

Figures 1.19 and 1.20 show that with an increase in the number of stages, the performance index can be slightly improved. However, the portion of the control policy after the maximum bound is not very important. The very low sensitivity of this part of the control policy makes an accurate determination of the optimal control difficult. The most important part of the policy is the switching time, and it is essential to obtain the switching time from 0 to 0.5 at time $t = 0.5796$ as accurately as possible.

The benefits of using stages of varying lengths for establishing the optimal control policy is obvious from this example. In this chapter we outline the algorithm

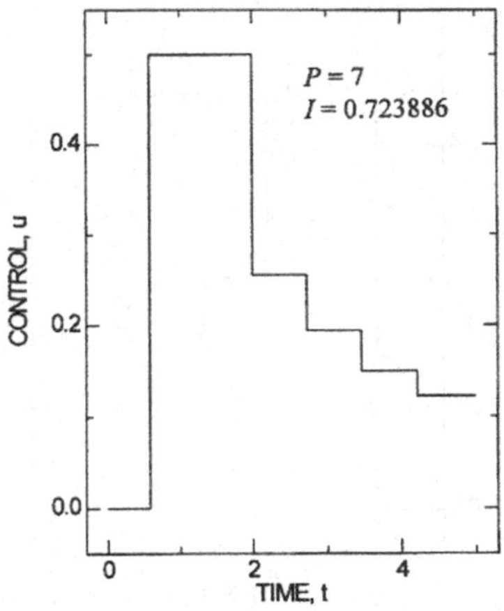


**Fig. 1.18** Optimal control policy obtained with $P = 7$ time stages of varying length

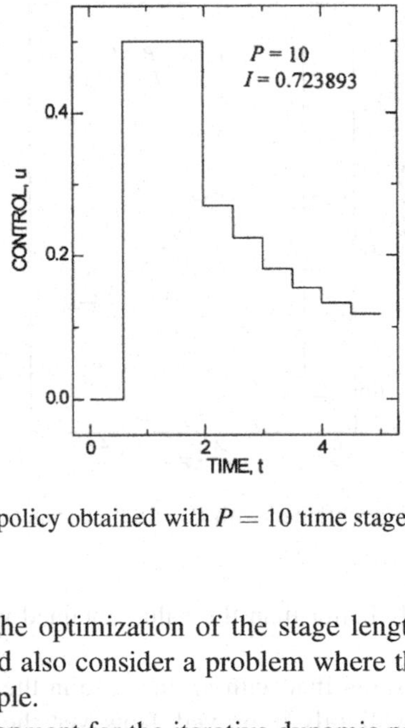


**Fig. 1.19** Optimal control policy obtained with $P = 10$ time stages of varying length

needed to incorporate the optimization of the stage lengths into the iterative dynamic programming and also consider a problem where the final time is specified as in the previous example.

The algorithm development for the iterative dynamic programming to allow the stages to be of varying length was first done by Bojkov and Luus (1996) [184]. Here, the stage lengths needed to constitute another control variable and the formulation are quite straightforward. Consider an optimal control problem where the state equation is

$$\frac{d\mathbf{x}}{dt} = \mathbf{f}(\mathbf{x}, \mathbf{u}) \tag{1.141}$$

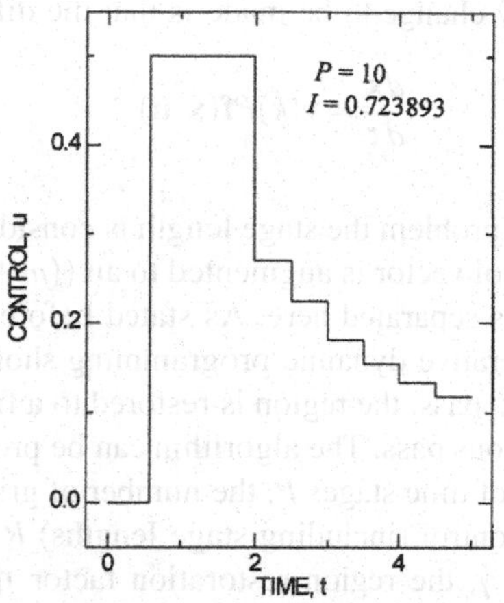


**Fig. 1.20** Optimal control policy obtained with $P = 15$ time stages of varying length

with the initial state $\mathbf{x}(0)$ given. The $(m \times 1)$ control vector is bounded by

$$\alpha_j \le u_j(t) \le \beta_j, \quad j = 1, 2, \cdots, m. \tag{1.142}$$

The performance index to be maximized is a scalar function of the state at the final time $t_f$

$$I = \Phi(\mathbf{x}(t_f)) \tag{1.143}$$

where the final time $t_f$ is not specified.

Divide the time interval into $P$ time stages, with each of variable length

$$v(k) = t_k - t_{k-1}, \quad k = 1, 2, \cdots, P, \tag{1.144}$$

where

$$v(k) \ge 0, \quad k = 1, 2, \cdots, P. \tag{1.145}$$

Control is kept constant in each of these time intervals.

Now a normalized time variable $\tau$ is introduced by defining $d\tau$ in the time interval $t_{k-1} \le t < t_k$ through the relationship

$$dt = v(k)Pd\tau, \tag{1.146}$$

so that

$$t_k - t_{k-1} = v(k)P(\tau_k - \tau_{k-1}). \tag{1.147}$$

Therefore

$$\tau_k - \tau_{k-1} = \frac{1}{P} \tag{1.148}$$

and in the normalized time the stages are of equal length and the final time is $\tau = 1$, so that the algorithm for the iterative dynamic programming presented earlier can be used directly. The only change to be made is that the differential equation now becomes

$$\frac{d\mathbf{x}}{d\tau} = v(k)P\mathbf{f}(\mathbf{x},\mathbf{u}) \tag{1.149}$$

at the kth time interval.

For this free final-time problem the stage length is considered an additional control variable and the control vector is augmented to an $((m+1)\times 1)$ vector, but for presentation clarity, this is separated here. As stated before, it is clear that for the greatest efficiency, the iterative dynamic programming should be used in a multi-pass fashion, where after a pass, the region is restored to a fraction $\eta$ of its value at the beginning of the previous pass. The algorithm can be presented in eight steps:

1. Choose the number of time stages $P$, the number of grid points $N$, the number of allowable values for control (including stage lengths) $R$ at each grid point, the region contraction factor $\gamma$, the region restoration factor $\eta$, initial values for the control and stage lengths, the initial region sizes, the number of iterations to be used in every pass, and the number of passes.

2. By choosing $N$ values for the control and stage length, which are evenly distributed inside the specified regions, integrate Eq. (1.149) from $\tau = 0$ to $\tau = 1$ to generate $N$ trajectories. The $N$ values for $\mathbf{x}$ at the beginning of each time stage constitute the $N$ grid points at each stage.

3. Starting at stage $P$, and corresponding to the normalized time $\tau = (P-1)/P$,for each grid point generate $R$ sets of values for the control and stage length:

$$\mathbf{u}(P) = \mathbf{u}^{*j}(P) + \mathbf{D}\mathbf{r}^{j}(P) \tag{1.150}$$

$$v(P) = v^{*j}(P) + \omega s^{j}(P) \tag{1.151}$$

where $\mathbf{D}$ is an $m\times m$ diagonal matrix with different random numbers between $-1$ and 1 along the diagonal and $\omega$ is another random number between $-1$ and 1; $\mathbf{u}^{*j}(P)$ is the best value for control, and $v^{*j}(P)$ is the best value for the stage length, both obtained for that particular grid point in the previous iteration. Integrate Eq. (1.149) from $(P-1)/P$ to $\tau = 1$ once with each of the $R$ allowable values for the control and stage length to yield $\mathbf{x}(t_f)$, so that the performance index can be evaluated. Compare the performance index $R$ values and choose the control and stage length which give the maximum value. The corresponding control and stage lengths are stored for use in Step 4.

4. Go back to stage $P-1$, with the corresponding time $\tau = (P-2)/P$. For each grid point generate $R$ allowable sets for the control and stage lengths. Integrate Eq. (1.149) from $(P-2)/P$ to $(P-1)/P$ once with each of the $R$ sets. To continue integration, choose the control and the stage length from step 3 that corresponds to the grid point that is closest to the $\mathbf{x}$ at $\tau = (P-1)/P$. Now compare the performance index $R$ values and store the control policy and the stage length that yield the maximum value.

5. Go back to stage $P-2$, and continue the procedure as in the previous step. Continue until stage 1 corresponds to $\tau = 0$ with the given initial state as the grid

point is reached. Compare the performance index $R$ values to give the best control and stage length for this stage. The best control policy and the stage length for each stage in the sense of maximizing the performance index from the allowable choices have now been determined.

6. In preparation for the next iteration, reduce the size of the allowable regions

$$\mathbf{r}^{j+1}(k) = \gamma \mathbf{r}^{j}(k), \quad k = 1, 2, \cdots, P \tag{1.152}$$

$$s^{j+1}(k) = \gamma s^{j}(k), \quad k = 1, 2, \cdots, P \tag{1.153}$$

where $\gamma$ is the region reduction factor and $j$ is the iteration index. Use the best control policy and the stage lengths from Step 5 as the midpoint for the next iteration.

7. Increment the iteration index $j$ by 1 and go to Step 2 to generate another set of grid points. Continue for the specified number of iterations.

8. Increment the pass number index by 1, set the iteration index $j$ to 1 and go to step 2. Continue for the specified number of passes, and examine the results.

A reasonable number of iterations per pass is 20 or 30 with the region contraction factor $\gamma = 0.95$, as this indicated that the region is covered well. For many problems, an optimum with reasonable accuracy is not found in a single pass. Therefore the region should be restored to a reasonable value before the iterations are performed again. An easy way to restore the region size is to take as its value a fraction $\eta$ of its value at the beginning of the previous pass. To help in deciding what value for $\eta$ to use, refer to Table 1.3, which shows how the initial region is reduced through the choice of $\eta$ if 20 or 30 passes are made. If 30 passes are used and $\eta$ is 0.95, then the initial region size after 30 passes becomes 0.2146 times its value at the beginning of the first pass; if $\eta$ is 0.85, then the initial region size after 30 passes becomes 0.0076 times its value at the beginning of the first pass. Therefore, a balance is sought between a premature collapse of the region size and the speed of convergence.

This method for restoring the region size after every pass is not the best method, as it depends very much on the initial region size choice. However, this method is used for all the examples in this book. Recently, Luus (1998b) [182] showed that the extent by which a variable changes during a pass can be used as the value for the region size for the next pass and in a steady state optimization this method yielded a significant improvement in the convergence rate, and enabled a complex batch fed reactor problem to be optimized with great accuracy [183].

### *1.4.4 State Constraints*

In optimal control problems constraints are frequently imposed on state variables. For example, if one of the state variables is the reactor temperature, then it should be obvious that there should be an upper limit on the temperature. If a state variable is the reaction mixture volume, this volume should not surpass the physical volume of the reactor. Such constraints take the form of inequalities which need to be incorporated into the system model. Another occasion when state constraints are encountered is when a system transfers from one operation state to another. For example,

**Table 1.3** Effect of the region restoration factor $\eta$ on the region size after 20 and 30 passes

| Region restoration factor | Number of passes | |
|---|---|---|
| $\eta$ | 20 | 30 |
| 0.95 | 0.3584 | 0.2146 |
| 0.94 | 0.2901 | 0.1563 |
| 0.93 | 0.2342 | 0.1134 |
| 0.92 | 0.1887 | 0.0820 |
| 0.91 | 0.1516 | 0.0591 |
| 0.90 | 0.1216 | 0.0424 |
| 0.89 | 0.0972 | 0.0303 |
| 0.88 | 0.0776 | 0.0216 |
| 0.87 | 0.0617 | 0.0153 |
| 0.86 | 0.0490 | 0.0108 |
| 0.85 | 0.0388 | 0.0076 |

when starting a reactor, the system start conditions are usually not the same as the desired conditions for the system's operation. In these situations the constraints on the system state variables are in the form of equality constraints at the final time $t_f$, which is the point at which some of the state variables need to have specified desired values at time $t_f$. The methods to deal with these two types of state constraints are somewhat different; so, here they are considered separately.

Consider a system described by the differential equation

$$\frac{d\mathbf{x}}{dt} = \mathbf{f}(\mathbf{x},\mathbf{u}) \tag{1.154}$$

where the initial state $\mathbf{x}(0)$ is given. The state vector $\mathbf{x}$ is an $(n \times 1)$ vector and $u$ is an $(m \times 1)$ control vector bounded by

$$\alpha_j \leq u_j \leq \beta_j, \quad j = 1,2,\cdots,m. \tag{1.155}$$

At the final time $t_f$, there are $k$ state variables that must be at the desired values, i.e.,

$$h_i = x_i(t_f) - x_i^d = 0, \quad i = 1,2,\cdots,k \leq n. \tag{1.156}$$

The optimal control problem is to find the control policy $\mathbf{u}(t)$ at the time interval $0 \leq t < t_f$ that minimizes the performance index

$$I = \Phi(\mathbf{x}(t_f)) \tag{1.157}$$

where the final time $t_f$ is specified. This problem differs from those considered earlier as there are k equality constraints that must be satisfied. Numerically, the equality constraints are expressed explicitly as

$$|x_i(t_f) - x_i^d| \leq \varepsilon_i, \quad i = 1,2,\cdots,k, \tag{1.158}$$

where $\varepsilon_i$ is the tolerance for the $i^{th}$ constraint and is sufficiently small. It is also necessary to know how the choice of the tolerance $\varepsilon_i$ affects the value of the performance index $I$ at the optimum.

Construct an augmented performance index to take account of the $k$ equality constraints by adding a quadratic penalty function containing shifting terms to the performance index.

$$J = I + \theta \sum_{i=1}^{k} (h_i - s_i)^2, \tag{1.159}$$

where $\theta$ is a positive penalty function factor and the shifting terms $s_i$ are chosen so that computationally Eq. (1.156) is satisfied and the performance index in Eq. (1.157) is minimized as the iterative procedure progresses. The use of such shifting terms in the quadratic penalty function was first explored by Luus (1996) [185] for steady state optimization. It is noted that

$$(h_i - s_i)^2 = h_i^2 - 2s_i h_i + s_i^2, \tag{1.160}$$

so that near the optimum when Eq. (1.156) is almost satisfied, the term $h_i^2$ is negligibly small with respect to the other two terms and the minimization of the augmented performance index in Eq. (1.159) becomes equivalent to minimizing

$$J = I - 2\theta \sum_{i=1}^{k} s_i h_i, \tag{1.161}$$

since the term involving the square of the shifting term is simply a constant. Thus, the product $-2\theta s_i$ gives the Lagrange multiplier or sensitivity of the performance index to the change in the $i^{th}$ equality constraint in Eq. (1.156). This aspect of the shifting term was pointed out by Luus (1996) [185], when using penalty functions in dealing with difficult equality constraints in steady state optimization. The use of shifting terms was illustrated with iterative dynamic programming by Luus (1996) [186], and Luus and Storey (1997) [187]. When prior information on the sensitivity is not available, the shifting term $s_i$ is initially chosen to be equal to zero, and then, after every pass of the iterative dynamic programming, it is updated according to

$$s_i^{q+1} = s_i^q - h_i, \tag{1.162}$$

where $s_i^q$ is the value of $s_i$ at pass number $q$. The shifting terms can be updated after every iteration instead of after every pass if a very small number of passes is used. To show the computational aspects of the use of the quadratic penalty function with shifting terms, consider the following example.

*Example 1.8.* Consider a very simple example with a final state constraint that can be readily solved using Pontryagin's maximum principle. This problem was first considered by Goh and Teo (1988) [188] and then optimized using iterative dynamic programming by Luus (1991) [189] and Dadebo and McAuley (1995) [190]. The system is described by

$$\frac{dx_1}{dt} = u \tag{1.163}$$

$$\frac{dx_2}{dt} = x_1^2 + u^2 \tag{1.164}$$

with the initial state $\mathbf{x}(0) = [1,0]^T$, and final time $t_f = 1$. There are no constraints on the scalar control variable $u$, but there is an equality constraint to be satisfied at the final time

$$h_1 = x_1(t_f) - 1 = 0. \quad (1.165)$$

It is necessary to find the unconstrained control policy that minimizes the performance index

$$I = x_2(t_f). \quad (1.166)$$

The augmented performance index to be minimized is therefore chosen as

$$J = I + \theta(h_1 - s_1)^2. \quad (1.167)$$

To determine how closely the equality constraint is satisfied, the absolute value of the deviation is examined from the desired state at the final time $|x_1(t_f) - 1|$.

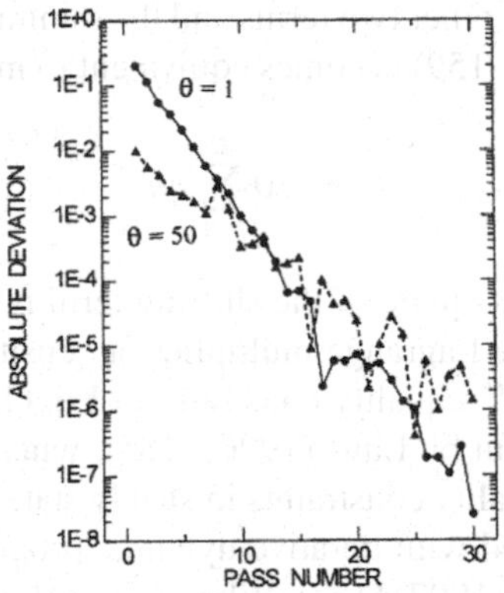


**Fig. 1.21** Effect of the penalty function factor $\theta$ on convergence for Example 1.8

To solve this problem using iterative dynamic programming, $P = 10$ stages are chosen, with an $N = 1$ grid point, $R = 5$ randomly chosen points, 30 passes with the region restoration factor $\eta = 0.70$, and 5 iterations with the region reduction factor $\gamma = 0.70$ in each pass. The initial value for control was zero, the initial region size was 1.0, and the initial value for the shifting term $s_1$ was zero. The goal is to choose a piecewise linear control policy to minimize the augmented performance index $J$.

With both $\theta = 1$ and $\theta = 50$, a rapid convergence to a minimum value of $I = 0.92424$ was achieved with the absolute difference from the desired state of less than $10^{-6}$ over 30 passes, as shown in Figure 1.21. The computation time for the 30 passes was 0.8$s$ on a Core $i$5 computer. From the further results shown in Table 1.4, it can be seen that the product $-2\theta s_1$ is constant and equal to $-I$ in the range $0.7 \leq \theta \leq 30$. In this $\theta$ range, the minimum value of the performance index is $I = 0.924235$. For a penalty function factor greater than 30, the deviation from the desired state is larger and also the performance index is larger than the optimum value. The control policy and the state trajectory are shown in Figure 1.22.

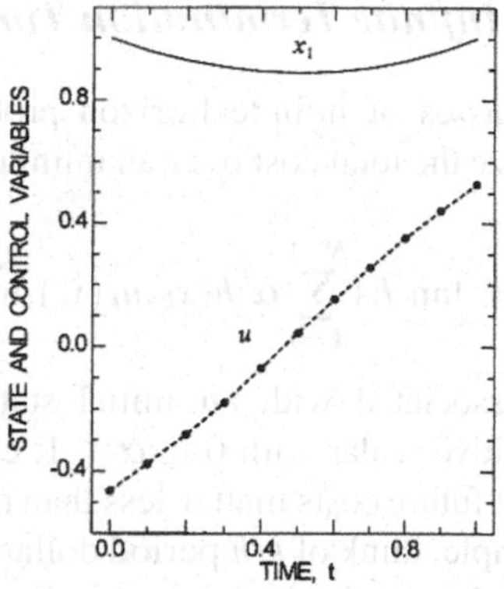


Fig. 1.22 Effect of the penalty function factor $\theta$ on convergence for Example 1.8

**Table 1.4** Effect of the penalty function factor $\theta$ on convergence for Example 1.8

| Penalty factor $\theta$ | Deviation from desired state $\lvert x_1(1)-1\rvert$ | Performance index $I=x_2(1)$ | Shifting term $s_1$ | Sensitivity factor $-2\theta s_1$ |
|---|---|---|---|---|
| 0.7 | $5.210\times10^{-7}$ | 0.924235 | 0.660319 | −0.92445 |
| 0.8 | $2.585\times10^{-7}$ | 0.924235 | 0.577780 | −0.92445 |
| 1.0 | $2.485\times10^{-8}$ | 0.924235 | 0.462224 | −0.92445 |
| 3.0 | $5.404\times10^{-8}$ | 0.924235 | 0.154075 | −0.92445 |
| 10 | $1.226\times10^{-7}$ | 0.924235 | 0.046223 | −0.92446 |
| 30 | $3.077\times10^{-8}$ | 0.924235 | 0.015407 | −0.92442 |
| 50 | $1.454\times10^{-6}$ | 0.924239 | 0.009243 | −0.92430 |
| 80 | $1.317\times10^{-6}$ | 0.925144 | 0.005847 | −0.93552 |
| 100 | $2.257\times10^{-5}$ | 0.925661 | 0.004719 | −0.94380 |

## 1.5 MOMSDM under an Infinite Termination Time

In this section, an introduction to infinite horizon problems is given. These problems differ from those considered so far in two respects: (a) The number of stages is infinite. (b) The system is stationary, so the system equation, the cost per stage, and the random disturbance statistics do not change from one stage to the next. The assumption of an infinite number of stages is rarely satisfied in practice, but is a reasonable approximation for problems involving a finite but very large number of stages. The assumption of stationarity is often satisfied in practice, and in other cases it is useful when the system parameters vary slowly with time. Infinite horizon problems are interesting because their analysis is elegant and insightful, and the implementation of optimal policies is often instructive. On the other hand, infinite horizon problems generally require more sophisticated analysis than their finite horizon counterparts because of the need to analyze the limiting behavior as the horizon tends to infinity. This analysis at times reveals surprising possibilities. The treatment of these cases here, however, is limited to finite-state problems.

### 1.5.1 System with an Infinite Termination Time

There are four principal classes of infinite horizon problems. In the first three classes, the aim is to minimize the total cost over an infinite number of stages, given by

$$J_\pi(x_0) = \lim_{N\to\infty} E\{\sum_{k=0}^{N-1} \alpha^k h(x_k, u_k(x_k), w_k)\}. \quad (1.168)$$

$J_\pi(x_0)$ denotes the costs associated with an initial state $x_0$ and a policy $\pi = \{u_0, u_1, \cdots\}$, and $\alpha$ is a positive scalar with $0 < \alpha \leq 1$, called the discount factor. The meaning of $\alpha < 1$ is that future costs matter less than the same costs incurred in the present time. As an example, think of $kth$ period dollars depreciated to an initial period dollars by a factor of $(1+r)$ where $r$ is a rate of interest; here $\alpha = 1/(1+r)$. An important concern in total cost problems is that the limit in the definition of $J_\pi(x_0)$ is finite. In the first two of the following classes of problems, this is guaranteed through various assumptions in the problem structure as well as the discount factor. Generally, a typical MOMSDM formulation under an implicit termination time is given as follows,

$$\begin{cases} \max J_1(\mathbf{x}(\infty), \pi), J_2(\mathbf{x}(\infty), \pi), \cdots, J_n(\mathbf{x}(\infty), \pi) \\ \text{s.t.} \begin{cases} \mathbf{x}(k+1) = f_k(\mathbf{x}(k), \mathbf{u}(k)), \\ \mathbf{x}(0) \in I, \\ g_j(x, \xi) \leq 0, \ j = 1, 2, \cdots, p, \\ \mathbf{x}(k) \in X, \ k = 1, 2, \cdots, \\ \mathbf{u}(k) \in U, \ k = 1, 2, \cdots. \end{cases} \end{cases} \quad (1.169)$$

where $\mathbf{x}(k)$ and $\mathbf{u}(k)$ are state and decision variables; $I$ is the initial state space; $X$ is the state space; and $U$ is the decision space.

### 1.5.2 Typical Problems under an Infinite Termination Time

In the third class, the analysis is adjusted to deal with the infinite costs of some of policies. In the fourth class, the sum need not be finite for any policy, and for this reason, the cost is appropriately redefined.

(a) Stochastic shortest path problems. Here, $\alpha = 1$ but there is a special cost-free termination state; once the system reaches that state no further costs are incurred. Assume a problem structure in which termination is inevitable. Thus, the horizon is in effect finite, but its length is random and may be affected by the policy being used. These problems are considered in the next section and the analysis provides the foundation for the analysis of the other types of problems considered in this chapter.

(b) Discounted problems with bounded cost per stage. $\alpha < 1$ and the absolute cost per stage $|h(x,u,w)|$ I is bounded from above by some constant; this makes the cost $J_\pi(x_0)$ well defined because it is the infinite sum of a sequence of numbers that are bounded in absolute value by the decreasing geometric progression $\{\alpha^k M\}$.

(c) Discounted and undiscounted problems with unbounded cost per stage. Here the discount factor may not be less than 1, and the cost per stage may be unbounded. These problems require a complicated analysis because the possibility of infinite cost for some of the policies is explicitly dealt with.

(d) Average cost per problem. Minimization of the total cost $J_\pi(x_0)$ makes sense only if $J_\pi(x_0)$ is finite for at least some admissible policies $\pi$ and some initial states $x_0$. Frequently, however, it turns out that $J_\pi(x_0) = \infty$ for every policy $\pi$ and initial state $x_0$ (think of a case where $\alpha = 1$, and the cost for every state and control is positive). It turns out that in many such problems the average cost per stage, defined by

$$lim_{N\to\infty} = \frac{1}{N}E\{\sum_{k=0}^{N-1} h(x_k, u_k(x_k), w_k)\} \tag{1.170}$$

is well defined and finite.

### 1.5.3 Analysis of Infinite Horizon Results

There are several analytical and computational issues regarding infinite horizon problems. Many of these revolve around the relation between the optimal cost-to-go function $J^*$ of the infinite horizon problem and the optimal cost-to-go functions of the corresponding N-stage problems. In particular, consider the case $\alpha = 1$ and let $J_N(x)$ denote the optimal cost of the problem involving $N$ stages, initial state $x$, cost per stage $g(x,u,w)$, and zero terminal cost. The optimal N-stage cost is generated after $N$ iterations of the DP algorithm

$$J_{k+1}(x) = \min_{u\in U(x)} E\{g(x,u,w) + J_k(J(x,u,w))\}, \quad k = 0, 1, \cdots, \tag{1.171}$$

starting from the initial condition $J_0(x) = 0$ for all $x$ (note here that we have reversed the time indexing to suit our purposes). Since the infinite horizon cost of a given policy is, by definition, the limit of the corresponding N-stage costs as $N \to \infty$, it is natural to speculate that:

(1) The optimal infinite horizon cost is the limit of the corresponding N-stage optimal costs as $N \to \infty$, that is,

$$J^*(x) = \lim_{N\to\infty} J_N(x) \tag{1.172}$$

for all states $x$. This relation is extremely valuable computationally and analytically, and, fortunately, it typically holds. However, there are some unusual exceptions for problems in the category above, and this illustrates that infinite horizon problems should be approached with some care.

(2) The following limiting form of the DP algorithm should hold for all states $x$,

$$J^*(x) = \min_{u\in U(x)} E\{g(x,u,w) + J^*(J(x,u,w))\} \tag{1.173}$$

as suggested by Eqs. (1.171) and (1.172). This is not really an algorithm, but rather a system of equations (one equation per state), which has as a solution the costs-to-go of all the states. It can also be viewed as a functional equation for the cost-to-go function $J^*$, and is called Bellman's equation. Fortunately again, an appropriate form of this equation holds for every type of infinite horizon problem of interest.

(3) If $u(x)$ attains the minimum in the right-hand side of Bellman's equation for each $x$, then the policy $\{u_0, u_1, \cdots\}$ should be optimal. This is true for most infinite horizon problems of interest and in particular, for all the models discussed in this chapter. Most of the analysis of infinite horizon problems revolves around the above three issues and also around the issue of the efficient computation of $J^*$ and an optimal policy. The next three sections provide a discussion of these issues for some of the simpler infinite horizon problems, all of which involve a finite state space.

# 2

# Elements of Fuzzy-Like MOMSDM

The need to address uncertainty in multiple objective multistage decision making (MOMSDM) is widely recognized, as uncertainties exist in a variety of system components. As a result, the inherent complexity and uncertainty existing in real-world MOMSDM problems has essentially placed them beyond conventional deterministic optimization methods. Fuzzy set theory was developed for solving problems in which descriptions of the activities and observations are imprecise, vague, and uncertain. The term "fuzzy" refers to a situation in which there are no well-defined boundaries to the set of activities or observations to which the descriptions apply. This fuzziness exists in many situations in the practical applications of MOMSDM, such as in equipment failure rate, effective monthly working days, transportation and stay time, arrival rate of vehicles, and facilities service rate. This leads to the concept of fuzzy multiple objective multistage decision making (F-MOMSDM). In some other cases, especially for large scale construction projects, the practical problems are often faced with a hybrid uncertain environment where fuzziness and randomness co-exist in a decision making process. For example, over the years, the advancement in engineering technology has provided more alternatives for construction operations and, subsequently, decision-makers are faced with more imprecise information than ever before. Therefore, fuzzy random variables that can take into account both fuzziness and randomness are favored by decision-makers to describe the uncertainty and vague information encountered in reality. This has lead to the development of fuzzy random multiple objective multistage decision making (FR-MOMSDM). From the point of view of functional analysis, both the F-MOMSDM and FR-MOMSDM are essentially different from the MOMSDM under a deterministic environment. Generally, the F-MOMSDM and FR-MOMSDM are called by a joint name, fuzzy-like multiple objective multistage decision making. This chapter presents some basic concepts for the elements in fuzzy-like MOMSDM.

## 2.1 Fuzzy Theory

Fuzzy is a concept in which the meaningful content, value, or boundaries of the application can vary considerably according to context or conditions, instead of being

J. Xu and Z. Zeng, *Fuzzy-Like Multiple Objective Multistage Decision Making*, Studies in Computational Intelligence 533,
DOI: 10.1007/978-3-319-03398-3_2, 

fixed (Haack 1996 [296]). This generally means the concept is vague, and lacking a fixed, precise meaning, without, however, being meaningless altogether. It has a meaning, or multiple meanings (it has different semantic associations). But these can become clearer only through further elaboration and specification, including a closer definition of the context in which they are used. Fuzzy concepts "lack clarity and are difficult to test or operationalize" (Markusen 2003 [283]). The basic concepts and terminology central to the study of fuzzy theory and fuzzy systems include fuzzy sets, fuzzy membership functions, and fuzzy variables. For easy reference, these concepts and terminology, and the brief descriptions adopted from Zadeh (1965, 1975a, 1975b, 1968a, 1973) [22, 556, 290, 201, 256], Dubois and Prade (1980) [571], Lee (1990) [570], and Zimmermann (1991) [573], are presented in this chapter.

### *2.1.1 Fundamental Concepts of Fuzzy Sets*

Fuzzy set theory (FST) is a powerful tool for dealing with complexities existing in real world problems. Although stochastic methods are used in modeling uncertainties due to randomness, there are some kinds of uncertainties that may not have a random or stochastic nature. The error or imprecision caused by discretizing continuous variables in the modeling process is a good example of a type of uncertainty that is not related to randomness. Such an imprecision can be dealt with by FST in which a fuzzy number represents an interval rather than a single valued point (Mousavi et al. (2004) [172]). Fuzzy sets are a generalization of conventional set theory, which was introduced by Zadeh in 1965 [22], and is a mathematical way to represent vagueness in everyday life.

Actually, in our daily life, some events may have clear and crisp boundaries, while some may have vague and indistinct boundaries. Typical example for events that have crisp boundaries can be found in distinguishing adults from juveniles. In most countries, the legal age for an adult should be not less than 18 years old. Let the classical set $A$ be the set of adults and the classical set $J$ be the set of juveniles as expressed below:

$$A = \{p(x)|x \geq 18, x \in U\}, \quad J = \{p(x)|x < 18, x \in U\} \tag{2.1}$$

where $U = [0, +\infty]$ is the universal space of personal ages, $p(x)$ is a person whose age is $x$, and 18 is a crisp boundary point that distinguishes the adults and juveniles. If $x \geq 18$, the person $p(x)$ belongs to the set $A$, otherwise $p(x)$ belongs to the set $J$. Being different from a classical set, a fuzzy set has a vague boundary instead of a crisp one. This means the boundary of a fuzzy set is extended to be an interval with gradual degrees which could be characterized by membership functions. In real life problems, it is common to meet linguistic terms, such as "it is expensive" or "demand is low". Since these expressions contain human subjectivity, it is difficult to define clear and crisp boundaries for the corresponding fuzzy set. In this situation, the membership function of a fuzzy set becomes a useful tool for describing the degree of an object belonging to a fuzzy set. For example, let $H$ be the fuzzy set for "the temperature is high", $M$ be the fuzzy set for "the temperature is medium", and

$L$ be the fuzzy set for "the temperature is low". The membership functions for the three fuzzy sets are defined as follows:

$$\mu_H(x) = \begin{cases} 0, & x < 24 \\ \frac{x-24}{28-24}, & 24 \leq x < 28 \\ 1, & x \geq 28 \end{cases} \tag{2.2}$$

$$\mu_M(x) = \begin{cases} 0, & x < 10 \\ \frac{x-10}{18-10}, & 10 \leq x < 18 \\ 1, & 18 \leq x < 22 \\ \frac{x-26}{22-26}, & 22 \leq x < 26 \\ 0, & x \geq 26 \end{cases} \tag{2.3}$$

$$\mu_L(x) = \begin{cases} 1, & x < 5 \\ \frac{x-15}{5-15}, & 5 \leq x < 15 \\ 0, & x \geq 15 \end{cases} \tag{2.4}$$

where $\mu_H(x)$, $\mu_M(x)$, and $\mu_L(x)$ are the membership functions for the fuzzy sets $H$, $M$, and $L$ respectively. Based on the membership functions, the range (i.e., $\mu_L(x) \geq 0$) of low temperature is $[-\infty, 15]°C$, and the most possible range (i.e., $\mu_L(x) = 1$) for low temperature is $[-\infty, 5]°C$, the range (i.e., $\mu_M(x) \geq 0$) for medium temperature is $[10, 26]°C$, and the most possible range (i.e., $\mu_M(x) = 1$) for medium temperature is $[18, 22]°C$, the range (i.e., $\mu_H(x) \geq 0$) for high temperature is $[24, +\infty]°C$, and the most possible range (i.e., $\mu_H(x) = 1$) for high temperature is $[28, +\infty]°C$. It can be seen that there is an overlap range for the fuzzy sets $H$, $M$, and $L$, so they have no crisp boundaries between them. Figure shows the difference between a classical set and a fuzzy set. Figure 2.1 shows the difference between a classical set and a fuzzy set.

**Definition 2.1.** *(Fuzzy set) (Zadeh 1965 [22]) Assume $U$ to be the universal space, which is a set of all elements, and $x$ is a generic element in $U$. A set $F$ in $U$ is a fuzzy set if and only if it is a set of ordered pairs:*

$$F = \{(x, \mu_F(x)) | x \in U\}, \tag{2.5}$$

where $\mu_F(x)$ is defined as the membership function of $x$ in $F$, which is a function that maps each element of $F$ to the interval $[0, 1]$.

Based on the above definition, it is not difficult to recognize that the fuzzy set is actually an extension of the classical set. When the value of the membership function of a fuzzy set is restricted to 0 or 1, the fuzzy set reduces to a classical set. On the contrary, when the value of the membership function of a classical set is extended to continuous values within the interval $[0, 1]$, the classical set is a fuzzy set.

*Example 2.1.* (Discrete Fuzzy Sets) Let $x$ be the number of customers in a restaurant, then the fuzzy set "the number of customers is close to 20" could be expressed as below

$$F = \{(17, 0.3), (18, 0.6), (19, 0.9), (20, 1), (21, 0.9), (22, 0.6), (23, 0.3)\}.$$

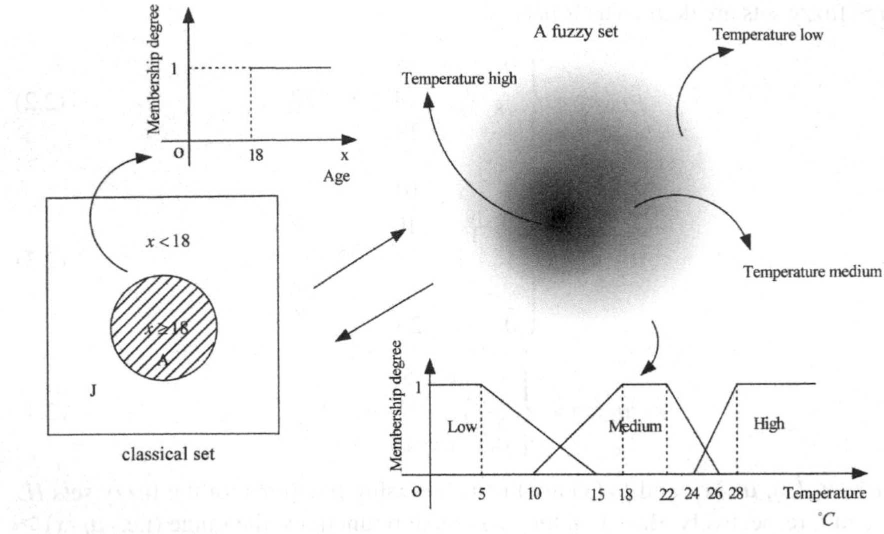


**Fig. 2.1** Difference between a classical set and a fuzzy set

*Example 2.2.* (Continuous Fuzzy Sets) The fuzzy set "the annual run-off of a river, i.e., $v$, is around 5 million $m^3$" could be expressed by using the following continuous membership function:

$$\mu_L(x) = \begin{cases} 0, & x < 4.5 \\ \frac{x-4.5}{5-4.5}, & 4.5 \le x < 5 \\ \frac{x-5.5}{5-5.5}, & 5 \le x < 5.5 \\ 0, & x \ge 5.5 \end{cases}$$

An interesting fact of the fuzzy set is that while the expression of a fuzzy set, such as "close to", "more or less", "about" and so on, is quite vague, the definition of its membership function is a very precise mathematical function, which can clearly describe the degree of membership for the fuzzy set. In other words, this means the vagueness in any fuzzy set has a precise mathematical description.

While the membership function of a fuzzy set is precise, the determination of a specific membership function for a fuzzy set is essentially subjective. In fact, different people have different opinions on the same linguistic term. For example, the fuzzy set "temperature is high" may vary significantly among different people. The reason for this subjectivity is because of the unclear definition of abstract concepts which cannot be described using randomness. The essential difference between fuzzy set theory and probability theory is in the way the nature of an event is described. This does not mean any event can be described by both fuzziness and randomness, as this depends on the attributes of the event. If the set of an event has crisp boundaries, it belongs to randomness, otherwise if a set has vague boundaries, it belongs to fuzziness. Therefore, for simplification, it could be said that the fuzziness describes a subjective uncertainty and randomness describes an objective uncertainty.

To further understand fuzzy set theory, the following related concepts are presented.

**Definition 2.2.** *(Support) (Zadeh 1965 [22]) A crisp set in a fuzzy set F supports the fuzzy set if and only if the values of its membership function are all positive.*

Let $S_F$ denote the support of a fuzzy set $F$ in the universal space $U$, the mathematical expression of $S_F$ is defined below,

$$S_F := \{x \in U | \mu_F(x) > 0\}. \tag{2.6}$$

Figure 2.2 shows the support for the fuzzy set "the temperature is high".

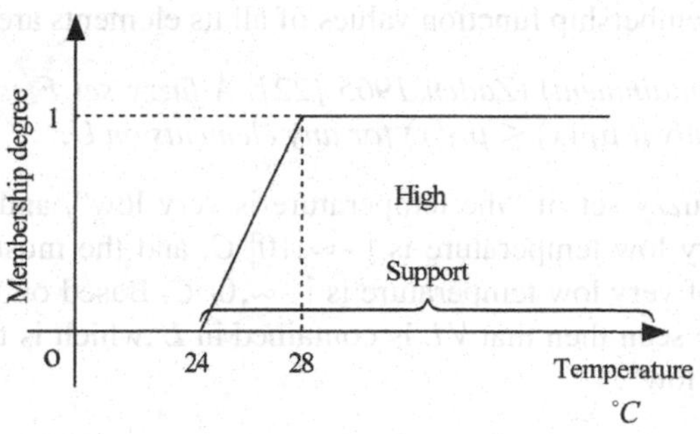


**Fig. 2.2** Support for the fuzzy set "the temperature is high"

**Definition 2.3.** *(Fuzzy Singleton) (Zadeh 1965 [22]) A fuzzy set is a fuzzy singleton if and only if its support is a single point in the universal space U.*

The fuzzy Singleton is an extreme case where a fuzzy set reduces to a classical set. When the mean value of all the elements in the fuzzy set is finite, such that their membership function values reach a maximum value, it is defined as the center of a fuzzy set. On the other hand, if there is positive (or negative) infinity, then the center is defined as the smallest (largest) among all elements that achieve a maximum membership value. Thus, if the support of a fuzzy singleton is finite, its support is also the center of this fuzzy set.

**Definition 2.4.** *(Normality) (Zadeh 1965 [22]) A fuzzy set is normal if and only if its largest membership function value achieves* 1.

If the membership function value of the support of a fuzzy singleton is 1, the fuzzy singleton is normal, and the support is the center of this fuzzy set.

**Definition 2.5.** *($\alpha$-Level Set and Strong $\alpha$-Level Set) (Zadeh 1965 [22]) A crisp set is an $\alpha$-cut or $\alpha$-level set of a fuzzy set F if and only if the membership function values of all the elements in the crisp set in F are greater than or equal to $\alpha$, i.e.,*

$$F_\alpha = \{x \in U | \mu_F(x) \geq \alpha\}. \tag{2.7}$$

*A crisp set is a strong $\alpha$-cut or strong $\alpha$-level set of a fuzzy set F if and only if the membership function values of all the elements in the crisp set in F are greater than $\alpha$, i.e.,*

$$F_\alpha = \{x \in U | \mu_F(x) > \alpha\}. \tag{2.8}$$

As shown in Figure 2.1, the $\alpha$-level set of the fuzzy set "the temperature is medium" ($\alpha = 0$) $F_0 = [10, 26]°C$, and the strong $\alpha$-level set of it $F_0 = (10, 26)°C$.

**Definition 2.6.** *(Convexity) (Zadeh 1965 [22]) A fuzzy set F in U is convex if and only if its $\alpha$-cut set is a convex set for any $\alpha$ in the interval $(0, 1)$.*

There are also some extreme cases of fuzzy set. For example, a fuzzy set is empty if and only if the membership function values of all its elements are zero on $U$.

**Definition 2.7.** *(Containment) (Zadeh 1965 [22]) A fuzzy set F is contained in a fuzzy set G if and only if $\mu_F(x) \leq \mu_G(x)$ for any elements on U.*

Let $VL$ be the fuzzy set of "the temperature is very low", and the range (i.e., $\mu_{VL}(x) \geq 0$) of very low temperature is $[-\infty, 10]°C$, and the most possible range (i.e., $\mu_{VL}(x) = 1$) of very low temperature is $[-\infty, 0]°C$. Based on the containment definition, it can be seen then that $VL$ is contained in $L$ ,which is the fuzzy set for "the temperature is low".

### 2.1.2 *Fuzzy Membership Function*

Membership functions in different systems may appear in a different form. In engineering applications, the most typical and commonly used membership functions, including triangular membership function and trapezoidal membership function, are explained in detail as follows.

In some engineering practical problems, the some parameter data are evaluated by a group of experts and expressed using linguistic terms, such as "the value is within the range $[a, b]$, and the most possible value is $c$". In this situation, a triangular membership function is employed to describe this type of uncertainty. The mathematical expression for a typical triangular membership function is as follows,

$$\mu_F(x) = \begin{cases} 0, & x < a \\ \frac{x-a}{c-a}, & a \leq x < c \\ 1, & x = c \\ \frac{x-b}{c-b}, & c \leq x < b \\ 0, & x \geq b \end{cases} \tag{2.9}$$

Figure 2.3 shows an example of triangular membership function.

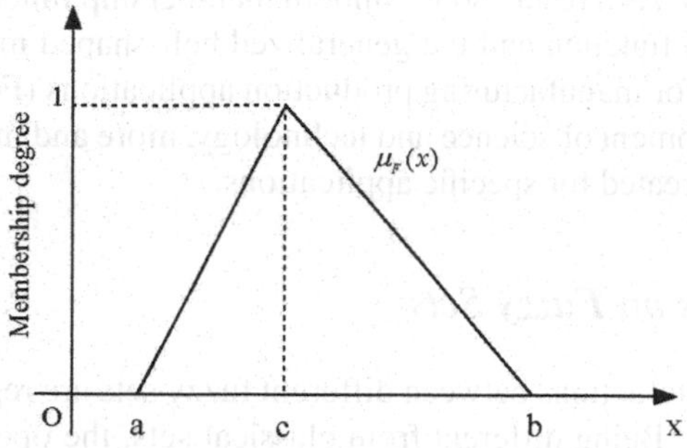


**Fig. 2.3** Example of triangular membership function

In some cases, the most possible value c could be extended to an interval $[c,d]$, which is contained in $[a,b]$. This means that experts express linguistic terms such as "the value is within the range $[a,b]$, and the most possible value is within the range $[c,d]$." To deal with this situation, a trapezoidal membership function, which could be regarded as an extension of triangular membership function, is employed as follows,

$$\mu_F(x) = \begin{cases} 0, & x < a \\ \frac{x-a}{c-a}, & a \le x < c \\ 1, & c \le x < d \\ \frac{x-b}{d-b}, & d \le x < b \\ 0, & x \ge b \end{cases} \tag{2.10}$$

Figure 2.4 shows an example of the trapezoidal membership function.

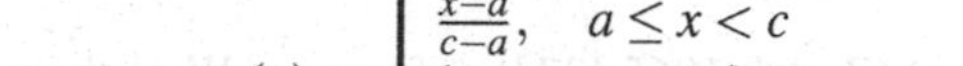

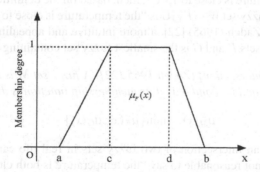


**Fig. 2.4** Example of trapezoidal membership function

While the triangular and trapezoidal membership functions have been widely used in practice because of their simple expressions and computational efficiency, they are not smooth functions due to the lack of smoothness in their turning points between line segments. As a result, some smooth membership functions, such as the Gaussian membership function and the generalized bell-shaped membership function, were developed for manufacturing production applications (Feng 2010 [460]). With the rapid development of science and technology, more and more membership functions have been created for specific applications.

### 2.1.3 Operations on Fuzzy Sets

In mathematics, the interactions between different fuzzy sets are regarded as the operations on fuzzy sets. Being different from classical sets, the operations on fuzzy sets (i.e., complement, union, and intersection) are defined based on the membership function instead of the set itself. The definition of complement, union, and intersection on fuzzy sets are presented as follows.

**Definition 2.8.** *(Complement) (Zadeh 1965 [22]) A fuzzy set $\bar{F}$ is the complement of a fuzzy set F if and only if its membership function is defined as*

$$\mu_{\bar{F}}(x) = 1 - \mu_F(x). \tag{2.11}$$

Based on the definition of the complement operation on a fuzzy set, it is relatively easy to determine that the complement of the fuzzy set "the temperature is close to 25°$C$" is actually that "the temperature is not close to 25°$C$".

**Definition 2.9.** *(Union) (Zadeh 1965 [22]) A fuzzy set W is the union of two fuzzy sets F and G if and only if its membership function is defined as*

$$\mu_W(x) = \max\{\mu_F(x), \mu_G(x)\} \tag{2.12}$$

Let $F$ be the fuzzy set for "the temperature is close to 25°$C$ and $G$ be the fuzzy set for "the temperature is close to 15°$C$", then, based on the definition for the union of fuzzy sets, the fuzzy set $W = F \bigcup G$ is "the temperature is close to 25°$C$ or 15°$C$". As pointed out by Zadeh (1965) [22], a more intuitive and appealing definition for the union of fuzzy sets $F$ and $G$ is the smallest fuzzy set containing both $F$ and $G$.

**Definition 2.10.** *(Intersection) (Zadeh 1965 [22]) A fuzzy set M is the intersection of two fuzzy sets F and G if and only if its membership function is defined as*

$$\mu_M(x) = \min\{\mu_F(x), \mu_G(x)\} \tag{2.13}$$

An example of an intersection of two fuzzy sets in real life can sometimes be in conflict, as it is not reasonable to say "the temperature is both close to 25°$C$ and 15°$C$. As pointed out by Zadeh (1965) [22], a more intuitive and appealing definition for the intersection of fuzzy sets $F$ and $G$ is the largest fuzzy set that is contained in both $F$ and $G$.

Figure 2.5 illustrates the 3 type of operations on the fuzzy sets $F$ and $G$. As pointed out above, sometimes the max and min operations in the above definitions may encounter some difficulties in the analysis of real life problems. As a result, some other operators have been proposed in research, such as the S-norm and T-norm operators (Dubois and Prade 1980 [571]) for fuzzy set union and intersection operations, respectively (Cheng 1998 [572]). Research on operations for fuzzy sets is ongoing as the theory is adapted for more application areas.

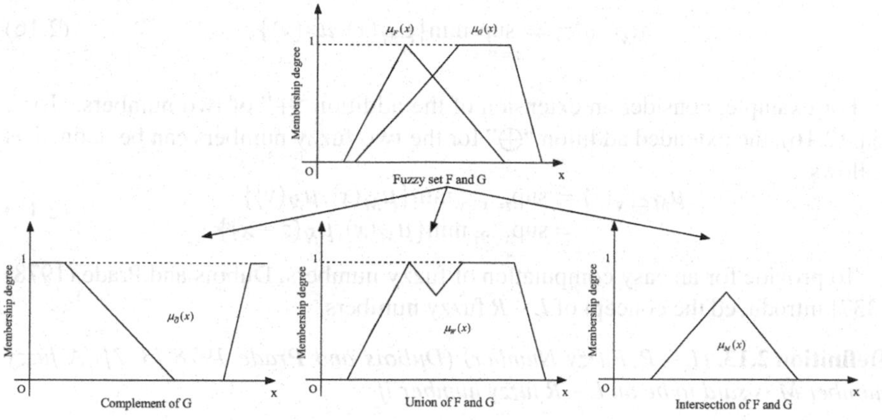


**Fig. 2.5** Operations on fuzzy sets

### *2.1.4 Extension Principle and Fuzzy Number*

The extension principle introduced by Zadeh (1965) [22] provides a general way to extend non-fuzzy mathematical concepts to the fuzzy framework.

**Definition 2.11.** *(Extension Principle) (Zadeh 1965 [22]) Let $f: X \to Y$ be a mapping from a set X to a set Y. Then, the extension principle allows us to define the fuzzy set $\tilde{B}$ in Y induced by the fuzzy set $\tilde{A}$ in X through f as follows:*

$$\tilde{B} = \{(y, \mu_{\tilde{B}}(y)) | y = f(x), x \in X\} \tag{2.14}$$

*with*

$$\mu_B(y) \triangleq \mu_{f(A)}(y) = \begin{cases} \sup_{y=f(x)} \mu_A(x) & \textit{if } f^{-1}(y) \neq \varnothing \\ 0 & \textit{if } f^{-1}(y) = \varnothing, \end{cases} \tag{2.15}$$

where $f^{-1}(y)$ is the inverse image of $y$, and $\varnothing$ means the empty set.

Because mathematical programming deals with uncertainty, among fuzzy sets, fuzzy numbers which are linguistically-expressed such as "approximately *m*" or "about *n*" play important roles. Before defining the fuzzy numbers, the definitions for the convex and normalized fuzzy sets are outlined. A fuzzy set $\tilde{A}$ is said to be convex if any $\alpha$-level set $A_\alpha$ of $\tilde{A}$ is convex, and a fuzzy set $\tilde{A}$ is said to be normal if there is $x$ such that $\mu_{\tilde{A}}(x) = 1$.

**Definition 2.12.** *(Fuzzy Number) (Dubois and Prade 1987 [578]) A fuzzy number is a convex normalized fuzzy set of the real line $\Re$ whose membership function is piecewise continuous.*

Using Zadeh's extension principle [22], the binary operation "$*$" in $\Re$ can be extended to the binary operation "$\circledast$" for fuzzy numbers $\tilde{M}$ and $\tilde{N}$ as

$$\mu_{\tilde{M}\circledast\tilde{N}}(z) = \sup_{z=x*y} \min\{\mu_{\tilde{M}}(x), \mu_{\tilde{N}}(y)\}. \tag{2.16}$$

For example, consider an extension of the addition "+" of two numbers. Using Eq. (2.16), the extended addition "$\bigoplus$" for the two fuzzy numbers can be defined as follows:

$$\begin{aligned} \mu_{\tilde{M}\bigoplus\tilde{N}}(z) &= \sup_{z=x+y} \min\{\mu_{\tilde{M}}(x), \mu_{\tilde{N}}(y)\} \\ &= \sup_{x\in\Re} \min\{\mu_{\tilde{M}}(x), \mu_{\tilde{N}}(z-x)\}. \end{aligned} \tag{2.17}$$

To provide for an easy computation of fuzzy numbers, Dubois and Prade (1978) [337] introduced the concept of $L-R$ fuzzy numbers.

**Definition 2.13.** *($L-R$ Fuzzy Number) (Dubois and Prade 1978 [337]) A fuzzy number $\tilde{M}$ is said to be an $L-R$ fuzzy number if*

$$\mu_{\tilde{M}}(x) = \begin{cases} L(\frac{m-x}{\alpha}) & \text{if } x \le m \\ R(\frac{x-m}{\beta}) & \text{if } x \ge m \end{cases} \tag{2.18}$$

where $m$ is the mean value of $\tilde{M}$, and $\alpha$ and $\beta$ are positive numbers which represent the left and right spreads of the fuzzy number, respectively; a function $L$ is a left shape function satisfying (i) $L(x) = L(-x)$; (ii) $L(0) = 1$, (iii) $L(x)$ is nonincreasing on $[0,\infty)$; and a right shape function $R$ is similarly defined as $L$.

Using its mean, left and right spreads, and shape functions, an $L-R$ fuzzy number $\tilde{M}$ is symbolically written as

$$\tilde{M} = (m, \alpha, \beta)_{LR}. \tag{2.19}$$

For two $L-R$ fuzzy numbers $\tilde{M} = (m, \alpha, \beta)_{LR}$ and $N = (n, \gamma, \delta)_{LR}$, the extended addition $\tilde{M} \bigoplus \tilde{N}$ is calculated as

$$(m, \alpha, \beta)_{LR} \bigoplus (n, \gamma, \delta)_{LR} = (m+n, \alpha+\gamma, \beta+\delta)_{LR}, \tag{2.20}$$

and scalar multiplication for an $L-R$ fuzzy numbers $\tilde{M} = (m, \alpha, \beta)_{LR}$ and a scalar value $\lambda$ is also given as

$$\lambda \otimes (m, \alpha, \beta)_{LR} = \begin{cases} (\lambda m, \lambda\alpha, \lambda\beta)_{LR} & \text{if } \lambda > 0 \\ (\lambda m, -\lambda\alpha, -\lambda\beta)_{LR} & \text{if } \lambda < 0, \end{cases} \tag{2.21}$$

where $\otimes$ represents the extended multiplication. The other operations are similarly given, and for further information on this issue, refer to Dubois and Prade (1978) [337].

### 2.1.5 Concept of Fuzzy Variable

From a mathematical perspective, a fuzzy variable which can describe an evaluation of an object or an event is a value which could lie in a probable range defined by quantitative limits or parameters, and can be usefully described with imprecise categories. Here, the concept of a fuzzy variable is introduced in a way that parallels the characterization of a non-fuzzy variable as expressed by Zadeh (1975b) [290].

Basic knowledge about fuzzy variables, including the definition, the measure and the expected value are introduced in the following.

**Definition 2.14.** *(Fuzzy Variable) (Zadeh 1975b [290]) A fuzzy variable is characterized by a triple* $(X, U, R(X;u))$*, in which* $X$ *is the name of the variable;* $U$ *is a universe of discourse (finite or infinite set);* $u$ *is a generic name for the elements of* $U$*; and* $R(X;u)$ *is a fuzzy subset of* $U$ *which represents a fuzzy restriction on the values of* $u$ *imposed by* $X$*. [As in the case of non-fuzzy variables,* $R(X;u)$ *is usually abbreviated to* $R(X)$ *or* $R(u)$ *or* $R(x)$*, where* $x$ *denotes a generic name for the values of* $X$*, and* $R(X;u)$ *is referred to as the restriction on* $u$ *or the restriction imposed by* $X$*.] The non-restricted non-fuzzy variable* $u$ *constitutes the base variable for* $X$*.*

The assignment equation for X has the form

$$x = u : R(X) \tag{2.22}$$

and represents an assignment of a value $u$ to $x$ subject to the restriction $R(X)$.

The degree to which this equation is satisfied is referred to as the compatibility of $u$ with $R(X)$ and is denoted by $c(u)$. By definition,

$$c(u) = \mu_{R(X)}(u), \quad u \in U \tag{2.23}$$

where $\mu_{R(X)}(u)$ is the grade of membership of $u$ in the restriction $R(X)$.

*Remark 2.1.* It is important to observe that the compatibility of $u$ is not the same as the probability of $u$. Thus, the compatibility of $u$ with $R(X)$ is merely a measure of the degree to which $u$ satisfies the restriction $R(X)$, and has no relation as to how probable or improbable $u$ is.

*Remark 2.2.* In terms of the valise analogy, a fuzzy variable may be likened to a tagged valise with soft sides, with $X$ representing the name on the tag, $U$ corresponding to a list of objects which can be put in a valise, and $R(X)$ representing a sub-list of $U$ in which each object $u$ is associated with a number $c(u)$ representing the degree of ease with which $u$ can be fitted in valise $X$.

In order to simplify the notation, it is convenient to use the same symbol for both $X$ and $x$, relying on the context for disambiguation. This is done in the following example.

*Example 2.3.* Consider a fuzzy variable named budget, with $U = [0, \infty]$ and $R(X)$ defined by (see Figure 2.6).

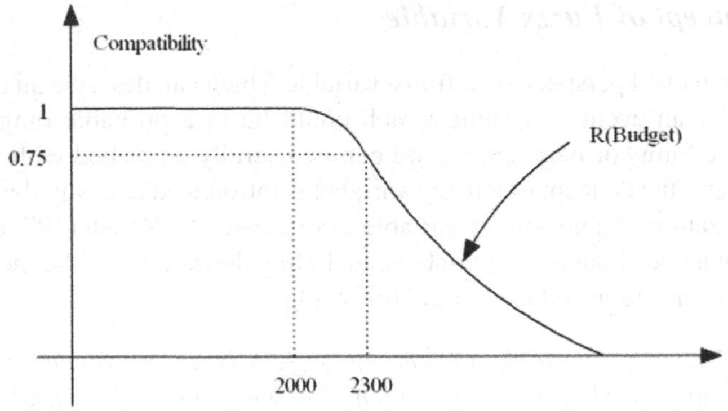


**Fig. 2.6** Compatibility function of budget

$$R(budget) = \int_0^{2000} 1/u + \int_{2000}^{\infty} [1 + (\frac{u-2000}{200})^2]^{-1}/u. \tag{2.24}$$

Then, in the assignment equation

$$budget = 2300 : R(budget), \tag{2.25}$$

the compatibility of 2300 with the restriction imposed by budget is

$$c(2300) = \mu_{R(budget)}(2300) = 0.75. \tag{2.26}$$

As in the case of nonfuzzy variables, if $X_1, \cdots, X_n$ are fuzzy variables in $U_1, \cdots, U_n$, respectively, then $X \triangleq (X_1, \cdots, X_n)$ is an $n$-ary composite (joint) variable in $U = U_1 \times \cdots \times U_n$. Correspondingly, in the $n$-ary assignment equation

$$(x_1, \cdots, x_n) = (u_1, \cdots, u_n) : R(X_1, \cdots, X_n), \tag{2.27}$$

$x_i$, $i = 1, \cdots, n$, is a generic name for the values of $X_i$; $u_i$ is a generic name for the elements of $U_i$; and $R(X) \triangleq R(X_1, \cdots, X_n)$ is an $n$-ary fuzzy relation in $U$ which represents the restriction imposed by $X \triangleq (X_1, \cdots, X_n)$. The compatibility of $(u_1, \cdots, u_n)$ with $R(X_1, \cdots, X_n)$ is defined by

$$c(u_1, \cdots, u_n) = \mu_{R(X)}(u_1, \cdots, u_n), \tag{2.28}$$

where $\mu_{R(X)}$ is the membership function of the restriction on $u \triangleq (u_1, \cdots, u_n)$.

*Example 2.4.* Suppose that $U_1 = U_2 = (-\infty, \infty)$, $X_1 \triangleq$ horizontal proximity; $X_2 \triangleq$ vertical proximity; and the restriction on $u$ is expressed by

$$R(X) = \int_{U_1 \times U_2} (1 + u_1^2 + u_2^2)^{-1}/(u_1, u_2). \tag{2.29}$$

Then the compatibility of the value $u = (2, 1)$ in the assignment equation

$$(x_1,x_2) = (2,1) : R(X) \tag{2.30}$$

is given by

$$c(2,1) = \mu_{R(X)}(2,1) = 0.16. \tag{2.31}$$

*Remark 2.3.* In terms of the valise analogy (see Remark 2.2), an $n$-ary composite fuzzy variable may be likened to a soft valise named $X$ with $n$ compartments named $X_1,\cdots,X_n$. The compatibility function $c(u_1,\cdots,u_n)$ represents the degree of ease with which objects $u_1,\cdots,u_n$ can be put into respective compartments $X_1,\cdots,X_n$ simultaneously.

A basic question that arises in connection with an $n$-ary assignment equation relates to its decomposition into a sequence of $n$ unary assignment equations. In the case of fuzzy variables, the process of decomposition is somewhat more involved, which is discussed after the marginal and conditioned restrictions are defined.

Fuzzy set theory is well developed and has been applied to a wide variety of real problems. Here the definition proposed by Zadeh (1965) [22] is adopted. The fuzzy variable was first introduced by Kaufmann (1975) [475], then it appeared in Nahmias (1978) [238]. The membership functions for fuzzy variables were introduced in Dubois and Prade (1980) [571], who further proposed possibility theory (Dubois and Prade 1988 [221]). However, the traditional fuzzy measures for fuzzy events are not able to express the preference of decision makers. Thus, here the fuzzy measure, Me (Xu and Zhou 2011 [281]), is introduced which embeds an optimistic-pessimistic parameter to determine the combined attitudes of decision makers.

For example, the duration for each activity is a typical uncertain variable, which can fluctuate because of many factors, such as, the weather, equipment properties, labor efficiency, execution errors of decision makers, supply conditions of materials, coordination problems among stake holders, and other uncertain factors. Let $\tilde{D}_i$ be the normal duration for Activity $i$, $\tilde{d}_i$, the crashed duration for Activity $i$, and $Y_i$, the crashing time for Activity $i$. In practice, the decision makers may give a statement such as "it is possible that the normal duration for Activity $i$ is within an optimistic and pessimistic range, where the optimistic margin is 15 months, and the pessimistic margin is 23 months, and the most likely value of the normal duration is 19 months," which can be translated into a triangular fuzzy number $\tilde{D}_i = (15,19,23)$. If the crashing time is 3 months, then the crashed duration for Activity $i$ can be calculated as a triangular fuzzy number $\tilde{d}_i = (12,16,20)$ as shown in Figure 2.7.

In order to transform the fuzzy numbers into crisp values, the expected value operator for the fuzzy measure $Me$ (Xu and Zhou 2011 [281]) is introduced to deal with the uncertainty in this book. The expected value of a triangular fuzzy number is as follows,

$$E^{Me}[\xi] = \begin{cases} \frac{\lambda}{2}r_1 + \frac{r_2}{2} + \frac{1-\lambda}{2}r_3, & if \quad r_3 \le 0 \\ \frac{\lambda}{2}(r_1+r_2) + \frac{\lambda r_3^2 - (1-\lambda)r_2^2}{2(r_3-r_2)}, & if \quad r_2 \le 0 \le r_3 \\ \frac{\lambda}{2}(r_3+r_2) + \frac{(1-\lambda)r_2^2 - \lambda r_1^2}{2(r_2-r_1)}, & if \quad r_1 \le 0 \le r_2 \\ \frac{(1-\lambda)r_1 + r_2 + \lambda r_3}{2}, & if \quad 0 \le r_1 \end{cases}$$

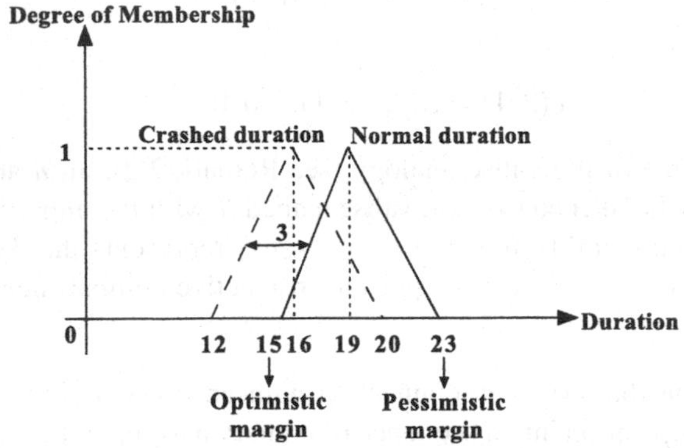


**Fig. 2.7** The membership function of duration for each activity

where $\lambda$ is the optimistic-pessimistic index to determine the combined attitudes of decision makers.

As all fuzzy variables are non-negative triangular fuzzy numbers in real world problems (Figure 2.7 shows an example), the problems in this book belong to the case $0 \leq r_1$, i.e., $E^{Me}[\xi] = \frac{(1-\lambda)r_1+r_2+\lambda r_3}{2}$.

## 2.2 Fuzzy MOMSDM

The general concepts of the fuzzy MOMSDM are developed to take into account the fuzziness and ambiguity of human decision making. Fuzzy uncertainty can appear in coefficients, termination time, state and decision variables, constraints, and objectives. How to deal with the fuzziness when modeling, what the relationship is between the fuzzy objectives and constraints and how to calculate the expected value of the fuzzy parameters are explained in this section.

### *2.2.1 Fuzzy Termination Time*

In practice, however, a crisp (non-fuzzy) termination time may often not represent the real perception of the planning (control) horizon appropriate for the problem being considered (Kacprzyk 1997 [162]). For example, in virtually all (longer term) planning problems even if the termination time (planning horizon) is stated as, say, 25 months, then it is tacitly assumed that this is just a rough estimate, and the process should terminate (e.g., with the attainment of some socioeconomic goals) in more or less than 10 months or, say, in not much more than 25 months. On the other hand, in many cases a small increase in the termination time can greatly improve the outcome or performance of the process, while a small decrease may have a negligible effect or none at all (Kacprzyk 1997 [162]).

This suggests that it may be expedient to allow for a less crisp, and "softer" definition of the control process termination time by allowing for its formulation as a fuzzy set such as 'more or less than 25 months', 'much less than 10 stages'. etc (Kacprzyk 1997 [162]).

The idea of such a fuzzy termination time was proposed in 1977 by Fung and Fu [171], and Kacprzyk (1997) [162]. Though these two approaches were similar conceptually, and with respect to the general problem perception and formulation, they differed in detail and solution procedures. This section considers Kacprzyk's concept, as it is much more consistent with the general line of reasoning adopted here.

If the termination time $\tilde{T}$ is a fuzzy variable, where its membership function is $\mu_{\tilde{T}}(x) = M(x)$, then the general model for the MOMSDM under a fuzzy termination time can be expressed as,

$$\begin{cases} \max \{J_1(\mathbf{x}(0),\mathbf{x}(\tilde{T}),\pi), J_2(\mathbf{x}(0),\mathbf{x}(\tilde{T}),\pi),\cdots,J_n(\mathbf{x}(0),\mathbf{x}(\tilde{T}),\pi)\} \\ \text{s.t.} \begin{cases} \mathbf{x}(k+1) = f_k(\mathbf{x}(k),\mathbf{u}(k)), \\ \mathbf{x}(0) \in I, \\ \mu_{\tilde{T}} = M(x), \\ g_j(\mathbf{x}(\cdot),\mathbf{u}(\cdot)) \le 0,\ j = 1,2,\cdots,p, \\ \mathbf{x}(k) \in X,\ k = 1,2,\cdots,\tilde{T}, \\ \mathbf{u}(k) \in U,\ k = 1,2,\cdots,\tilde{T}. \end{cases} \end{cases} \tag{2.32}$$

where $\mathbf{x}(k)$ is the state vector in Stage $k$; $\mathbf{u}(k)$ is the decision vector in Stage $k$; $J_i(\mathbf{x}(0),\mathbf{x}(\tilde{T}),\pi)$ is the objective function for the fuzzy MOMSDM; $g_j(\mathbf{x}(\cdot),\mathbf{u}(\cdot))$ is the general constraint function for the fuzzy MOMSDM; $f_k(\mathbf{x}(k),\mathbf{u}(k))$ is a state transition function of the fuzzy MOMSDM; $\pi = (\mathbf{u}(0),\mathbf{u}(1),\cdots,\mathbf{u}(\tilde{T}))$ is the policy for the whole decision making process.

### 2.2.2 *Fuzzy State and Decision Variables*

In some real problems, the MOMSDM state and decision variables are not crisp or deterministic values. Although stochastic methods are used to model some uncertainties in the state and decision variables due to randomness, there are some situations which do not necessarily have a random or stochastic nature. The good example is reservoir operations. Reservoir storage volume is an important variable, the discretization of which has a pronounced effect on the computational efforts. The error or imprecision caused by discretizing continuous variables in the modeling process is a good example of a type of uncertainty that is not related to randomness. Such an imprecision can be dealt with using fuzzy set theory in which a fuzzy number represents an interval rather than a single valued point. The error caused by storage volume discretization can be examined by considering it as a fuzzy state variable (Mousavi et al. 2004 [172]). Figure 2.8 shows an example of the fuzzy state variable, in which the error caused by storage volume discretization floats within the range of $[-0.053, 0.038]$ in a stage.

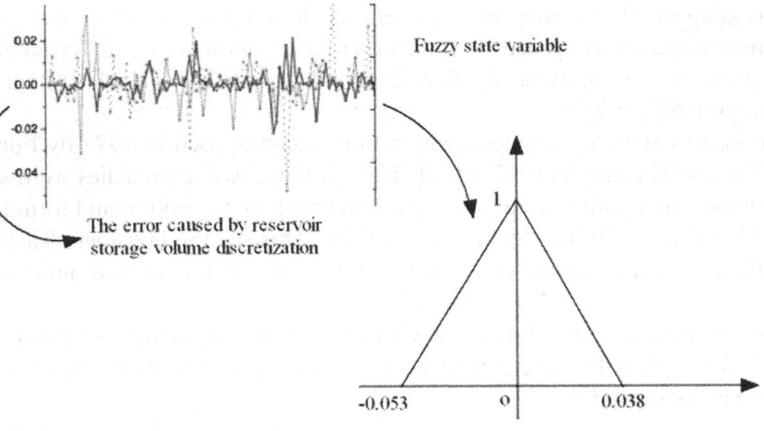


**Fig. 2.8** An example of the fuzzy state variable

Let $\tilde{\mathbf{x}}(k)$ be the fuzzy state vector in Stage $k$, $\tilde{\mathbf{u}}(k)$, the fuzzy decision vector in Stage $k$, then the general MOMSDM model with fuzzy state and decision variables can be expressed as,

$$\begin{cases} \max \{J_1(\tilde{\mathbf{x}}(0),\tilde{\mathbf{x}}(T),\pi),J_2(\tilde{\mathbf{x}}(0),\tilde{\mathbf{x}}(T),\pi),\cdots,J_n(\tilde{\mathbf{x}}(0),\tilde{\mathbf{x}}(T),\pi)\} \\ \text{s.t.} \begin{cases} \tilde{\mathbf{x}}(k+1) \cong f_k(\tilde{\mathbf{x}}(k),\tilde{\mathbf{u}}(k)), \\ \tilde{\mathbf{x}}(0) \in \tilde{I}, \\ T = a, \\ g_j(\tilde{\mathbf{x}}(\cdot),\tilde{\mathbf{u}}(\cdot)) \le 0, \ j = 1,2,\cdots,p, \\ \tilde{\mathbf{x}}(k) \in \tilde{X}, \ k = 1,2,\cdots,T, \\ \tilde{\mathbf{u}}(k) \in \tilde{U}, \ k = 1,2,\cdots,T. \end{cases} \end{cases} \tag{2.33}$$

where $a$ is the stage amount; $\tilde{I}$ is the fuzzy initial state space; $\tilde{X}$ is the fuzzy state space; $\tilde{U}$ is the fuzzy decision space; $J_i(\tilde{\mathbf{x}}(0),\tilde{\mathbf{x}}(T),\pi)$ is the objective function of the fuzzy MOMSDM; $g_j(\tilde{\mathbf{x}}(\cdot),\tilde{\mathbf{u}}(\cdot))$ is the general constraint function of the fuzzy MOMSDM; $f_k(\tilde{\mathbf{x}}(k),\tilde{\mathbf{u}}(k))$ is a state transition function of the fuzzy MOMSDM; $\pi = (\tilde{\mathbf{u}}(0),\tilde{\mathbf{u}}(1),\cdots,\tilde{\mathbf{u}}(T))$ is the policy of the whole decision making process.

### *2.2.3 Fuzzy Coefficients*

In conventional optimization problems, the coefficients are all assumed to be real numbers. However, uncertainty always occurs in the real world. The occurrence of fuzziness in the real world is inevitable due to unexpected situations (Wu 2008 [160]).For example, one of the key sources of uncertainty in any production-distribution system is product demand. Market demand is composed of firm orders and demand forecasts. The firm orders are known at the beginning of each planning horizon, while the foreseen demand is based on such factors as historical sales, suppliers of other sources or competitors. These factors make the demand fuzzy in nature as the capacity data is not a crisp value because during the production processes unforeseen events such as breakdowns, faulty production or preparation delays can occur. Also, planning costs cannot be measured easily since they require significant human perception for their valuation. For instance, in some manufacturing environments, the planners use an average production cost per hour which has been calculated from unitary production costs, which makes these costs fuzzy. The costs of the underuse or overuse of productive resources cannot be exact because of unknown elements, such as changed labor requirements, production mishaps or shortcomings. The demand backlog cost is also fuzzy in its composition as this cost consists not only of the administrative cost of managing the backlog of orders but also the costs due to a potential loss of clients. This type of cost is commonly estimated using human experience (Mula et al. 2007 [161]). As a result, the developing of a MOMSDM with fuzzy coefficients is necessary.

Let $\tilde{\xi}$ be the fuzzy coefficient vector in the objective functions and constraints, then the general model of the MOMSDM with fuzzy coefficients can be expressed as,

$$\begin{cases} \max \{J_1(\mathbf{x}(0),\mathbf{x}(T),\pi,\tilde{\xi}), J_2(\mathbf{x}(0),\mathbf{x}(T),\pi,\tilde{\xi}),\cdots,J_n(\mathbf{x}(0),\mathbf{x}(T),\pi,\tilde{\xi})\} \\ \text{s.t.} \begin{cases} \mathbf{x}(k+1) \cong f_k(\mathbf{x}(k),\mathbf{u}(k),\tilde{\xi}), \\ \mathbf{x}(0) \in I, \\ T = a, \\ g_j(\mathbf{x}(\cdot),\mathbf{u}(\cdot),\tilde{\xi}) \leq 0, \ j = 1,2,\cdots,p, \\ \mathbf{x}(k) \in X, \ k = 1,2,\cdots,T, \\ \mathbf{u}(k) \in U, \ k = 1,2,\cdots,T. \end{cases} \end{cases} \tag{2.34}$$

where $\mathbf{x}(k)$ is the state vector in Stage $k$, $\mathbf{u}(k)$, the decision vector in Stage $k$; $a$ is the stage amount; $I$ is the initial state space; $X$ is the state space; $U$ is the decision space; $J_i(\mathbf{x}(0),\mathbf{x}(T),\pi,\tilde{\xi})$ is the objective function of the fuzzy MOMSDM; $g_j(\mathbf{x}(\cdot),\mathbf{u}(\cdot),\tilde{\xi})$ is a general constraint function of the fuzzy MOMSDM; $f_k(\mathbf{x}(k),\mathbf{u}(k),\tilde{\xi})$ is a state transition function of the fuzzy MOMSDM; $\pi = (\mathbf{u}(0),\mathbf{u}(1),\cdots,\mathbf{u}(T))$ is the policy for the whole decision making process.

### 2.2.4 Fuzzy Goals and Constraints

Much of the decision-making in the real world takes place in an environment in which the goals, the constraints and the consequences of possible actions are not known precisely. The concept of decision-making in a fuzzy environment was introduced by Bellman and Zadeh [285] in 1970 and referred to as a decision process in which the goals and/or the constraints, but not necessarily the system under control, are fuzzy in nature. This means that the goals and/or the constraints constitute classes of alternatives whose boundaries are not sharply defined.

An example of a fuzzy constraint is: "The price of the construction material should not be *substantially* higher than $\alpha$," where $\alpha$ is a specified constant. Similarly, an example of a fuzzy goal is: "the total cost should be in the *vicinity* of $x_0$," where $x_0$ is a constant. The italicized words are the sources of fuzziness in these examples.

Fuzzy goals and fuzzy constraints can be defined precisely as fuzzy sets in the space of alternatives. A fuzzy decision, then, may be viewed as an intersection of the given goals and constraints. A maximizing decision is defined as a point in the space of alternatives at which the membership function of a fuzzy decision attains its maximum value.

Zimmermann (1978) [29] introduces the concept of fuzzy set theory into linear programming. Assuming that the membership functions for fuzzy sets are linear, he showed that, by employing the principle of the fuzzy decision by Bellman and Zadeh (1970) [285], a linear programming problem with a fuzzy goal and fuzzy constraints can be solved using standard linear programming techniques. For the MOMSDM with fuzzy goals and constraints, the general form of model is presented as,

$$\begin{cases} \{J_1(\mathbf{x}(0),\mathbf{x}(T),\pi) \gtrsim z_1, J_2(\mathbf{x}(0),\mathbf{x}(T),\pi) \gtrsim z_2, \cdots, J_n(\mathbf{x}(0),\mathbf{x}(T),\pi) \gtrsim z_n\} \\ \text{s.t.} \begin{cases} \mathbf{x}(k+1) = f_k(\mathbf{x}(k),\mathbf{u}(k)), \\ \mathbf{x}(0) \in I, \\ T = a, \\ g_j(\mathbf{x}(\cdot),\mathbf{u}(\cdot)) \lesssim 0, \ j = 1,2,\cdots,p, \\ \mathbf{x}(k) \in X, \ k = 1,2,\cdots,T, \\ \mathbf{u}(k) \in U, \ k = 1,2,\cdots,T. \end{cases} \end{cases} \tag{2.35}$$

where the symbols $\gtrsim$ and $\lesssim$ denote a relaxed or fuzzy version of the ordinary inequality $\geqslant$ and $\leqslant$, respectively; $\mathbf{x}(k)$ is the fuzzy state vector in Stage $k$, $\mathbf{u}(k)$, the fuzzy decision vector in Stage $k$; $a$ is the stage amount; $I$ is the initial state space; $X$ is the state space; $U$ is the decision space; $J_i(\mathbf{x}(0),\mathbf{x}(T),\pi)$ is the objective function of the fuzzy MOMSDM; $g_j(\mathbf{x}(\cdot),\mathbf{u}(\cdot))$ is general constraint function of the fuzzy MOMSDM; $f_k(\mathbf{x}(k),\mathbf{u}(k))$ is the state transition function of the fuzzy MOMSDM; $\pi = (\mathbf{u}(0),\mathbf{u}(1),\cdots,\mathbf{u}(T))$ is the policy of the whole decision making process.

From the decision maker's preference, the fuzzy goals and the fuzzy constraints mean that the objective functions need to be essentially larger than or equal to a certain level $z_i$, and that the constraint values should be substantially smaller than or equal to 0, respectively.

## 2.3 Fuzzy Random Theory

Fuzzy random variables were introduced to model and analyze 'imprecisely valued' measurable functions associated with the sample space of a random experiment, when the imprecision in the values of these functions is formalized in terms of fuzzy sets [105]. Different approaches to this concept have been developed in the literature, the most widely considered being one introduced by Kwakernaak (1978, 1979) [94, 95] and Kruse and Meyer (1987) [479], and the one by Puri and Ralescu (1986) [93]. This section reviews these main approaches and refers to relevant studies in probability and statistics in which they have been used.

### 2.3.1 General Concepts of Fuzzy Random Variable

Fuzzy random variables represent a well-formalized concept underlying many recent probabilistic and statistical studies involving data obtained from a random experiment when these data are assumed to be fuzzy set valued [105]. Fuzzy random variables have been considered in the setting of a random experiment to model: (1) either a fuzzy perception/observation of a mechanism (the so-called 'original random variable') associating a real value with each experimental outcome; (2) or an essentially fuzzy-valued mechanism, that is, a mechanism associating a fuzzy value with each experimental outcome.

For the first situation, Kwakernaak (1978, 1979) [94, 95] introduced a mathematical model which was later formalized in a clearer way by Kruse and Meyer (1987) [479]. In Kwakernaak/Kruse and Meyer's approach, a fuzzy random variable is viewed as a fuzzy perception/observation/report of a classical real-valued random variable. The definition is stated as follows.

**Definition 2.15.** *(Fuzzy Random Variable) (Kruse and Meyer 1987 [479]) Given a probability space $(\Omega, \mathscr{A}, P)$, a mapping*

$$\chi : \Omega \to \mathscr{F}_c(\mathbb{R})$$

*is said to be a fuzzy random variable (or FRV for short) if for all $\alpha \in [0,1]$ the two real-valued mappings*

$$\inf \chi_\alpha : \Omega \to \mathbb{R}, \quad \sup \chi_\alpha : \Omega \to \mathbb{R}$$

*(defined so that for all $\omega \in \Omega$ we have that $\chi_\alpha(\omega) = [\inf(\chi(\omega))_\alpha, \sup(\chi(\omega))_\alpha]$) are real-valued random variables.*

Probabilistic and statistical studies for FRVs in Kwakernaak/Kruse and Meyer's approach usually concern either 'crisp' parameters of the 'original' random variable or fuzzy-valued parameters defined on the basis of Zadeh's extension principle. More precisely, Kruse and Meyer (1987) [479] have defined

**Definition 2.16.** *(Kruse and Meyer 1987 [479]) If $\theta(X)$ is a real-valued parameter of a random variable $X : \Omega \to \mathbb{R}$ associated with the probability space $(\Omega, \mathscr{A}, P)$, and $\varepsilon(\Omega, \mathscr{A}, P)$ denotes the class of all possible 'originals' of a FRV $\chi : \Omega \to \mathscr{F}_c(\mathbb{R})$ associated with $(\Omega, \mathscr{A}, P)$, then the induced fuzzy parameter of $\chi$ corresponds to*

$$\theta(\chi) : \mathbb{R} \to [0,1]$$

*such that for all $t \in \mathbb{R}$*

$$\theta(\chi)(t) = \sup_{X \in \varepsilon(\Omega, \mathscr{A}, P) | \theta(X) = t} \inf_{\omega \in \Omega} \{\chi(\omega)(X(\omega))\}.$$

As an example of an induced fuzzy parameter, if $\theta(\chi) = E(X|P)$, then $\theta(\chi)$ corresponds to the so-called fuzzy expected value of $\chi$ which is the fuzzy set in $\mathscr{F}_c(\mathbb{R})$ such that for each $\alpha \in [0,1]$

$$(\theta(\chi))_\alpha = [E(\inf \chi_\alpha|P), E(\sup \chi_\alpha|P)].$$

It should be emphasized that the notion of the (induced) fuzzy parameter is essentially different from the classical one, since classical parameters are involved explicitly in the expression for the distribution of the associated variable, whereas the fuzzy-valued parameters are not necessarily included in it.

For the second situation to be modelled, Puri and Ralescu (1986) [93] introduced another mathematical approach. In Puri and Ralescu's approach, a fuzzy random variable is viewed as a mechanism associating a fuzzy set in $\mathscr{F}(\mathbb{R}^p)$ with each experimental outcome. The definition is stated as follows.

**Definition 2.17.** *(Puri and Ralescu 1986 [93]) Given a probability space* $(\Omega, \mathscr{A}, P)$, *a mapping*

$$\chi : \Omega \to \mathscr{F}(\mathbb{R}^p)$$

*is said to be a fuzzy random variable (also called random fuzzy set, and referred to as FRV for short) if for all* $\alpha \in [0,1]$ *the set-valued mappings*

$$\chi_\alpha : \Omega \to \mathscr{K}(\mathbb{R}^p),$$

*defined so that for all* $\omega \in \Omega$

$$\chi_\alpha(\omega) = (\chi(\omega))_\alpha,$$

*are random sets (that is, Borel-measurable mappings with the Borel* $\sigma$*-field generated by the topology associated with the Hausdorff metric on* $\mathscr{K}(\mathbb{R}^p)$*).*

*Example 2.5.* Let $(\Omega, \mathscr{A}, Pr)$ be a probability space. If $\Omega = \{\omega_1, \omega_2, \cdots, \omega_n\}$ and $\mu_1, \mu_2, \cdots, \mu_n$ are fuzzy variables in $\mathscr{F}$, then the function

$$\tilde{\tilde{\xi}}(\omega) = \begin{cases} \mu_1, & if \omega = \omega_1, \\ \mu_2, & if \omega = \omega_2, \\ \cdots, & \\ \mu_n, & if \omega = \omega_n \end{cases}$$

is clearly a fuzzy random variable, see Figure 2.9.

*Example 2.6.* In a resource-constrained multiple project scheduling problem, steel is a resource used in warehouse and factory construction activities such as in transport and power system projects in large-scale water conservancy and hydropower construction projects. As the price of steel is high, managers want to reduce the use of steel in order to reduce the cost, while also hoping that the reduction in steel use will not compromise the quality of the warehouses and factories. In this case, managers give the range of quantities in accordance with an expected value, i.e., viz $\tilde{\tilde{r}} = (50, \rho, 65)$ (units: ton) with $\rho \sim \mathscr{N}(1,1)$, see Figures 2.10 and 2.11. This means that the quantity of steel resources is a fuzzy variable taking a random parameter, i.e., a fuzzy random variable. Similarly, managers expect the processing time for a transport and power system project to be about 11 months with a normally distributed variable, $\tilde{\tilde{D}} = (10, \rho, 12)$ with $\rho \sim \mathscr{N}(0,1)$. Therefore, a situation arises where fuzzy variables take random parameters throughout the project. So, it becomes necessary to apply a fuzzy random variable to deal with these uncertain parameters of combined fuzziness and randomness.

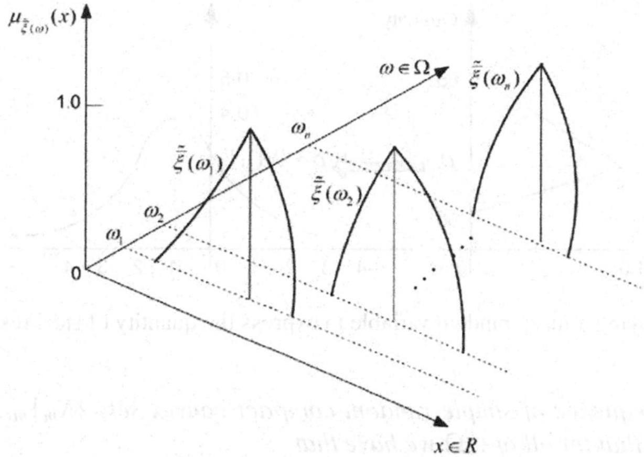


**Fig. 2.9** Representation of a fuzzy random variable

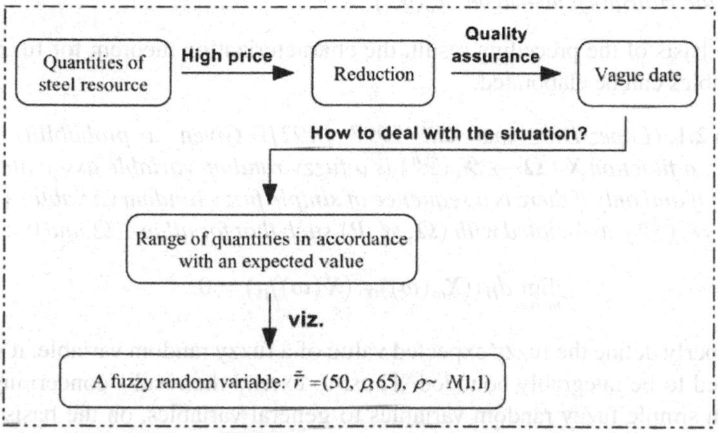


**Fig. 2.10** Flowchart of why the quantity of steel resource is a fuzzy random variable

## 2.3.2 *Characterization of Fuzzy Random Variables*

The main purpose of this section is to discuss the equivalence between the descriptive and constructive definitions of fuzzy random variables. The descriptive definition is based on the measurability of the e-level functions, and the constructive definition considers a fuzzy random variable as the Hausdorff limit (for each level) of a sequence of simple fuzzy random variables. This result is based on the following equivalence for random compact convex sets.

**Lemma 2.1.** *(Lopez-Diaz and Gil 1997 [492]) Given a probability space* $(\Omega, \mathscr{A}, P)$, *a set-valued function* $X : \Omega \to \mathscr{K}_c(\mathbb{R}^n)$ *is a random compact convex set if and only*

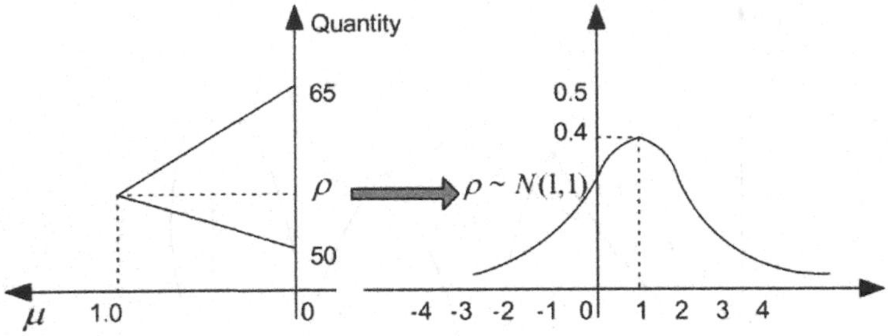


**Fig. 2.11** Employing a fuzzy random variable to express the quantity of steel resource

*if there is a sequence of simple random compact convex sets* $\{X_m\}_m$, $X_m : \Omega \to \mathcal{K}_c(\mathbb{R}^n)$, *such that for all* $\omega \in \Omega$ *we have that*

$$\lim_{m\to\infty} d_H(X_m(\omega), X(\omega)) = 0,$$

$d_H$ *being the Hausdorff metric on* $\mathcal{K}_c(\mathbb{R}^n)$.

On the basis of the preceding result, the characterization theorem for fuzzy random variables can be elaborated.

**Theorem 2.1.** *(Lopez-Diaz and Gil 1997 [492]) Given a probability space* $(\Omega, \mathscr{A}, P)$, *a function* $X : \Omega \to \mathscr{F}_c(\mathbb{R}^n)$ *is a fuzzy random variable associated with* $(\Omega, \mathscr{A}, P)$ *if and only if there is a sequence of simple fuzzy random variables* $\{X_m\}_m$, $X_m : \Omega \to \mathscr{F}_c(\mathbb{R}^n)$, *associated with* $(\Omega, \mathscr{A}, P)$ *such that for all* $\omega \in \Omega$ *and* $0 < \alpha \leqslant 1$

$$\lim_{m\to\infty} d_H((X_m(\omega))_\alpha, (X(\omega))_\alpha) = 0.$$

To properly define the fuzzy expected value of a fuzzy random variable, it should be assumed to be integrably bounded. Thus, to extend the results concerning integrals from simple fuzzy random variables to general variables, on the basis of the constructive definition, approximation conclusions like those discussed above have to be obtained for integrably bounded fuzzy random variables. In fact, the set of all simple fuzzy random variables from $\Omega$ to $\mathscr{F}_c(\mathbb{R}^n)$ is "dense" (in a Hausdorff sense and for each level in $(0,1]$) in the class of integrably bounded fuzzy random variables from $\Omega$ to $\mathscr{F}_c(\mathbb{R}^n)$. However, in contrast to the case of fuzzy random variables, the necessary condition in the Theorem 2.1 is not always sufficient for integrably bounded variables, since it is not sufficient to particularize the study to classical random variables. A counterexample involving fuzzy random variables and illustrating the last assertion is now elaborated.

*Example 2.7.* (Counter example)(Lopez-Diaz and Gil 1997 [492]) Let $\Omega$ be the real interval $[1, \infty)$, $\mathscr{A}$ the class of Lebesgue measurable sets on $\Omega$, and let $\mu$ be the Lebesgue measure on $\mathscr{A}$. Consider the measurable function $f : \Omega \to \mathbb{R}^+$, $f(\omega) = \omega^{-2}$, and let $P$ be the probability measure on $\mathscr{A}$ such that $P(A) = \int_A f d\mu$ for each $A \in \mathscr{A}$ (i.e., $f = dP/d\mu$ is the Radon-Nikodym derivative of $P$ with respect to $\mu$), so that $(\Omega, \mathscr{A}, P)$ is a probability space. If we define the fuzzy valued function $X : \Omega \to \mathscr{F}_c(\mathbb{R}^n)$, $(X(\omega))_\alpha = [1 + ((\omega - 1)/2)\alpha, \omega - ((\omega - 1)/2)\alpha]$, $0 < \alpha \leqslant 1$ (that is $X(\omega)$ is a symmetrical triangular fuzzy set with $X_0(\omega) = [1, \omega]$). $X$ is a fuzzy random variable, since for all $x \in \mathbb{R}$ and for all $0 \leqslant \alpha \leqslant 1$, the function $d_\alpha^x(\omega) = d(x, (X(\omega))_\alpha)$ is a continuous real function (see Himmelberg 1975 [495]; Hiai and Umegaki 1977 [493]). Nevertheless, $X$ is not integrably bounded, since (see Hiai and Umegaki 1977 [493]) the function $H(\omega) = \sup_{x \in X_0(\omega)} |x| = \omega$ is such that $H \notin L^1(P)$. On the other hand, the function $g : \Omega \to \mathbb{R}+$, $g(\omega) = \omega - 1$, is nonnegative $\mathscr{A}$-measurable, so that there exists a nondecreasing sequence of nonnegative simple $\mathscr{A}$-measurable functions $g_m : \Omega \to \mathbb{R}^+$ such that $\lim_{m\to\infty} g_m(\omega) = g(\omega)$ for all $\omega \in \Omega$. The fuzzy-valued functions $X_m : \Omega \to \mathscr{F}_c(R)$, $(X_m(\omega))_\alpha = [1 + (g_m(\omega)/2)\alpha,\ g_m(\omega) + 1 - (g_m(\omega)/2)\alpha]$, $0 < \alpha \leqslant 1$ (that is $X_m(\omega)$ is a symmetrical triangular fuzzy set with $(X_m)_0(\omega) = [1, g_m(\omega) + 1]$), are clearly simple fuzzy random variables, and they satisfy that, for all $0 < \alpha \leqslant 1$, $d_H((X_m(\omega))_\alpha, (X(\omega))_\alpha) \leqslant (g(\omega) - g_m(\omega))(1 + \alpha/2)$ that converges to 0 as $m \to \infty$.

There is a difference in the equivalence between the descriptive and the constructive definitions of integrably bounded fuzzy random variables. The descriptive definition is based on the measurability of the $\alpha$-level functions along with the integrable boundedness of the closed convex hull of the support, and the constructive definition considers an integrably bounded fuzzy random variable as the Hausdorff limit (for each level) of a sequence of simple fuzzy random variables dominated by an integrable real-valued function. This result is based on the following characterization theorem for integrably bounded fuzzy random variables.

**Theorem 2.2.** *(Lopez-Diaz and Gil 1997 [492]) Given a probability space* $(\Omega, \mathscr{A}, P)$*, a function* $X : \Omega \to \mathscr{F}_c(\mathbb{R}^n)$ *is an integrably bounded fuzzy random variable associated with* $(\Omega, \mathscr{A}, P)$ *if and only if there is a sequence of simple fuzzy random variables* $\{X_m\}_m$*,* $X_m : \Omega \to \mathscr{F}_c(\mathbb{R}^n)$*, associated with* $(\Omega, \mathscr{A}, P)$ *and a function* $h : \Omega \to \mathbb{R}$*,* $h \in L^1(P)$*, such that* $\sup_{x \in \{X_m\}_0(\omega)} |x| \leqslant h(\omega)$*, for all* $m \in \mathbb{N}$*, and* $\lim_{m\to\infty} d_H((X_m(\omega))_\alpha, (X(\omega))_\alpha) = 0$*, for all* $\omega \in \Omega$ *and for all* $0 < \alpha \leqslant 1$*.*

The constructive definition is of practical use to develop studies concerning the fuzzy integral of fuzzy random variables.

In particular, the constructive definition for the fuzzy integral suggested by Puri and Ralescu (1986) [93] can be now properly justified on the basis of Theorem 2.2. Thus, the fuzzy integral of integrably bounded fuzzy random variables can be defined as the Hausdorff limit (for each level) of a sequence of fuzzy integrals of simple fuzzy random variables, as certified in the following result,

**Proposition 2.1.** *(Lopez-Diaz and Gil 1997 [492]) Given a probability space* $(\Omega, \mathscr{A}, P)$*, and an integrably bounded fuzzy random variable* $X : \Omega \to \mathscr{F}_c(\mathbb{R}^n)$ *associated with it, then there is a sequence of simple fuzzy random variables associated with* $(\Omega, \mathscr{A}, P)$ *such that*

$$\lim_{m\to\infty} d_H((E_p(X_m))_\alpha, (E_p(X))_\alpha) = 0$$

*for all* $0 < \alpha \leqslant 1$*.*

### 2.3.3 Variance and Covariance of Fuzzy Random Variables

Let $(\Omega, \mathscr{A}, P)$ be a complete probability space. A fuzzy random variable (f.r.v., for short) is a Borel measurable function $X : (\Omega, \mathscr{A}) \to (E, d_\infty))$. If $X$ is an f.r.v. then $[X]^r = [X^-(r), X^+(r)]$, $r \in (0,1]$, is a random closed interval set and $X^-(r)$, $X^+(r)$ are realvalued r.v.'s. (Feng 1999 [494]). Since for every $1 \leqslant p \leqslant \infty$,

$$\begin{aligned}|d_p(u,v) - d_p(u_0,v_0)| &\leqslant d_p(u,u_0) + d_p(v,v_0)\\ &\leqslant d_\infty(u,u_0) + d_\infty(v,v_0)\end{aligned}$$

and

$$\begin{aligned}&|\langle u,v\rangle - \langle u_0,v_0\rangle|\\ &\leqslant |\langle u,v\rangle - \langle u_0,v\rangle| + |\langle u_0,v\rangle - \langle u_0,v_0\rangle|\\ &\leqslant 2(\|v\|_\infty d_1(u,u_0) + \|u_0\|_\infty d_1(v,v_0))\\ &\leqslant 2(\|v\|_\infty d_\infty(u,u_0) + \|u_0\|_\infty (v,v_0)),\end{aligned}$$

so the mappings

$$d_p(u,v) : (E, d_\infty) \times (E, d_\infty) \to R : (u,v) \mapsto d_p(u,v)$$

and

$$\langle u,v\rangle : (E, d_\infty) \times (E, d_\infty) \to R : (u,v) \mapsto \langle u,v\rangle$$

are continuous, hence if $X$ and $Y$ are f.r.v.'s then $d_p(X,Y)$, $\langle X,Y\rangle$ and $\|X\|_p$ are real-valued r.v.'s.

An f.r.v. $X$ is called integrably bounded if $E\|X\|_\infty < \infty$ and the expected value $EX$ is defined as the unique fuzzy number which satisfies the property: $[EX]^r = E[X]^r = [EX^-(r), EX^+(r)], 0 < r \leqslant 1$.

In the present case, if $E\|X\|_p < \infty$, for some $p : 1 \leqslant p < \infty$, then the expected value of $X$ still exists. To see this, let $E\|X\|_1 < \infty (\|X\|_1 \leqslant \|X\|_p$, for $1 \leqslant p < \infty)$, and it is easy to see that $E|X^\pm(r)| < \infty$ for all $r \in (0,1]\backslash A$, where $\lambda(A) = 0$, $\lambda$ is the Lebesgue measure. It is already known that a unique $u \in E$ exists such that $[u]^r = [EX^-(r), EX^+(r)]$, $r \in (0,1]$, so $EX = u$ is defined.

Let $1 \leqslant p < \infty$. $X$ is called a $p$-order f.r.v. provided $E\|X\|_p^p < \infty$. The family of all $p$-order f.r.v.'s $X$ and $Y$ is denoted by $\mathscr{L}_p(E)$ ($\mathscr{L}_p$, for short). Any two f.r.v.'s $X$ and $Y$ are called equivalent if $P(X \neq Y) = 0$. The all equivalent elements in $\mathscr{L}_p$ are identified. If $X, Y \in \mathscr{L}_p$, then $d_p^p(X,Y)$ is an integrably real-valued r.v., so define

$$D_p(X,Y) = (Ed_p^p(X,Y))^{1/p}, \quad X,Y \in \mathscr{L}_p.$$

Then $(\mathscr{L}_p, D_p)$ is a metric space. The following theorem and corollary are also important.

**Theorem 2.3.** *(Feng et al. 2001 [474]) Let $(X_n, n \geqslant 1)$ be a sequence in $\mathscr{L}_p$, $1 \leqslant p < \infty$. The following conditions are equivalent.*

*(i) $X \in \mathscr{L}_p$ and $D_p(X_n, X) \to 0$;*

*(ii) $(X_n, n \geqslant 1)$ is a Cauchy sequence in $\mathscr{L}_p$, i.e. $D_p(X_m, X_n) \to 0$, as $m$, $n \to \infty$;*

*(iii) $(\|X_n\|_p^p, n \geqslant 1)$ is uniformly integrable and $d_p(X_n, X)$ converges in probability to zero, as $n \to \infty$.*

**Corollary 2.1.** *(Feng et al. 2001 [474])* $(\mathscr{L}_p, D_p)$ *is a complete metric space.*

In the next discussion $p = 2$. $\mathscr{L}_2 = \{X | X \text{ is an f.r.v. and } E\|X\|_2^2 < \infty\}$. If $X, Y \in \mathscr{L}_2$, then $\langle X, Y\rangle$ is an integrably real-valued r.v. Set

$$D_*(X,Y) = (Ed_*^2(X,Y))^{1/2},$$

where $d_*$ is defined by

$$d_*(u,v) = \sqrt{\langle u,u\rangle - 2\langle u,v\rangle + \langle v,v\rangle}. \tag{2.36}$$

$D_*$ is still a metric in $\mathscr{L}_2$ and equivalent to the metric $D_2$. This is apparent from the following equation

$$\begin{aligned} d_2^2(u,v) &\leqslant \textstyle\int_0^1 [(u^-(r) - v^-(r))^2 \\ &\quad + (u^+(r) - v^+(r))^2]dr \\ &= \langle u,u\rangle - 2\langle u,v\rangle + \langle v,v\rangle = d_*^2(u,v) \\ &\leqslant 2\textstyle\int_0^1 h^2([u]^r, [v]^r)dr = 2d_2^2(u,v). \end{aligned}$$

for any $(X, X_n, n \geqslant 1) \subset \mathscr{L}_2$, we have

$$\lim_{n\to\infty} D_2(X_n, X) = 0 \Leftrightarrow \lim_{n\to\infty} D_*(X_n, X) = 0.$$

**Lemma 2.2.** *(Feng et al. 2001 [474]) (i) The mapping* $E\langle\cdot,\cdot\rangle : (\mathscr{L}_2, D_2) \times (\mathscr{L}_2, D_2) \to R$; $(X,Y) \mapsto E\langle X,Y\rangle$, *is continuous.*

*(ii)* $(X_n, n \geqslant 1)$ *is a Cauchy sequence in* $(\mathscr{L}_2, D_2)$ *if and only if as* $n, m \to \infty$, *the limit* $E\langle X_n, X_m\rangle$ *exists.*

The concepts of f.r.v. variance and covariance are now explained. The expected value of an f.r.v. is defined as a fuzzy number by Puri and Ralescu (1986) [93], but the variance of an f.r.v. must be an accurate measure of the spread or dispersion of the f.r.v. about its mean and the covariance or the correlation coefficient of the two f.r.v.'s must measure their linear interdependence. They should have no fuzziness.

**Definition 2.18.** *(Feng et al. 2001 [474]) Let f.r.v.'s X and Y be in* $\mathscr{L}_2$. *The covariance of X and Y is defined as*

$$\begin{aligned} Cov(X,Y) = \tfrac{1}{2}\textstyle\int_0^1 &(Cov(X^-(r), Y^-(r)) \\ &+ Cov(X^+(r), Y^+(r)))dr. \end{aligned} \tag{2.37}$$

*The variance of X is defined by*

$$DX = Cov(X,X). \tag{2.38}$$

*The normalized covariance is given by*

$$\rho(X,Y) = \frac{Cov(X,Y)}{\sqrt{DX}\sqrt{DY}} \tag{2.39}$$

*and is called the correlation coefficient of X and Y. If* $\rho(X,Y) = 0$ *then the f.r.v.'s X and Y are said to be uncorrelated.*

Note that if $X \in \mathscr{L}_2$, then $E(X^{\pm}(r))^2 < \infty$ for all $r \in (0,1]\backslash A$, where $\lambda(A) = 0$, $\lambda$ is the Lebesgue measure. On the other hand, if $X$, $Y \in \mathscr{L}_2$, then $\langle X,Y \rangle$ is an integrably real-valued r.v. and $\langle EX,EY \rangle$ exists. So Eq. (2.37) makes sense. The variance and covariance of f.r.v.'s have many properties which are similar to the ones of real-valued r.v.'s.

**Theorem 2.4.** *(Feng et al. 2001 [474]) Let f.r.v.'s $X$ and $Y$ be in $\mathscr{L}_2$. Then*
*(i)* $Cov(X,Y) = \frac{1}{2}(E\langle X,Y \rangle - \langle EX,EY \rangle)$, $DX = \frac{1}{2}D_*^2(X,EX)$;
*(ii)* $Cov(\lambda X + u, kY + v) = \lambda k\, Cov(X,Y)$, *where* $u$, $v \in E$ *and* $\lambda, k \in R$, $\lambda k \geqslant 0$;
*(iii)* $D(\lambda X + u) = \lambda^2 DX$;
*(iv)* $D(X+Y) = DX + DY + 2Cov(X,Y)$;
*(v)* $|\rho(X,Y)| \leqslant 1$ *and* $\rho(X,Y) = 1$ *if and only if* $Y + \lambda EX = EY + \lambda X$, *a.s.*, $\rho(X,Y) = -1$ *if and only if* $Y + \lambda X = EY + \lambda EX$, *a.s., where* $\lambda = \sqrt{DY/DX}$;
*(vi) (Chebyshev inequality)* $P(d_2(X,EX) > \varepsilon) \leqslant (2DX)/\varepsilon^2$, *for any* $\varepsilon > 0$.

It can be clearly seen that if $X$ and $Y$ are independent, i.e. $\sigma(X)$ and $\sigma(Y)$ are independent where $\sigma(X)$ is the smallest $\sigma$-field with respect to which the f.r.v. $X$ is measurable, then for every $r \in (0,1]$, $X^{\pm}(r)$ and $Y^{\pm}(r)$ are independent. So from the definition there is $Cov(X,Y) = 0$ and $E\langle X,Y \rangle = \langle EX,EY \rangle$ from the property (i), which shows that the two independent f.r.v.'s $X$ and $Y$ in $\mathscr{L}_2$ are uncorrelated.

If $(X,Y)$ is a two-dimensional Gaussian f.r.v. then $X = EX + \hat{\xi}$, and $Y = EY + \hat{\eta}$, where $(\xi,\eta)$ is a two-dimensional Gaussian real-valued r.v. with $E\xi = E\eta = 0$. Since $\rho(X,Y) = \rho(\xi,\eta)$, the independency and the lack of correlation of $X$ and $Y$ are equivalent.

All the above sufficiently shows that the variance and covariance that were defined are reasonable, calculable and applicable. Next some examples are provided to compute the variance of an f.r.v. and to show an application of statistical estimation when samples or prior information are fuzzy.

*Example 2.8.* (Feng et al. 2001 [474]) Consider an f.r.v. $X$ of discrete type, i.e. $P(X = u_k) = p_k$, $k = 1,2,\cdots$. where $u_k : R \to [0,1]$ are continuous with compact support. The infinite sum of the fuzzy numbers is defined by

$$(\sum_{k=1}^{\infty} v_k)(x) = \sup(\inf_{k \geqslant 1}(v_k(y_k))),$$

where the supremum is taken over all sequences $\{y_1, y_2, \cdots\}$ such that $x = \sum_{j=1}^{\infty} y_j$. Let $X \in \mathscr{L}_2$, then $EX = \sum_{k=1}^{\infty} p_k u_k$. It is easy to check that $E\langle X,X \rangle = \sum_{k=1}^{\infty} p_k \langle u_k, u_k \rangle$. Using Theorem 2.4 (i) and the fact that $\langle EX,EX \rangle = \sum_{k=1}^{\infty}\sum_{j=1}^{\infty} p_k p_j \langle u_k, u_j \rangle$, so

$$\begin{aligned} DX &= \tfrac{1}{2}(E\langle X,X \rangle - \langle EX,EX \rangle) \\ &= \tfrac{1}{2}(\textstyle\sum_{k=1}^{\infty} p_k \langle u_k,u_k \rangle - \sum_{k=1}^{\infty}\sum_{j=1}^{\infty} p_k p_j \langle u_k,u_j \rangle). \end{aligned}$$

*Example 2.9.* (Feng et al. 2001 [474])

$$X(\omega)=\begin{cases}\frac{x-\xi}{\eta} & if \xi<x<\xi+\eta,\\ 1 & if x=\xi+\eta,\\ \frac{\xi+\eta+\zeta-x}{\zeta} & if \xi+\eta<x<\xi+\eta+\zeta,\\ 0 & elsewhere,\end{cases}$$

where $\xi$, $\eta$ and $\zeta$ are real-valued r.v.'s with $\eta>0$ and $\zeta>0$, a.s. Fix $\omega\in\Omega$, $X(\omega)$ is a continuous triangular fuzzy number. It is relatively easy then to show that $[X]^r=[\xi+r\eta,\xi+\eta+(1-r)\zeta]$, for $r\in(0,1]$. Therefore,

$$D(X^-(r))=D\xi+r^2D\eta+2rCov(\xi,\eta),$$

$$\begin{aligned}D(X^+(r))=D\xi+D\eta+(1-r)^2D\zeta\\ +2Cov(\xi,\eta)+2(1-r)Cov(\xi,\zeta)\\ +2(1-r)Cov(\eta,\zeta).\end{aligned}$$

Hence from Eq. (2.37),

$$\begin{aligned}DX=D\xi+\tfrac{2}{3}D\eta+\tfrac{1}{6}D\zeta+\tfrac{3}{2}Cov(\xi,\eta)\\ +\tfrac{1}{2}Cov(\xi,\zeta)+\tfrac{1}{2}Cov(\eta,\zeta).\end{aligned}$$

If $\eta=\zeta$, a.s., i.e. $X(\omega)$ is a symmetric triangular fuzzy number, then

$$DX=D\xi+\frac{4}{3}D\eta+2Cov(\xi,\eta).$$

*Example 2.10.* (Feng et al. 2001 [474]) A simple random sample $X_1,X_2,\cdots,X_n$ is taken from an f.r.v. $X$ with $EX=u$ and $DX=\sigma^2$, i.e. $X_1,X_2,\cdots,X_n$ are independent and identically distributed with $X$. Let

$$\bar{X}=\frac{1}{n}\sum_{i=1}^{n}X_i,\quad M_2=\frac{1}{2n}\sum_{i=1}^{n}d_*^2(X_i,\bar{X}),$$

$$S^2=\frac{1}{2(n-1)}\sum_{i=1}^{2}d_*^2(X_i,\bar{X})$$

denote the sample average, the sample second central moment and the sample variance, respectively. Then

(i) $\bar{X}$ converges in probability to $u$ in $d_2$, as $n\to\infty$.

(ii) $\bar{X}$ is the minimum variance linear unbiased estimator of the expected value $u$ of $X$.

(iii) $S^2$ is an unbiased estimator of the variance $\sigma^2$ of $X$.

In fact, from Theorem 2.4 (iii) and (iv),

$$E\bar{X}=\frac{1}{n}\sum_{i=1}^{n}EX_i=u,$$

$$D\bar{X} = \frac{1}{n^2}\sum_{i=1}^{n} DX_i = \frac{\sigma^2}{n}$$

thus the Chebyshev inequality (Theorem 2.4 (vi)) yields that in probability $d_2(\bar{X},u) \to 0$, as $n \to \infty$. If $Y = \sum_{i=1}^{n} a_i X_i$ is a linear unbiased estimator of the expected value $u$ of $X$, since $\sum_{i=1}^{n} a_i = 1$ and $DY = \sum_{i=1}^{n} a_i^2\sigma^2 \geqslant (1/n)\sigma^2 = DX$, hence (ii) holds. In order to show (iii), note that $E\langle X_i, X_i\rangle = 2\sigma^2 + \langle u,u\rangle$ and $E\langle \bar{X}, \bar{X}\rangle = (2\sigma^2/n) + \langle u,u\rangle$ from Theorem 2.4 (i), then by 2.36,

$$\begin{aligned} EM_2 &= \tfrac{1}{2n}E\textstyle\sum_{i=1}^{n}(\langle X_i,X_i\rangle - 2\langle Xi,\bar{X}\rangle + \langle\bar{X},\bar{X}\rangle) \\ &= \tfrac{1}{2n}E(\textstyle\sum_{i=1}^{n}\langle X_i,X_i\rangle - \langle\bar{X},\bar{X}\rangle) = \tfrac{n-1}{n}\sigma^2 \end{aligned}$$

hence $ES^2 = E((n/n-1)M_2) = \sigma^2$.

### 2.3.4 Expected Value of Fuzzy Random Variables

To calculate the expected value of fuzzy random variables, a hybrid crisp approach is developed to integrate the decision maker's optimistic-pessimistic attitudes in adopting real world practice. Take the triangular fuzzy random variable as an example, this method first transforms fuzzy random parameters into $(r,\sigma)$-level trapezoidal fuzzy variables, which are subsequently defuzzified using an expected value operator with an optimistic-pessimistic index. The following describes the transformation process for converting fuzzy random variables into $(r,\sigma)$-level trapezoidal fuzzy variables.

With no loss of generality, the fuzzy random variable is denoted by $\tilde{\bar{\kappa}} = ([\kappa]_L, \varphi(\omega), [\kappa]_R)$, where $\varphi(\omega)$ is a random variable with a probability density function $p_\varphi(x)$. It is assumed that $\varphi(\omega)$ follows a Normal distribution $\mathcal{N}(\zeta, \nu^2)$, then $p_\varphi(x) = \frac{1}{\sqrt{2\pi\nu}}e^{-\frac{(x-\zeta)^2}{2\nu^2}}$. Let $\sigma$ be any given probability level of a random variable, $r$ be any given possibility level of a fuzzy variable. The parameters $\sigma \in [0, \sup p_\varphi(x)]$ and $r \in [\frac{[\kappa]_R - [\kappa]_L}{[\kappa]_R - [\kappa]_L + \varphi_\sigma^R - \varphi_\sigma^L}, 1]$ reflect the decision-maker's degree of optimism. For an easy description, $\sigma$ is called the *probability* level and $r$ , the *possibility* level. A stepwise transformation method is presented below:

***Step.1.*** Estimate the parameters $[\kappa]_L$, $[\kappa]_R$, $\zeta$, and $\nu$ from collected data and professional experience using statistical methods.

***Step.2.*** Obtain the decision-maker's degree of optimism, i.e., the values of a *probability* level $\sigma \in [0, \sup p_\varphi(x)]$ and *possibility* level $r \in [\frac{[\kappa]_R - [\kappa]_L}{[\kappa]_R - [\kappa]_L + \varphi_\sigma^R - \varphi_\sigma^L}, 1]$, which often can be determined through a group decision making approach (Herrera et al. 1996 [66]; Wu and Xu 2012 [65]).

***Step.3.*** Let $\varphi_\sigma$ be the $\sigma$-level sets (or $\sigma$-cuts) of the random variable $\varphi(\omega)$, i.e. $\varphi_\sigma = [\varphi_\sigma^L, \varphi_\sigma^R] = \{x \in R | p_\varphi(x) \geqslant \sigma\}$, then compute the values for $\varphi_\sigma^L$ and $\varphi_\sigma^R$. They are:

$$\varphi_\sigma^L = \inf\{x \in R | p_\varphi(x) \geqslant \sigma\} = \inf p_\varphi^{-1}(\sigma) = \zeta - \sqrt{-2\nu^2 \ln(\sqrt{2\pi}\nu\sigma)}, \quad (2.40)$$

$$\varphi_\sigma^R = \sup\{x \in R | p_\varphi(x) \geqslant \sigma\} = \sup p_\varphi^{-1}(\sigma) = \zeta + \sqrt{-2\nu^2 \ln(\sqrt{2\pi}\nu\sigma)}. \quad (2.41)$$

***Step.4.*** Transform the fuzzy random variable $\tilde{\bar{\kappa}} = ([\kappa]_L, \varphi(\omega), [\kappa]_R)$ into the $(r,\sigma)$-level trapezoidal fuzzy variable $\tilde{\kappa}_{(r,\sigma)}$ using the following equation:

$$\tilde{\bar{\kappa}} \to \tilde{\kappa}_{(r,\sigma)} = ([\kappa]_L, \underline{\kappa}, \overline{\kappa}, [\kappa]_R),$$

where

$$\underline{\kappa} = [\kappa]_R - r([\kappa]_R - \varphi_\sigma^L) = [\kappa]_R - r([\kappa]_R - \zeta + \sqrt{-2\nu^2 \ln(\sqrt{2\pi}\nu\sigma)}), \quad (2.42)$$

$$\overline{\kappa} = [\kappa]_L + r(\varphi_\sigma^R - [\kappa]_L) = [\kappa]_L + r(\zeta + \sqrt{-2\nu^2 \ln(\sqrt{2\pi}\nu\sigma)} - [\kappa]_L). \quad (2.43)$$

The membership function of $\tilde{\kappa}_{(r,\sigma)}$ is as below:

$$\mu_{\tilde{\kappa}_{(r,\sigma)}}(x) = \begin{cases} 0 & if \quad x < [\kappa]_L, x > [\kappa]_R, \\ \frac{x-[\kappa]_L}{\underline{\kappa}-[\kappa]_L} & if \quad [\kappa]_L \leqslant x < \underline{\kappa}, \\ 1 & if \quad \underline{\kappa} \leqslant x \leqslant \overline{\kappa}, \\ \frac{[\kappa]_R - x}{[\kappa]_R - \overline{\kappa}} & if \quad \overline{\kappa} < x \leqslant [\kappa]_R. \end{cases}$$

As shown, the *possibility* level (i.e., $r$) should be within $[\frac{[\kappa]_R-[\kappa]_L}{[\kappa]_R-[\kappa]_L+\varphi_\sigma^R-\varphi_\sigma^L}, 1]$ in order to deal with the real world problems. The decision maker can choose the value of $r$ accordingly. When the *possibility* level increases, it enlarges the distance between $\underline{\kappa}$ and $\overline{\kappa}$ in the $(r,\sigma)$-level trapezoidal fuzzy variable, which means the decision maker holds a pessimistic attitude about the accuracy of data. Contrarily, when the *possibility* level decreases, it reduces the distance between $\underline{\kappa}$ and $\overline{\kappa}$ in the $(r,\sigma)$-level trapezoidal fuzzy variable, which means the decision maker holds an optimistic attitude about the accuracy of data.

To calculate the expected value of the above $(r,\sigma)$-level trapezoidal fuzzy variables, a new fuzzy measure with an optimistic-pessimistic adjusting index is introduced to characterize real life problems. The definition of this fuzzy measure *Me*, which is a convex combination of *Pos* and *Nec*, can be found in Xu and Zhou (2011) [281], and the basic knowledge for the measures *Pos* and *Nec* can be seen in Dubois and Prade (1988) [221].

Let $\tilde{\kappa} = (r_1, r_2, r_3, r_4)$ denote a trapezoidal fuzzy variable. In fact, in real world problems, especially in inventory problems in large-scale construction projects, the case is often encountered when $r_1 > 0$. Based on the definition and the properties of the expected value operator of the fuzzy variable using measure *Me* (Xu and Zhou 2011 [281]), if the fuzzy random variable $\tilde{\bar{\kappa}}$ is transformed into the $(r,\sigma)$-level trapezoidal fuzzy variable $\tilde{\kappa}_{(r,\sigma)} = ([\kappa]_L, \underline{\kappa}, \overline{\kappa}, [\kappa]_R)$, where $[\kappa]_L > 0$, then the expected value of $\tilde{\kappa}_{(r,\sigma)}$ should be

$$E^{Me}[\tilde{\kappa}_{(r,\sigma)}] = \frac{1-\lambda}{2}([\kappa]_L + \underline{\kappa}) + \frac{\lambda}{2}(\overline{\kappa} + [\kappa]_R). \quad (2.44)$$

The linearity of $E^{Me}$ has also been proved in Xu and Zhou (2011) [281], thus it is suitable to employ this expected value operator to deal with the linear objective functions and the constraints that contain the $(r,\sigma)$-level trapezoidal fuzzy variable.

## 2.4 Fuzzy Random MOMSDM

In some practical MOMSDM processes, the decision making problems often take place in a hybrid uncertain environment. The uncertain events in this environment are characterized by both fuzzy uncertainty and randomness, the so-called "twofold" uncertainty. In this case, fuzzy random theory becomes an effective tool for dealing with these issues in MOMSDM processes. Fuzzy random uncertainty appears in many real world applications, such as inventory management, facilities planning, and transportation assignment. This section will explain the principle of the fuzzy random MOMSDM.

### *2.4.1 Motivation for Employing Fuzzy Random Uncertainty*

The need to address uncertainty in the MOMSDM is widely recognized, as uncertainties exist in a variety of system components in a dynamic process. As a result, the inherent complexity and stochastic uncertainty existing in real-world multiple objective multistage decision-making have essentially placed them beyond conventional deterministic optimization methods.

A typical example can be found in the water resources allocation problem. Previous studies about water resources allocation under market mechanisms usually consider water demand as deterministic. However, this is not consistent with reality. Since the decision making of the water manager of each sub-area in the lower level for water use is dependent on the rate structures, they have the incentive to reduce water use in order to reduce the revenues that should be paid to the regional authority in the upper level. At the same time, it is also difficult to avoid the influence of the reduction in water use on the income of water users in each sub-area, so in these cases, decisions on water demand tend to fluctuate due to uncertainty of judgment, lack of evidence, insufficient information, and the dynamic environment of the regional water system. Sometimes it is difficult for the water manager to give a crisp description, so these parameters are assumed to be flexible or imprecise in nature. For instance, because it is very difficult to estimate the exact value of water demand, it is vaguely defined, with water managers giving a range, in which the most possible value is regarded as a random variable, i.e., viz $(a,\rho,b)$. Based on the statistical characteristics of the irrigation district, it is found that the most possible value for the water demand approximately follows a normal distribution, i.e., $\rho \sim \mathcal{N}(\mu,\sigma^2)$. In this situation, the triangular fuzzy random variable $(a,\rho,b)$, where $\rho \sim \mathcal{N}(\mu,\sigma^2)$ is applied to deal with these uncertain parameters by combining fuzziness and randomness. Therefore, it is appropriate to consider water demand as a fuzzy random variable (see Figure 2.12).

The regional authority needs to make a decision about the allocation of scarce water supplies to competing users. At the beginning, the future uncertainty of the river flow is a concern and, at a future time, when the uncertainties of water flow are quantified, resource allocation action can be taken. The future availability of this water supply is uncertain because of the variability of the stream flow, which is judged using three different scenarios: low level, medium level, and high level, and are

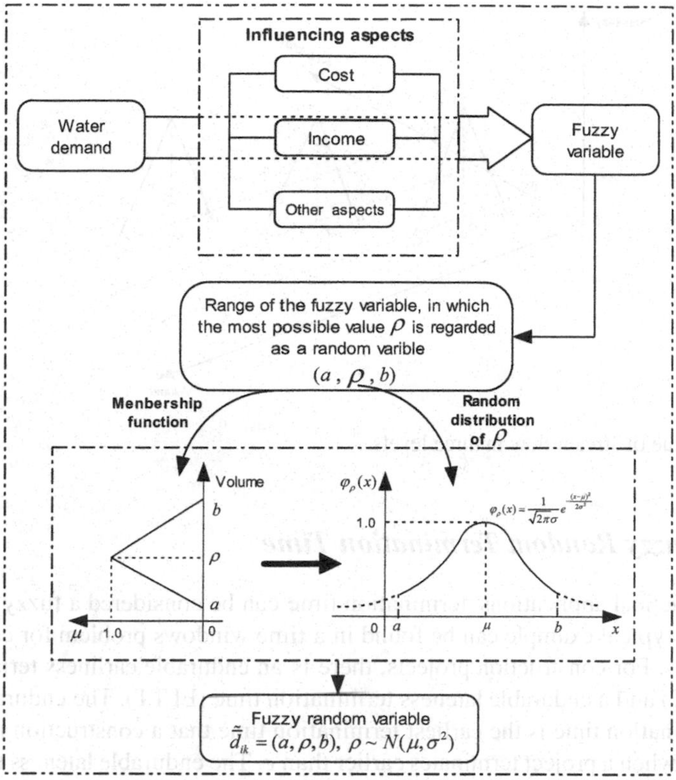


**Fig. 2.12** Flowchart of why water demand is a fuzzy random variable

presented as probability distributions [222]. Based on the statistical data characteristics, the probabilities of the three levels are $p_1$, $p_2$, and $p_3$ respectively. However, for each flow level, it is very difficult for the regional water authority to determine the specific stream flow data, so these are vaguely defined. To collect stream flow data at each level, interviews were conducted with different experts with the stream flow of level $t$ being described in linguistic terms as "between $a_t$ and $c_t$, and with a most possible value of $b_t$." where $t$ is denoted as the level of water flow (for $t = 1, 2, 3$ corresponding to low level, medium level and high level). It should be noted that the experts gave linguistic explanations based on their observations of the data over time. Therefore, the stream flow at each level can be conveniently described using triangular fuzzy sets, as shown in Figure 2.13. In summary, the stream flow volume (i.e., available water for distribution) is considered a fuzzy random variable.

$$\tilde{\bar{Q}} = \begin{cases} (a_1, b_1, c_1) & \text{with probability } p_1 \\ (a_2, b_2, c_2) & \text{with probability } p_2 \\ (a_3, b_3, c_3) & \text{with probability } p_3 \end{cases} \tag{2.45}$$

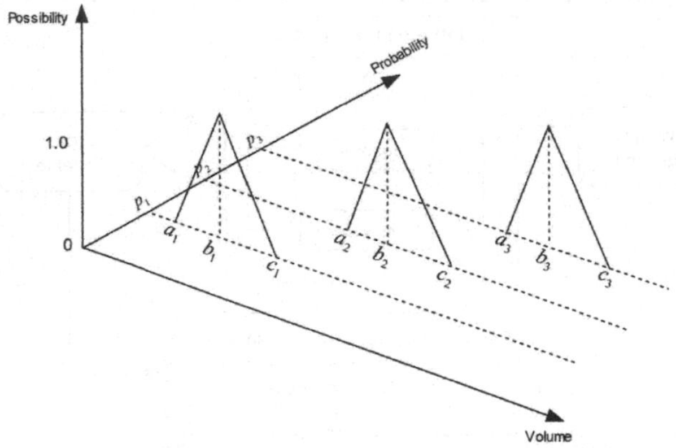


**Fig. 2.13** Value of stream flow in three levels

### 2.4.2 *Fuzzy Random Termination Time*

In some practical applications, termination time can be considered a fuzzy random variable. A typical example can be found in a time windows problem for construction projects. For construction projects, there is an endurable earliness termination time (EETT) and a endurable lateness termination time (ELTT). The endurable earliness termination time is the earliest termination time that a construction manager can endure when a project terminates earlier than $e$. The endurable lateness termination time is the latest termination time that a construction manager can endure when a project terminates later than $l$. The expected termination time (i.e., $\rho(\omega)$) can be influenced by many different factors, such as the season, the weather and other stochastic factors, and usually follows a random distribution. In this case, fuzzy random theory is useful tool when dealing with such uncertain information.

If the termination time $\tilde{\bar{T}} = (e, \rho(\omega), l)$ is a fuzzy variable, then the general model for the MOMSDM under a fuzzy random termination time can be expressed as,

$$\begin{cases} \max \, \{J_1(\mathbf{x}(0), \mathbf{x}(\tilde{\bar{T}}), \pi), J_2(\mathbf{x}(0), \mathbf{x}(\tilde{\bar{T}}), \pi), \cdots, J_n(\mathbf{x}(0), \mathbf{x}(\tilde{\bar{T}}), \pi)\} \\ \text{s.t.} \begin{cases} \mathbf{x}(k+1) = f_k(\mathbf{x}(k), \mathbf{u}(k)), \\ \mathbf{x}(0) \in I, \\ \tilde{\bar{T}} = (e, \rho(\omega), l), \\ g_j(\mathbf{x}(\cdot), \mathbf{u}(\cdot)) \leq 0, \; j = 1, 2, \cdots, p, \\ \mathbf{x}(k) \in X, \; k = 1, 2, \cdots, \tilde{\bar{T}}, \\ \mathbf{u}(k) \in U, \; k = 1, 2, \cdots, \tilde{\bar{T}}. \end{cases} \end{cases} \tag{2.46}$$

where $\mathbf{x}(k)$ is a state vector in Stage $k$; $\mathbf{u}(k)$ is a decision vector in Stage $k$; $J_i(\mathbf{x}(0), \mathbf{x}(\tilde{\bar{T}}), \pi)$ is the objective function for the fuzzy random MOMSDM; $g_j(\mathbf{x}(\cdot), \mathbf{u}(\cdot))$ is the general constraint function for the fuzzy random MOMSDM; $f_k(\mathbf{x}(k), \mathbf{u}(k))$ is a state transition function for the fuzzy MOMSDM; $\pi = (\mathbf{u}(0), \mathbf{u}(1), \cdots, \mathbf{u}(\tilde{\bar{T}}))$ is the policy for the whole decision making process.

### 2.4.3 Fuzzy Random State and Decision Variables

A dynamic system usually contains hybrid inherent uncertainties and some of these uncertainties reflect on the state variables or decision variables. In some practical applications, the uncertainty of the state and decision variables can be expressed as fuzzy random uncertainty. The fuzzy random state variable characterizes the "two fold" uncertainty of the status of a dynamic system. The fuzzy random decision variable features the statistics and ambiguity of the output of the previous stage of the system and the input for the next stage in the system as shown in Figure 2.14.

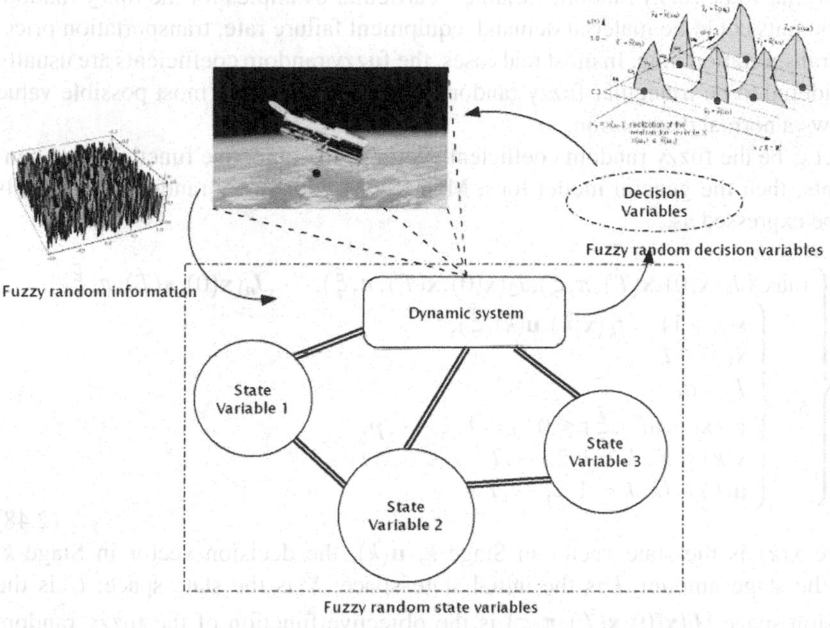


**Fig. 2.14** Fuzzy random state and decsion variables in dynamic system

Let $\tilde{\bar{\mathbf{x}}}(k)$ be the fuzzy random state vector in Stage $k$, $\tilde{\bar{\mathbf{u}}}(k)$, the fuzzy random decision vector in Stage $k$, then the general model of a MOMSDM with fuzzy random state and decision variables can be expressed as,

$$\begin{cases} \max \{J_1(\tilde{\bar{\mathbf{x}}}(0),\tilde{\bar{\mathbf{x}}}(T),\pi), J_2(\tilde{\bar{\mathbf{x}}}(0),\tilde{\bar{\mathbf{x}}}(T),\pi), \cdots, J_n(\tilde{\bar{\mathbf{x}}}(0),\tilde{\bar{\mathbf{x}}}(T),\pi)\} \\ \text{s.t.} \begin{cases} \tilde{\bar{\mathbf{x}}}(k+1) \simeq f_k(\tilde{\bar{\mathbf{x}}}(k),\tilde{\bar{\mathbf{u}}}(k)), \\ \tilde{\bar{\mathbf{x}}}(0) \in \tilde{\bar{I}}, \\ T = a, \\ g_j(\tilde{\bar{\mathbf{x}}}(\cdot),\tilde{\bar{\mathbf{u}}}(\cdot)) \leq 0, \ j = 1,2,\cdots,p, \\ \tilde{\bar{\mathbf{x}}}(k) \in \tilde{\bar{X}}, \ k = 1,2,\cdots,T, \\ \tilde{\bar{\mathbf{u}}}(k) \in \tilde{\bar{U}}, \ k = 1,2,\cdots,T. \end{cases} \end{cases} \tag{2.47}$$

where $a$ is the stage amount; $\tilde{\bar{I}}$ is the fuzzy initial state space; $\tilde{\bar{X}}$ is the fuzzy state space; $\tilde{\bar{U}}$ is the fuzzy decision space; $J_i(\tilde{\bar{\mathbf{x}}}(0), \tilde{\bar{\mathbf{x}}}(T), \pi)$ is the objective function of the fuzzy random MOMSDM; $g_j(\tilde{\bar{\mathbf{x}}}(\cdot), \tilde{\bar{\mathbf{u}}}(\cdot))$ is the general constraint function of the fuzzy random MOMSDM; $f_k(\tilde{\bar{\mathbf{x}}}(k), \tilde{\bar{\mathbf{u}}}(k))$ is the state transition function of the fuzzy random MOMSDM; $\pi = (\tilde{\bar{\mathbf{u}}}(0), \tilde{\bar{\mathbf{u}}}(1), \cdots, \tilde{\bar{\mathbf{u}}}(T))$ is the policy for the whole decision making process.

### 2.4.4 *Fuzzy Random Coefficients*

Sometimes it is necessary to consider the coefficients in the objective functions or constraints to be fuzzy random variables. Particular examples for the fuzzy random coefficients could be material demand, equipment failure rate, transportation price, and transportation time. In most real cases, the fuzzy random coefficients are usually considered to be triangular fuzzy random numbers, where the most possible value follows a normal distribution.

Let $\tilde{\bar{\xi}}$ be the fuzzy random coefficient vector in the objective functions and constraints, then the general model for a MOMSDM with fuzzy random coefficients can be expressed as,

$$
\begin{cases}
\max \{J_1(\mathbf{x}(0), \mathbf{x}(T), \pi, \tilde{\bar{\xi}}), J_2(\mathbf{x}(0), \mathbf{x}(T), \pi, \tilde{\bar{\xi}}), \cdots, J_n(\mathbf{x}(0), \mathbf{x}(T), \pi, \tilde{\bar{\xi}})\} \\
\text{s.t.} \begin{cases}
\mathbf{x}(k+1) = f_k(\mathbf{x}(k), \mathbf{u}(k), \tilde{\bar{\xi}}), \\
\mathbf{x}(0) \in I, \\
T = a, \\
g_j(\mathbf{x}(\cdot), \mathbf{u}(\cdot), \tilde{\bar{\xi}}) \leq 0, \ j = 1, 2, \cdots, p, \\
\mathbf{x}(k) \in X, \ k = 1, 2, \cdots, T, \\
\mathbf{u}(k) \in U, \ k = 1, 2, \cdots, T.
\end{cases}
\end{cases}
\tag{2.48}
$$

where $\mathbf{x}(k)$ is the state vector in Stage $k$, $\mathbf{u}(k)$, the decision vector in Stage $k$; $a$ is the stage amount; $I$ is the initial state space; $X$ is the state space; $U$ is the decision space; $J_i(\mathbf{x}(0), \mathbf{x}(T), \pi, \tilde{\bar{\xi}})$ is the objective function of the fuzzy random MOMSDM; $g_j(\mathbf{x}(\cdot), \mathbf{u}(\cdot), \tilde{\bar{\xi}})$ is the general constraint function of the fuzzy random MOMSDM; $f_k(\mathbf{x}(k), \mathbf{u}(k), \tilde{\bar{\xi}})$ is the state transition function of the fuzzy random MOMSDM; $\pi = (\mathbf{u}(0), \mathbf{u}(1), \cdots, \mathbf{u}(T))$ is the policy for the whole decision making process.

## 2.5 Methodology for Fuzzy-Like MOMSDM

A methodology is usually a guideline for solving a problem, with specific components such as phases, tasks, methods, techniques and tools. The methodology for a fuzzy-like MOMSDM describes the research paradigm, the nature of the trade space for time, the functional space of the fuzzy-like MOMSDM, and the future prospects. Generally speaking, the methodology for fuzzy-like MOMSDM does not describe the specific methods despite the attention given to the nature and kinds of processes to be followed in a given procedure or in attaining an objective.

### *2.5.1 Research Paradigm of Fuzzy-Like MOMSDM*

The word research is derived from the Middle French "recherche", which means "to go about seeking", with the term itself being derived from an old French term "recerchier" a compound word from "re-" + "cerchier", or "sercher", meaning 'search'. The earliest recorded use of the term was in 1577 [220].

Research has been defined in a number of different ways:

(1) A broad definition of research was given by Shuttleworth (2008) [218] - "In the broadest sense of the word, the definition of research includes any gathering of data, information and facts for the advancement of knowledge."

(2) Another definition of research was given by Creswell (2008) [219] who states - "Research is a process of steps used to collect and analyze information to increase our understanding of a topic or an issue". Research consists of three steps: the posing of a question, the collection of data to answer the question, and the presentation of an answer to the question.

(3) The Merriam-Webster Online Dictionary defines research in more detail as "a studious inquiry or examination; especially: an investigation or experimentation aimed at the discovery and interpretation of facts, the revision of accepted theories or laws in the light of new facts, or a practical application of such new or revised theories or laws."

In this section, the research paradigm for the fuzzy-like MOMSDM is defined to allow for a systematic investigation into the theory of the fuzzy-like MOMSDM to elaborate existing applications and to provide new knowledge for future applications. This investigation seeks to establish or confirm facts, reaffirm the results of previous work, solve new or existing problems, support theorems, or develop new theories in the fields of fuzzy-like MOMSDM.

Research in the field of fuzzy-like MOMSDM has often been conducted using the hourglass model structure of research. The hourglass model starts with a broad spectrum for the research, then focuses in on the required information using the methodology of the project (like the neck of the hourglass), then expands the research in the form of discussion and results. The major steps in conducting research are:

(1) Identification of research problem
(2) Literature review
(3) Establishment of the model system for the fuzzy-like MOMSDM
(4) Design of an algorithm for the model
(5) Practical application of the model
(6) Reporting and evaluation of the research

Figure 2.15 shows the framework for the research paradigm for the fuzzy-like MOMSDM. The steps generally represent the overall process, however they should be viewed as an ever-changing process rather than a fixed set of steps. Most research begins with a general statement of the problem, or rather, the purpose for engaging in the study. The literature review identifies the flaws or holes in the

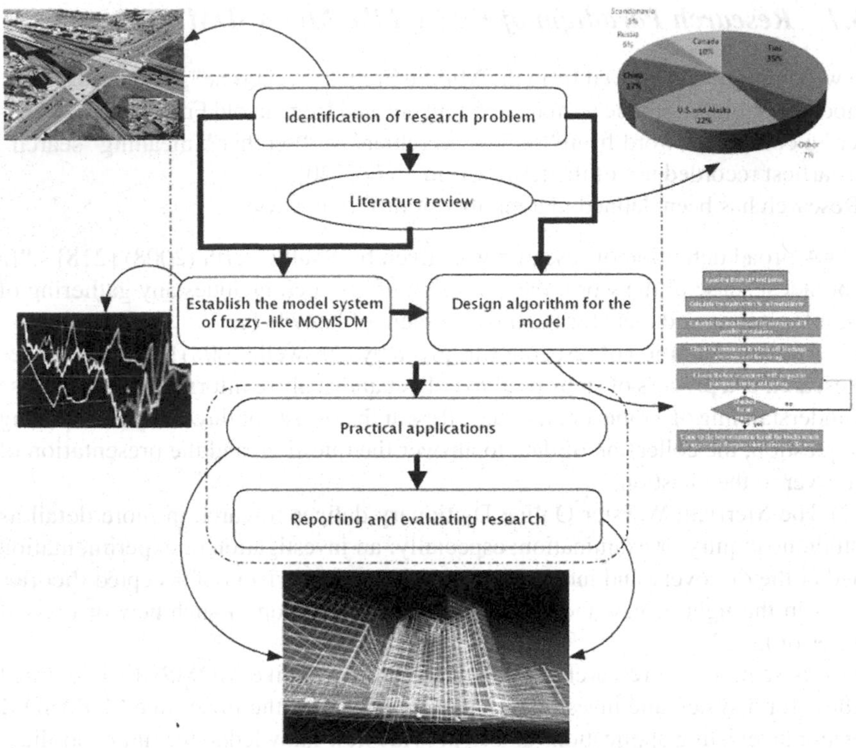


**Fig. 2.15** Framework of research paradigm of fuzzy-like MOMSDM

previous research, which provides a justification for the study. Often, a literature review is conducted in a given subject area before a research question is identified. A gap in the current literature, as identified by a researcher, then engenders the research question. The establishment of the model system for the fuzzy-like MOMSDM is based on the problem statement. Then an algorithm is designed to solve the model. Practical applications can demonstrate the effectiveness and efficiency of the developed method. Finally, the researcher may report and evaluate the research and discuss avenues for further research.

### 2.5.2 *The Nature of Space-Time Trade-Off*

The essential ideal of the fuzzy-like MOMSDM is to balance the space-time trade-off in a dynamic system as Figure 2.16 shows. For example, in computer science, a space-time or time-memory trade-off is a situation where the memory use can be reduced at the cost of slower program execution (and, conversely, the computation time can be reduced at the cost of increased memory use). As the relative costs of CPU cycles, RAM space, and hard drive space change, hard drive space has for some

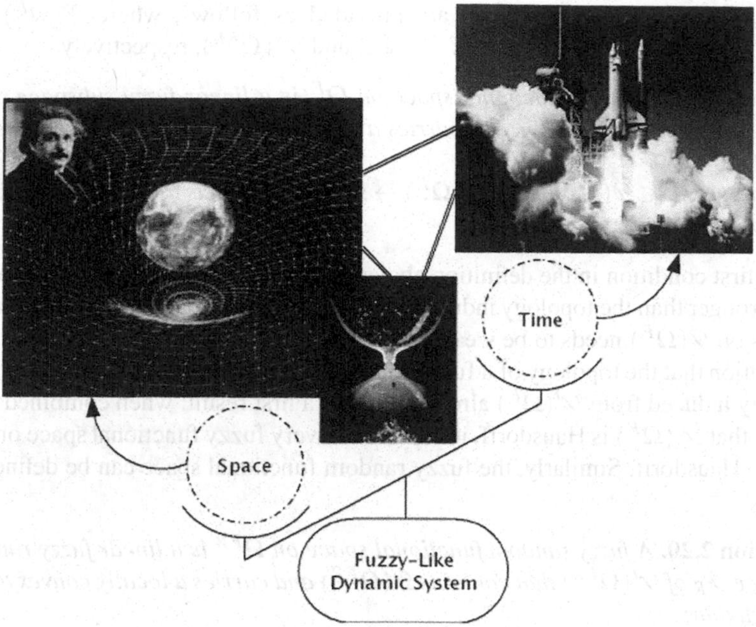


**Fig. 2.16** Space-time trade-off in a dynamic system

time been getting cheaper at a much faster rate than other components of computers, so appropriate choices for space-time trade-offs have changed radically. Often, by exploiting a space-time trade-off, a program can be made to run much faster.

A space-time trade-off can be applied to a data storage problem. If data is stored uncompressed, it takes up more space but can be accessed in less time than if the data were stored compressed (since compressing the data reduces the amount of space it takes, but it takes time to run the decompression algorithm). Depending on the particular instance of the problem, either way is practical.

In most cases of a fuzzy-like MOMSDM, the optimal way is to trade space for time. Since most construction projects need to be completed within a limited duration, the objective functions for these problems usually require a minimization of the construction time. Thus, the complete multistage optimization process is to trade space for time. Similar examples can be found in many practical application fields.

### 2.5.3 Functional Space of Fuzzy-Like MOMSDM

In this part, the concept of functional space on the fuzzy state or decision space and the fuzzy random state or decision space is introduced. Let $\Omega^F$ denote an open subset of the fuzzy state or decision space and $\Omega^{FR}$ denote an open subset of the fuzzy random state or decision space. If $\Omega^F = \emptyset$ or $\Omega^{FR} = \emptyset$, there are difficulties or changes are required, thus it is assumed that $\Omega^F$ or $\Omega^{FR}$ is nonempty.

The most important definitions are provided as follows, where $\mathscr{D}(\Omega^F)$ and $\mathscr{D}(\Omega^{FR})$ are viewed as subspaces of $\mathscr{D}'(\Omega^F)$ and $\mathscr{D}'(\Omega^{FR})$, respectively.

**Definition 2.19.** *A fuzzy functional space on $\Omega^F$ is a linear fuzzy subspace $\mathscr{F}$ of $\mathscr{D}'(\Omega^F)$ that contains $\mathscr{D}(\Omega^F)$ and carries a locally convex topology such that:*
*1. $\mathscr{D}(\Omega^F) \subseteq \mathscr{F} \subseteq \mathscr{D}'(\Omega^F)$ and*
*2. for every $\varphi \in \mathscr{D}(\Omega^F)$, $m_\varphi : \mathscr{D}'(\Omega^F) \to \mathscr{D}'(\Omega^F)$ restricts to a continuous linear map from $\mathscr{F}$ into $\mathscr{F}$.*

The first condition in the definition above implies that the topology on $\mathscr{F}$ needs to be stronger than the topology induced from $\mathscr{D}'(\Omega^F)$ and that the topology that $\mathscr{F}$ induces on $\mathscr{D}(\Omega^F)$ needs to be weaker than the intrinsic topology on $\mathscr{D}(\Omega^F)$. The observation that the topology of a fuzzy functional space on $\Omega^F$ is stronger than the topology induced from $\mathscr{D}'(\Omega^F)$ already leads to a first result: when combined with the fact that $\mathscr{D}'(\Omega^F)$ is Hausdorff, it shows that every fuzzy functional space on $\Omega^F$ must be Hausdorff. Similarly, the fuzzy random functional space can be defined as follows.

**Definition 2.20.** *A fuzzy random functional space on $\Omega^{FR}$ is a linear fuzzy random subspace $\mathscr{F}_R$ of $\mathscr{D}'(\Omega^{FR})$ that contains $\mathscr{D}(\Omega^{FR})$ and carries a locally convex topology such that:*
*1. $\mathscr{D}(\Omega^{FR}) \subseteq \mathscr{F}_R \subseteq \mathscr{D}'(\Omega^{FR})$ and*
*2. for every $\varphi_R \in \mathscr{D}(\Omega^{FR})$, $m_{\varphi_R} : \mathscr{D}'(\Omega^{FR}) \to \mathscr{D}'(\Omega^{FR})$ restricts to a continuous linear map from $\mathscr{F}_R$ into $\mathscr{F}_R$.*

Although functional spaces are the most important objects in the theory, spaces are also encountered that are 'almost' functional spaces.

**Definition 2.21.** *A fuzzy semi-functional space on $\Omega^F$ is a linear fuzzy subspace $\mathscr{F}$ of $\mathscr{D}'(\Omega^F)$ carrying a locally convex topology such that:*
*1. $\mathscr{F} \subseteq \mathscr{D}'(\Omega^F)$ and*
*2. for every $\varphi \in \mathscr{D}(\Omega^F)$, $m_\varphi : \mathscr{D}'(\Omega^F) \to \mathscr{D}'(\Omega^F)$ restricts to a continuous linear map from $\mathscr{F}$ into $\mathscr{F}$.*

Note that in comparison with the definition for a fuzzy functional space on $\Omega^F$, there is only one difference: for a fuzzy semi-functional space $\mathscr{F}$ on $\Omega^F$ we do not require that $\mathscr{D}(\Omega^F)$ is a subspace of $\mathscr{F}$ (with continuous inclusion). This extra property that fuzzy functional spaces are required to have turns out to be essential for only a limited, but important, part of the theory. Despite the fact that fuzzy functional spaces are the main concept, the part of the theory for which this extra assumption is not required will be largely formulated in terms of fuzzy semi-functional spaces to allow for a smoother treatment of the theory.

It is true that every fuzzy functional space on $\Omega^F$ is a fuzzy semi-functional space on $\Omega^F$. Also, by the same argument used for fuzzy functional spaces, it can be seen that every fuzzy semi-functional space is Hausdorff. The fact that every fuzzy functional space on $\Omega^F$ is a dense subspace of $\mathscr{D}'(\Omega^F)$ does not generalize to fuzzy semi-functional spaces, or for the definition of the fuzzy random semi-functional spaces.

**Definition 2.22.** *A fuzzy random semi-functional space on $\Omega^{FR}$ is a linear fuzzy random subspace $\mathscr{F}_R$ of $\mathscr{D}'(\Omega^{FR})$ carrying a locally convex topology such that:*
*1. $\mathscr{F} \subseteq \mathscr{D}'(\Omega^{FR})$ and*
*2. for every $\varphi \in \mathscr{D}(\Omega^{FR})$, $m_{\varphi_R} : \mathscr{D}'(\Omega^{FR}) \to \mathscr{D}'(\Omega^{FR})$ restricts to a continuous linear map from $\mathscr{F}_R$ into $\mathscr{F}_R$.*

### 2.5.4 Future Prospect

Fuzzy-like MOMSDM is an effective methodology for handling decision making problems. This book attempts to integrate the MOMSDM and fuzzy-like theory and explain their practical applications. While practical applications are the most important consideration here, the theory, especially for the functional space of fuzzy-like MOMSDM, still needs further future research. Future discussions could focus on fixed compact support, properties, compact support, locality, semi-locality, positive powers, duals, and the normalizing of the functional space of the fuzzy-like MOMSDM.

From the viewpoint of the models, fuzzy-like theory is applied to all elements of the MOMSDM, and since fuzzy-like theory is a rapidly growing subject area, the mathematical analysis of these models' properties may be an interesting area in future research.

From the perspective of the algorithms, this book provides quite a few algorithms to deal with the fuzzy-like MOMSDM. Since there are increasingly more new application problems, the improvement and modification of these algorithms need to be further explored in future research.

From the application aspect, more practical problems with increasing complexity are arising in the modern technological world. The description of these problems using multiple objective multistage decision making may be an interesting study in the future.

To sum up, this book provides the principles and methodologies for the fuzzy-like MOMSDM. The practical applications of the fuzzy-like MOMSDM are also discussed in detail. However, future research into this method may focus on new theorems, models and algorithms as new and more complex application problems arise.

# 3

# Fuzzy MOMSDM for Dynamic Machine Allocation

Large scale manufacturing or construction systems are characterized by various complicated machines for different types of interrelated jobs. Machine breakdown and preventative maintenance lead to losses in production capability in manufacturing and construction industries [1]. The complexity in production planning and the construction process focuses on a search for an ideal machine utilization level [2, 3]. In fact, proper machine allocation and management can meaningfully increase productivity or construction throughput. Therefore, it is important to improve the effectiveness of machine allocation that significantly contributes to the success of a project. In this chapter, a dynamic machine allocation problem (DMAP) is presented. The modeling process for the DMAP is explained in detail. Then a model analysis of the DMAP is discussed in which a theoretical algorithm is designed for small scale problems. To solve large scale problems, a dynamic programming based particle swarm optimization algorithm is developed. Finally, a case study from the Shuibuya Hydropower Project is used as an application example.

## 3.1 Statement of DMAP

This section, based on an analysis of manufacturing and construction systems, presents a dynamic allocation of machines problem with uncertain machine breakdowns for maximizing total production or construction throughput within a limited duration.

### *3.1.1 Problem Description*

In the manufacturing and construction industries, machines are an important production and construction element. Every machine is unreliable in the sense that it degrades with age, and usage, and in the end ultimately fails [4]. Machine failure with uncertainty has a significant impact on business performance and is a difficult problem for managers. It is commonly agreed nowadays that periodic preventive maintenance policies can be very successful in improving machine reliability [5, 6].

J. Xu and Z. Zeng, *Fuzzy-Like Multiple Objective Multistage Decision Making*, Studies in Computational Intelligence 533,
DOI: 10.1007/978-3-319-03398-3_3, © Springer International Publishing Switzerland 2014

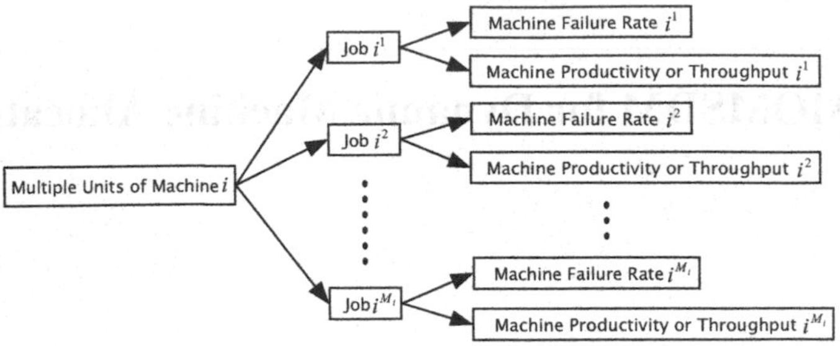


**Fig. 3.1** Relationship of machine, jobs, machine failure rate, and machine productivity or throughput

To minimize maintenance-related costs, the period of preventive maintenance may range from one week to several months depending on the type of machine and the kind of work [7]. Although periodic preventive maintenance policies can be very successful in improving machine reliability, they still cannot prevent machine failure between periodic preventive maintenances.

Normally, in a production or construction project, each type of machine can do several different jobs. Also, machines degrade in different situations when doing different jobs, and machine productivity or throughput is determined by the mode of production or working conditions (e.g., travel distance, rolling resistance for a truck), which varies from job to job. Therefore, machine failure rate and machine productivity or throughput may be different for different kinds of jobs processed. The relationship between machine, jobs, machine failure rate and machine productivity or throughput is illustrated in Figure 3.1. As shown, multiple units of machine type $i$ ($i = 1,2,\cdots,m$) are assigned to multiple kinds of jobs. Note that $i^\theta$ ($\theta = 1,2,\cdots,M_i$) is used as an index for a particular existing job $w$ ($w = 1,2,\cdots,|W|$) that may potentially be assigned to machine type $i$. Specifically, let $W = \{w|w = 1,2,\cdots,|W|\}$ denote the set of all jobs, because the same job may be assigned to different types of machine in the same time stage; hence, $i^\theta \in \{w|w \in W\}$. Moreover, each of these assignments is associated with a corresponding machine failure rate ($p^k_{i^\theta}$) and a productivity or throughput rate ($\tilde{Q}_{i^\theta}$). The objective is to dynamically allocate all types of machine to the available jobs under uncertain machine failures and productivity or throughput in every time stage while maximizing total production or construction throughput produced by the machines in a whole period. Actual examples can be found in both the manufacturing and construction industries. As an example, the machines in a large scale iron and steel plant can produce both iron and steel. The differences between the production processes of iron and steel are in the proportion of carbon and the smelting temperatures which lead to different machine failure rates [8], as shown in Figure 3.2. In rockfill dam construction projects, similar cases can also be found. As there are always many borrow areas in a rockfill dam construction project, dump trucks are used to transport different types of rockfill between different borrow areas and the dam. The machine failure rates and the machine throughput can be different

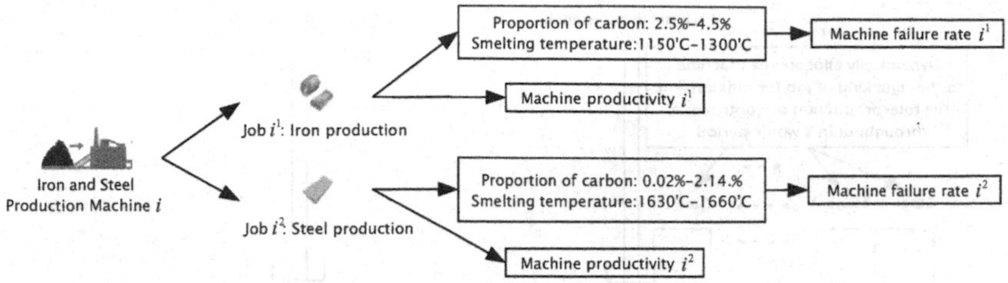


**Fig. 3.2** Relationship of machine, jobs, machine failure rate, and machine productivity in iron and steel production

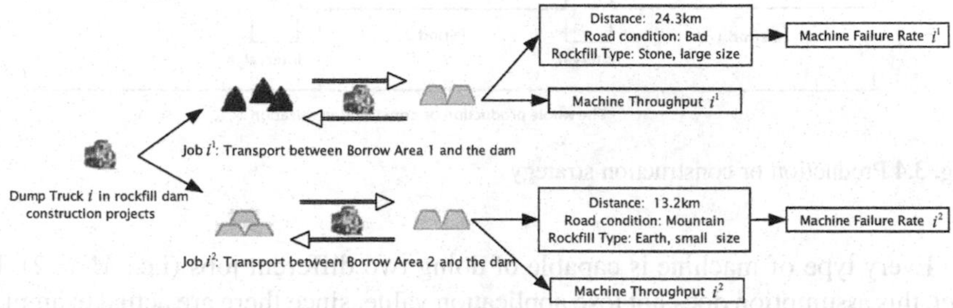


**Fig. 3.3** Relationship of machine, jobs, machine failure rate, and throughput in rockfill dam construction projects

depending on the distances, road conditions, the type of rockfill [9], user skill and previous maintenance effectiveness, as shown in Figure 3.3.

In fact, most production and construction systems use the following strategy to improve production or construction throughput within some limited duration. The entire production or construction duration is divided into several periods. Preventative maintenance is done during the intervals between all adjacent periods. Therefore, a machine that breaks down in a certain period cannot resume work until preventative maintenance has been performed. Moreover, every period is divided into several stages. Because the span of a period may range from one week to several months, depending on the type of machine and mode of work, the span of a stage may accordingly range from one day to several weeks. The problem is how to dynamically allocate all machines to the available jobs under uncertain machine failures in every stage while maximizing the production or construction throughput produced by the machines in one whole period. This production or construction strategy is illustrated in Figure 3.4.

To better understand the problem, the following assumptions and hypothesis are made in this chapter:

(1) Once equipment is fixed or maintained, it returns to an "as good as new" condition.

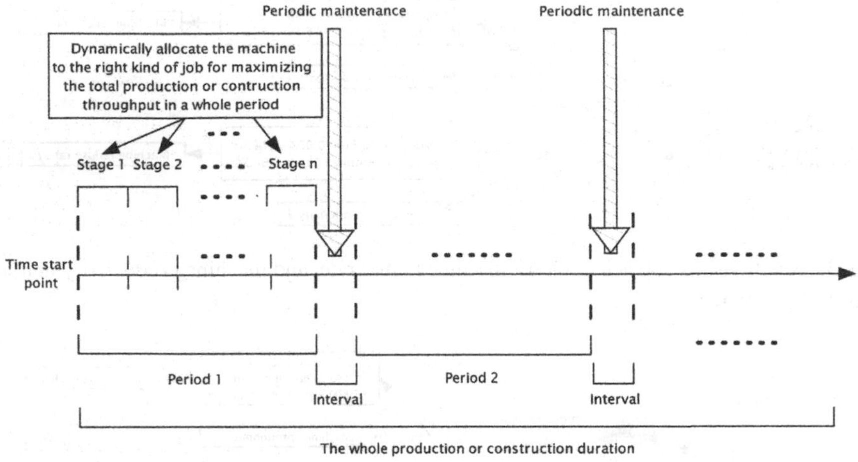


**Fig. 3.4** Production or construction strategy

(2) Every type of machine is capable of doing two different jobs (i.e., $M_i = 2$). In fact, this assumption does not lose application value, since there are actual examples as shown in Figures 3.2 and 3.3.
(3) Jobs for every type of machine are assumed to have the same conditions for machine failure rates and machine productivity or throughput. This means the sequence of several jobs for one type of machine does not affect machine failure rate and machine productivity or throughput.
(4) Preventive maintenance is done during the interval between all adjacent periods. Therefore, any machine that breaks down in a certain period is unable to be used until preventive maintenance has been done.
(5) The machine failure time follows a stochastic distribution.
(6) The machine productivity or throughput is regarded as a fuzzy variable.

### *3.1.2 Distribution of Machine Failure Time*

Before the model presentation, one important thing needs to be considered: the distribution of machine failure time. Weibull distributions have been widely used in reliability engineering for failure time distribution formulation. Prior studies by Dogramaci and Fraiman (2004) [10], Das et al. (2007) [11], McCool (2006) [12], Seifried (2004) [13], Patankar and Mitra (1995) [14], and Friebelova and Friebel (2006) [15] all considered that machine failure time follows a Weibull distribution, including another study which looked at shovel machines used in an open cast coal mine [16]. Hegab and Smith (2007) [18], based on data characteristics, also found that the Weibull distribution best represents the machine failure time in microtunneling projects. To further demonstrate, actual machine failure time data from the Xiaolangdi Hydropower Project were collected from the Yellow River Water and Hydroelectric Power Development Corporation (YRWHDC). The probability plots

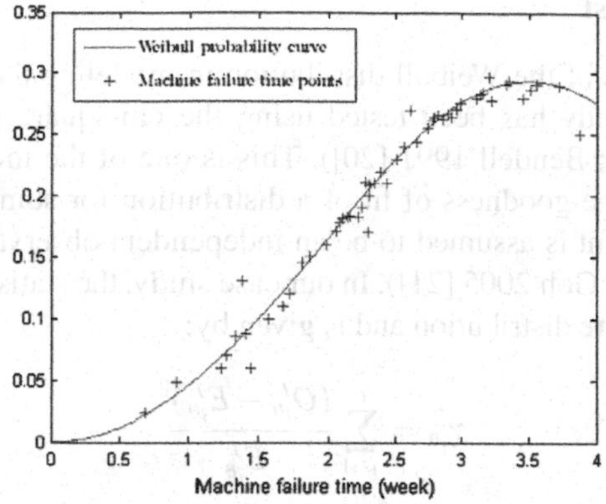


**Fig. 3.5** Weibull probability plot of a certain type of dump truck in Xiaolangdi Hydropower Project

for the machine failure times for each type of construction machines were also obtained. Through observation and comparison, the Weibull distribution was finally determined to be the best fit for characterizing machine failure time in the Xiaolangdi Hydropower Project. Figure 3.5 shows one of the Weibull probability plots for a certain type of dump truck at the Xiaolangdi Hydropower Project. In another case study, actual data for equipment failure times at the Shuibuya Hydropower Project were also collected by Hubei Qingjiang Shuibuya Project Construction Company. The equipment failure time was modeled using a two-parameter Weibull distribution. In order to justify the appropriateness of the Weibull distribution in modeling the equipment failure time, a goodness of fit test was employed. The two parameters (scale and shape parameters) were also estimated using the maximum likelihood method.

### Two-Parameter Weibull

The probability density function (PDF) of the two-parameter Weibull distribution of Equipment $i$ allocated to Job $i^\theta$ $(\theta = 1, 2, \cdots, M_i)$ is expressed as:

$$f_{i\theta}(t) = \lambda_{i\theta}\gamma_{i\theta}(\lambda_{i\theta}t)^{\gamma_{i\theta}-1}\exp[-(\lambda_{i\theta}t)^{\gamma_{i\theta}}] \tag{3.1}$$

where $\gamma_{i\theta} > 0$ and $\lambda_{i\theta} > 0$ are the shape and scale parameters for Equipment $i$ allocated to Job $i^\theta$; $t$ = the equipment failure time of every type of equipment allocated to every job; $i$ = the equipment type index, where $i = 1, 2, \cdots, N$; $\theta$ = the job selection index, where $\theta = 1, 2, \cdots, M_i$; $M_i$ = the number of jobs assigned to Equipment $i$; $i^\theta$ = the index of jobs handled by Equipment $i$.

### Goodness-of-Fit Test

The appropriateness of the Weibull distribution in modeling the equipment failure time in our case study has been tested using the chi-square goodness-of-fit test (Conover 1980 [19]; Bendell 1991 [20]). This is one of the most commonly used tests to determine the goodness of fit of a distribution for some observed data, in which each data point is assumed to be an independent observation of the random variable $t$ (Chua and Goh 2005 [21]). In our case study, the statistic, $\chi^2_{i\theta}$, for this test follows the chi-square distribution and is given by:

$$\chi^2_{i\theta} = \sum_{j=1}^{c} \frac{(O^j_{i\theta} - E^j_{i\theta})^2}{E^j_{i\theta}} \tag{3.2}$$

where $O^j_{i\theta}$ is the number of observed data in class $j$ for Equipment $i$ allocated to Job $i^\theta$, and $E^j_{i\theta}$ is the expected number of observed data in that class as given by the Weibull distribution so that:

$$E^j_{i\theta} = q^j_{i\theta} \sum_{j=1}^{c} O^j_{i\theta} \tag{3.3}$$

where class $j = \{t | \alpha^j_{i\theta} \leqslant t < \beta^j_{i\theta}\}$, $c$ is the number of classes, and $q^j_{i\theta}$ is the probability that $t = t_j \in \{t | \alpha^j_{i\theta} \leqslant t < \beta^j_{i\theta}\}$, given by:

$$q^j_{i\theta} = \int_{\alpha^j_{i\theta}}^{\beta^j_{i\theta}} f_{i\theta}(t)dt = \exp[-(\lambda_{i\theta}\alpha^j_{i\theta})^{\gamma_{i\theta}}] - \exp[-(\lambda_{i\theta}\beta^j_{i\theta})^{\gamma_{i\theta}}] \tag{3.4}$$

where $\alpha^j_{i\theta}$ and $\beta^j_{i\theta}$ are the left and right boundary points of class $j$ for Equipment $i$ allocated to Job $i^\theta$, respectively.

The classes for the test have been designed carefully to ensure that the assumptions of the test are not violated. For example, it is necessary to make the number of classes (i.e., $c$) as large as possible while ensuring that the value of $E^j_{i\theta}$ (i.e., the expected number of observed data in class $j$ as given by the Weibull distribution) is not too low, i.e., no less than 5 (Roxy 2001 [36]). If the expected number of observed data in a class is less than 5, then the class is merged with adjacent classes to increase the $E^j_{i\theta}$ (Montgomery and Runger 1999 [37]). Table 3.1 shows the results of the goodness-of-fit test on the data for 6 types of equipment in the case study (note that the model here is applicable to multiple jobs. In the case study, each type of equipment is arranged to handle 2 jobs). Thus, the Weibull distribution was selected because $\chi^2_{i\theta} < \chi^2_{c-2-1,\alpha}$ for all the 6 types of equipment which represents a good fit to the data.

### Maximum Likelihood Estimation

To perform the maximum likelihood estimation, the likelihood function of equipment failure time for Equipment $i$ allocated to Job $i^\theta$ is expressed as:

$$L_{i\theta}(\lambda_{i\theta},\gamma_{i\theta}) = \prod_{s=1}^{n_{i\theta}} \lambda_{i\theta}\gamma_{i\theta}(\lambda_{i\theta}t_s)^{\gamma_{i\theta}-1}\exp[-(\lambda_{i\theta}t_s)^{\gamma_{i\theta}}] \tag{3.5}$$

where $s$ = the sample index; $n_{i\theta}$ = the sample size; $t_s$ = the observed data of sample $s$. Then the log-likelihood function is

$$\ln L_{i\theta}(\lambda_{i\theta},\gamma_{i\theta}) = n_{i\theta}(\ln\gamma_{i\theta} + \gamma_{i\theta}\ln\lambda_{i\theta}) + (\gamma_{i\theta}-1)\sum_{s=1}^{n_{i\theta}}\ln t_s - \lambda_{i\theta}^{\gamma_{i\theta}}\sum_{s=1}^{n_{i\theta}} t_s^{\gamma_{i\theta}} \tag{3.6}$$

Based on Eq. (3.6), the maximum likelihood equations over $\lambda_{i\theta}$ and $\gamma_{i\theta}$ are

$$\frac{\partial \ln L_{i\theta}(\lambda_{i\theta},\gamma_{i\theta})}{\partial \lambda_{i\theta}} = 0 \quad\Rightarrow\quad n_{i\theta} - \lambda_{i\theta}^{\gamma_{i\theta}}\sum_{s=1}^{n_{i\theta}} t_s^{\gamma_{i\theta}} = 0 \tag{3.7}$$

$$\frac{\partial \ln L_{i\theta}(\lambda_{i\theta},\gamma_{i\theta})}{\partial \gamma_{i\theta}} = 0 \quad\Rightarrow\quad \frac{n_{i\theta}}{\gamma_{i\theta}} + n_{i\theta}\ln\lambda_{i\theta} + \sum_{s=1}^{n_{i\theta}}\ln t_s - \lambda_{i\theta}^{\gamma_{i\theta}}\sum_{s=1}^{n_{i\theta}}[t_s^{\gamma_{i\theta}}(\ln\lambda_{i\theta} + \ln t_s)] = 0 \tag{3.8}$$

The scale and shape parameters $\lambda_{i\theta}$ and $\gamma_{i\theta}$ can be obtained by solving Eqs. (3.7) and (3.8) using the Newton-Raphson method (Skitmore et al. 2007 [39]). The estimated results for the two parameters are shown in Table 3.1.

**Table 3.1** Results of goodness-of-fit test on the data and estimated parameters for their Weibull distributions

| Type index | Equipment | Job index | Sample size | Number of classes | Chi-square test | | Two-parameter Weibull | |
|---|---|---|---|---|---|---|---|---|
| $i$ | *(Dump truck)* | $i^\theta$ | $\sum_{j=1}^{c} O_{i\theta}^{j}$ | $c$ | $\chi_{i\theta}^2$ | $\chi_{c-2-1,\alpha}^2$ ($\alpha = 0.05$) | $\lambda_{i\theta}$ (scale) | $\gamma_{i\theta}$ (shape) |
| 1 | Terex TA25 (13.5$m^3$/23$t$) | $1^1$ | 189 | 13 | 12.634 | 18.307 | 0.301 | 3.121 |
| | | $1^2$ | 197 | 14 | 15.298 | 19.675 | 0.251 | 3.012 |
| 2 | Terex TA28 (17$m^3$/28$t$) | $2^1$ | 166 | 12 | 9.676 | 16.919 | 0.271 | 3.324 |
| | | $2^2$ | 142 | 10 | 11.575 | 14.067 | 0.291 | 3.423 |
| 3 | K29N-6.4 (11$m^3$/20$t$) | $3^1$ | 159 | 12 | 8.151 | 16.919 | 0.254 | 3.327 |
| | | $3^2$ | 154 | 11 | 10.922 | 15.507 | 0.271 | 3.308 |
| 4 | K30N-8.4 (16$m^3$/26$t$) | $4^1$ | 183 | 11 | 9.278 | 15.507 | 0.275 | 3.526 |
| | | $4^2$ | 211 | 13 | 14.979 | 18.307 | 0.253 | 3.304 |
| 5 | Perlini DP366 (20$m^3$/36$t$) | $5^1$ | 129 | 10 | 10.708 | 14.067 | 0.267 | 3.315 |
| | | $5^2$ | 133 | 10 | 9.812 | 14.067 | 0.308 | 3.294 |
| 6 | Perlini DP755 (30$m^3$/65$t$) | $6^1$ | 158 | 11 | 12.454 | 15.507 | 0.242 | 3.256 |
| | | $6^2$ | 161 | 11 | 8.489 | 15.507 | 0.268 | 3.157 |

### 3.1.3 *Failure Probability-Work Time Equation*

The problem is to determine the relationship between the machine failure probability and the machine failure time. Let $t$ denote the machine failure time of every type of machine allocated to every job, and $p_{i\theta}^{k}$, the machine failure probability for Machine $i$ allocated to Job $i^\theta$ during the $(k+1)$th stage, where $k \leqslant t \leqslant k+1$, $k = 0, 1, \cdots, n-1$, and $\theta = 1, 2$. Let $f_{i\theta}(t)$ be the probability density function (PDF) of the Weibull distribution for Machine $i$ allocated to Job $i^{\theta}$ ($\theta = 1,2$), then, according to probability theory, the value of $p_{i\theta}^{k}$ can be determined:

$$p_{i\theta}^{k} = P_{i\theta}\left[k \leqslant t \leqslant k+1\right] = \int_{k}^{k+1} f_{i\theta}(t)dt \tag{3.9}$$

where the value of $t$ is measured by the stage unit; $P_{i\theta}\left[k \leqslant t \leqslant k+1\right] =$ the probability that $t$ lies in the interval $[k, k+1]$; $k =$ index of stage, where $k = 0,1,\cdots,n-1$; $n =$ the total number of stages; $i =$ the machine type index, where $i = 1,2,\cdots,m$; $m =$ the total number of machine types; $\theta =$ the job selection index, where $\theta = 1,2$; $i^{\theta} =$ the index of jobs handled by Machine $i$. When $k = 1$, the value of $p_{i\theta}^{1}$ is just the area of the shaded part as shown in Figure 3.6.

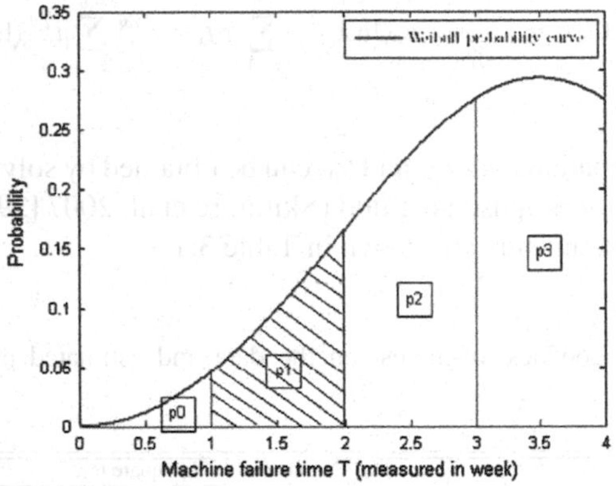


**Fig. 3.6** Probability density function of a Weibull distribution and the machine failure probability in the (k+1)th stage

For a practical application, the conception of mean time to work (MTTW) is usually used for calculations in production or construction projects. Note that $t$ denotes the machine failure time (which is also the machine work time for failed individuals) for every type of machine allocated to every job. Here we define $T_{i\theta}^{k}$ to be the machine work time for every unit of Machine $i$ allocated to Job $i^{\theta}$ during the $(k+1)$th stage, and is expressed as:

$$T_{i\theta}^{k} = \begin{cases} k+1, & if \quad \omega = \omega_1 \\ t, & if \quad \omega = \omega_2 \end{cases} \tag{3.10}$$

where $\omega_1 =$ machine does not fail during the $(k+1)$th stage; $\omega_2 =$ machine fails during the $(k+1)$th stage; $P(\omega_1) = 1 - p_{i\theta}^{k}$ and $P(\omega_2) = p_{i\theta}^{k}$; $t$ follows a Weibull distribution, the value of $t$ is measured by the stage unit, and $k \leqslant t \leqslant k+1$; $k+1 =$ the machine work time for the operating units for Machine $i$ allocated to Job $i^{\theta}$

during the $(k+1)$th stage; $t$ = the machine work (failure) time for the failed units for Machine $i$ allocated to Job $i^\theta$ during the $(k+1)$th stage. $T_{i\theta}^k$ is a variable that takes twofold randomness, and is called a bi-random variable [280].

The expected value $E[T_{i\theta}^k]$ denotes the mean time for the work for all the units of Machine $i$ allocated to Job $i^\theta$ during the $(k+1)$th stage (i.e., $k \leqslant t \leqslant k+1$). From this then, the following equation can be derived:

$$E[T_{i\theta}^k] = (1-p_{i\theta}^k)(k+1) + p_{i\theta}^k \frac{\int_k^{k+1} t f_{i\theta}(t)dt}{\int_k^{k+1} f_{i\theta}(t)dt}, \tag{3.11}$$

where $\frac{\int_k^{k+1} t f_{i\theta}(t)dt}{\int_k^{k+1} f_{i\theta}(t)dt}$ = the mean time for work (failure) for the failed units of Machine $i$ allocated to Job $i^\theta$ during the $(k+1)$th stage; $k+1$ = the machine work time for the operating units of Machine $i$ allocated to Job $i^\theta$ during the $(k+1)$th stage; $1-p_{i\theta}^k$ and $p_{i\theta}^k$ are the rates of the operating units and the failed units of Machine $i$ allocated to Job $i^\theta$ during the $(k+1)$th stage, respectively.

According to Eq. (3.9), Eq. (3.11) can be simplified as:

$$E[T_{i\theta}^k] = (1-p_{i\theta}^k)(k+1) + \int_k^{k+1} t f_{i\theta}(t)dt \tag{3.12}$$

Here, Eq. (3.12) is called the failure probability-work time (FPWT) equation.

Based on Eqs. (3.9) and (3.11), if the Weibull distribution of the machine failure time for every type of machine allocated to every job is determined, the following values can be easily calculated: 1) the machine failure probability of Machine $i$ allocated to Job $i^\theta$ during the $(k+1)$th stage (i.e., $p_{i\theta}^k$); 2) the mean time to work (MTTW) for all units of Machine $i$ allocated to Job $i^\theta$ during the $(k+1)$th stage (i.e., $E[T_{i\theta}^k]$).

### *3.1.4 Fuzziness of Machine Productivity or Throughput*

Since it was first proposed by Zadeh (1965) [22], and further developed by many researchers such as Dubois and Prade (1988) [221], and Nahmias (1978) [238], fuzzy theory has been a useful tool when dealing with ambiguous information, while stochastic theory is suitable for the stochastic influencing factors. The machine productivity or throughput is a typical uncertain variable, which can fluctuate in each stage because of many factors, such as user skill, weather conditions, festival and holiday events, and other uncertain factors. Table 3.2 lists the machine throughput for one type of dump truck [Perlini DP366 ($20m^3/36t$)] at the Xiaolangdi Hydropower Project from December 1995 to March 1996. Let $\tilde{Q}_{i\theta}$ denote the machine productivity or throughput of Machine $i$ allocated to Job $i^\theta$. Based on the data shown in Table 3.2, it is known that the machine throughput for each stage is "between 4118 and 4153 $m^3/week \cdot unit$, and the most likely value is 4132 $m^3/week \cdot unit$," which can be translated into a triangular fuzzy number $\tilde{Q}_{i\theta} = (4118, 4132, 4153)$ as shown in Figure 3.7.

**Table 3.2** Machine throughput of Perlini DP366 for each month in Xiaolangdi Hydropower Project

| Year | 1995 | | | | 1996 | | | | | | | | | | | |
|---|---|---|---|---|---|---|---|---|---|---|---|---|---|---|---|---|
| Month | December | | | | January | | | | February | | | | March | | | |
| Week index | 1 | 2 | 3 | 4 | 1 | 2 | 3 | 4 | 1 | 2 | 3 | 4 | 1 | 2 | 3 | 4 |
| Machine throughput ($m^3/week \cdot unit$) | 4132 | 4118 | 4127 | 4140 | 4132 | 4123 | 4132 | 4136 | 4125 | 4132 | 4153 | 4138 | 4132 | 4130 | 4121 | 4148 |

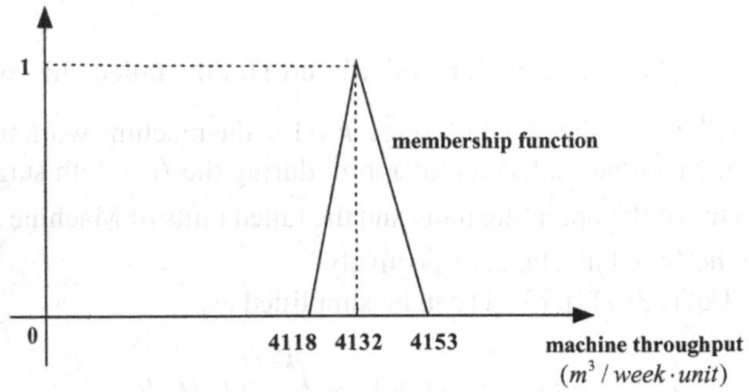


**Fig. 3.7** Probability density function of a Weibull distribution and the machine failure probability in the (k+1)th stage

## 3.2 Modelling Process of DMAP

This section discusses the modelling process of the DMAP. The key features of the termination time of the DMAP, the decreasing state transition equation, the initial and constraint conditions, and the objective functions with fuzzy coefficients are detailed.

### *3.2.1 Termination Time of DMAP*

In Chapter 1, it was discussed that the multiple objective multistage decision making model can be take three kinds of cases as follows:

(1) A MOMSDM under a specified termination time

If the termination time $T$ is fixed in advance (i.e., $T = a$, where $a$ is the stage amount), this means the number of stages of the decision making process has been previously determined . In this case, the decision making is called a MOMSDM under a specified termination time, and its general model is expressed as:

$$\begin{cases} \max \{J_1(\mathbf{x}(0),\mathbf{x}(T),\pi),J_2(\mathbf{x}(0),\mathbf{x}(T),\pi),\cdots,J_n(\mathbf{x}(0),\mathbf{x}(T),\pi)\} \\ \text{s.t.} \begin{cases} \mathbf{x}(k+1) = f_k(\mathbf{x}(k),\mathbf{u}(k)), \\ \mathbf{x}(0) \in I, \\ T = a, \\ g_j(\mathbf{x}(\cdot),\mathbf{u}(\cdot)) \leq 0,\ j=1,2,\cdots,p, \\ \mathbf{x}(k) \in X,\ k=1,2,\cdots,T, \\ \mathbf{u}(k) \in U,\ k=1,2,\cdots,T. \end{cases} \end{cases} \tag{3.13}$$

*Example 3.1.* A typical example problem of a MOMSDM under a specified termination time can be found in construction management, production operations, and transportation planning.

(2) A MOMSDM under an implicit termination time

In a more general case, the termination time $T$ can be assumed to be determined implicitly by a subsidiary condition of the form $\mathbf{x}(T) \in U(\pi)$, where $U(\pi)$ is the termination state space. In this case, the process terminates when the state of the system under control enters, for the first time, a specified subset of the state space (i.e., $U(\pi)$). Thus, its general model is expressed as:

$$\begin{cases} \max \{J_1(\mathbf{x}(0),\mathbf{x}(T),\pi),J_2(\mathbf{x}(0),\mathbf{x}(T),\pi),\cdots,J_n(\mathbf{x}(0),\mathbf{x}(T),\pi)\} \\ \text{s.t.} \begin{cases} \mathbf{x}(k+1) = f_k(\mathbf{x}(k),\mathbf{u}(k)), \\ \mathbf{x}(0) \in I, \\ \mathbf{x}(T) \in U(\pi), \\ g_j(\mathbf{x}(\cdot),\mathbf{u}(\cdot)) \leq 0,\ j=1,2,\cdots,p, \\ \mathbf{x}(k) \in X,\ k=1,2,\cdots,T, \\ \mathbf{u}(k) \in U,\ k=1,2,\cdots,T. \end{cases} \end{cases} \tag{3.14}$$

*Example 3.2.* First discussed by Bellman and Zadeh (1970) [285], practical applications for this model have been widely employed in infrastructure management, equipment allocation and project scheduling. Taking a hydropower construction project as an example, usually, in a hydropower construction project, the concrete production system has several groups of concrete mixing buildings which are located at different sites in the dam area. In order to construct these concrete mixing buildings efficiently, several construction teams are dispatched in multiple stages. The state of the system is the number of finished concrete mixing buildings.

(3) A MOMSDM under an infinite termination time

Formulations with an infinite termination time are mostly used for cases with long planning horizons, such as water resource management (Nandalal and Bogardi 2007 [168]), urban planning and development, and outer space exploration. The number of stages in this case is assumed to be $T = \infty$, therefore, its general model is expressed as:

$$\begin{cases} \max \{J_1(\mathbf{x}(0),\mathbf{x}(T),\pi), J_2(\mathbf{x}(0),\mathbf{x}(T),\pi),\cdots, J_n(\mathbf{x}(0),\mathbf{x}(T),\pi)\} \\ \text{s.t.} \begin{cases} \mathbf{x}(k+1) = f_k(\mathbf{x}(k),\mathbf{u}(k)), \\ \mathbf{x}(0) \in I, \\ T = \infty, \\ g_j(\mathbf{x}(\cdot),\mathbf{u}(\cdot)) \leq 0,\ j = 1,2,\cdots,p. \\ \mathbf{x}(k) \in X,\ k = 1,2,\cdots, \\ \mathbf{u}(k) \in U,\ k = 1,2,\cdots. \end{cases} \end{cases} \tag{3.15}$$

The problem of an infinite termination time was first formulated and solved by Kacprzyk et al. (1981) [169] and Kacprzyk and Staniewski (1983) [170]; see also Kacprzyk (1983a) [284].

*Example 3.3.* One of the most representative applications of this model can be found in reservoir operations. Reservoirs have to be operated at best practice to achieve the maximum benefits. For many years the rule curves, which define the ideal reservoir storage levels in each season or month, have been the essential operational tool. Reservoir operators are expected to maintain these pre-fixed water levels as closely as possible while generally trying to satisfy various water needs downstream. If the reservoir storage levels are above the target or desired levels, the release rates are increased. Conversely, if the levels are below the targets, the release rates are decreased. Sometimes operation rules are defined to include not only storage target levels, but also the various storage allocation zones, such as water conservation, flood control, spill or surcharge, buffer, and inactive or dead storage zones. These zones may also vary throughout the year and the advised release range for each zone is provided by the rules. The desired storage levels and allocation zones mentioned above which are usually defined based on historical operating practice and experience. Having only these target levels for each reservoir, the reservoir operator has considerable responsibility in the day-to-day operations with respect to the appropriate trade-off between storage levels and discharge deviations from ideal conditions.

### *3.2.2 Decreasing State Transition Equation*

Let $x_i(k)$ denote the number of Machine $i$ that are available at the beginning of the $(k+1)^{th}$ stage (i.e., $t = k$), and $u_{i^\theta}(k)$ ($\theta = 1,2$), the number of Machine $i$ allocated to Job $i^\theta$ at the beginning of the $(k+1)^{th}$ stage (i.e., $t = k$). The relationship equation over $x_i(k)$, $u_{i^\theta}(k)$ and $x_i(k+1)$ is expressed as:

$$x_i(k+1) = \sum_{\theta=1}^{2} (1 - p^k_{i^\theta}) u_{i^\theta}(k), \quad \forall i \in \Psi, \quad \forall k \in \Phi \tag{3.16}$$

$$x_i(k) = \sum_{\theta=1}^{2} u_{i^\theta}(k), \quad \forall i \in \Psi, \quad \forall k \in \Phi \tag{3.17}$$

where $(1 - p^k_{i^\theta})u_{i^\theta}(k)$ = number of operating units of Machine $i$ allocated to Job $i^\theta$ at the end of the $(k+1)^{th}$ stage (i.e., $t = k+1$). Since the end of the $(k+1)^{th}$ stage is the beginning of the $(k+2)^{th}$ stage, the number of Machine $i$ that are available at the beginning of the $(k+2)^{th}$ stage (i.e., $t=k+1$) $x_i(k+1)$ should be the sum of $(1-p_{i\theta}^k)u_{i\theta}(k)$ for $\theta=1,2$, (See Eq. (3.16)). On the other hand, all the available machines at the beginning of every stage should be allocated to do jobs. Thus, Eq. (3.17) is obtained. For model analysis convenience, based on Eqs. (3.16) and (3.17), the relationship equation can be also expressed as:

$$x_i(k+1)=(1-p_{i1}^k)u_{i1}(k)+(1-p_{i2}^k)(x_i(k)-u_{i1}(k)),\quad \forall i\in\Psi,\quad \forall k\in\Phi \quad (3.18)$$

Figure 3.8 shows the detailed dynamic allocation process for Machine $i$.

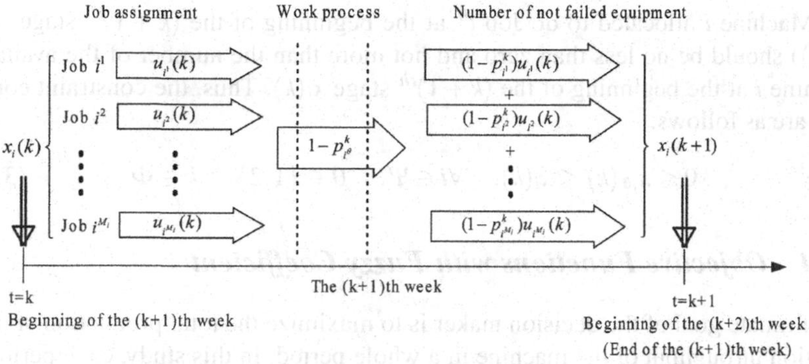


**Fig. 3.8** Dynamic allocation process for Machine $i$

### *3.2.3 Initial and Constraint Conditions*

Serial multistage decision problems can be classified into three categories as follows.

1. Initial value problem. If the value of the initial state variable, $x_i(0)$, is prescribed, the problem is called an initial value problem.
2. Final value problem. If the value of the final state variable, $x_i(T)$ is prescribed, the problem is called a final value problem. Notice that a final value problem can be transformed into an initial value problem by reversing the directions of $x_i(k)$, $k=1,2,\cdots,T$.
3. Boundary value problem. If the values of both the input and output variables are specified, the problem is called a boundary value problem.

For the DMAP discussed in this chapter, the multistage decision making problem is considered to be an initial value problem.

**Initial Conditions**

Initial conditions specify the number of the available machine at the beginning of the whole period. Let $a_i$ denote the initial number of Machine $i$ at the beginning of the whole period. We present the initial conditions as follows:

$$x_i(0) = a_i, \quad \forall i \in \Psi \tag{3.19}$$

where $x_i(0)$ = number of Machine $i$ that are available at the beginning of the first stage (i.e., $t = 0$); and $i$ = index of machine type, where $i = 1, 2, \cdots, m$.

**Inequality Constraint Conditions**

The Machine $i$ allocated to do Job $i^\theta$ at the beginning of the $(k+1)^{th}$ stage (i.e., $u_{i^\theta}(k)$) should be no less than zero and not more than the number of the available Machine $i$ at the beginning of the $(k+1)^{th}$ stage $x_i(k)$. Thus, the constraint conditions are as follows:

$$0 \leqslant u_{i^\theta}(k) \leqslant x_i(k), \quad \forall i \in \Psi, \quad \theta \in \{1,2\}, \quad k \in \Phi \tag{3.20}$$

### 3.2.4 Objective Functions with Fuzzy Coefficient

The ultimate goal of the decision maker is to maximize the total production or construction throughput of the machine in a whole period. In this study, each period is divided into $n$ stages. Let $J_i^n$ denote the total production or construction throughput produced by Machine $i$ in a whole period. Then the objective function can be expressed as:

$$J_i^n = J_i^n(u_{i^\theta}(\cdot)) = \sum_{\theta=1}^{2} [\tilde{Q}_{i^\theta} \sum_{k=0}^{n-1} \zeta_{k,i^\theta}^{rand} u_{i^\theta}(k)] \tag{3.21}$$

i.e.,

$$J_i^n = J_i^n(x_i(\cdot), u_{i^1}(\cdot)) = \tilde{Q}_{i^1} \sum_{k=0}^{n-1} \zeta_{k,i^1}^{rand} u_{i^1}(k) + \tilde{Q}_{i^2} \sum_{k=0}^{n-1} \zeta_{k,i^2}^{rand} [x_i(k) - u_{i^1}(k)] \tag{3.22}$$

where $\tilde{Q}_{i^\theta}$ = the machine productivity or throughput of Machine $i$ allocated to Job $i^\theta$; $k$ = the stage index, where $k = 0, 1, \cdots, n-1$; $i$ = the machine type index, where $i = 1, 2, \cdots, m$; $\tilde{Q}_{i^\theta} \sum_{k=0}^{n-1} \zeta_{k,i^\theta}^{rand} u_{i^\theta}(k)$ = the production or construction throughput produced by Machine $i$ in a whole period allocated to Job $i^\theta$; note that $\zeta_{k,i^\theta}^{rand}$ ($\theta = 1, 2$) is a conversion factor for the production or construction throughput produced by Machine $i$ allocated to Job $i^\theta$ during the $(k+1)^{th}$ stage (i.e., $k \leqslant t \leqslant k+1$).

The reason why it is necessary to introduce $\zeta_{k,i^\theta}^{rand}$ is that machine failures do not happen at the end of every stage, and often happen in the middle period of a stage. In fact, this is the problem with machine failure time. As a result, the production or construction throughput produced per stage by Machine $i$ allocated to Job $i^\theta$ during the $(k+1)^{th}$ stage must be less than $\tilde{Q}_{i^\theta} u_{i^\theta}(k)$, since not all Machine $i$ allocated to Job $i^\theta$ during the $(k+1)^{th}$ stage can function for a whole stage because possible machine failures may occur in the middle period of the stage. Therefore, it is necessary to introduce a conversion factor $\zeta^{rand}_{k,i^\theta}$ $(\theta = 1,2)$ the role of which is to balance the difference between $\tilde{Q}_{i^\theta} u_{i^\theta}(k)$ and the actual value of the production or construction throughput produced by Machine $i$ allocated to Job $i^\theta$ during the $(k+1)^{th}$ stage.

Then, the expression $\zeta^{rand}_{k,i^\theta}$ is introduced. Let $A^1_{i^\theta}(k), A^2_{i^\theta}(k), \cdots, A^{u_{i^\theta}(k)}_{i^\theta}(k)$ be the actual values for the machine productivity or throughput of Machine $i$ allocated to Job $i^\theta$ during the $(k+1)^{th}$ stage respectively, $0 \leqslant A^j_{i^\theta}(k) \leqslant \tilde{Q}_{i^\theta}$, where $j = 1,2,\cdots,u_{i^\theta}(k)$. Thus, the expression of the conversion factor $\zeta^{rand}_{k,i^\theta}$ is presented as:

$$\zeta^{rand}_{k,i^\theta} = \frac{\sum_{j=1}^{u_{i^\theta}(k)} A^j_{i^\theta}(k)}{\tilde{Q}_{i^\theta} u_{i^\theta}(k)} \tag{3.23}$$

Note that the current form of Eq. (3.23) is not very useful since the allocation decision must be made at the beginning of a whole period. At that time, the machine failures do not always happen, so it is impossible to know the value of $A^j_{i^\theta}(k)$. Eq. (3.23), therefore, needs some transformations.

Since $T^k_{i^\theta}$ is the machine work time for every unit of Machine $i$ allocated to Job $i^\theta$ during the $(k+1)^{th}$ stage, where the value of $T^k_{i^\theta}$ is measured by the stage unit, then we have $k \leqslant T^k_{i^\theta} \leqslant k+1$, and:

$$(T^k_{i^\theta} - k)\tilde{Q}_{i^\theta} = A^j_{i^\theta}(k) \tag{3.24}$$

Thus, Eq. (3.23) can be transformed into the following form:

$$\zeta^{rand}_{k,i^\theta} = \frac{(T^k_{i^\theta} - k)\tilde{Q}_{i^\theta} u_{i^\theta}(k)}{\tilde{Q}_{i^\theta} u_{i^\theta}(k)}, \tag{3.25}$$

i.e.,

$$\zeta^{rand}_{k,i^\theta} = T^k_{i^\theta} - k \tag{3.26}$$

Finally, we can formulate a discrete-time uncertain optimal model for the DMAP as follows:

$$\max J^n_i(x_i(\cdot), u_{i^1}(\cdot)) = \tilde{Q}_{i^1} \sum_{k=0}^{n-1} \zeta^{rand}_{k,i^1} u_{i^1}(k) + \tilde{Q}_{i^2} \sum_{k=0}^{n-1} \zeta^{rand}_{k,i^2} [x_i(k) - u_{i^1}(k)]$$

$$s.t. \begin{cases} x_i(k+1) = (1-p^k_{i^1})u_{i^1}(k) + (1-p^k_{i^2})(x_i(k) - u_{i^1}(k)), & \forall i \in \Psi, \quad \forall k \in \Phi, \\ 0 \leqslant u_{i^\theta}(k) \leqslant x_i(k), \quad \forall i \in \Psi, \quad \theta \in \{1,2\}, \quad k \in \Phi, \\ x_i(0) = a_i, \quad \forall i \in \Psi. \end{cases} \tag{3.27}$$

### 3.2.5 FEVM of DMAP

It is often quite difficult to handle the DMAP when it involves uncertain information. The fuzzy coefficients involved in the DMAP may make the optimal control model very hard to solve [583]. One strategy is to employ a transformation method to convert the current triangular fuzzy number into its deterministic equivalent. In this section, a crisp approach is used to integrate the decision maker's optimistic-pessimistic attitudes in adopting real world practice. The fuzzy coefficients in the model are defuzzified using an expected value operator with an optimistic-pessimistic index. To calculate the expected value of the triangular fuzzy numbers, a new measure, Me, which has an optimistic-pessimistic adjusting index, is introduced to characterize real life problems. The definition of this fuzzy measure Me and its relative expected value operator $E^{Me}$ can be found in Xu and Zhou (2011) [281]. For example, let $\tilde{Q}_{i\theta} = (r_1, r_2, r_3)$ be any triangular fuzzy number in the DMAP. In fact, in real world problems, especially decision problems in manufacturing and construction industries, the case is often encountered where $r_1 > 0$, then the expected value of $\tilde{Q}_{i\theta}$ should be,

$$E^{Me}[\tilde{Q}_{i\theta}] = \frac{(1-\eta)}{2} r_1 + \frac{1}{2} r_2 + \frac{\eta}{2} r_3, \tag{3.28}$$

where $\eta$ is the optimistic-pessimistic index to determine the combined attitude of a decision maker.

To transform the uncertain variable $\zeta^{rand}_{k,i^\theta}$ into a deterministic variable, the expected value model (EVM) [280] is proposed here to form an equivalent crisp model.

Based on the properties of the expected value operator, Eq. (3.26) is transformed into the following form:

$$E[\zeta^{rand}_{k,i^\theta}] = E[T^k_{i\theta}] - k. \tag{3.29}$$

From this, Eq. (3.29) can be used to calculate the expected value of the conversion factor for the production or construction throughput produced by Machine $i$ allocated to Job $i^\theta$ during the $(k+1)^{th}$ stage, i.e., $E[\zeta^{rand}_{k,i^\theta}]$ $(\theta = 1,2)$.

After proper transformation, the expected value of the objective function (3.22) can be transformed as:

$$\begin{aligned} E[J^n_i(x_i(\cdot), u_{i^1}(\cdot))] = {} & E^{Me}[\tilde{Q}_{i1}] \Sigma^{n-1}_{k=0} E[\zeta^{rand}_{k,i^1}] u_{i^1}(k) + E^{Me}[\tilde{Q}_{i2}] \Sigma^{n-1}_{k=0} E[\zeta^{rand}_{k,i^2}] \\ & [x_i(k) - u_{i^1}(k)], \end{aligned} \tag{3.30}$$

and the expected value model of the DMAP can be rewritten as:

$$\begin{cases} \max E[J^n_i(x_i(\cdot), u_{i^1}(\cdot))] = E^{Me}[\tilde{Q}_{i1}] \Sigma^{n-1}_{k=0} E[\zeta^{rand}_{k,i^1}] u_{i^1}(k) + E^{Me}[\tilde{Q}_{i2}] \Sigma^{n-1}_{k=0} E[\zeta^{rand}_{k,i^2}] \\ \qquad [x_i(k) - u_{i^1}(k)] \\ \text{s.t.} \begin{cases} x_i(k+1) = (1-p^k_{i1}) u_{i^1}(k) + (1-p^k_{i2})(x_i(k) - u_{i^1}(k)), \quad \forall i \in \Psi, \quad \forall k \in \Phi, \\ 0 \leqslant u_{i^\theta}(k) \leqslant x_i(k), \quad \forall i \in \Psi, \quad \theta \in \{1,2\}, \quad k \in \Phi, \\ x_i(0) = a_i, \quad \forall i \in \Psi. \end{cases} \end{cases} \tag{3.31}$$

The fuzzy coefficients in the objective functions have been transformed into fuzzy expected values, which reflect the expectation of the decision makers. The above model is called the fuzzy expected value model of the DMAP in this book.

## 3.3 Model Analysis of DMAP

In order to solve the expected value model above, the support function and Hamilton function are introduced to derive the theorem vital for finding the optimal solution to our problem. But before the theorem can be proved, a series of mathematical statements and definitions are introduced.

### *3.3.1 Expanded Discrete State Attainable Set*

Firstly, it is necessary to consider the expected value of the objective function (3.30). To prove the crucial theorem, a function which is similar to (3.30) and useful in the following process of proofs must be defined. We describe this function below.

Let $L_i^k$ be a function defined by

$$L_i^k = -\{E^{Me}[\tilde{Q}_{i1}]\sum_{j=0}^{k-1} E[\zeta_{j,i1}^{rand}]u_{i1}(j) + E^{Me}[\tilde{Q}_{i2}]\sum_{j=0}^{k-1} E[\zeta_{j,i2}^{rand}][x_i(j) - u_{i1}(j)]\},$$

$\forall i \in \Psi$, where $\Psi = \{1,2,\cdots,m\}$, $k = 1,2,\cdots,n$.

Then, the expanded state variable and expanded state vector need to be defined, so it is much more convenient to discuss the problem in Euclidean space.

**Definition 3.1.** *A variable $x_i^0(k)$ is called an expanded state variable if and only if*

$$x_i^0(0) = 0, \quad x_i^0(k) = L_i^k, \quad \forall i \in \Psi, \quad k = 1,2,\cdots,n.$$

**Definition 3.2.** *A vector $y_i(k)$ is called an expanded state vector if and only if*

$$y_i(k) = (x_i^0(k), x_i(k)), \quad \forall i \in \Psi, \quad k = 0,1,\cdots,n.$$

*Remark 3.1.* Let $\mathbf{Y_i}$ be the space of all $\mathbf{y_i}(k)$ $(k = 0,1,\cdots,n)$, then $\mathbf{Y_i} \subset \mathbf{R^2}$, $\forall i \in \Psi$.

To simplify the expression, the relationship is described. Let $f_i^k$ be a function defined by

$$x_i(k+1) = f_i^k(x_i(k), u_{i1}(k)) = (p_{i2}^k - p_{i1}^k)u_{i1}(k) + (1 - p_{i2}^k)x_i(k), \quad \forall i \in \Psi,$$

where $k = 0,1,\cdots,n-1$.

Then, let $\mathbf{F_i^k}$ be a function defined by

$$\mathbf{F_i^k} = (L_i^{k+1}, f_i^k), \quad \forall i \in \Psi, \quad k = 0,1,\cdots,n-1.$$

Therefore, $\mathbf{y_i}(0) = (0, x_i(0)) = (0, a_i)$, $\mathbf{y_i}(k+1) = \mathbf{F_i^k}(\mathbf{y_i}(k), u_{i1}(k)), \forall i \in \Psi, k = 0,1,\cdots,n-1$.

Based on the above, it is possible to define the state attainable set and the expanded state attainable set.

The state attainable set of the $2^{th}$ stage $\mathbf{X}_\mathbf{i}^\mathbf{1}(x_i(0))$ which is derived from the initial condition $x_i(0) = a_i$ is defined by

$$\mathbf{X}_\mathbf{i}^\mathbf{1}(x_i(0)) = \{x_i(1)|x_i(1) = f_i^0(x_i(0), u_{i1}(0)), 0 \leqslant u_{i1}(0) \leqslant x_i(0)\}, \quad \forall i \in \Psi.$$

**Definition 3.3.** *(Xu and Zeng 2012 [198])* $\boldsymbol{X}_i^\boldsymbol{k}(x_i(k-1))$ *is the state attainable set of the* $(k+1)$*th stage which is derived from* $x_i(k-1)$ *if and only if*

$$\boldsymbol{X}_i^\boldsymbol{k}(x_i(k-1)) = \{x_i(k)|x_i(k) = f_i^{k-1}(x_i(k-1), u_{i1}(k-1)), 0 \leqslant u_{i1}(k-1) \leqslant x_i(k-1)\},$$

$\forall x_i(k-1) \in \boldsymbol{X}_i^{\boldsymbol{k-1}}, \forall i \in \Psi, k = 2, 3, \cdots, n.$

The expanded state attainable set of the $2^{nd}$ stage $\mathbf{Y}_\mathbf{i}^\mathbf{1}(\mathbf{y}_\mathbf{i}(0))$ which is in the corresponding expanded space $\mathbf{Y}_\mathbf{i}$ is defined by

$$\mathbf{Y}_\mathbf{i}^\mathbf{1}(\mathbf{y}_\mathbf{i}(0)) = \{\mathbf{y}_\mathbf{i}(1)|\mathbf{y}_\mathbf{i}(1) = \mathbf{F}_\mathbf{i}^\mathbf{0}(\mathbf{y}_\mathbf{i}(0), u_{i1}(0)), 0 \leqslant u_{i1}(0) \leqslant x_i(0)\}, \quad \forall i \in \Psi.$$

**Definition 3.4.** *(Xu and Zeng 2012 [198])* $\boldsymbol{Y}_i^\boldsymbol{k}(\boldsymbol{y}_i(k-1))$ *is the expanded state attainable set of the* $(k+1)$*th stage which is derived from* $\mathbf{y}_i(k-1)$ *if and only if*

$$\boldsymbol{Y}_i^\boldsymbol{k}(\boldsymbol{y}_i(k-1)) = \{\boldsymbol{y}_i(k)|\boldsymbol{y}_i(k) = \boldsymbol{F}_i^{\boldsymbol{k-1}}(\boldsymbol{y}_i(k-1), u_{i1}(k-1)), 0 \leqslant u_{i1}(k-1) \leqslant x_i(k-1)\},$$

$\forall \boldsymbol{y}_i(k-1) \in \boldsymbol{Y}_i^{\boldsymbol{k-1}}, \forall i \in \Psi, k = 2, 3, \cdots, n.$

All of the state attainable set of the $(k+1)$th stage $\mathbf{X}_\mathbf{i}^\mathbf{k}$ is defined by

$$\mathbf{X}_\mathbf{i}^\mathbf{k} = \bigcup\{\mathbf{X}_\mathbf{i}^\mathbf{k}(x_i(k-1))|x_i(k-1) \in \mathbf{X}_\mathbf{i}^{\mathbf{k-1}}\}, \quad \forall i \in \Psi, \quad k = 2, 3, \cdots, n.$$

All of the expanded state attainable set of the $(k+1)$th stage $\mathbf{Y}_\mathbf{i}^\mathbf{k}$ is defined by

$$\mathbf{Y}_\mathbf{i}^\mathbf{k} = \bigcup\{\mathbf{Y}_\mathbf{i}^\mathbf{k}(\mathbf{y}_\mathbf{i}(k-1))|\mathbf{y}_\mathbf{i}(k-1) \in \mathbf{Y}_\mathbf{i}^{\mathbf{k-1}}\}, \quad \forall i \in \Psi, \quad k = 2, 3, \cdots, n.$$

Up to now, a series of concepts about the solution space have been outlined. With this completed the conditional function of optimal value and the function of optimal value are now defined to allow for a further preparation of the proof.

### *3.3.2 Conditional Function of Optimal Construction Throughput*

The conditional function of optimal value of the $2^{nd}$ stage $fov_i^1(x_i(1)|\mathbf{y}_\mathbf{i}(0))$ is defined by

$$fov_i^1(x_i(1)|\mathbf{y}_\mathbf{i}(0)) = \min\{L_i^1(u_{i1}(0))|0 \leqslant u_{i1}(0) \leqslant x_i(0), f_i^0(x_i(0), u_{i1}(0)) = x_i(1)\},$$

$\forall i \in \Psi.$

The function of optimal value of the $2^{nd}$ stage $FOV_i^1(x_i(1))$ is defined by

$$FOV_i^1(x_i(1)) = \min\{L_i^1(u_{i^1}(0)) | 0 \leqslant u_{i^1}(0) \leqslant x_i(0), f_i^0(x_i(0), u_{i^1}(0)) = x_i(1)\},$$

$\forall i \in \Psi$.

Here, the value of the conditional function of optimal value $fov_i^1(x_i(1)|\mathbf{y_i}(0))$ is calculated under the condition that the value of the expanded state vector $\mathbf{y_i}(0)$ is fixed. Since $\mathbf{y_i}(0) = (0, x_i(0)) = (0, a_i)$ is the only fixed value in the $2^{th}$ stage, hence, the conditional function of optimal value of the $2^{th}$ stage $fov_i^1(x_i(1)|\mathbf{y_i}(0))$ is the same as the function of optimal value of the $2^{th}$ stage $FOV_i^1(x_i(1))$ in this special case, i.e. $fov_i^1(x_i(1)|\mathbf{y_i}(0)) = FOV_i^1(x_i(1))$. But there are differences between the two function above when $k = 2, 3, \cdots, n$.

**Definition 3.5.** *(Xu and Zeng 2012 [198]) $fov_i^k(x_i(k)|\mathbf{y}_i(k-1))$ is the conditional function of optimal value of the $(k+1)$th stage if and only if*

$$fov_i^k(x_i(k)|\mathbf{y}_i(k-1)) = \min\left\{ L_i^k(u_{i^1}(k-1)) \left| \begin{array}{l} 0 \leqslant u_{i^1}(k-1) \leqslant x_i(k-1), \\ L_i^k = x_i^0(k-1) + \{E[\zeta_{k-1,i^1}^{rand}]E^{Me}[\tilde{Q}_{i^1}]u_{i^1}(k-1) + \\ E[\zeta_{k-1,i^2}^{rand}]E^{Me}[\tilde{Q}_{i^2}][x_i(k-1) - u_{i^1}(k-1)]\}, \\ f_i^{k-1}(x_i(k-1), u_{i^1}(k-1)) = x_i(k) \end{array} \right. \right\},$$

$\forall i \in \Psi, k = 2, 3, \cdots, n$.

**Lemma 3.1.** *The conditional function of optimal value of the $(k+1)$th stage $fov_i^k(x_i(k)|\mathbf{y}_i(k-1))$ is differentiable of its argument $x_i(k)$, $\forall i \in \Psi$, $k = 1, 2, \cdots, n$.*

*Proof.* According to Def. 3.5 and the statement of $fov_i^1(x_i(1)|\mathbf{y_i}(0))$, $\forall i \in \Psi$, $k = 1, 2, \cdots, n$, we get

$$fov_i^k(x_i(k)|\mathbf{y_i}(k-1)) = \min\left\{ L_i^k(u_{i^1}(k-1)) \left| \begin{array}{l} 0 \leqslant u_{i^1}(k-1) \leqslant x_i(k-1), \\ L_i^k = x_i^0(k-1) + \{E[\zeta_{k-1,i^1}^{rand}]E^{Me}[\tilde{Q}_{i^1}]u_{i^1}(k-1) + \\ E[\zeta_{k-1,i^2}^{rand}]E^{Me}[\tilde{Q}_{i^2}][x_i(k-1) - u_{i^1}(k-1)]\}, \\ f_i^{k-1}(x_i(k-1), u_{i^1}(k-1)) = x_i(k) \end{array} \right. \right\}.$$

Since

$$\begin{aligned} &L_i^k(u_{i^1}(k-1)) \\ &= -\{E^{Me}[\tilde{Q}_{i^1}]\Sigma_{j=0}^{k-1}E[\zeta_{j,i^1}^{rand}]u_{i^1}(j) + E^{Me}[\tilde{Q}_{i^2}]\Sigma_{j=0}^{k-1}E[\zeta_{j,i^2}^{rand}][x_i(j) - u_{i^1}(j)]\} \\ &= -\{[E^{Me}[\tilde{Q}_{i^1}]\Sigma_{j=0}^{k-2}E[\zeta_{j,i^1}^{rand}]u_{i^1}(j) - E^{Me}[\tilde{Q}_{i^2}]\Sigma_{j=0}^{k-2}E[\zeta_{j,i^2}^{rand}]u_{i^1}(j)] + E^{Me}[\tilde{Q}_{i^2}] \\ &\Sigma_{j=0}^{k-2}E[\zeta_{j,i^2}^{rand}]x_i(j)\} + [E[\zeta_{k-1,i^1}^{rand}]E^{Me}[\tilde{Q}_{i^1}] - E[\zeta_{k-1,i^2}^{rand}]E^{Me}[\tilde{Q}_{i^2}]]u_{i^1}(k-1) \\ &+E[\zeta_{k-1,i^2}^{rand}]E^{Me}[\tilde{Q}_{i^2}]x_i(k-1), \end{aligned}$$

and $x_i(k-1)$ is fixed by the value of $\mathbf{y_i}(k-1)) = (x_i^0(k-1), x_i(k-1))$, so $u_{i^1}(k-1)$ and $x_i(k)$ have one to one correspondence as a result of the equation $f_i^{k-1}(x_i(k-1), u_{i^1}(k-1)) = x_i(k)$. Hence, when $0 \leqslant u_{i^1}(k-1) \leqslant x_i(k-1)$,

$$
\begin{aligned}
& fov_i^k(x_i(k)|\mathbf{y_i}(k-1)) \\
= & \min\left\{ L_i^k(u_{i^1}(k-1)) \left| \begin{array}{l} 0 \leqslant u_{i^1}(k-1) \leqslant x_i(k-1), \\ L_i^k = x_i^0(k-1) + \{E[\zeta_{k-1,i^1}^{rand}]E^{Me}[\tilde{Q}_{i^1}]u_{i^1}(k-1) + \\ E[\zeta_{k-1,i^2}^{rand}]E^{Me}[\tilde{Q}_{i^2}][x_i(k-1) - u_{i^1}(k-1)]\}, \\ f_i^{k-1}(x_i(k-1), u_{i^1}(k-1)) = x_i(k) \end{array} \right. \right\} \\
= & \min\left\{ \begin{array}{l} -\sum_{j=0}^{k-2}\{[E[\zeta_{j,i^1}^{rand}]E^{Me}[\tilde{Q}_{i^1}] - E[\zeta_{j,i^2}^{rand}]E^{Me}[\tilde{Q}_{i^2}]] \\ u_{i^1}(j) + E[\zeta_{j,i^2}^{rand}]E^{Me}[\tilde{Q}_{i^2}]x_i(j)\} \end{array} \left| \begin{array}{l} 0 \leqslant u_{i^1}(j) \leqslant x_i(j), \\ j = 0, \cdots, k-2 \end{array} \right. \right\} \\
& + [E[\zeta_{k-1,i^1}^{rand}]E^{Me}[\tilde{Q}_{i^1}] - E[\zeta_{k-1,i^2}^{rand}]E^{Me}[\tilde{Q}_{i^2}]]u_{i^1}(k-1) + E[\zeta_{k-1,i^2}^{rand}]E^{Me}[\tilde{Q}_{i^2}] \\
& x_i(k-1).
\end{aligned}
$$

If $E[\zeta_{j,i^1}^{rand}]E^{Me}[\tilde{Q}_{i^1}] - E[\zeta_{j,i^2}^{rand}]E^{Me}[\tilde{Q}_{i^2}] > 0$, the minimum value of $L_i^k(u_{i^1}(k-1))$ can be arrived when $u_{i^1}(j) = x_i(j)$, where $j = 0, \cdots, k-2$, thus

$$
\begin{aligned}
& \min\left\{ \begin{array}{l} -\sum_{j=0}^{k-2}\{[E[\zeta_{j,i^1}^{rand}]E^{Me}[\tilde{Q}_{i^1}] - E[\zeta_{j,i^2}^{rand}]E^{Me}[\tilde{Q}_{i^2}]] \\ u_{i^1}(j) + E[\zeta_{j,i^2}^{rand}]E^{Me}[\tilde{Q}_{i^2}]x_i(j)\} \end{array} \left| \begin{array}{l} 0 \leqslant u_{i^1}(j) \leqslant x_i(j), \\ j = 0, \cdots, k-2 \end{array} \right. \right\} \\
= & -\sum_{j=0}^{k-2} E[\zeta_{j,i^1}^{rand}]E^{Me}[\tilde{Q}_{i^1}]x_i(j).
\end{aligned}
$$

If $E[\zeta_{j,i^1}^{rand}]E^{Me}[\tilde{Q}_{i^1}] - E[\zeta_{j,i^2}^{rand}]E^{Me}[\tilde{Q}_{i^2}] < 0$, the minimum value of $L_i^k(u_{i^1}(k-1))$ can be arrived when $u_{i^1}(j) = 0$, where $j = 0, \cdots, k-2$, thus

$$
\begin{aligned}
& \min\left\{ \begin{array}{l} -\sum_{j=0}^{k-2}\{[E[\zeta_{j,i^1}^{rand}]E^{Me}[\tilde{Q}_{i^1}] - E[\zeta_{j,i^2}^{rand}]E^{Me}[\tilde{Q}_{i^2}]] \\ u_{i^1}(j) + E[\zeta_{j,i^2}^{rand}]E^{Me}[\tilde{Q}_{i^2}]x_i(j)\} \end{array} \left| \begin{array}{l} 0 \leqslant u_{i^1}(j) \leqslant x_i(j), \\ j = 0, \cdots, k-2 \end{array} \right. \right\} \\
= & -\sum_{j=0}^{k-2} E[\zeta_{j,i^2}^{rand}]E^{Me}[\tilde{Q}_{i^2}]x_i(j).
\end{aligned}
$$

If $E[\zeta_{j,i^1}^{rand}]E^{Me}[\tilde{Q}_{i^1}] - E[\zeta_{j,i^2}^{rand}]E^{Me}[\tilde{Q}_{i^2}] = 0$, then

$$
\begin{aligned}
& \min\left\{ \begin{array}{l} -\sum_{j=0}^{k-2}\{[E[\zeta_{j,i^1}^{rand}]E^{Me}[\tilde{Q}_{i^1}] - E[\zeta_{j,i^2}^{rand}]E^{Me}[\tilde{Q}_{i^2}]] \\ u_{i^1}(j) + E[\zeta_{j,i^2}^{rand}]E^{Me}[\tilde{Q}_{i^2}]x_i(j)\} \end{array} \left| \begin{array}{l} 0 \leqslant u_{i^1}(j) \leqslant x_i(j), \\ j = 0, \cdots, k-2 \end{array} \right. \right\} \\
= & -\sum_{j=0}^{k-2} E[\zeta_{j,i^2}^{rand}]E^{Me}[\tilde{Q}_{i^2}]x_i(j).
\end{aligned}
$$

To sum up, we have

$$
\begin{aligned}
& fov_i^k(x_i(k)|\mathbf{y_i}(k-1)) \\
= & \left\{ \begin{array}{l} -\sum_{j=0}^{k-2} E[\zeta_{j,i^1}^{rand}]E^{Me}[\tilde{Q}_{i^1}]x_i(j) + [E[\zeta_{k-1,i^1}^{rand}]E^{Me}[\tilde{Q}_{i^1}] - E[\zeta_{k-1,i^2}^{rand}]E^{Me}[\tilde{Q}_{i^2}]] \\ u_{i^1}(k-1) + E[\zeta_{k-1,i^2}^{rand}]E^{Me}[\tilde{Q}_{i^2}]x_i(k-1), \ E[\zeta_{j,i^1}^{rand}]E^{Me}[\tilde{Q}_{i^1}] - E[\zeta_{j,i^2}^{rand}] \\ E^{Me}[\tilde{Q}_{i^2}] > 0; \\ -\sum_{j=0}^{k-2} E[\zeta_{j,i^2}^{rand}]E^{Me}[\tilde{Q}_{i^2}]x_i(j) + [E[\zeta_{k-1,i^1}^{rand}]E^{Me}[\tilde{Q}_{i^1}] - E[\zeta_{k-1,i^2}^{rand}]E^{Me}[\tilde{Q}_{i^2}]] \\ u_{i^1}(k-1) + E[\zeta_{k-1,i^2}^{rand}]E^{Me}[\tilde{Q}_{i^2}]x_i(k-1), \ E[\zeta_{j,i^1}^{rand}]E^{Me}[\tilde{Q}_{i^1}] - E[\zeta_{j,i^2}^{rand}] \\ E^{Me}[\tilde{Q}_{i^2}] \leqslant 0. \end{array} \right.
\end{aligned}
\tag{3.32}
$$

From the above expression, we can know that $fov_i^k(x_i(k)|\mathbf{y_i}(k-1))$ is a linear function and is differentiable of its argument $x_i(k-1)$. Let

$$
\begin{aligned}
& g_i^1(x_i(k)) \\
&= [E[\zeta_{k-1,i^1}^{rand}]E^{Me}[\tilde{Q}_{i^1}] + E[\zeta_{k-1,i^2}^{rand}]E^{Me}[\tilde{Q}_{i^2}]]x_i(k-1) + [E[\zeta_{k-1,i^1}^{rand}]E^{Me}[\tilde{Q}_{i^1}] - E[\zeta_{k-1,i^2}^{rand}] \\
&\quad E^{Me}[\tilde{Q}_{i^2}]]u_{i^1}(k-1) - [E[\zeta_{k,i^1}^{rand}]E^{Me}[\tilde{Q}_{i^1}] - E[\zeta_{k,i^2}^{rand}]E^{Me}[\tilde{Q}_{i^2}]]u_{i^1}(k) - E[\zeta_{k,i^2}^{rand}] \\
&\quad E^{Me}[\tilde{Q}_{i^2}]x_i(k),
\end{aligned}
$$

$$
\begin{aligned}
& g_i^2(x_i(k)) \\
&= 2E[\zeta_{k-1,i^2}^{rand}]E^{Me}[\tilde{Q}_{i^2}]x_i(k-1) + [E[\zeta_{k-1,i^1}^{rand}]E^{Me}[\tilde{Q}_{i^1}] - E[\zeta_{k-1,i^2}^{rand}]E^{Me}[\tilde{Q}_{i^2}]]u_{i^1}(k-1) - \\
&\quad [E[\zeta_{k,i^1}^{rand}]E^{Me}[\tilde{Q}_{i^1}] - E[\zeta_{k,i^2}^{rand}]E^{Me}[\tilde{Q}_{i^2}]]u_{i^1}(k) - E[\zeta_{k,i^2}^{rand}]E^{Me}[\tilde{Q}_{i^2}]x_i(k).
\end{aligned}
$$

According to Eq. (3.32), we get

$$
\begin{aligned}
& fov_i^k(x_i(k)|\mathbf{y_i}(k-1)) \\
&= \begin{cases} fov_i^{k+1}(x_i(k+1)|\mathbf{y_i}(k)) + g_i^1(x_i(k)), & E[\zeta_{j,i^1}^{rand}]E^{Me}[\tilde{Q}_{i^1}] - E[\zeta_{j,i^2}^{rand}]E^{Me}[\tilde{Q}_{i^2}] > 0; \\ fov_i^{k+1}(x_i(k+1)|\mathbf{y_i}(k)) + g_i^2(x_i(k)), & E[\zeta_{j,i^1}^{rand}]E^{Me}[\tilde{Q}_{i^1}] - E[\zeta_{j,i^2}^{rand}]E^{Me}[\tilde{Q}_{i^2}] \leqslant 0. \end{cases}
\end{aligned}
$$

We have proved that $fov_i^{k+1}(x_i(k+1)|\mathbf{y_i}(k))$ is a linear function and is differentiable of its argument $x_i(k)$. On the other hand, it is obvious that $g_i^1(x_i(k))$ and $g_i^2(x_i(k))$ are both linear function and differentiable of their argument $x_i(k)$. Therefore, the linear combinations

$$fov_i^{k+1}(x_i(k+1)|\mathbf{y_i}(k)) + g_i^1(x_i(k))$$

and

$$fov_i^{k+1}(x_i(k+1)|\mathbf{y_i}(k)) + g_i^2(x_i(k))$$

are both differentiable of their argument $x_i(k)$. That means $fov_i^k(x_i(k)|\mathbf{y_i}(k-1))$ is differentiable of its argument $x_i(k)$ whenever $E[\zeta_{j,i^1}^{rand}]E^{Me}[\tilde{Q}_{i^1}] - E[\zeta_{j,i^2}^{rand}]E^{Me}[\tilde{Q}_{i^2}] > 0$ or $E[\zeta_{j,i^1}^{rand}]E^{Me}[\tilde{Q}_{i^1}] - E[\zeta_{j,i^2}^{rand}]E^{Me}[\tilde{Q}_{i^2}] \leqslant 0$. The proof is completed. □

**Definition 3.6.** *(Xu and Zeng 2012 [198]) $FOV_i^k(x_i(k))$ is the function of optimal value of the $(k+1)$th stage if and only if*

$$
FOV_i^k(x_i(k)) = \min \left\{ L_i^k(\cdot) \left| \begin{array}{l} 0 \leqslant u_{i^1}(0) \leqslant x_i(0), 0 \leqslant u_{i^1}(1) \leqslant x_i(1), \cdots, \\ 0 \leqslant u_{i^1}(k-1) \leqslant x_i(k-1), \\ f_i^0(x_i(0), u_{i^1}(0)) = x_i(1), f_i^1(x_i(1), u_{i^1}(1)) = x_i(2), \cdots, \\ f_i^{k-1}(x_i(k-1), u_{i^1}(k-1)) = x_i(k) \end{array} \right. \right\},
$$

$\forall i \in \Psi, k = 2, 3, \cdots, n.$

**Lemma 3.2.** *The function of optimal value of the $(k+1)$th stage $FOV_i^k(x_i(k))$ is differentiable of its argument $x_i(k)$, $\forall i \in \Psi$, $k = 1, 2, \cdots, n$.*

*Proof.* According to Def. 3.6 and the statement of $FOV_i^1(x_i(1))$, $\forall i \in \Psi$, $k = 1, 2, \cdots, n$, we get

$$FOV_i^k(x_i(k)) = \min\left\{ L_i^k(\cdot) \left| \begin{array}{l} 0 \leqslant u_{i^1}(0) \leqslant x_i(0), 0 \leqslant u_{i^1}(1) \leqslant x_i(1), \cdots, \\ 0 \leqslant u_{i^1}(k-1) \leqslant x_i(k-1), \\ f_i^0(x_i(0), u_{i^1}(0)) = x_i(1), f_i^1(x_i(1), u_{i^1}(1)) = x_i(2), \cdots, \\ f_i^{k-1}(x_i(k-1), u_{i^1}(k-1)) = x_i(k) \end{array} \right. \right\}.$$

Since

$$L_i^k = -\sum_{j=0}^{k-1}\{[E[\zeta_{j,i^1}^{rand}]E^{Me}[\tilde{Q}_{i^1}] - E[\zeta_{j,i^2}^{rand}]E^{Me}[\tilde{Q}_{i^2}]]u_{i^1}(j) + E[\zeta_{j,i^2}^{rand}]E^{Me}[\tilde{Q}_{i^2}]x_i(j)\},$$

if $E[\zeta_{j,i^1}^{rand}]E^{Me}[\tilde{Q}_{i^1}] - E[\zeta_{j,i^2}^{rand}]E^{Me}[\tilde{Q}_{i^2}] > 0$, the minimum value of $L_i^k(\cdot)$ can be arrived when $u_{i^1}(j) = x_i(j)$, where $j = 0, \cdots, k-1$, thus

$$FOV_i^k(x_i(k)) = -\sum_{j=0}^{k-1}\{E[\zeta_{j,i^1}^{rand}]E^{Me}[\tilde{Q}_{i^1}]x_i(j)\};$$

if $E[\zeta_{j,i^1}^{rand}]E^{Me}[\tilde{Q}_{i^1}] - E[\zeta_{j,i^2}^{rand}]E^{Me}[\tilde{Q}_{i^2}] < 0$, the minimum value of $L_i^k(\cdot)$ can be arrived when $u_{i^1}(j) = 0$, where $j = 0, \cdots, k-1$, thus

$$FOV_i^k(x_i(k)) = -\sum_{j=0}^{k-1}\{E[\zeta_{j,i^2}^{rand}]E^{Me}[\tilde{Q}_{i^2}]x_i(j)\};$$

if $E[\zeta_{j,i^1}^{rand}]E^{Me}[\tilde{Q}_{i^1}] - E[\zeta_{j,i^2}^{rand}]E^{Me}[\tilde{Q}_{i^2}] = 0$, then

$$FOV_i^k(x_i(k)) = -\sum_{j=0}^{k-1}\{E[\zeta_{j,i^2}^{rand}]E^{Me}[\tilde{Q}_{i^2}]x_i(j)\}.$$

To sum up, we get

$$\begin{aligned} &FOV_i^k(x_i(k)) \\ &= \begin{cases} -\sum_{j=0}^{k-1}\{E[\zeta_{j,i^1}^{rand}]E^{Me}[\tilde{Q}_{i^1}]x_i(j)\}, & E[\zeta_{j,i^1}^{rand}]E^{Me}[\tilde{Q}_{i^1}] - E[\zeta_{j,i^2}^{rand}]E^{Me}[\tilde{Q}_{i^2}] > 0; \\ -\sum_{j=0}^{k-1}\{E[\zeta_{j,i^2}^{rand}]E^{Me}[\tilde{Q}_{i^2}]x_i(j)\}, & E[\zeta_{j,i^1}^{rand}]E^{Me}[\tilde{Q}_{i^1}] - E[\zeta_{j,i^2}^{rand}]E^{Me}[\tilde{Q}_{i^2}] \leqslant 0. \end{cases} \end{aligned} \tag{3.33}$$

From the above expression, we can know that $FOV_i^k(x_i(k))$ is a linear function and is differentiable of its argument $x_i(k-1)$. So it is obvious that $FOV_i^{k+1}(x_i(k+1))$ is a linear function and is differentiable of its argument $x_i(k)$. According to Eq. (3.33), we get

$$\begin{aligned} &FOV_i^k(x_i(k)) \\ &= \begin{cases} FOV_i^{k+1}(x_i(k+1)) + E[\zeta_{k,i^1}^{rand}]E^{Me}[\tilde{Q}_{i^1}]x_i(k), E[\zeta_{j,i^1}^{rand}]E^{Me}[\tilde{Q}_{i^1}] - E[\zeta_{j,i^2}^{rand}]E^{Me}[\tilde{Q}_{i^2}] > 0; \\ FOV_i^{k+1}(x_i(k+1)) + E[\zeta_{k,i^2}^{rand}]E^{Me}[\tilde{Q}_{i^2}]x_i(k), E[\zeta_{j,i^1}^{rand}]E^{Me}[\tilde{Q}_{i^1}] - E[\zeta_{j,i^2}^{rand}]E^{Me}[\tilde{Q}_{i^2}] \leqslant 0. \end{cases} \end{aligned}$$

Since $E[\zeta_{k,i^1}^{rand}]E^{Me}[\tilde{Q}_{i^1}]x_i(k)$ and $E[\zeta_{k,i^2}^{rand}]E^{Me}[\tilde{Q}_{i^2}]x_i(k)$ are both linear function and differentiable of their argument $x_i(k)$. Therefore, the linear combinations

$$FOV_i^{k+1}(x_i(k+1)) + E[\zeta_{k,i^1}^{rand}]E^{Me}[\tilde{Q}_{i^1}]x_i(k)$$

and

$$FOV_i^{k+1}(x_i(k+1)) + E[\zeta_{k,i^2}^{rand}]E^{Me}[\tilde{Q}_{i^2}]x_i(k)$$

are both differentiable of their argument $x_i(k)$. That means $FOV_i^k(x_i(k))$ is differentiable of its argument $x_i(k)$ whenever $E[\zeta_{j,i^1}^{rand}]E^{Me}[\tilde{Q}_{i^1}] - E[\zeta_{j,i^2}^{rand}]E^{Me}[\tilde{Q}_{i^2}] > 0$ or $E[\zeta_{j,i^1}^{rand}]E^{Me}[\tilde{Q}_{i^1}] - E[\zeta_{j,i^2}^{rand}]E^{Me}[\tilde{Q}_{i^2}] \leqslant 0$. The proof is completed. □

Based on the above, the conditional lower bound surface and the lower bound surface are defined.

The conditional lower bound surface of the $2^{nd}$ stage is defined as follows:

Let $\Pi_i^1(\mathbf{y_i}(0))$ be the conditional lower bound surface of the $2^{nd}$ stage which is defined by

$$\Pi_i^1(\mathbf{y_i}(0)) = \{\mathbf{y_i}(1)|x_i^0(1) = fov_i^1(x_i(1)|\mathbf{y_i}(0)), x_i(1) \in \mathbf{X_i^1}(x_i(0))\}, \quad \forall i \in \Psi.$$

Let $\Pi_i^1$ be the lower bound surface of the $2^{nd}$ stage which is defined by

$$\Pi_i^1 = \{\mathbf{y_i}(1)|x_i^0(1) = FOV_i^1(x_i(1)), x_i(1) \in \mathbf{X_i^1}\}, \quad \forall i \in \Psi.$$

Since it is known $fov_i^1(x_i(1)|\mathbf{y_i}(0)) = FOV_i^1(x_i(1))$, it is obvious that $\Pi_i^1(\mathbf{y_i}(0)) = \Pi_i^2$ in this special case. However, the situations are different when $k = 2, 3, \cdots, n$, as seen in the following definitions.

**Definition 3.7.** *(Xu and Zeng 2012 [198]) $\Pi_i^k(\mathbf{y}_i(k-1))$ is the conditional lower bound surface of the $(k+1)$th stage if and only if*

$$\Pi_i^k(\mathbf{y}_i(k-1)) = \{\mathbf{y}_i(k)|x_i^0(k) = fov_i^k(x_i(k)|\mathbf{y}_i(k-1)), x_i(k) \in \boldsymbol{X}_i^k(x_i(k-1))\},$$

$\forall \mathbf{y}_i(k-1) \in \boldsymbol{Y}_i^{k-1}, \forall i \in \Psi, k = 2, 3, \cdots, n.$

**Definition 3.8.** *(Xu and Zeng 2012 [198]) $\Pi_i^k$ is the lower bound surface of the $(k+1)$th stage if and only if*

$$\Pi_i^k = \{\mathbf{y}_i(k)|x_i^0(k) = FOV_i^k(x_i(k)), x_i(k) \in \boldsymbol{X}_i^k\}, \quad \forall i \in \Psi, \quad k = 2, 3, \cdots, n.$$

According to Def. 3.5, 3.6 and the above statement for $fov_i^1(x_i(1)|\mathbf{y_i}(0))$ and $FOV_i^1(x_i(1))$, it is obvious that

$$FOV_i^k(x_i(k)) \leqslant fov_i^k(x_i(k)|\mathbf{y_i}(k-1)), \quad \forall \mathbf{y_i}(k-1) \in \mathbf{Y_i^{k-1}}, \quad \forall i \in \Psi, \quad k = 1, 2, \cdots, n. \tag{3.34}$$

Assume that $\{u_{i^1}^*(0), \cdots, u_{i^1}^*(n-1)\}$ is the optimal solution sequence of the number of Machine $i$ allocated to Job $i^1$ at the beginning of every stage; $\{x_i^*(0) = a_i, x_i^*(1), \cdots, x_i^*(n)\}$ is the optimal solution sequence of the number of Machine $i$ that are available at the beginning of every stage; $\{\mathbf{y_i^*}(0), \cdots, \mathbf{y_i^*}(n)\}$ is the optimal solution sequence of the corresponding expanded vectors, so then the following lemmas can be elaborated.

**Lemma 3.3.** *The point $\mathbf{y}_i^*(k)$ is on the lower bound surface of the $(k+1)$ stage $\Pi_i^k$, $\forall i \in \Psi$, $k = 1, 2, \cdots, n$.*

*Proof.* When $k = n, \forall i \in \Psi$, assume that $\mathbf{y_i^*}(n) \notin \Pi_i^n$, then there exists sequence $\{\bar{u}_{i1}(0), \cdots, \bar{u}_{i1}(n-1)\}$ such that the corresponding expanded vector $\bar{\mathbf{y}}_\mathbf{i}(n) = (\bar{L}_i^n, \bar{x}_i(n))$ satisfies

$$\bar{x}_i(n) = x_i^*(n),$$

$$\bar{L}_i^n = -\sum_{j=0}^{n-1}\{E[\zeta_{j,i1}^{rand}]E^{Me}[\tilde{Q}_{i1}]\bar{u}_{i1}(j) + E[\zeta_{j,i2}^{rand}]E^{Me}[\tilde{Q}_{i2}][x_i(j) - \bar{u}_{i1}(j)]\},$$

$$\bar{L}_i^n < FOV_i^n(\bar{x}_i(n)), \quad \bar{x}_i(n) = x_i^*(n).$$

Thus,

$$\bar{L}_i^n < FOV_i^n(x_i^*(n)) = -\max E[J_i^n] = -E[J_i^{n*}],$$

where $J_i^{n*}$ is the optimal total production or construction throughput produced by Machine $i$ in a whole period. That means the sequence $\{\bar{u}_{i1}(0), \cdots, \bar{u}_{i1}(n-1)\}$ is better than the optimal solution sequence $\{u_{i1}^*(0), \cdots, u_{i1}^*(n-1)\}$, but it is impossible.

When $1 \leqslant k \leqslant n-1$, assume that $\mathbf{y_i^*}(k) \notin \Pi_i^k$, then there exists sequence $\{\bar{u}_{i1}(0), \cdots, \bar{u}_{i1}(k-1)\}$ such that the corresponding expanded vector $\bar{\mathbf{y}}_\mathbf{i}(k) = (\bar{L}_i^k, \bar{x}_i(k))$ satisfies

$$\bar{x}_i(k) = x_i^*(k),$$

$$\bar{L}_i^k = -\sum_{j=0}^{k-1}\{E[\zeta_{j,i1}^{rand}]E^{Me}[\tilde{Q}_{i1}]\bar{u}_{i1}(j) + E[\zeta_{j,i2}^{rand}]E^{Me}[\tilde{Q}_{i2}][x_i(j) - \bar{u}_{i1}(j)]\},$$

$$\bar{L}_i^k < FOV_i^k(\bar{x}_i(k)) = FOV_i^k(x_i^*(k)) = -E[J_i^{k*}] = -\Sigma_{j=0}^{k-1}\{E[\zeta_{j,i1}^{rand}]E^{Me}[\tilde{Q}_{i1}]u_{i1}^*(j) + E[\zeta_{j,i2}^{rand}]E^{Me}[\tilde{Q}_{i2}][x_i^*(j) - u_{i1}^*(j)]\}. \tag{3.35}$$

Now consider the sequence $\{\bar{u}_{i1}(0), \cdots, \bar{u}_{i1}(k-1), u_{i1}^*(k), \cdots, u_{i1}^*(n-1)\}$, let

$$\hat{L}_i^n = \bar{L}_i^k + \sum_{j=k}^{n-1} -\{E[\zeta_{j,i1}^{rand}]E^{Me}[\tilde{Q}_{i1}]u_{i1}^*(j) + E[\zeta_{j,i2}^{rand}]E^{Me}[\tilde{Q}_{i2}][x_i^*(j) - u_{i1}^*(j)]\}. \tag{3.36}$$

From Eq. (3.35) and (3.36), we have

$$\hat{L}_i^n < \sum_{j=0}^{n-1} -\{E[\zeta_{j,i1}^{rand}]E^{Me}[\tilde{Q}_{i1}]u_{i1}^*(j) + E[\zeta_{j,i2}^{rand}]E^{Me}[\tilde{Q}_{i2}][x_i^*(j) - u_{i1}^*(j)]\} = -E[J_i^{n*}].$$

That means the sequence $\{\bar{u}_{i1}(0), \cdots, \bar{u}_{i1}(k-1), u_{i1}^*(k), \cdots, u_{i1}^*(n-1)\}$ is better than the optimal solution sequence $\{u_{i1}^*(0), \cdots, u_{i1}^*(n-1)\}$, but it is impossible. The proof is completed. □

From Lemma 3.3, it is not difficult to prove the following lemma.

**Lemma 3.4.** *The point $\mathbf{y}_i^*(k)$ is on the conditional lower bound surface of the $(k+1)$th stage $\Pi_i^k(\mathbf{y}_i(k-1))$, $\forall i \in \Psi$, $k = 1, \cdots, 4$.*

*Proof.* Let $\mathbf{y}_i^*(k) = (L_i^{k*}, x_i^*(k))$, where

$$L_i^{k*} = -\sum_{j=0}^{k-1}\{E[\zeta_{j,i1}^{rand}]E^{Me}[\tilde{Q}_{i1}]u_{i1}^*(j) + E[\zeta_{j,i2}^{rand}]E^{Me}[\tilde{Q}_{i2}][x_i^*(j) - u_{i1}^*(j)]\},$$

$\forall i \in \Psi, k = 1, 2, \cdots, n$. According to Def. 3.5 and the statement of $fov_i^1(x_i(1)|\mathbf{y}_i(0))$, we have

$$L_i^{k*} \geqslant fov_i^k(x_i^*(k)|\mathbf{y}_i^*(k-1)),$$

but according to Lemma 3.3, we get

$$L_i^{k*} = FOV_i^k(x_i^*(k)). \tag{3.37}$$

On the other hand, by Eq. (3.34), we get

$$fov_i^k(x_i^*(k)|\mathbf{y}_i^*(k-1)) \geqslant FOV_i^k(x_i^*(k)).$$

Hence,

$$fov_i^k(x_i^*(k)|\mathbf{y}_i^*(k-1)) = L_i^{k*}. \tag{3.38}$$

The above equation means the point $\mathbf{y}_i^*(k)$ is on the conditional lower bound surface of the $(k+1)$th stage $\Pi_i^k(\mathbf{y}_i(k-1))$, $\forall i \in \Psi$, $k = 1, 2, \cdots, n$. The proof is completed. □

### 3.3.3 Support Function of Surface

Now the support function of the surface is introduced in the following definition.

**Definition 3.9.** *(Xu and Zeng 2012 [198]) Let $S_i^k(x_i(k)|y_i(k-1))$ be a finite function which is continuous for all real values $x_i(k)$ and is defined on the state attainable set of the $(k+1)$th stage $\mathbf{X}_i^k(x_i(k-1))$ which is derived from $x_i(k-1)$, then $S_i^k(x_i(k)|y_i(k-1))$ is called the support function of the conditional lower bound surface of the $(k+1)$th stage $\Pi_i^k(y_i(k-1))$ on the point $\hat{x}_i(k)$, if and only if there exists a real number $\alpha_i^k$ such that:*

$$S_i^k(x_i(k)|y_i(k-1)) - \alpha_i^k \leqslant fov_i^k(x_i(k)|y_i(k-1)), \quad \forall x_i(k) \in \mathbf{X}_i^k(x_i(k-1)), \tag{3.39}$$

$$S_i^k(\hat{x}_i(k)|y_i(k-1)) - \alpha_i^k = fov_i^k(\hat{x}_i(k)|y_i(k-1)), \tag{3.40}$$

$\forall i \in \Psi$, $k = 1, 2, \cdots, n$.

**Lemma 3.5.** *Let $S_i^k(x_i(k)|y_i^*(k-1))$ be the support function of the conditional lower bound surface of the $(k+1)$th stage $\Pi_i^k(y_i^*(k-1))$ on the point $x_i^*(k)$, if $S_i^k(x_i(k)|y_i^*(k-1))$ is differentiable of its argument $x_i(k)$, and $x_i^*(k)$ is the interior point of $\mathbf{X}_i^k(x_i^*(k-1))$ then we have*

$$\frac{dS_i^k(x_i^*(k)|y_i^*(k-1))}{dx_i(k)} = \frac{dfov_i^k(x_i^*(k)|y_i^*(k-1))}{dx_i(k)} = \frac{dFOV_i^k(x_i^*(k))}{dx_i(k)},$$

$\forall i \in \Psi$, $k = 1, 2, \cdots, n$.

*Proof.* By Eq. (3.39) and (3.40), we get

$$S_i^k(x_i(k)|\mathbf{y_i^*}(k-1)) - fov_i^k(x_i(k)|\mathbf{y_i^*}(k-1)) \leqslant \alpha_i^k,$$
$$S_i^k(x_i^*(k)|\mathbf{y_i^*}(k-1)) - fov_i^k(x_i^*(k)|\mathbf{y_i^*}(k-1)) = \alpha_i^k,$$

i.e. $S_i^k(x_i(k)|\mathbf{y_i^*}(k-1)) - fov_i^k(x_i(k)|\mathbf{y_i^*}(k-1))$ arrives the maximum value $\alpha_i^k$ at the point $x_i^*(k)$, $\forall i \in \Psi, k = 1, 2, \cdots, n$. Since $S_i^k(x_i(k)|\mathbf{y_i^*}(k-1))$ is differentiable of its argument $x_i(k)$, and by Lemma 3.1, we can know that the conditional function of optimal value of the $(k+1)$th stage $fov_i^k(x_i(k)|\mathbf{y_i^*}(k-1))$ is differentiable of its argument $x_i(k)$, hence, we can get

$$\frac{dS_i^k(x_i^*(k)|\mathbf{y_i^*}(k-1))}{dx_i(k)} - \frac{dfov_i^k(x_i^*(k)|\mathbf{y_i^*}(k-1))}{dx_i(k)} = 0, \quad \forall i \in \Psi, \quad k = 1, 2, \cdots, n,$$

i.e.

$$\frac{dS_i^k(x_i^*(k)|\mathbf{y_i^*}(k-1))}{dx_i(k)} = \frac{dfov_i^k(x_i^*(k)|\mathbf{y_i^*}(k-1))}{dx_i(k)}, \quad \forall i \in \Psi, \quad k = 1, 2, \cdots, n.$$

From Eq. (3.37) and (3.38), we have

$$FOV_i^k(x_i^*(k)) - fov_i^k(x_i^*(k)|\mathbf{y_i^*}(k-1)) = 0,$$

from Eq. (3.34), we get

$$FOV_i^k(x_i(k)) - fov_i^k(x_i(k)|\mathbf{y_i^*}(k-1)) \leqslant 0,$$

i.e. $FOV_i^k(x_i(k)) - fov_i^k(x_i(k)|\mathbf{y_i^*}(k-1))$ arrives the maximum value 0 at the point $x_i^*(k)$, $\forall i \in \Psi, k = 1, 2, \cdots, n$. By Lemma 3.2, we can know that the function of optimal value of the $(k+1)$th stage $FOV_i^k(x_i(k))$ is differentiable of their argument $x_i(k)$. Hence, we can get

$$\frac{dfov_i^k(x_i^*(k)|\mathbf{y_i^*}(k-1))}{dx_i(k)} - \frac{dFOV_i^k(x_i^*(k))}{dx_i(k)} = 0, \quad \forall i \in \Psi, \quad k = 1, 2, \cdots, n,$$

i.e.

$$\frac{dfov_i^k(x_i^*(k)|\mathbf{y_i^*}(k-1))}{dx_i(k)} = \frac{dFOV_i^k(x_i^*(k))}{dx_i(k)}, \quad \forall i \in \Psi, \quad k = 1, 2, \cdots, n.$$

The proof is completed. □

Then, the generalized Hamilton function is introduced in the following definition.

**Definition 3.10.** *(Xu and Zeng 2012 [198]) Let $S_i^k(x_i(k)|y_i(k-1))$ be the support function of the conditional lower bound surface of the $(k+1)$th stage $\Pi_i^k(y_i(k-1))$ on the point $x_i(k)$, then $H_i^k(x_i(k), u_{i1}(k), S_i^{k+1})$ is the generalized Hamilton function if and only if*

$$
\begin{aligned}
H_i^k(x_i(k), u_{i1}(k), S_i^{k+1}) = {} & \{E[\zeta_{k,i1}^{rand}]E^{Me}[\tilde{Q}_{i1}]u_{i1}(k) + E[\zeta_{k,i2}^{rand}]E^{Me}[\tilde{Q}_{i2}][x_i(k) - u_{i1}(k)]\} \\
& + S_i^{k+1}(x_i(k+1)|y_i(k)),
\end{aligned}
\tag{3.41}
$$

$\forall i \in \Psi, k = 0, 1, \cdots, n-1.$

Based on Lemma 3.4, the above statements and definitions, the most important lemma can be proved as outlined below.

**Lemma 3.6.** *It is said that $S_i^{k+1}(x_i(k+1)|\mathbf{y}_i^*(k))$ is the support function of the conditional lower bound surface of the $(k+1)$th stage $\Pi_i^{k+1}(\mathbf{y}_i^*(k))$ on the point $x_i^*(k+1)$, if and only if $H_i^k(x_i^*(k), u_{i1}(k), S_i^{k+1})$ arrives at the maximum value on point $u_{i1}^*(k)$, where $0 \leqslant u_{i1}^*(k) \leqslant x_i^*(k)$, $\forall i \in \Psi, k = 0, 1, \cdots, n-1$.*

*Proof.* Let $S_i^{k+1}(x_i(k+1)|\mathbf{y}_i^*(k))$ be the support function of the conditional lower bound surface of the $(k+1)$th stage $\Pi_i^{k+1}(\mathbf{y}_i^*(k))$ on the point $x_i^*(k+1)$, then

$$
-fov_i^{k+1}(x_i(k+1)|\mathbf{y}_i^*(k)) + S_i^{k+1}(x_i(k+1)|\mathbf{y}_i^*(k)) \leqslant \alpha_i^{k+1}, \quad \forall x_i(k+1) \in \mathbf{X}_i^{k+1}(x_i^*(k)).
\tag{3.42}
$$

Let $0 \leqslant \bar{u}_{i1}(k) \leqslant x_i(k)$, and $\bar{x}_i(k+1) = f_i^k(x_i^*(k), \bar{u}_{i1}(k)) = (1-p_{i1}^k)\bar{u}_{i1}(k) + (1-p_{i2}^k)[x_i^*(k) - \bar{u}_{i1}(k)]$, by Def. 3.5 and the statement of $fov_i^1(x_i(1)|\mathbf{y}_i(0))$, we get

$$
\begin{aligned}
& -x_i^{0*}(k) + \{E[\zeta_{k,i1}^{rand}]E^{Me}[\tilde{Q}_{i1}]\bar{u}_{i1}(k) + E[\zeta_{k,i2}^{rand}]E^{Me}[\tilde{Q}_{i2}][x_i^*(k) - \bar{u}_{i1}(k)]\} \\
& \leqslant -fov_i^{k+1}(\bar{x}_i(k+1)|\mathbf{y}_i^*(k)).
\end{aligned}
$$

With Eq. (3.42), and by adding $S_i^{k+1}(\bar{x}_i(k+1)|\mathbf{y}_i^*(k))$ to both sides of the above inequality, we can get

$$
\begin{aligned}
& -x_i^{0*}(k) + \{E[\zeta_{k,i1}^{rand}]E^{Me}[\tilde{Q}_{i1}]\bar{u}_{i1}(k) + E[\zeta_{k,i2}^{rand}]E^{Me}[\tilde{Q}_{i2}][x_i^*(k) - \bar{u}_{i1}(k)]\} + \\
& S_i^{k+1}(\bar{x}_i(k+1)|\mathbf{y}_i^*(k)) \leqslant \alpha_i^{k+1},
\end{aligned}
$$

i.e.

$$
H_i^k(x_i^*(k), u_{i1}(k), S_i^{k+1}(x_i(k+1)|\mathbf{y}_i^*(k))) \leqslant \alpha_i^{k+1} + x_i^{0*}(k), \quad 0 \leqslant u_{i1}(k) \leqslant x_i^*(k).
$$

On the other hand, by Eq. (3.38) in Lemma 3.4, we have

$$
\begin{aligned}
& fov_i^{k+1}(x_i^*(k+1)|\mathbf{y}_i^*(k)) \\
& = x_i^{0*}(k) - \{E[\zeta_{k,i1}^{rand}]E^{Me}[\tilde{Q}_{i1}]u_{i1}^*(k) + E[\zeta_{k,i2}^{rand}]E^{Me}[\tilde{Q}_{i2}][x_i^*(k) - u_{i1}^*(k)]\}.
\end{aligned}
\tag{3.43}
$$

From Def. 3.9, we have

$$-fov_i^{k+1}(x_i^*(k+1)|\mathbf{y_i^*}(k)) + S_i^{k+1}(x_i^*(k+1)|\mathbf{y_i^*}(k)) \leqslant \alpha_i^{k+1},$$

i.e.

$$H_i^k(x_i^*(k), u_{i^1}^*(k), S_i^{k+1}(x_i^*(k+1)|\mathbf{y_i^*}(k))) = \alpha_i^{k+1} + x_i^{0*}(k).$$

This means $H_i^k(x_i^*(k), u_{i^1}(k), S_i^{k+1}(x_i(k+1)|\mathbf{y_i^*}(k)))$ arrives the maximum value $\alpha_i^{k+1} + x_i^{0*}(k)$ on the point $u_{i^1}^*(k)$, $\forall i \in \Psi, k = 0, 1, \cdots, n-1$.

Contrarily, if $H_i^k(x_i^*(k), u_{i^1}(k), S_i^{k+1}(x_i(k+1)|\mathbf{y_i^*}(k)))$ arrives the maximum value on the point $u_{i^1}^*(k)$, then there exists $\beta_i^k$ such that

$$\{E[\zeta_{k,i^1}^{rand}]E^{Me}[\tilde{Q}_{i^1}]u_{i^1}(k) + E[\zeta_{k,i^2}^{rand}]E^{Me}[\tilde{Q}_{i^2}][x_i^*(k) - u_{i^1}(k)]\} + S_i^{k+1}(x_i(k+1)|\mathbf{y_i^*}(k)) \leqslant \beta_i^k, \tag{3.44}$$

where $0 \leqslant u_{i^1}(k) \leqslant x_i^*(k)$,

$$\{E[\zeta_{k,i^1}^{rand}]E^{Me}[\tilde{Q}_{i^1}]u_{i^1}^*(k) + E[\zeta_{k,i^2}^{rand}]E^{Me}[\tilde{Q}_{i^2}][x_i^*(k) - u_{i^1}^*(k)]\} + S_i^{k+1}(x_i^*(k+1)|\mathbf{y_i^*}(k)) = \beta_i^k. \tag{3.45}$$

Let $\alpha_i^{k+1} = \beta_i^k - x_i^{0*}(k)$, from Eq. (3.43) and (3.45) we have

$$-fov_i^{k+1}(x_i^*(k+1)|\mathbf{y_i^*}(k)) + S_i^{k+1}(x_i^*(k+1)|\mathbf{y_i^*}(k)) = \beta_i^k - x_i^{0*}(k) = \alpha_i^{k+1}.$$

On the other hand, by Eq. (3.44), we get

$$-x_i^{0*}(k) + \{E[\zeta_{k,i^1}^{rand}]E^{Me}[\tilde{Q}_{i^1}]u_{i^1}(k) + E[\zeta_{k,i^2}^{rand}]E^{Me}[\tilde{Q}_{i^2}][x_i^*(k) - u_{i^1}(k)]\} + S_i^{k+1}(x_i(k+1)|\mathbf{y_i^*}(k)) \leqslant \alpha_i^{k+1},$$

where $0 \leqslant u_{i^1}(k) \leqslant x_i^*(k)$.

From Def. 3.5 and the statement of $fov_i^1(x_i(1)|\mathbf{y_i}(0))$ and Eq. (3.38) we get

$$-fov_i^{k+1}(x_i(k+1)|\mathbf{y_i^*}(k)) + S_i^{k+1}(x_i(k+1)|\mathbf{y_i^*}(k)) \leqslant \alpha_i^{k+1}, \quad \forall x_i(k+1) \in \mathbf{X_i^{k+1}}(x_i^*(k)).$$

According to Def 3.9, we can know that $S_i^{k+1}(x_i(k+1)|\mathbf{y_i^*}(k))$ is the support function of the conditional lower bound surface of the $(k+1)$th stage $\Pi_i^{k+1}(\mathbf{y_i^*}(k))$ on the point $x_i^*(k+1)$, $\forall i \in \Psi, k = 0, 1, \cdots, n-1$. The proof is completed. □

**Theorem 3.1.** *Let $\{u_{i^1}^*(0), \cdots, u_{i^1}^*(n-1)\}$ be the optimal solution sequence of the number of Machine i allocated to Job $i^1$ at the beginning of every stage, and $\{x_i^*(0) = a_i, x_i^*(1), \cdots, x_i^*(n)\}$ be the optimal solution sequence of the number of Machine i that are available at the beginning of every stage. Assumed that $x_i^*(k)$ is an interior point of $\mathbf{X_i^k}(x_i^*(k-1))$, where $k = 0, 1, \cdots, n$. Then there is a function $S_i^k(x_i(k)|\mathbf{y}_i^*(k-1))$ which is the support function of the conditional lower bound surface of the $(k+1)$th stage $\Pi_i^k(\mathbf{y}_i^*(k-1))$ on the point $x_i^*(k)$, $\forall i \in \Psi, k = 1, 2, \cdots, n$, such that,*

*(i)*

$$\frac{dS_i^k(x_i^*(k)|\mathbf{y}_i^*(k-1))}{dx_i(k)} = \frac{\partial H_i^k(x_i^*(k),u_{i1}^*(k),S_i^{k+1}(x_i^*(k+1)|\mathbf{y}_i^*(k)))}{\partial x_i(k)}, \quad \forall i \in \Psi, \tag{3.46}$$

*where* $k = 1,2,\cdots,n-1$*;*
*(ii)*

$$H_i^k(x_i^*(k),u_{i1}^*(k),S_i^{k+1*}) = \max[H_i^k(x_i^*(k),u_{i1}(k),S_i^{k+1})|0 \leqslant u_{i1}(k) \leqslant x_i^*(k)], \quad \forall i \in \Psi, \tag{3.47}$$

*where* $S_i^{k+1*} = S_i^{k+1}(x_i^*(k+1)|\mathbf{y}_i^*(k))$*,* $S_i^{k+1} = S_i^{k+1}(x_i(k+1)|\mathbf{y}_i^*(k))$*,* $k = 0,1,\cdots,n-1$*;*
*(iii)*

$$\frac{dS_i^n(x_i^*(n)|\mathbf{y}_i^*(n-1))}{dx_i(n)} = 0, \quad \forall i \in \Psi. \tag{3.48}$$

*Proof.* By Eq. (3.34), (3.37) and (3.38), we can get

$$FOV_i^k(x_i(k)) \leqslant fov_i^k(x_i(k)|\mathbf{y}_i^*(k-1)), \quad \forall i \in \Psi, \quad k = 1,2,\cdots,n,$$

$$FOV_i^k(x_i^*(k)) = fov_i^k(x_i^*(k)|\mathbf{y}_i^*(k-1)), \quad \forall i \in \Psi, \quad k = 1,2,\cdots,n.$$

By Def. 3.9, we can know that $FOV_i^k(x_i(k))$ is the support function of the conditional lower bound surface of the $(k+1)$th stage $\Pi_i^k(\mathbf{y}_i^*(k-1))$ on the point $x_i^*(k)$ (here $\alpha_i^k = 0$). Now we only need to prove the function $FOV_i^k(x_i^*(k))$ satisfies (i), (ii), (iii), so it is just the very support function $S_i^k(x_i(k)|\mathbf{y}_i^*(k-1))$ which we want to find.

Let $\hat{\mathbf{X}}_\mathbf{i}^\mathbf{k}$ be a subset of $\mathbf{X}_\mathbf{i}^\mathbf{k}$, we define $\hat{\mathbf{X}}_\mathbf{i}^\mathbf{k}$ as follows,

$$\hat{\mathbf{X}}_\mathbf{i}^\mathbf{k} = \{x_i(k)|x_i(k) = f_i^{k-1}(x_i(k-1),u_{i1}^*(k-1)),x_i(k-1) \in \mathbf{X}_\mathbf{i}^\mathbf{k-1}\}.$$

Let $\hat{F}_i^k(x_i(k))$ be a function defined on the subset $\hat{\mathbf{X}}_\mathbf{i}^\mathbf{k}$ as follows,

$$\begin{aligned} &\hat{F}_i^k(x_i(k)) \\ &= \min\{FOV_i^{k-1}(x_i(k-1)) - \{E[\zeta_{k-1,i^1}^{rand}]E^{Me}[\tilde{Q}_{i1}]u_{i1}^*(k-1) + E[\zeta_{k-1,i^2}^{rand}]E^{Me}[\tilde{Q}_{i2}] \\ &\quad [x_i(k-1) - u_{i1}^*(k-1)]\}|f_i^{k-1}(x_i(k-1),u_{i1}^*(k-1)) = x_i(k),x_i(k-1) \in \mathbf{X}_\mathbf{i}^\mathbf{k-1}\}. \end{aligned}$$

From the above definition, we can get

$$FOV_i^k(x_i(k)) \leqslant \hat{F}_i^k(x_i(k)), \quad \forall x_i(k) \in \hat{\mathbf{X}}_\mathbf{i}^\mathbf{k}, \tag{3.49}$$

$$FOV_i^k(x_i^*(k)) = \hat{F}_i^k(x_i^*(k)).$$

By Eq. (3.37), we have

$$FOV_i^{k+1}(x_i^*(k+1)) \\ = FOV_i^k(x_i^*(k)) - \{E[\zeta_{k,i^1}^{rand}]E^{Me}[\tilde{Q}_{i1}]u_{i^1}^*(k) + E[\zeta_{k,i^2}^{rand}]E^{Me}[\tilde{Q}_{i2}][x_i^*(k) - u_{i^1}^*(k)]\}. \tag{3.50}$$

According to Eq. (3.49) and the definition of $\hat{F}_i^k(x_i(k))$, we get

$$FOV_i^{k+1}(x_i(k+1)) \leqslant \hat{F}_i^{k+1}(x_i(k+1)) \leqslant FOV_i^k(x_i(k)) - \{E[\zeta_{k,i^1}^{rand}]E^{Me}[\tilde{Q}_{i1}]u_{i^1}^*(k) + \\ E[\zeta_{k,i^2}^{rand}]E^{Me}[\tilde{Q}_{i2}][x_i(k) - u_{i^1}^*(k)]\}, \tag{3.51}$$

$\forall x_i(k) \in \mathbf{X_i^k}, x_i(k+1) \in \hat{\mathbf{X}}_\mathbf{i}^{\mathbf{k+1}}$.

Since, $x_i(k+1) = f_i^k(x_i(k), u_{i1}(k))$, we transform Eq. (3.50) and (3.51) as follows

$$\{E[\zeta_{k,i^1}^{rand}]E^{Me}[\tilde{Q}_{i^1}]u_{i^1}^*(k) + E[\zeta_{k,i^2}^{rand}]E^{Me}[\tilde{Q}_{i^2}][x_i^*(k) - u_{i^1}^*(k)]\} + FOV_i^{k+1}(f_i^k(x_i^*(k), u_{i^1}^*(k))) \\ -FOV_i^k(x_i^*(k)) = 0, \tag{3.52}$$

$$\{E[\zeta_{k,i^1}^{rand}]E^{Me}[\tilde{Q}_{i^1}]u_{i^1}^*(k) + E[\zeta_{k,i^2}^{rand}]E^{Me}[\tilde{Q}_{i^2}][x_i(k) - u_{i^1}^*(k)]\} + FOV_i^{k+1}(f_i^k(x_i(k), u_{i^1}^*(k))) - \\ FOV_i^k(x_i(k)) \leqslant 0, \tag{3.53}$$

$\forall x_i(k) \in \mathbf{X_i^k}$.

Let

$$\Upsilon_i(x_i(k)) = \{E[\zeta_{k,i^1}^{rand}]E^{Me}[\tilde{Q}_{i1}]u_{i^1}^*(k) + E[\zeta_{k,i^2}^{rand}]E^{Me}[\tilde{Q}_{i2}][x_i(k) - u_{i^1}^*(k)]\} + \\ FOV_i^{k+1}(f_i^k(x_i(k), u_{i^1}^*(k))) - FOV_i^k(x_i(k)).$$

From Eq. (3.52) and (3.53), we can know that $\Upsilon_i(x_i(k))$ arrives the maximum value 0 on the point $x_i^*(k)$. It is obvious that

$$\{E[\zeta_{k,i^1}^{rand}]E^{Me}[\tilde{Q}_{i1}]u_{i^1}^*(k) + E[\zeta_{k,i^2}^{rand}]E^{Me}[\tilde{Q}_{i2}][x_i(k) - u_{i^1}^*(k)]\}$$

and

$$f_i^k(x_i(k), u_{i^1}^*(k)) = (1 - p_{i^1}^k)u_{i^1}^*(k) + (1 - p_{i^2}^k)[x_i(k) - u_{i^1}^*(k)]$$

are both differentiable of their argument $x_i(k)$. From Lemma 3.2, we can know that $FOV_i^k(x_i(k))$ is differentiable of its argument $x_i(k)$, so $\Upsilon_i(x_i(k))$ is differentiable of its argument $x_i(k)$ as a result of the fact that it is a linear combination of $\{E[\zeta_{k,i^1}^{rand}]E^{Me}[\tilde{Q}_{i1}]u_{i^1}^*(k) + E[\zeta_{k,i^2}^{rand}]E^{Me}[\tilde{Q}_{i2}]\,[x_i(k) - u_{i^1}^*(k)]\}$, $FOV_i^{k+1}(x_i(k+1))$ and $FOV_i^k(x_i(k))$. Hence, we can get

$$\frac{d\Upsilon_i(x_i^*(k))}{dx_i(k)} = 0, \quad \forall i \in \Psi. \tag{3.54}$$

By Def. 3.10, we have

$$\Upsilon_i(x_i(k)) = H_i^k(x_i(k), u_{i^1}^*(k), FOV_i^{k+1}(f_i^k(x_i(k), u_{i^1}^*(k)))) - FOV_i^k(x_i(k)).$$

By Eq. (3.54), we get

$$\frac{dFOV_i^k(x_i^*(k))}{dx_i(k)} = \frac{\partial H_i^k(x_i^*(k), u_{i1}^*(k), FOV_i^{k+1}(x_i^*(k+1)))}{\partial x_i(k)}, \tag{3.55}$$

$\forall i \in \Psi$, $k = 1,2,\cdots,n-1$. This means $FOV_i^k(x_i(k))$ satisfies (i), $\forall i \in \Psi, k = 1,2,\cdots,n-1$.

Since we have proved previously that $FOV_i^k(x_i(k))$ is the support function of the conditional lower bound surface of the $(k+1)$th stage $\Pi_i^k(\mathbf{y_i^*}(k-1))$ on the point $x_i^*(k)$, according to Lemma 3.6, we can know that $FOV_i^{k+1}(x_i(k+1))$ satisfies (ii), $\forall i \in \Psi, k = 0,1,\cdots,n-1$.

From Def. 3.6, Lemma 3.3 and Eq. (3.37), we have

$$FOV_i^n(x_i(n)) \geqslant \min[FOV_i^n(x_i(n)) | x_i(n) \in \mathbf{X_i^n}] = L_i^{k*} = FOV_i^n(x_i^*(n)),$$

i.e. $FOV_i^n(x_i(n))$ arrives the minimum value at the point $x_i^*(n)$, so we can get

$$\frac{dFOV_i^n(x_i^*(n))}{dx_i(n)} = 0, \quad \forall i \in \Psi.$$

Hence, $FOV_i^n(x_i(n))$ satisfies (iii). The proof is completed. □

### 3.3.4 Existence of Linear Support Function

However, the Hamilton function in the Theorem 3.1 may not be linear. To solve the expected value model discussed previously, a linear form of the Hamilton function needs to be defined as follows:

$$\begin{aligned} H_i^k(x_i(k), u_{i1}(k), \lambda_i(k+1)) = {} & \{E[\zeta_{k,i^1}^{rand}]E^{Me}[\tilde{Q}_{i^1}]u_{i^1}(k) + E[\zeta_{k,i^2}^{rand}]E^{Me}[\tilde{Q}_{i^2}][x_i(k) - \\ & u_{i^1}(k)]\} + \lambda_i(k+1)[(p_{i^2}^k - p_{i^1}^k)u_{i^1}(k) + (1-p_{i^2}^k)x_I(k)], \end{aligned} \tag{3.56}$$

$\forall i \in \Psi, k = 0,1,\cdots,n-1$.
i.e.

$$\begin{aligned} H_i^k(x_i(k), u_{i1}(k), \lambda_i(k+1)) = {} & \{E[\zeta_{k,i^1}^{rand}]E^{Me}[\tilde{Q}_{i1}]u_{i1}(k) + E[\zeta_{k,i^2}^{rand}]E^{Me}[\tilde{Q}_{i^2}][x_i(k) \\ & -u_{i1}(k)]\} + \lambda_i(k+1) f_i^k(x_i(k), u_{i1}(k)), \end{aligned}$$

where

$$f_i^k(x_i(k), u_{i1}(k)) = x_i(k+1) = (p_{i2}^k - p_{i1}^k)u_{i1}(k) + (1-p_{i2}^k)x_i(k),$$

$\forall i \in \Psi, k = 0,1,\cdots,n-1$.

Compared with Def. 3.10, this means the support function $S_i^{k+1}(x_i(k+1)|\mathbf{y_i}(k))$ is of linear form as follows:

$$S_i^{k+1}(x_i(k+1)) = \lambda_i(k+1)x_i(k+1), \quad \forall i \in \Psi, \quad k = 0,1,\cdots,n-1,$$

i.e.

$$S_i^{k+1}(f_i^k(x_i(k),u_{i1}(k))) = \lambda_i(k+1)f_i^k(x_i(k),u_{i1}(k)), \quad \forall i \in \Psi, \quad k=0,1,\cdots,n-1.$$

Now, the problem is whether there is a linear support function $S_i^{k+1}(x_i(k+1)) = \lambda_i(k+1)x_i(k+1)$ or not. This indicates that the existence of the linear support function needs to be proved. Thus, the following lemmas are needed.

**Lemma 3.7.** *The state attainable set of the $(k+1)$th stage $\mathbf{X_i^k}(x_i(k-1))$ which is derived from $x_i(k-1)$ is a convex set, $\forall i \in \Psi, k=1,2,\cdots,n$.*

*Proof.* Let $\delta \in [0,1]$, $\forall x_i^1(k), x_i^2(k) \in \mathbf{X_i^k}(x_i(k-1))$, we want to prove that $\delta x_i^1(k) + (1-\delta)x_i^2(k) \in \mathbf{X_i^k}(x_i(k-1))$, $\forall i \in \Psi, k=1,2,\cdots,n$. In fact, by Def. 3.3 and the statement of $\mathbf{X_i^2}(x_i(1))$, we have

$$\begin{aligned}
&\delta x_i^1(k) + (1-\delta)x_i^2(k)\\
&= \delta f_i^{k-1}(x_i(k-1),u_{i1}^1(k-1)) + (1-\delta)f_i^{k-1}(x_i(k-1),u_{i1}^2(k-1))\\
&= \delta[(p_{i2}^k - p_{i1}^k)u_{i1}^1(k-1) + (1-p_{i2}^k)x_i(k-1)] + (1-\delta)[(p_{i2}^k - p_{i1}^k)u_{i1}^2(k-1)+\\
&\quad (1-p_{i2}^k)x_i(k-1)]\\
&= [(p_{i2}^k - p_{i1}^k)[\delta u_{i1}^1(k-1) + (1-\delta)u_{i1}^2(k-1)] + (1-p_{i2}^k)x_i(k-1)]\\
&= f_i^{k-1}(x_i(k-1), \delta u_{i1}^1(k-1) + (1-\delta)u_{i1}^2(k-1)),
\end{aligned} \tag{3.57}$$

and

$$0 \leqslant \delta u_{i1}^1(k-1) \leqslant \delta x_i(k-1), \quad 0 \leqslant (1-\delta)u_{i1}^2(k-1) \leqslant (1-\delta)x_i(k-1).$$

Hence,

$$0 \leqslant \delta u_{i1}^1(k-1) + (1-\delta)u_{i1}^2(k-1) \leqslant x_i(k-1). \tag{3.58}$$

By Def. 3.3 and the statement of $\mathbf{X_i^2}(x_i(1))$, Eq. (3.57) and (3.58), we can know that

$$\delta x_i^1(k) + (1-\delta)x_i^2(k) \in \mathbf{X_i^k}(x_i(k-1)), \quad \forall i \in \Psi, k=1,2,\cdots,n.$$

This means $\mathbf{X_i^k}(x_i(k-1))$ is a convex set, $\forall i \in \Psi, k=1,2,\cdots,n$. The proof is completed. □

**Definition 3.11.** *(Xu and Zeng 2012 [198]) Let $z \in \boldsymbol{R^2}$, if $\forall \delta \in [0,1], \forall \boldsymbol{y_i^1}(k), \boldsymbol{y_i^2}(k) \in \boldsymbol{Y_i^k}(\boldsymbol{y_i}(k-1))$, then the set $\boldsymbol{Y_i^k}(\boldsymbol{y_i}(k-1)) \subset \boldsymbol{R^2}$ is called convex in direction $z$, if and only if there exists $\gamma \geqslant 0$, such that $\delta \boldsymbol{y_i^1}(k) + (1-\delta)\boldsymbol{y_i^2}(k) - \gamma z \in \boldsymbol{Y_i^k}(\boldsymbol{y_i}(k-1))$.*

**Lemma 3.8.** *The expanded state attainable set of the $(k+1)$th stage $\boldsymbol{Y_i^k}(\boldsymbol{y_i}(k-1))$ which is derived from $\boldsymbol{y_i}(k-1)$ is convex in direction $\boldsymbol{e} = (1,0)$, $\forall i \in \Psi, k = 1,2,\cdots,n$.*

*Proof.* Let $\delta \in [0,1]$, $\forall \mathbf{y_i^1}(k), \mathbf{y_i^2}(k) \in \mathbf{Y_i^k}(\mathbf{y_i}(k-1))$, by Def. 3.11, we want to prove that there exists $\gamma \geqslant 0$, such that $\delta \mathbf{y_i^1}(k) + (1-\delta)\mathbf{y_i^2}(k) - \gamma \mathbf{e} \in \mathbf{Y_i^k}(\mathbf{y_i}(k-1))$, $\forall i \in \Psi, k = 1,2,\cdots,n$. In fact, by Def. 3.4 and the statement of $\mathbf{Y_i^2}(\mathbf{y_i}(1))$, we have

$$\begin{aligned}
&\delta \mathbf{y_i^1}(k) + (1-\delta)\mathbf{y_i^2}(k) - \gamma \mathbf{e} \\
&= \delta \mathbf{F_i^{k\text{-}1}}(\mathbf{y_i}(k-1), u_{i1}^1(k-1)) + (1-\delta)\mathbf{F_i^{k\text{-}1}}(\mathbf{y_i}(k-1), u_{i1}^2(k-1)) - (\gamma, 0) \\
&= \delta(L_i^k(u_{i1}^1(k-1)), f_i^{k-1}(x_i(k-1), u_{i1}^1(k-1))) + (1-\delta) \\
&\quad (L_i^k(u_{i1}^2(k-1)), f_i^{k-1}(x_i(k-1), u_{i1}^2(k-1))) - (\gamma, 0) \\
&= (L_i^k(\delta u_{i1}^1(k-1) + (1-\delta)u_{i1}^2(k-1)) - \gamma, f_i^{k-1}(x_i(k-1), \delta u_{i1}^1(k-1) + (1-\delta) \\
&\quad u_{i1}^2(k-1))).
\end{aligned} \tag{3.59}$$

Let $\gamma = 0$, then we have

$$\begin{aligned}
&(L_i^k(\delta u_{i1}^1(k-1) + (1-\delta)u_{i1}^2(k-1)) - \gamma, f_i^{k-1}(x_i(k-1), \delta u_{i1}^1(k-1) + (1-\delta)u_{i1}^2(k-1))) \\
&= (L_i^k(\delta u_{i1}^1(k-1) + (1-\delta)u_{i1}^2(k-1)), f_i^{k-1}(x_i(k-1), \delta u_{i1}^1(k-1) + (1-\delta)u_{i1}^2(k-1))) \\
&= \mathbf{F_i^{k\text{-}1}}(\mathbf{y_i}(k-1), \delta u_{i1}^1(k-1) + (1-\delta)u_{i1}^2(k-1)).
\end{aligned}$$

From Def. 3.4 and the statement of $\mathbf{Y_i^2}(\mathbf{y_i}(1))$ and Eq. (3.58), we get

$$\delta \mathbf{y_i^1}(k) + (1-\delta)\mathbf{y_i^2}(k) - \gamma \mathbf{e} = \mathbf{F_i^{k\text{-}1}}(\mathbf{y_i}(k-1), \delta u_{i1}^1(k-1) + (1-\delta)u_{i1}^2(k-1)) \in \mathbf{Y_i^k}(\mathbf{y_i}(k-1)).$$

This means $\mathbf{Y_i^k}(\mathbf{y_i}(k-1))$ $\mathbf{y_i}(k-1)$ is convex in direction $\mathbf{e} = (1,0)$, $\forall i \in \Psi, k = 1,2,\cdots,n$. The proof is completed. □

**Lemma 3.9.** *If the expanded state attainable set of the $(k+1)$th stage $\mathbf{Y_i^k}(\mathbf{y_i}(k-1))$ which is derived from $\mathbf{y_i}(k-1)$ is convex in direction $\mathbf{e} = (1,0)$, then the conditional function of the optimal value of the $(k+1)$th stage $fov_i^k(x_i(k)|\mathbf{y_i}(k-1))$ is a convex function, $\forall i \in \Psi, k = 1,2,\cdots,n$.*

*Proof.* Assumed that $\mathbf{Y_i^k}(\mathbf{y_i}(k-1))$ is convex in direction $\mathbf{e} = (1,0)$. $\forall i \in \Psi, k = 1,2,\cdots,n$, let $x_i^1(k), x_i^2(k) \in \mathbf{X_i^k}(x_i(k-1))$, chose $x_i^{0,1}(k), x_i^{0,2}(k)$, such that $\mathbf{y_i^1}(k) = (x_i^{0,1}(k), x_i^1(k)), \mathbf{y_i^2}(k) = (x_i^{0,2}(k), x_i^2(k))$, $\mathbf{y_i^1}(k)$, $\mathbf{y_i^2}(k) \in \mathbf{Y_i^k}(\mathbf{y_i}(k-1))$. From Def. 3.11, $\forall \delta \in [0,1]$, there exists $\gamma \geqslant 0$, such that

$$\delta \mathbf{y_i^1}(k) + (1-\delta)\mathbf{y_i^2}(k) - \gamma \mathbf{e} \in \mathbf{Y_i^k}(\mathbf{y_i}(k-1)),$$

i.e.

$$\delta x_i^1(k) + (1-\delta)x_i^2(k) \in \mathbf{X_i^k}(x_i(k-1)).$$

Hence, $\mathbf{X_i^k}(x_i(k-1))$ is a convex set. Now let $\overline{x_i^1(k)}, \overline{x_i^2(k)} \in \mathbf{X_i^k}(x_i(k-1))$, let

$$\overline{x_i^{0,1}(k)} = fov_i^k(\overline{x_i^1(k)}|\mathbf{y_i}(k-1)), \quad \overline{x_i^{0,2}(k)} = fov_i^k(\overline{x_i^2(k)}|\mathbf{y_i}(k-1)),$$

$$\overline{\mathbf{y_i^1}(k)} = (\overline{x_i^{0,1}(k)}, \overline{x_i^1(k)}), \quad \overline{\mathbf{y_i^2}(k)} = (\overline{x_i^{0,2}(k)}, \overline{x_i^2(k)}),$$

then we have $\overline{\mathbf{y_i^1}(k)}, \overline{\mathbf{y_i^2}(k)} \in \mathbf{Y_i^k}(\mathbf{y_i}(k-1))$. Since $\mathbf{Y_i^k}(\mathbf{y_i}(k-1))$ is convex in direction $\mathbf{e} = (1,0)$, $\forall \delta \in [0,1]$, there exists $\bar{\gamma} \geqslant 0$, such that

$$\delta \overline{\mathbf{y_i^1}(k)} + (1-\delta)\overline{\mathbf{y_i^2}(k)} - \bar{\gamma}\mathbf{e} \in \mathbf{Y_i^k}(\mathbf{y_i}(k-1)),$$

i.e.

$$\begin{aligned}&\delta fov_i^k(\overline{x_i^1(k)}|\mathbf{y_i}(k-1)) + (1-\delta) fov_i^k(\overline{x_i^2(k)}|\mathbf{y_i}(k-1)) - \bar{\gamma}\\ &\geqslant fov_i^k(\delta \overline{x_i^1(k) + (1-\delta)x_i^2(k)}|\mathbf{y_i}(k-1)).\end{aligned}$$

Since $\bar{\gamma} \geqslant 0$, we have

$$\begin{aligned}&\delta fov_i^k(\overline{x_i^1(k)}|\mathbf{y_i}(k-1)) + (1-\delta) fov_i^k(\overline{x_i^2(k)}|\mathbf{y_i}(k-1))\\ &\geqslant fov_i^k(\delta \overline{x_i^1(k) + (1-\delta)x_i^2(k)}|\mathbf{y_i}(k-1)).\end{aligned}$$

This means $fov_i^k(x_i(k)|\mathbf{y_i}(k-1))$ is a convex function, $\forall i \in \Psi, k = 1,2,\cdots,n$. The proof is completed. □

Based on Lemma 3.7, the following lemma can be proved.

**Lemma 3.10.** *If the conditional function of the optimal value of the* $(k+1)$*th stage* $fov_i^k(x_i(k)|\mathbf{y}_i(k-1))$ *is convex on* $\mathbf{X_i^k}(x_i(k-1))$*, and* $\hat{x}_i(k)$ *is any interior point of* $\mathbf{X_i^k}(x_i(k-1))$*, then there exists a linear function* $S_i^k(x_i(k)) = \lambda_i(k)x_i(k)$ *which is the support function of the conditional lower bound surface of the* $(k+1)$*th stage* $\Pi_i^k(\mathbf{y}_i(k-1))$ *on point* $\hat{x}_i(k)$*,* $\forall i \in \Psi, k = 1,2,\cdots,n$*.*

*Proof.* Let $\overline{\mathbf{Y_i^k}}$ be a subset of $\mathbf{Y_i^k}(\mathbf{y_i}(k-1))$, $\forall i \in \Psi, k = 1,2,\cdots,n$, we define it as below

$$\overline{\mathbf{Y_i^k}} = \left\{\mathbf{y_i}(k) \left| \begin{array}{l} fov_i^k(x_i(k)|\mathbf{y_i}(k-1)) \leqslant x_i^0(k) \leqslant \sup[fov_i^k(x_i(k)|\mathbf{y_i}(k-1))|\\ x_i(k) \in \mathbf{X_i^k}(x_i(k-1))], x_i(k) \in \mathbf{X_i^k}(x_i(k-1)) \end{array}\right.\right\}.$$

Firstly, we want to prove $\overline{\mathbf{Y_i^k}}$ is a convex set. Let $\mathbf{y_i^1}(k) = (x_i^{0,1}(k), x_i^1(k))$, $\mathbf{y_i^2}(k) = (x_i^{0,2}(k), x_i^2(k))$, $\mathbf{y_i^1}(k), \mathbf{y_i^2}(k) \in \overline{\mathbf{Y_i^k}}$, we define

$$\mathbf{y_i^k}(\delta) = \delta \mathbf{y_i^1}(k) + (1-\delta)\mathbf{y_i^2}(k), \quad 0 \leqslant \delta \leqslant 1.$$

According to Lemma 3.7, $\mathbf{X_i^k}(x_i(k-1))$ is a convex set. Let $x_i^1(k), x_i^2(k) \in \mathbf{X_i^k}(x_i(k-1))$, then

$$\delta x_i^1(k) + (1-\delta)x_i^2(k) \in \mathbf{X_i^k}(x_i(k-1)).$$

On the other hand, it is obvious that

$$x_i^{0,1}(k) \geqslant fov_i^k(x_i^1(k)|\mathbf{y_i}(k-1)), \quad x_i^{0,2}(k) \geqslant fov_i^k(x_i^2(k)|\mathbf{y_i}(k-1)),$$

since $fov_i^k(x_i(k)|\mathbf{y_i}(k-1))$ is convex, then we can get

$$\delta x_i^{0,1}(k) + (1-\delta)x_i^{0,2} \geqslant fov_i^k(\delta x_i^1(k) + (1-\delta)x_i^2(k)|\mathbf{y_i}(k-1)).$$

Since

$$x_i^{0,1}(k) \leqslant \sup[fov_i^k(x_i(k)|\mathbf{y_i}(k-1))|x_i(k) \in \mathbf{X_i^k}(x_i(k-1))],$$
$$x_i^{0,2}(k) \leqslant \sup[fov_i^k(x_i(k)|\mathbf{y_i}(k-1))|x_i(k) \in \mathbf{X_i^k}(x_i(k-1))],$$

it is obvious that

$$\delta x_i^{0,1}(k) + (1-\delta)x_i^{0,2} \leqslant \sup[fov_i^k(x_i(k)|\mathbf{y_i}(k-1))|x_i(k) \in \mathbf{X_i^k}(x_i(k-1))].$$

This means $\mathbf{y_i^k}(\delta) \in \overline{\mathbf{Y_i^k}}$, so $\overline{\mathbf{Y_i^k}}$ is a convex set.

According to the theorems of convex set, $\forall \hat{\mathbf{y}}_\mathbf{i}(k) = (\hat{x}_i^0(k), \hat{x}_i(k))$ which is a boundary point of $\overline{\mathbf{Y_i^k}}$, there exists a support plane $\mathbf{P_i^k}$ which is defined as below

$$\mathbf{P_i^k} = \{\mathbf{y_i}(k)|\mathbf{y_i}(k) = (x_i^0(k), x_i(k)), a_i^0(k)x_i^0(k) + a_i(k)x_i(k) = b_i(k)\},$$

where $a_i^0(k), a_i(k)$ and $b_i(k)$ are all nonnegative real values. Then we have

$$a_i^0(k)\hat{x}_i^0(k) + a_i(k)\hat{x}_i(k) = b_i(k),$$
$$a_i^0(k)x_i^0(k) + a_i(k)x_i(k) \geqslant b_i(k), \quad \forall \mathbf{y_i}(k) \in \overline{\mathbf{Y_i^k}}.$$

Now let $\hat{x}_i(k)$ be an interior point of $\mathbf{X_i^k}(x_i(k-1))$, let $\hat{\mathbf{y}}_\mathbf{i}(k) = (fov_i^k(\hat{x}_i(k)|\mathbf{y_i}(k-1)), \hat{x}_i(k))$, then by the definition of $\overline{\mathbf{Y_i^k}}$, we can see that $\hat{\mathbf{y}}_\mathbf{i}(k)$ is a boundary point of $\overline{\mathbf{Y_i^k}}$, so we have

$$a_i^0(k)fov_i^k(\hat{x}_i(k)|\mathbf{y_i}(k-1)) + a_i(k)\hat{x}_i(k) = b_i(k),$$
$$a_i^0(k)fov_i^k(x_i(k)|\mathbf{y_i}(k-1)) + a_i(k)x_i(k) \geqslant b_i(k), \quad \forall x_i(k) \in \mathbf{X_i^k}(x_i(k-1)).$$

Here, it must be that $a_i^0(k) \neq 0$, otherwise we have

$$a_i(k)\hat{x}_i(k) = b_i(k),$$
$$a_i(k)x_i(k) \geqslant b_i(k), \quad \forall x_i(k) \in \mathbf{X_i^k}(x_i(k-1)).$$

This means $a_i(k)x_i(k) - b_i(k) = 0$ is just the support plane of $\mathbf{X_i^k}(x_i(k-1))$ on the point $\hat{x}_i(k)$, i.e. $\hat{x}_i(k)$ is a boundary point of $\mathbf{X_i^k}(x_i(k-1))$, but it is impossible since we have assumed that $\hat{x}_i(k)$ is an interior point of $\mathbf{X_i^k}(x_i(k-1))$.

As $a_i^0(k) \neq 0$, let $\alpha_i^k = -\frac{b_i(k)}{a_i^0(k)}$, $\lambda_i(k) = -\frac{a_i(k)}{a_i^0(k)}$, then we can get

$$\lambda_i(k)\hat{x}_i(k) - \alpha_i^k = fov_i^k(\hat{x}_i(k)|\mathbf{y_i}(k-1)),$$
$$\lambda_i(k)x_i(k) - \alpha_i^k \leqslant fov_i^k(x_i(k)|\mathbf{y_i}(k-1)), \quad \forall x_i(k) \in \mathbf{X_i^k}(x_i(k-1)).$$

From Def. 3.9, we can get

$$S_i^k(x_i(k)) = \lambda_i(k)x_i(k), \quad \forall i \in \Psi, \quad k = 1, 2, \cdots, n.$$

This is just the linear support function of the conditional lower bound surface of the $(k+1)$th stage $\Pi_i^k(\mathbf{y_i}(k-1))$ on the point $\hat{x}_i(k)$. The proof is completed. □

Now the theorem for the existence of the linear support function can be proved.

**Theorem 3.2.** *If $\hat{x}_i(k)$ is an interior point of $\mathbf{X}_i^k(x_i(k-1))$, then there exists a $\lambda_i(k)$ such that $S_i^k(x_i(k)) = \lambda_i(k)x_i(k)$ is a linear support function of the conditional lower bound surface of the $(k+1)$th stage $\Pi_i^k(y_i(k-1))$ on the point $\hat{x}_i(k)$ $\forall i \in \Psi, k = 1,2,\cdots,n$.*

*Proof.* According to Lemma 3.8, 3.9 and 3.10, we can prove the theorem of the existence of the linear support function as below.

From Lemma 3.8, we can know that the expanded state attainable set of the $(k+1)$th stage $\mathbf{Y}_\mathbf{i}^\mathbf{k}(\mathbf{y}_\mathbf{i}(k-1))$ which is derived from $\mathbf{y}_\mathbf{i}(k-1)$ is convex in direction $\mathbf{e} = (1,0)$, $\forall i \in \Psi, k = 1,2,\cdots,n$. Hence, the conditional function of optimal value of the $(k+1)$th stage $fov_i^k(x_i(k)|\mathbf{y}_\mathbf{i}(k-1))$ is a convex function according to Lemma 3.9. From Lemma 3.10, we can know that there exists a $\lambda_i(k)$ such that $S_i^k(x_i(k)) = \lambda_i(k)x_i(k)$ is a linear support function of the conditional lower bound surface of the $(k+1)$th stage $\Pi_i^k(\mathbf{y}_\mathbf{i}(k-1))$ on the point $\hat{x}_i(k)$ $\forall i \in \Psi, k = 1,2,\cdots,n$. The proof is completed. □

Since the existence of the linear support function is proved, the linear form of the Hamilton function can be defined as below.

**Definition 3.12.** *Let $S_i^{k+1}(x_i(k+1)) = \lambda_i(k+1)x_i(k+1)$ be the linear support function of the conditional lower bound surface of the $(k+1)$th stage $\Pi_i^{k+1}(y_i(k))$ on point $x_i(k+1)$, then $H_i^k(x_i(k),u_{i^1}(k),\lambda_i(k+1))$ is the linear Hamilton function if and only if*

$$
\begin{aligned}
H_i^k(x_i(k),u_{i^1}(k),\lambda_i(k+1)) = {} & \{E[\zeta_{k,i^1}^{rand}]E^{Me}[\tilde{Q}_{i^1}]u_{i^1}(k) + E[\zeta_{k,i^2}^{rand}]E^{Me}[\tilde{Q}_{i^2}][x_i(k) - u_{i^1}(k)]\} + \\
& \lambda_i(k+1)f_i^k(x_i(k),u_{i^1}(k)),
\end{aligned}
\tag{3.60}
$$

*where*

$$
f_i^k(x_i(k),u_{i^1}(k)) = x_i(k+1) = (p_{i^2}^k - p_{i^1}^k)u_{i^1}(k) + (1-p_{i^2}^k)x_i(k),
$$

$\forall i \in \Psi, k = 0,1,\cdots,n-1$.

### 3.3.5 The Maximal Theory of DMAP

Now the theorem which is crucial for finding the optimal solution of our problem based on the Theorems 3.1 and 3.2 can be proved.

**Theorem 3.3.** *Let $\{u_{i^1}^*(0),\cdots,u_{i^1}^*(n-1)\}$ be the optimal solution sequence of the number of Machine i allocated to Job $i^1$ at the beginning of every stage, and $\{x_i^*(0) = a_i, x_i^*(1),\cdots,x_i^*(n)\}$ be the optimal solution sequence of the number of Machine i that are available at the beginning of every stage. Assume that $x_i^*(k)$ is an interior point of $\mathbf{X}_i^k(x_i^*(k-1))$, where $k = 1,2,\cdots,n$. Then there exists a $\lambda_i^*(k)$, $\forall i \in \Psi, k = 1,2,\cdots,n$, such that,*

*(i)*

$$\lambda_i^*(k) = \frac{\partial H_i^k(x_i^*(k), u_{i1}^*(k), \lambda_i^*(k+1))}{\partial x_i(k)} = (1-p_{i2}^k)\lambda_i^*(k+1) + E[\zeta_{k,i2}^{rand}]E^{Me}[\tilde{Q}_{i2}], \tag{3.61}$$

$\forall i \in \Psi, k = 1, 2, \cdots, n-1$;
*(ii)*

$$H_i^k(x_i^*(k), u_{i1}^*(k), \lambda_i^*(k+1)) = \max[H_i^k(x_i^*(k), u_{i1}(k), \lambda_i^*(k+1) | 0 \leqslant u_{i1}(k) \leqslant x_i^*(k))], \tag{3.62}$$

$\forall i \in \Psi, k = 0, 1, \cdots, n-1$;
*(iii)*

$$\lambda_i^*(n) = 0, \quad \forall i \in \Psi. \tag{3.63}$$

*Proof.* By Theorem 3.2, since $x_i^*(k)$ is an interior point of $\mathbf{X_i^k}(x_i^*(k-1))$, we can know that there exists a $\lambda_i^*(k)$ such that $\hat{S}_i^k(x_i^*(k)) = \lambda_i^*(k)x_i^*(k)$ is a linear support function of the conditional lower bound surface of the $(k+1)$th stage $\Pi_i^k(\mathbf{y_i^*}(k-1))$ on the point $x_i^*(k)$, $\forall i \in \Psi, k = 1, 2, \cdots, n$. From Def. 3.12, we get

$$\frac{\partial H_i^k(x_i^*(k), u_{i1}^*(k), \lambda_i^*(k+1))}{\partial x_i(k)} = E[\zeta_{k,i2}^{rand}]E^{Me}[\tilde{Q}_{i2}] + \frac{d\hat{S}_i^{k+1}(x_i^*(k+1))}{dx_i(k+1)} \times \frac{\partial f_i^k(x_i^*(k), u_{i1}^*(k))}{\partial x_i(k)}.$$

According to Lemma 3.5, we have

$$\frac{d\hat{S}_i^{k+1}(x_i^*(k+1))}{dx_i(k+1)} = \frac{dFOV_i^{k+1}(x_i^*(k+1))}{dx_i(k+1)}. \tag{3.64}$$

This means

$$\frac{\partial H_i^k(x_i^*(k), u_{i1}^*(k), \lambda_i^*(k+1))}{\partial x_i(k)} = \frac{\partial H_i^k(x_i^*(k), u_{i1}^*(k), FOV_i^{k+1}(x_i^*(k+1)))}{\partial x_i(k)}. \tag{3.65}$$

From Eq. (3.55) in Theorem 3.1, we have

$$\frac{dFOV_i^k(x_i^*(k))}{dx_i(k)} = \frac{\partial H_i^k(x_i^*(k), u_{i1}^*(k), FOV_i^{k+1}(x_i^*(k+1)))}{\partial x_i(k)}. \tag{3.66}$$

By Eq. (3.64), (3.65) and (3.66), we get

$$\frac{\partial H_i^k(x_i^*(k), u_{i1}^*(k), \lambda_i^*(k+1))}{\partial x_i(k)} = \frac{d\hat{S}_i^k(x_i^*(k))}{dx_i(k)}. \tag{3.67}$$

Since

$$\frac{d\hat{S}_i^k(x_i^*(k))}{dx_i(k)} = \lambda_i^*(k),$$

so

$$\lambda_i^*(k) = \frac{\partial H_i^k(x_i^*(k), u_{i1}^*(k), \lambda_i^*(k+1))}{\partial x_i(k)} = (1-p_{i2}^k)\lambda_i^*(k+1) + E[\zeta_{k,i2}^{rand}]E^{Me}[\tilde{Q}_{i2}],$$

$\forall i \in \Psi$, $k = 1,2,\cdots,n-1$. This means $\lambda_i^*(k)$ satisfies (i).

Since

$$\hat{S}_i^k(x_i^*(k)) = \lambda_i^*(k)x_i^*(k)$$

is a linear support function of the conditional lower bound surface of the $(k+1)$th stage $\Pi_i^k(\mathbf{y}_i^*(k-1))$ on the point $x_i^*(k)$, according to Lemma 3.6, we can know that

$$\hat{S}_i^{k+1}(x_i^*(k+1)) = \lambda_i^*(k+1)x_i^*(k+1)$$

satisfies (ii), $\forall i \in \Psi, k = 0,1,\cdots,n-1$.

According to Theorem 3.1, we have proved that

$$\frac{dFOV_i^n(x_i^*(n))}{dx_i(n)} = 0, \quad \forall i \in \Psi.$$

By Eq. (3.64), we have

$$\frac{d\hat{S}_i^n(x_i^*(n))}{dx_i(n)} = \frac{dFOV_i^n(x_i^*(n))}{dx_i(n)}, \quad \forall i \in \Psi.$$

Since

$$\frac{d\hat{S}_i^n(x_i^*(n))}{dx_i(n)} = \lambda_i^*(n), \quad \forall i \in \Psi,$$

then we get

$$\lambda_i^*(n) = 0, \quad \forall i \in \Psi.$$

This means $\lambda_i^*(n)$ satisfies (iii). The proof is completed. □

### 3.3.6 *Theoretical Algorithm for DMAP*

Since the crucial theorem for finding the optimal solution of our problem has been derived, it is not difficult to form the theoretical algorithm (TA) to solve the expected value model (EVM) for the dynamic allocation of machines with uncertain failures to maximize the total production or construction throughput within a whole period. Based on Theorem 3.3, the steps for formulation of the theoretical algorithm are as follows.

- ***Step.1:*** According to Eq. (3.61) $\lambda_i^*(k) = (1-p_{i2}^k)\lambda_i^*(k+1) + E[\zeta_{k,i2}^{rand}]E^{Me}[\tilde{Q}_{i2}]$ and Eq. (3.63) $\lambda_i^*(n) = 0$ in Theorem 3.3, calculate $\lambda_i^*(k)$, $\forall i \in \Psi$, $k = 1,2,\cdots,n-1$.

Then, to find the optimal solution of the expected value model, the Hamilton function is applied in the Theorem 3.3.

- ***Step.2:*** Transform Eq. (3.62) in Theorem 3.3 as follows:

$$\begin{aligned}&H_i^k(x_i^*(k),u_{i1}^*(k),\lambda_i^*(k+1)) = \max_{0\leqslant u_{i1}(k)\leqslant x_i^*(k)} H_i^k(x_i^*(k),u_{i1}(k),\lambda_i^*(k+1))\\ &= [\lambda_i^*(k+1)(1-p_{i2}^k)+E[\zeta_{k,i2}^{rand}]E^{Me}[\tilde{Q}_{i2}]]x_i^*(k)+\max_{0\leqslant u_{i1}(k)\leqslant x_i^*(k)}\{[\lambda_i^*(k+1)(p_{i2}^k-p_{i1}^k)\\ &+(E[\zeta_{k,i1}^{rand}]E^{Me}[\tilde{Q}_{i1}]-E[\zeta_{k,i2}^{rand}]E^{Me}[\tilde{Q}_{i2}])]u_{i1}(k)\}\end{aligned} \tag{3.68}$$

$\forall i \in \Psi, k = 0,1,\cdots,n-1.$

Based on the deduction in Step. 2, information can be obtained through the simple analysis below.

- ***Step.3:*** According to Eq. (3.68), $\forall i \in \Psi, k = 0,1,\cdots,n-1$, the value of $u_{i1}^*(k)$ can be determined through easy analysis. To maximize $H_i^k(x_i^*(k),u_{i1}^*(k),\lambda_i^*(k+1))$, the last item which only needs to be considered is $\max_{0\leqslant u_{i1}(k)\leqslant x_i^*(k)}\{[\lambda_i^*(k+1)(p_{i2}^k-p_{i1}^k)+(E[\zeta_{k,i1}^{rand}]E^{Me}[\tilde{Q}_{i1}]-E[\zeta_{k,i2}^{rand}]E^{Me}[\tilde{Q}_{i2}])]u_{i1}(k)\}$. There are three different cases:
  Let

$$\phi_i(k+1) = \lambda_i^*(k+1)(p_{i2}^k-p_{i1}^k)+(E[\zeta_{k,i1}^{rand}]E^{Me}[\tilde{Q}_{i1}]-E[\zeta_{k,i2}^{rand}]E^{Me}[\tilde{Q}_{i2}]),$$

  then

$$\begin{aligned}&\max_{0\leqslant u_{i1}(k)\leqslant x_i^*(k)}\{[\lambda_i^*(k+1)(p_{i2}^k-p_{i1}^k)+(E[\zeta_{k,i1}^{rand}]E^{Me}[\tilde{Q}_{i1}]-E[\zeta_{k,i2}^{rand}]E^{Me}[\tilde{Q}_{i2}])]u_{i1}(k)\}\\ &= \max_{0\leqslant u_{i1}(k)\leqslant x_i^*(k)}\{\phi_i(k+1)u_{i1}(k)\},\end{aligned}$$

  ***Case.1:*** If $\phi_i(k+1)<0$, the value of $\phi_i(k+1)u_{i1}(k)$ must be nonpositive, and can arrive at its maximal value zero if and only if $u_{i1}^*(k)=0, \forall i\in\Psi, k=0,1,\cdots, n-1$.
  ***Case.2:*** If $\phi_i(k+1)>0$, by Eq. (3.20), since $0\leqslant u_{i1}(k)\leqslant x_i(k)$, $\phi_i(k+1)u_{i1}(k)$ must be positive. Therefore, it can arrive at its maximal value if and only if $u_{i1}^*(k)=x_i^*(k), \forall i\in\Psi, k=0,1,\cdots,n-1$.
  ***Case.3:*** If $\phi_i(k+1)=0$, then $\phi_i(k+1)u_{i1}(k)$ must be zero. So $u_{i1}^*(k)$ can be any value in $[0,x_i^*(k)]$, $\forall i\in\Psi, k=0,1,\cdots,n-1$.

Finally, the optimal solution to the problem is determined using Step. 4.

- ***Step.4:*** Since the value of $u_{i1}^*(k)$ ($\forall i\in\Psi, k=0,1,\cdots,n-1$) was determined in Step. 3, the value of $x_i^*(k)$ ($\forall i\in\Psi, k=\{1,2,\cdots,n-1\}$) can be calculated by solving Eq. (3.16).

The overall procedure for the theoretical algorithm is shown in Figure 3.9. In order to illustrate the efficiency and practicality of the theoretical algorithm, a numerical example is used here.

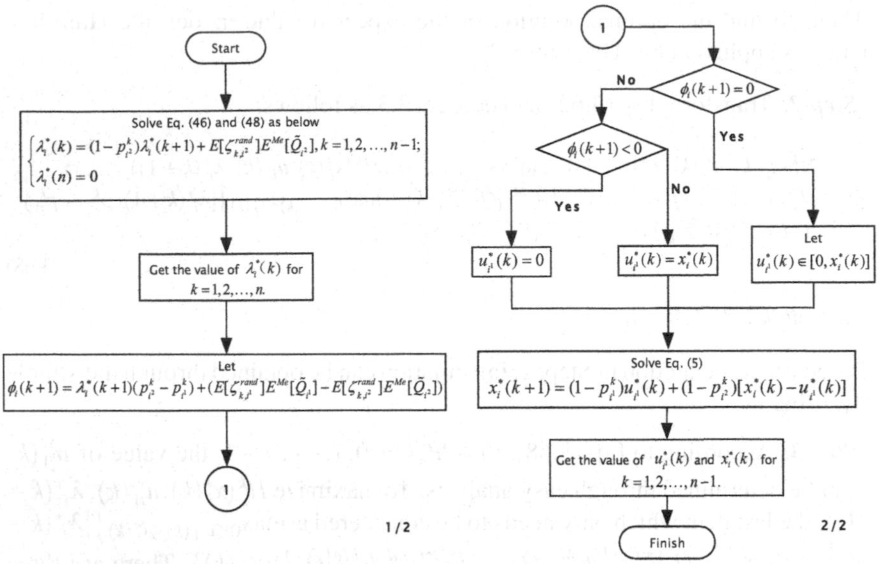


**Fig. 3.9** Overall procedure of the theoretical algorithm for DMAP

*Example 3.4.* The example project is a steel production project at Baoshan Iron & Steel Co., Ltd. In this case, our objective is to maximize the production throughput for 5 types of cold rolling mill in a 4-week period production project using the production strategy that is shown in Figure 3.4. All the detailed data for the steel production project were obtained from Baoshan Iron & Steel Co., Ltd. Every type of cold rolling mill is arranged to do 2 kinds of jobs, and the initial number of every type of cold rolling mill and the expected value of machine productivity (i.e., $E^{Me}[\tilde{Q}_{i\theta}]$, where the optimistic-pessimistic index $\eta$ is set to be 0.5) are shown in Table 4.2. The machine failure probability (i.e., $p^k_{i\theta}$), and the expected value of the conversion factor (i.e., $E[\zeta^{rand}_{k,i\theta}]$) are shown in Tables 3.3 and 3.4.

**Table 3.3** Job assignments of the 5 types of cold rolling mills in a steel production project

| Machine | Type index | Machine type | Initial number | Job index | Job assignment | Expected value of machine productivity |
|---|---|---|---|---|---|---|
| | $i$ | | $a_i$ (unit) | $i^\theta$ | | $E^{Me}[\tilde{Q}_{i\theta}]$ $(ton/week \cdot unit)$ |
| Cold rolling mill | 1 | 2030MM tandem cold mill | 11 | $1^1$ | Product cold rolled coil SPCC-SD | 334 |
| | | | | $1^2$ | Product cold rolled coil ST12-SB | 342 |
| | 2 | 1550MM tandem cold mill | 13 | $2^1$ | Product cold rolled sheet SPCC-B | 328 |
| | | | | $2^2$ | Product cold rolled sheet Q195 | 322 |
| | 3 | 1420MM reversing cold mill | 10 | $3^1$ | Product tinplate sheet JISG3303 | 297 |
| | | | | $3^2$ | Product tinplate sheet ASTMA599 | 296 |
| | 4 | 1220MM reversing cold mill | 12 | $4^1$ | Product galvanized steel strip S220GD+Z | 334 |
| | | | | $4^2$ | Product galvanized steel strip S280GD+Z | 329 |
| | 5 | 1730MM skin pass mill | 9 | $5^1$ | Product electrical steel B35A440 | 345 |
| | | | | $5^2$ | Product electrical steel B65A600 | 352 |

**Table 3.4** Detailed information of cold rolling mills in a steel production project

| Machine | Type index | Machine type | Job index | Machine failure probability during the $(k+1)$th week $p^k_{i\theta}$ | | | | Expected value of the conversion factor during the $(k+1)$th week $E[\zeta^{rand}_{k,i\theta}]$ | | | |
|---|---|---|---|---|---|---|---|---|---|---|---|
| | $i$ | | $i^\theta$ | k=0 | k=1 | k=2 | k=3 | k=0 | k=1 | k=2 | k=3 |
| Cold rolling mill | 1 | 2030MM tandem cold mill | $1^1$ | 0.0142 | 0.0929 | 0.2093 | 0.2764 | 0.9964 | 0.9627 | 0.9045 | 0.8630 |
| | | | $1^2$ | 0.0206 | 0.1146 | 0.2288 | 0.2731 | 0.9946 | 0.9527 | 0.8935 | 0.8625 |
| | 2 | 1550MM tandem cold mill | $2^1$ | 0.0208 | 0.1108 | 0.2180 | 0.2631 | 0.9944 | 0.9540 | 0.8984 | 0.8681 |
| | | | $2^2$ | 0.0104 | 0.0829 | 0.2112 | 0.2971 | 0.9975 | 0.9678 | 0.9053 | 0.8535 |
| | 3 | 1420MM reversing cold mill | $3^1$ | 0.0100 | 0.0852 | 0.2239 | 0.3125 | 0.9977 | 0.9672 | 0.8999 | 0.8451 |
| | | | $3^2$ | 0.0104 | 0.0855 | 0.2200 | 0.3059 | 0.9976 | 0.9669 | 0.9014 | 0.8485 |
| | 4 | 1220MM reversing cold mill | $4^1$ | 0.0117 | 0.0910 | 0.2252 | 0.3038 | 0.9972 | 0.9645 | 0.8984 | 0.8489 |
| | | | $4^2$ | 0.0125 | 0.0900 | 0.2144 | 0.2892 | 0.9970 | 0.9645 | 0.9029 | 0.8567 |
| | 5 | 1730MM skin pass mill | $5^1$ | 0.0073 | 0.0679 | 0.1935 | 0.2984 | 0.9984 | 0.9743 | 0.9149 | 0.8550 |
| | | | $5^2$ | 0.0094 | 0.0795 | 0.2099 | 0.3016 | 0.9978 | 0.9694 | 0.9064 | 0.8516 |

The results for the DMAP for the example project, as compared to the actual data from the corresponding project, are listed in Table 3.5. The growth rate in Table 3.5 is calculated from the following formula:

$$Growth\ rate = \frac{Algorithm\ result - Actual\ data\ from\ the\ project}{Actual\ data\ from\ the\ project} \times 100\%$$

As shown, the solution of the DEAP for Equipment $i$ is expressed by $u^*_{i\theta}(k)$ and $x^*_i(k)$. By comparing the algorithm result (total production throughput) with the actual data from the example project in Table 3.5, the findings indicate that there are differences between the solutions generated using the theoretical algorithm developed in this chapter and the actual production throughput. The net increase in the total production throughput for every type of machine ranged from 425 to 964 *tons* for the steel production project, and the rate of increase, from 4.18% to 7.56% for the steel production project. This improvement in production efficiency could bring a considerable economic benefit for each project, especially if they are large scale. While this finding is quite encouraging, at present it is not 100% sure that all these differences represent a net improvement because this mathematical model presented here was formulated with some assumptions. Some portion of these differences could result from possible modeling errors. To some extent, it is possible to conclude from the resulting data that the theoretical optimized values for total production throughput for the design variables are better than the actual values for each project. This means that the current practice for each project is not optimal. In fact, in the current practice, the decision makers usually make decisions on experience alone. Thus, the results here can be used to provide decision makers with a theoretical optimal dynamic equipment allocation plan for guiding current practice.

**Table 3.5** Results of theoretical algorithm for DMAP in the steel production project

| Machine | Type index $i$ | Machine type | Variable | Week index $k$ | | | | Total production throughput | | | |
|---|---|---|---|---|---|---|---|---|---|---|---|
| | | | | 0 | 1 | 2 | 3 | Algorithm result ($ton$) | Actual data from the project ($ton$) | Net increase ($ton$) | Rate of increase |
| Cold rolling mill | 1 | 2030MM tandem cold mill | $x_1^*(k)$ | 11 | 10 | 9 | 7 | 11,741 | 11,270 | 471 | 4.18% |
| | | | $u_{11}^*(k)$ | 0 | 10 | 9 | 0 | | | | |
| | | | $u_{12}^*(k)$ | 11 | 0 | 0 | 7 | | | | |
| | 2 | 1550MM tandem cold mill | $x_2^*(k)$ | 13 | 12 | 11 | 9 | 13,719 | 12,755 | 964 | 7.56% |
| | | | $u_{21}^*(k)$ | 0 | 0 | 11 | 9 | | | | |
| | | | $u_{22}^*(k)$ | 13 | 12 | 0 | 0 | | | | |
| | 3 | 1420MM reversing cold mill | $x_3^*(k)$ | 10 | 9 | 9 | 7 | 9,708 | 9,283 | 425 | 4.58% |
| | | | $u_{31}^*(k)$ | 10 | 9 | 0 | 0 | | | | |
| | | | $u_{32}^*(k)$ | 0 | 0 | 9 | 7 | | | | |
| | 4 | 1220MM reversing cold mill | $x_4^*(k)$ | 12 | 11 | 10 | 8 | 12,779 | 12,015 | 764 | 6.36% |
| | | | $u_{41}^*(k)$ | 12 | 11 | 0 | 8 | | | | |
| | | | $u_{42}^*(k)$ | 0 | 0 | 10 | 0 | | | | |
| | 5 | 1730MM skin pass mill | $x_5^*(k)$ | 9 | 8 | 8 | 6 | 10,174 | 9,537 | 637 | 6.68% |
| | | | $u_{51}^*(k)$ | 0 | 8 | 8 | 0 | | | | |
| | | | $u_{52}^*(k)$ | 9 | 0 | 0 | 6 | | | | |

## 3.4 Particle Swarm Optimization

The theoretical algorithm for the DMAP discussed above can be categorized into exact techniques. However, the theoretical algorithm is unable to solve large and hard instances measured in terms of, say, the number of machines and tasks. In terms of what current state-of-art algorithms can do, and considering the structure of the project networks as well as the number of machines and tasks, instances with a large number of machines and tasks cannot be solved optimally in a reasonable amount of time. It follows that the search for exact algorithms which are also formally efficient is all but futile and that it is necessary to instead search for effective heuristic algorithms to solve a general DMAP. To solve the expected value model, a dynamic programming-based particle swarm optimization (PSO) algorithm is proposed. The advantage of using a PSO algorithm over other techniques is that it is computationally tractable, easy-to-implement, and does not require any gradient information of an objective function but its value. Therefore, the PSO algorithm can be developed to solve the above problem. The key features of the algorithm are explained in detail including the general concept of PSO, the solution representation, the initializing method, the adjusting method, the decoding method, the fitness value function, and the parameter selection, followed by an explanation of algorithm framework.

### *3.4.1 General Concept of PSO*

PSO is a population-based self-adaptive search optimization technique first proposed by Kennedy and Eberhart (1995) [23]. It simulates a social behavior, such as birds flocking to a promising position, for certain objectives in a multidimensional space (Eberhart and Shi 2001 [24]; Clerc and Kennedy 2002 [25]). There have been many practical applications of PSO to problems in construction, including resource-constrained project scheduling (Zhang et al. 2006a [27]), preemptive scheduling under break and resource-constraints (Zhang et al. 2006b [28]), construction site unequal-area layout (Zhang and Wang 2008 [31]), and construction operations (Zhang et al. 2006c [30]).

Similar to other evolutionary algorithms, PSO conducts its search using a population (i.e., a swarm) of individuals (i.e., particles). In the simplest (and original) version of PSO, each particle represents a candidate position or solution that is moved through a problem space by two elastic forces. One attracts it with random magnitude towards the best location so far encountered by the particle, and the other, towards the best location encountered by any member of the swarm (Langdon and Poli 2007 [32]). Thus, global sharing of information takes place and particles profit from their own discoveries (i.e., personal bests) and the previous experience of all other companions (i.e., global bests) during the search process. Initialized with a swarm of random particles, PSO searches for the best position or optimum by updating the particles until reaching a relatively steady position or the iteration limit. In every iteration or generation, the personal bests and global bests are determined by evaluating the fitness or objectives of the current particles. Each particle is treated as a point in an $H$-dimension space and the status of a particle in the search space is characterized by its position and velocity (Kennedy and Eberhart 1995 [23]). In the case study here, the $H$-dimension position of Equipment $i$ for the $l$th particle in the $\tau$th generation can be denoted as $P_l^i(\tau) = [p_{l1}^i(\tau), p_{l2}^i(\tau), \cdots, p_{lH}^i(\tau)]$. Similarly, the velocity (i.e., distance change) of Equipment $i$, also a $H$-dimension vector, for the $l$th particle in the $\tau$th generation can be described as $V_l^i(\tau) = [v_{l1}^i(\tau), v_{l2}^i(\tau), \cdots, v_{lH}^i(\tau)]$. The following formulas (Kennedy and Eberhart 1995 [23]) represent the updating mechanism of a swarm of particles' status from the ones of the last generation during the search process:

$$V_l^i(\tau+1) = w(\tau)V_l^i(\tau) + c_p r_p (PBest_l^i - P_l^i(\tau)) + c_g r_g (GBest^i - P_l^i(\tau)) \tag{3.69}$$

$$P_l^i(\tau+1) = V_l^i(\tau+1) + P_l^i(\tau) \tag{3.70}$$

where $i =$ the equipment type index $= 1, 2, \cdots, N$; $l =$ the particle index $= 1, 2, \cdots, L$; $L =$ population size; $\tau =$ the iteration index $= 0, 1, \cdots, T$; $T =$ the iteration limit; $PBest_l^i = [pBest_{l1}^i, pBest_{l2}^i, \cdots, pBest_{lH}^i] =$ the personal best of the $l$th particle encountered after $\tau$ iterations for Equipment $i$; and $GBest^i = [gBest_1^i, gBest_2^i, \cdots, gBest_H^i] =$ the global best among all the swarm of particles achieved so far for Equipment $i$; $c_p$ and $c_g =$ the positive constants (namely, learning factors) that determine the relative weight that the global best has versus the personal best; $r_p$ and $r_g =$ the random numbers between 0 and 1; and $w(\tau) =$ the inertia weight used to control the impact of the previous velocities on the current velocity, influencing the trade-off between the global and local exploration abilities during the search.

Eq. (3.69) is used to calculate the particle's new velocity according to its previous velocity and the distance of its current position from its own best experience and the group's best experience. Eq. (3.70) is traditionally used to update the particle such that it flies toward a new position (Shi and Eberhart 1998a [34]).

### 3.4.2 Solution-Representation of PSO

The solution representation of the DEAP is one of the key elements for the effective implementation of PSO. An indirect representation is proposed here. In this study, every particle consists of $4 \times (M_i - 1)$ dimensions (i.e., $H = 4 \times (M_i - 1)$), and is divided into 4 parts which are expressed as:

$$P_l^i(\tau) = [p_{l1}^i(\tau), p_{l2}^i(\tau), \cdots, p_{l(4\times(M_i-1))}^i(\tau)] = [Y_{l1}^i(\tau), Y_{l2}^i(\tau), Y_{l3}^i(\tau), Y_{l4}^i(\tau)] \tag{3.71}$$

where $Y_{l(k+1)}^i(\tau)$ = the $(k+1)$th part of the $l$th particle in the $\tau$th generation for Equipment $i$, $k = 0, 1, 2, 3$. Note that every part of a particle is a $(M_i - 1)$-dimension vector, and can be denoted as:

$$Y_{l(k+1)}^i(\tau) = [y_{l(k+1)}^{i^1}(\tau), y_{l(k+1)}^{i^2}(\tau), \cdots, y_{l(k+1)}^{i^{M_i-1}}(\tau)] \tag{3.72}$$

where $y_{l(k+1)}^{i^\theta}(\tau)$ = the $\theta$th dimension of $Y_{l(k+1)}^i(\tau)$ for the $l$th particle in the $\tau$th generation; $k = 0, 1, 2, 3$; $\theta = 1, 2, \cdots, M_i - 1$. In order to be in accordance with the expression $P_l^i(\tau) = [p_{l1}^i(\tau), p_{l2}^i(\tau), \cdots, p_{lH}^i(\tau)]$, there is $y_{l(k+1)}^{i^\theta}(\tau) = p_{l(k\times(M_i-1)+\theta)}^i(\tau)$, where $H = 4 \times (M_i - 1)$.

Note that the $\theta$th dimension of $Y_{l(k+1)}^i(\tau)$ (i.e., $y_{l(k+1)}^{i^\theta}(\tau)$) represents the number of Equipment $i$ allocated to Job $i^\theta$ at the beginning of the $(k+1)$th week (i.e., $u_{i^\theta}(k)$). Based on the relationship equations (i.e., Eqs. (3.16) and (3.17)) and constraint conditions (i.e., Eq. (3.20)), the $\theta$th dimension of $Y_{l(k+1)}^i(\tau)$ (i.e., $y_{l(k+1)}^{i^\theta}(\tau)$), either initialized or updated, must be an integer that is limited to $[0, x_i(k)]$ (i.e., $[p_i^{min}, p_i^{max}]$) and satisfy Eqs. (3.16) and (3.17) so as to avoid infeasible positions, and thus expedite the PSO search.

**Initializing Method**

The element in the multidimensional particle position can be initialized as follows to avoid infeasible positions:

*Step 1:* Set $k = 0$.

*Step 2:* Set $\theta = 1$ and $y_{l(k+1)}^{i^0}(0) = 0$.

*Step 3:* Initialize $y_{l(k+1)}^{i^\theta}(0)$ by generating a random integer within $[0, x_i(k) - \Sigma_{\eta=0}^{\theta-1} y_{l(k+1)}^{i^\eta}(0)]$ (note that $x_i(0) = a_i$, where $a_i$ denotes the initial number of Equipment $i$ at the beginning of the whole period).

*Step 4:* If $\theta = M_i - 1$, based on Eq. (3.16), then $x_i(k+1) = (1 - p_{i^{M_i}}^k)[x_i(k) - \Sigma_{\eta=1}^{M_i-1} y_{l(k+1)}^{i^\eta}(0)] + \Sigma_{\eta=1}^{M_i-1}(1 - p_{i^\eta}^k) y_{l(k+1)}^{i^\eta}(0)$. Otherwise, $\theta = \theta + 1$ and return to step 3.

*Step 5:* If the stopping criterion is met, i.e., $k = 3$, then the initialization for the $l$th particle of Equipment $i$ is completed. Otherwise, $k = k+1$ and return to step 2.

**Adjustment Method**

After updating, the element in the multidimensional particle position can be adjusted as follows to avoid an infeasible position:

*Step 1:* Adjust $y^{i^\theta}_{l(k+1)}(\tau)$ $(\theta = 1,2,\cdots,M_i - 1; k = 0,1,2,3)$ to a nearest integer; Set $k = 0$.

*Step 2:* For $\theta = 1,2,\cdots,M_i - 1$, if $y^{i^\theta}_{l(k+1)}(\tau) < 0$, then $y^{i^\theta}_{l(k+1)}(\tau) = 0$; else if $y^{i^\theta}_{l(k+1)}(\tau) > x_i(k)$ (note that $x_i(0) = a_i$, where $a_i$ denotes the initial number of Equipment $i$ at the beginning of the whole period), then $y^{i^\theta}_{l(k+1)}(\tau) = x_i(k)$.

*Step 3:* If $\sum_{\theta=1}^{M_i-1} y^{i^\theta}_{l(k+1)}(\tau) > x_i(k)$, then let $U = \sum_{\theta=1}^{M_i-1} y^{i^\theta}_{l(k+1)}(\tau) - x_i(k)$ and select the largest $y^{i^{\theta^*}}_{l(k+1)}(\tau)$ $(\theta^* \in \{1,2,\cdots,M_i - 1\})$ (if several are equally the largest, then select one by random), $y^{i^{\theta^*}}_{l(k+1)}(\tau) = y^{i^{\theta^*}}_{l(k+1)}(\tau) - U$ and return to step 2. Otherwise, based on Eq. (3.16), then $x_i(k+1) = (1 - p^k_{i^{M_i}})[x_i(k) - \sum_{\theta=1}^{M_i-1} y^{i^\theta}_{l(k+1)}(\tau)] + \sum_{\theta=1}^{M_i-1}(1 - p^k_{i^\theta}) y^{i^\theta}_{l(k+1)}(\tau)$.

*Step 4:* If the stopping criterion is met, i.e., $k = 3$, then the adjustment for the $l$th particle in the $\tau$th generation of Equipment $i$ is completed. Otherwise, $k = k+1$ and return to step 2.

**Decoding Method**

The decoding method for this representation into the problem solution starts by transforming a particle into the corresponding number of Equipment $i$ allocated to Job $i^\theta$ at the beginning of every week. The overall procedure for decoding is detailed below:

*Step 1:* Decode the $\theta$th dimension of $Y^i_{l(k+1)}(\tau)$ (i.e., $y^{i^\theta}_{l(k+1)}(\tau)$) into the number of Equipment $i$ allocated to Job $i^\theta$ at the beginning of the $(k+1)$th week, i.e., $u_{i^\theta}(k) = y^{i^\theta}_{l(k+1)}(\tau)$, $\theta = 1,2,\cdots,M_i - 1$, $k = 0,1,2,3$.

*Step 2:* Set $k = 0$.

*Step 3:* Calculate the number of Equipment $i$ allocated to Job $i^{M_i}$ at the beginning of $(k+1)$th week by $u_{i^{M_i}}(k) = x_i(k) - \sum_{\theta=1}^{M_i-1} u_{i^\theta}(k)$ (note that $x_i(0) = a_i$). Then, based on Eq. (3.16), $x_i(k+1) = \sum_{\theta=1}^{M_i}(1 - p^k_{i^\theta}) u_{i^\theta}(k)$.

*Step 3:* If the stopping criterion is met (i.e., $k = 3$), then integrate $u_{i^\theta}(k)$ and $x_i(k)$ (for $\theta = 1,2,\cdots,M_i$, $k = 0,1,2,3$) in order to form a solution to the problem for Equipment $i$. Otherwise, $k = k+1$ and return to step 3.

Figure 3.10 illustrates the decoding method and mapping between PSO particles and solutions to the problem.

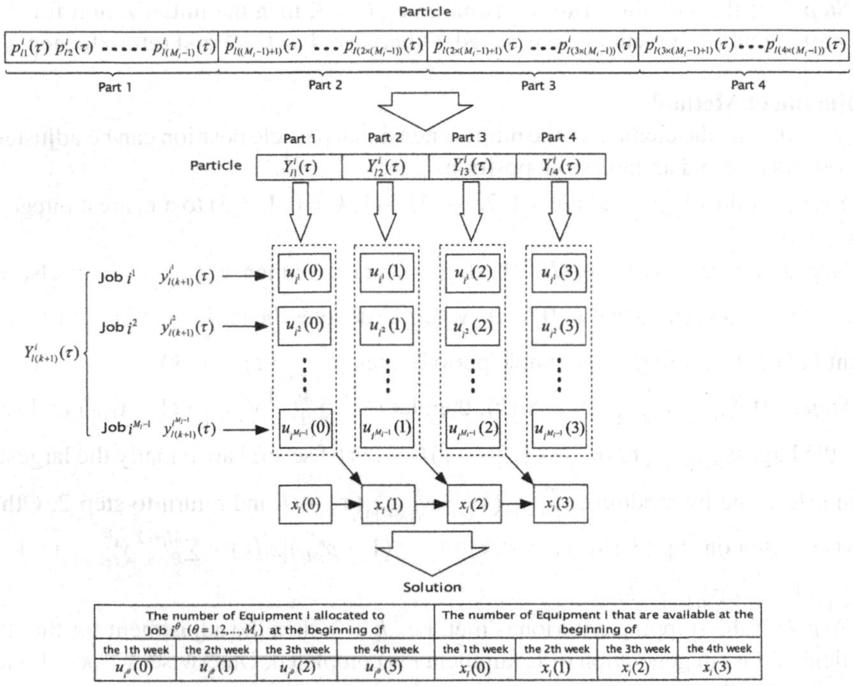


**Fig. 3.10** Decoding method and mapping between PSO particles and solutions to the problem

### 3.4.3 *Fitness Value*

The fitness value used to evaluate a particle is just the value of the objective function in Eq.(3.31), i.e.,

$$Fitness^i(P_l^i(\tau)) = \sum_{\theta=1}^{M_i-1}\{E[Q_{i\theta}]\sum_{k=0}^{3}E[\zeta_{k,i\theta}^{rand}]y_{l(k+1)}^{i\theta}(\tau)\} + E[Q_{iM_i}]\sum_{k=0}^{3}E[\zeta_{k,iM_i}^{rand}][x_i(k) - \sum_{\theta=1}^{M_i-1}y_{l(k+1)}^{i\theta}(\tau)] \quad (3.73)$$

where $y_{l(k+1)}^{i\theta}(\tau) = u_{i\theta}(k)$; $x_i(0) = a_i$; $x_i(k+1) = (1-p_{iM_i}^k)[x_i(k) - \sum_{\theta=1}^{M_i-1}y_{l(k+1)}^{i\theta}(\tau)] + \sum_{\theta=1}^{M_i-1}(1-p_{i\theta}^k)y_{l(k+1)}^{i\theta}(\tau)$, $\theta = 1,2,\cdots,M_i-1$; $k = 0,1,2,3$; $i = 1,2,\cdots,N$.

### 3.4.4 *Parameter Selection for PSO*

The PSO parameters are set based on the results of some preliminary experiments that were carried out to observe the behavior of an algorithm in different parameter settings. By comparing several sets of parameters, including population size, iteration number, acceleration constant, initial velocity, and inertia weight, the most reasonable parameters for each type of dump trucks were chosen. Since more particles require more evaluation runs, leading to a higher optimization cost (Trelea 2003 [40]), 50 particles were selected as the population size and 100 times as the iteration number for all types of construction equipment in the case study. The inertia weight $w(\tau)$ was set to be varying with the iteration as follows:

$$w(\tau) = w(T) + \frac{\tau - T}{1 - T}[w(1) - w(T)] \tag{3.74}$$

where $\tau$ = iteration index $= 1, 2, \cdots, T$; $T$ = iteration limit. Other parameters were selected by comparing the results and observing the dynamic search behaviors of the swarm. Table 3.4 summarizes the values of all parameters selected for the 6 types of dump trucks in the case study.

### 3.4.5 Framework of PSO for DMAP

Based on the above basic concepts (i.e., solution-representation, initializing method, adjustment method, decoding method, fitness value function and parameter selection), the framework to implement the PSO for the DEAP optimization is developed. The procedure for the PSO framework is shown as a flowchart in Figure 3.11. The initial positions of the $L$ particles for Equipment $i$, i.e., $P_l^i(0) = [p_{l1}^i(0), p_{l2}^i(0), \cdots, p_{l(4\times(M_i-1))}^i(0)]$ $(l = 1, 2, \cdots, L)$, are randomly generated using the initializing method described in the previous section to avoid an infeasible position. The initial velocities of the $L$ particles for Equipment $i$, i.e., $V_l^i(0) = [v_{l1}^i(0), v_{l2}^i(0), \cdots, v_{l(4\times(M_i-1))}^i(0)]$ $(l = 1, 2, \cdots, L)$, are also generated according to the data in Table 4. The particle-updating mechanism represented by Eqs. (3.69) and (3.70) is used to update the velocities and positions (i.e., $V_l^i(\tau)$ and $P_l^i(\tau)$) of the particles until the optimal solution is found. During the PSO search, each element in the multidimensional particle position (i.e., $p_{l(k+1)}^i(\tau)$) should be updated using the adjustment method described above when it is beyond $[p_i^{min}, p_i^{max}]$.

The initial or updated positions must be evaluated with respect to a fitness value (i.e., Eq. (3.73))in order to identify the personal best for each particle for Equipment $i$ (i.e., $PBest_l^i = [pBest_{l1}^i, pBest_{l2}^i, \cdots, pBest_{l(4\times(M_i-1))}^i]$) $(l = 1, 2, \cdots, L)$ and the global best for Equipment $i$ (i.e., $GBest^i = [gBest_1^i, gBest_2^i, \cdots, gBest_{4\times(M_i-1)}^i]$) in the swarm.

The PSO search terminates when either one of the following two stopping criteria is met: (1) convergence iteration limit (i.e., maximum number of iterations since the last updating of the global best); (2) total iteration limit (i.e., $T$). Then, based on the decoding method described above, the global best particle-represented solution for Equipment $i$ (i.e., $GBest^i = [gBest_1^i, gBest_2^i, \cdots, gBest_{4\times(M_i-1)}^i]$) can be transformed into the equipment allocation plan for Equipment $i$, which is the final solution to the problem.

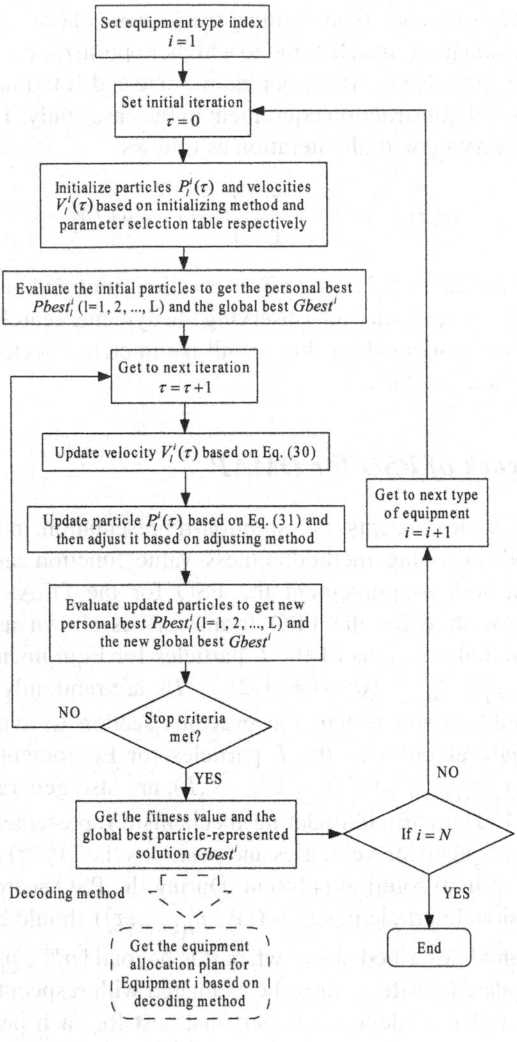


**Fig. 3.11** Procedure of the PSO framework

## 3.5 Case Studies: Earthwork Transportation Projects

This section introduces two actual construction projects, i.e., Xiaolangdi and Shuibuya Hydropower Projects, as examples for the DMAP, and employs the theoretical algorithm and a DP-based PSO to solve these problems, respectively. The project description and data collection are explained in detail and the specific features of each case are outlined and fully explained. The results and an analysis are also provided to allow for a full understanding of the solution methods.

### 3.5.1 Project Background

Two actual construction projects are used as practical application examples for the optimization methods. Here we introduce the background to the Xiaolangdi and Shuibuya Hydropower Projects.

**Xiaolangdi Hydropower Project**

The first project is the Xiaolangdi Hydropower Project, which is located at the exit of the last gorge in the middle reaches of the Yellow River, bordering Luoyang and Jiyuan cities, Henan Province, 130 *km* downstream of the Sanmenxia reservoir in China [35]. The maximum height of the main dam is 154 *m* with a crest length of 1667 *m*, total reservoir storage of 12.65 billion $m^3$, and an installed capacity of 1836 *MW*. The objectives of this project are flood control, sediment mitigation, water supply, irrigation and power generation. The construction was finished in 2001, and significant results have been achieved, especially in regulating flow and sediment load.

**Shuibuya Hydropower Project**

The Shuibuya Hydropower Project is in Badong County located in the middle reaches of the Qingjiang River in Sichuan province, China. The project is the first cascaded project on the Qingjiang mainstream and the third most important project after Geheyan and Gaobazhou in China. Once completed, it will be a major power source for the meeting of the peak load demand in the Central China Power Grid. The installed capacity and annual output of the Shuibuya Power Plant are 1,600 *MW* and 3.92 *GWh* respectively. The project has a powerful regulating ability with a normal pool level of 400 *m* and reservoir capacity of 4.58 billion $m^3$. The project consists of a concrete faced rockfill dam (CFRD), underground power house, chute spillway on the left bank, and the sluice tunnel on the right bank. The dam is 233 *m* high and is the highest of its kind in the world at present with a total volume of $15.64 \times 10^6\ m^3$.

### 3.5.2 Data Collection

The detailed data for the Xiaolangdi Hydropower Project were obtained from the Yellow River Water and Hydroelectric Power Development Corporation (YRWHDC) and the detailed data for the Shuibuya Hydropower Project were obtained from the Hubei Qingjiang Shuibuya Project Construction Company.

**Xiaolangdi Hydropower Project**

In the Xiaolangdi Hydropower Project, there are 4 borrow areas which are the main source of the rockfill. The location and detailed information of each borrow area is illustrated in Figure 3.12. In this case, 2 types of dump trucks and 4 types of

**Fig. 3.12** Layout of the borrow areas of Xiaolangdi Hydropower Project

hydraulic excavators are dynamically allocated in the construction project. Each type of these construction machines is arranged to do 2 kinds of jobs, and the initial number of each type of machine and the expected value of machine throughput (i.e., $E^{Me}[\tilde{Q}_{i\theta}]$, where the optimistic-pessimistic index $\eta$ is set to be 0.5) are shown in Table 3.6. To minimize maintenance-related costs, the decision makers use the construction strategy shown in Figure 3.4, in which the span of each stage is one week and the span of each period is one month. It has been explained that machine failure time follows a Weibull distribution, so based on the machine failure time data at the Xiaolangdi Hydropower Project, the different Weibull distributions were estimated for each type of the construction machine allocated to Job $i^\theta$ $(\theta = 1,2)$ during the $(k+1)$th week, where $k = 0,1,\cdots,3$, $i = 1,2,\cdots,6$. The machine failure probability and the expected value of the conversion factor are calculated using Eq. (3.9), (3.12) and (3.29). The machine failure probability (i.e., $p^k_{i\theta}$), the expected value of the conversion factor (i.e., $E[\zeta^{rand}_{k,i^\theta}]$) are shown in Table 3.7.

**Shuibuya Hydropower Project**

In this case, the objective is to maximize the construction throughput for 4 types of dump trucks in the project over a 4-week period using the construction strategy and the dynamic allocation method that shown in Figure 3.4. In the Shuibuya Hydropower Project, there are 2 borrow areas, 3 stockpile areas and 2 orphan banks. While each type of dump trucks can be arranged to do several different jobs, there are usually some physical constraints for job assignments since some jobs are equipment specified. In this case, there are massive schists and limestone in Borrow Area 2

**Table 3.6** Job assignments of the construction machines in Xiaolangdi Hydropower Project

| Type | Machine index | Machine type | Initial number | Job index | Job assignment | Expected value of machine throughput |
|---|---|---|---|---|---|---|
| | $i$ | | $a_i$ (unit) | $i^\theta$ | | $E^{Me}[\tilde{Q}_{i\theta}]$ ($m^3/week \cdot unit$) |
| Dump truck | 1 | Perlini DP366 ($20m^3/36t$) | 34 | $1^1$ | Transport between Borrow Area 1 and the dam | 4,010 |
| | | | | $1^2$ | Transport between Borrow Area 2 and the dam | 4,130 |
| | 2 | Perlini DP755 ($30m^3/65t$) | 41 | $2^1$ | Transport between Borrow Area 3 and the dam | 5,010 |
| | | | | $2^2$ | Transport between Borrow Area 4 and the dam | 5,170 |
| Hydraulic excavator | 3 | Hitachi EX400 ($2.2m^3$) | 28 | $3^1$ | Work in Borrow Area 1 | 4,280 |
| | | | | $3^2$ | Work in Borrow Area 3 | 4,170 |
| | 4 | Hitachi EX1100 ($5.1m^3$) | 32 | $4^1$ | Work in Borrow Area 2 | 4,810 |
| | | | | $4^2$ | Work in Borrow Area 4 | 4,740 |
| | 5 | Hitachi EX1800 ($10.3m^3$) | 27 | $5^1$ | Work in Borrow Area 3 | 5,160 |
| | | | | $5^2$ | Work in Borrow Area 4 | 5,290 |
| | 6 | Komatsu PC220 ($1.0m^3$) | 31 | $6^1$ | Work in Borrow Area 1 | 3,870 |
| | | | | $6^2$ | Work in Borrow Area 2 | 3,920 |

**Table 3.7** Detailed information of construction machines in Xiaolangdi Hydropower Project

| Machine | Type index | Machine type | Job index | Machine failure probability during the $(k+1)$th week $p^k_{i\theta}$ | | | | Expected value of the conversion factor during the $(k+1)$th week $E[\zeta^{rand}_{k,i\theta}]$ | | | |
|---|---|---|---|---|---|---|---|---|---|---|---|
| | $i$ | | $i^\theta$ | k=0 | k=1 | k=2 | k=3 | k=0 | k=1 | k=2 | k=3 |
| Dump truck | 1 | Perlini DP366 ($20m^3/36t$) | $1^1$ | 0.0133 | 0.1035 | 0.2497 | 0.3167 | 0.9968 | 0.9595 | 0.8865 | 0.8402 |
| | | | $1^2$ | 0.0204 | 0.1327 | 0.2776 | 0.3072 | 0.9949 | 0.9464 | 0.8706 | 0.8414 |
| | 2 | Perlini DP755 ($30m^3/65t$) | $2^1$ | 0.0099 | 0.0877 | 0.2338 | 0.3226 | 0.9977 | 0.9665 | 0.8956 | 0.8393 |
| | | | $2^2$ | 0.0152 | 0.1133 | 0.2633 | 0.3181 | 0.9964 | 0.9554 | 0.8794 | 0.8379 |
| Hydraulic excavator | 3 | Hitachi EX400 ($2.2m^3$) | $3^1$ | 0.0122 | 0.1035 | 0.2609 | 0.3306 | 0.9972 | 0.9601 | 0.8820 | 0.8324 |
| | | | $3^2$ | 0.0136 | 0.0984 | 0.2314 | 0.3003 | 0.9967 | 0.9611 | 0.8947 | 0.8498 |
| | 4 | Hitachi EX1100 ($5.1m^3$) | $4^1$ | 0.0176 | 0.1237 | 0.2732 | 0.3146 | 0.9957 | 0.9507 | 0.8737 | 0.8385 |
| | | | $4^2$ | 0.0153 | 0.1115 | 0.2570 | 0.3134 | 0.9963 | 0.9559 | 0.8823 | 0.8409 |
| | 5 | Hitachi EX1800 ($10.3m^3$) | $5^1$ | 0.0086 | 0.0765 | 0.2083 | 0.3053 | 0.9980 | 0.9708 | 0.9076 | 0.8502 |
| | | | $5^2$ | 0.0129 | 0.1030 | 0.2528 | 0.3214 | 0.9970 | 0.9599 | 0.8853 | 0.8376 |
| | 6 | Komatsu PC220 ($1.0m^3$) | $6^1$ | 0.0158 | 0.1053 | 0.2343 | 0.2936 | 0.9961 | 0.9578 | 0.8925 | 0.8526 |
| | | | $6^2$ | 0.0142 | 0.1065 | 0.2511 | 0.3138 | 0.9966 | 0.9581 | 0.8855 | 0.8414 |

and Stockpile Area 3, which needs to be transported to the dam. The massive schists and limestone are both extremely large size kinds of rockfill, and only a Terex TA28 ($17m^3/28t$) can transport them. Thus the Terex TA28 ($17m^3/28t$) is arranged to transport the massive schists and limestone from Borrow Area 2 to the dam as well as from Stockpile Area 3 to the dam. The job assignments for the other types of dump truck are similar. As a result, each type of dump truck is arranged to do 2 kinds of jobs as shown in Table 3.8 (i.e., $M_i$=2 for $i = 1,2,\cdots,6$). The initial number for each type of dump trucks at the beginning of the whole period and the expected value of the machine throughput (i.e., $E^{Me}[\tilde{Q}_{i\theta}]$, where the optimistic-pessimistic index $\eta$ is set to be 0.5) for each type of dump truck are shown in Table3.8. All necessary data for each type of dump truck, including the machine failure probability, the expected value of the conversion factor, were calculated or derived from actual data (See Tables 3.9 and 3.10).

**Table 3.8** Job assignments of the 6 types of dump trucks in Shuibuya Hydropower Project

| Type index $i$ | Equipment (*Dump truck*) | Initial number $a_i$ (*unit*) | Job index $i^\theta$ | Job assignment | Throughput (Expected value of equipment productivity) $E[Q_{i\theta}]$ ($m^3/week \cdot unit$) |
|---|---|---|---|---|---|
| 1 | Terex TA25 ($13.5m^3/23t$) | 50 | $1^a$ | Transport between Borrow Area 1 and the dam | 4000 |
| | | | $1^b$ | Transport between Stockpile Area 1 and the dam | 3700 |
| 2 | Terex TA28 ($17m^3/28t$) | 40 | $2^a$ | Transport between Borrow Area 2 and the dam | 4200 |
| | | | $2^b$ | Transport between Stockpile Area 3 and the dam | 4150 |
| 3 | K29N-6.4 ($11m^3/20t$) | 40 | $3^a$ | Transport between Stockpile Area 2 and the dam | 4500 |
| | | | $3^b$ | Transport between Orphan Bank 2 and the dam | 4600 |
| 4 | K30N-8.4 ($16m^3/26t$) | 50 | $4^a$ | Transport between Orphan Bank 1 and the dam | 4300 |
| | | | $4^b$ | Transport between Stockpile Area 2 and the dam | 4350 |
| 5 | Perlini DP366 ($20m^3/36t$) | 34 | $5^a$ | Transport between Borrow Area 1 and the dam | 4010 |
| | | | $5^b$ | Transport between Stockpile Area 1 and the dam | 4130 |
| 6 | Perlini DP755 ($30m^3/65t$) | 41 | $6^a$ | Transport between Stockpile Area 3 and the dam | 5010 |
| | | | $6^b$ | Transport between Borrow Area 2 and the dam | 5170 |

**Table 3.9** The detailed information of dump trucks in Shuibuya Hydropower Project

| Type index | Equipment | Job index | Equipment failure probability during the $(k+1)$th week $p^k_{i\theta}$ | | | | Mean time to work during the $(k+1)$th week $E[T^k_{i\theta}]$ | | | |
|---|---|---|---|---|---|---|---|---|---|---|
| $i$ | (*Dump truck*) | $i^\theta$ | k=0 | k=1 | k=2 | k=3 | k=0 | k=1 | k=2 | k=3 |
| 1 | Terex TA25 ($13.5m^3/23t$) | $1^1$ | 0.0231 | 0.1472 | 0.2971 | 0.3080 | 0.9942 | 1.9403 | 2.8603 | 3.8388 |
| | | $1^2$ | 0.0155 | 0.1020 | 0.2267 | 0.2879 | 0.9961 | 1.9591 | 2.8961 | 3.8560 |
| 2 | Terex TA28 ($17m^3/28t$) | $2^1$ | 0.0139 | 0.1065 | 0.2542 | 0.3176 | 0.9967 | 1.9583 | 2.8842 | 3.8392 |
| | | $2^2$ | 0.0146 | 0.1039 | 0.2396 | 0.3028 | 0.9964 | 1.9588 | 2.8905 | 3.8477 |
| 3 | K29N-6.4 ($11m^3/20t$) | $3^1$ | 0.0103 | 0.0929 | 0.2477 | 0.3322 | 0.9977 | 1.9646 | 2.8891 | 3.8332 |
| | | $3^2$ | 0.0132 | 0.1098 | 0.2707 | 0.3319 | 0.9969 | 1.9575 | 2.8769 | 3.8306 |
| 4 | K30N-8.4 ($16m^3/26t$) | $4^1$ | 0.0109 | 0.0848 | 0.2122 | 0.2952 | 0.9974 | 1.9669 | 2.9046 | 3.8543 |
| | | $4^2$ | 0.0103 | 0.0863 | 0.2244 | 0.3112 | 0.9976 | 1.9667 | 2.8995 | 3.8456 |
| 5 | Perlini DP366 ($20m^3/36t$) | $5^1$ | 0.0133 | 0.1035 | 0.2497 | 0.3167 | 0.9968 | 1.9595 | 2.8865 | 3.8402 |
| | | $5^2$ | 0.0204 | 0.1327 | 0.2776 | 0.3072 | 0.9949 | 1.9464 | 2.8706 | 3.8414 |
| 6 | Perlini DP755 ($30m^3/65t$) | $6^1$ | 0.0099 | 0.0877 | 0.2338 | 0.3226 | 0.9977 | 1.9665 | 2.8956 | 3.8393 |
| | | $6^2$ | 0.0152 | 0.1133 | 0.2633 | 0.3181 | 0.9964 | 1.9554 | 2.8794 | 3.8379 |

**Table 3.10** Expected value of the conversion factor $\zeta^{rand}_{k,i^\theta}$

| Type index | Equipment | Job index | Expected value of the conversion factor during the $(k+1)$th week $E[\zeta^{rand}_{k,i^\theta}]$ | | | |
|---|---|---|---|---|---|---|
| $i$ | (*Dump truck*) | $i^\theta$ | k=0 | k=1 | k=2 | k=3 |
| 1 | Terex TA25 ($13.5m^3/23t$) | $1^1$ | 0.9942 | 0.9403 | 0.8603 | 0.8388 |
| | | $1^2$ | 0.9961 | 0.9591 | 0.8961 | 0.8560 |
| 2 | Terex TA28 ($17m^3/28t$) | $2^1$ | 0.9967 | 0.9583 | 0.8842 | 0.8392 |
| | | $2^2$ | 0.9964 | 0.9588 | 0.8905 | 0.8477 |
| 3 | K29N-6.4 ($11m^3/20t$) | $3^1$ | 0.9977 | 0.9646 | 0.8891 | 0.8332 |
| | | $3^2$ | 0.9969 | 0.9575 | 0.8769 | 0.8306 |
| 4 | K30N-8.4 ($16m^3/26t$) | $4^1$ | 0.9974 | 0.9669 | 0.9046 | 0.8543 |
| | | $4^2$ | 0.9976 | 0.9667 | 0.8995 | 0.8456 |
| 5 | Perlini DP366 ($20m^3/36t$) | $5^1$ | 0.9968 | 0.9595 | 0.8865 | 0.8402 |
| | | $5^2$ | 0.9949 | 0.9464 | 0.8706 | 0.8414 |
| 6 | Perlini DP755 ($30m^3/65t$) | $6^1$ | 0.9977 | 0.9665 | 0.8956 | 0.8393 |
| | | $6^2$ | 0.9964 | 0.9554 | 0.8794 | 0.8379 |

### 3.5.3 Physical Constraints for Job Assignments

Some actual examples can be easily found in CFDR construction projects. In fact, a CFDR construction project has many dam areas, borrow areas, stockpile areas, and orphan banks as shown in Figure 3.13. Different borrow areas supply different kinds of rockfill, and different kinds of rockfill should be transported by different types of dump trucks and stored in different stockpile areas. That means one type of dump truck can be arranged to do several different jobs. But there are usually some physical constraints for job assignments since some jobs are equipment specified. For example, one type of dump truck is arranged to transport a specified kind of rockfill. This specified rockfill need to be transported from Borrow Area 1 to Dam Area 2 as well as from Stockpile Area 2 to Dam Area 2. As a result, the equipment failure rates and equipment productivity can be different depending on transport distances, road conditions and kind of rockfill (Varty et al. 1985 [9]). In addition, equipment failure rates also depend on user skill and previous maintenance effectiveness.

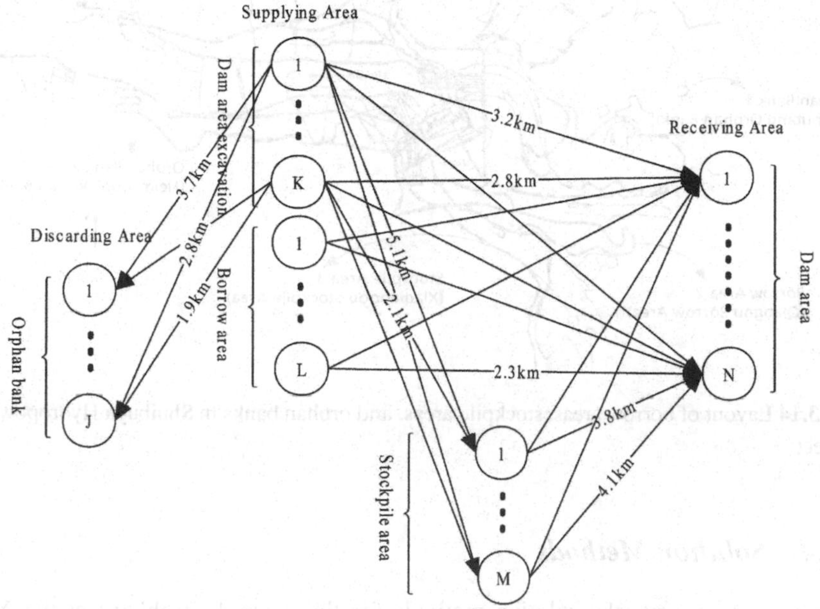


**Fig. 3.13** Earthwork transportation among dam areas, borrow areas, stockpile areas, and orphan banks

In the Shuibuya Hydropower Project, there are 2 borrow areas, 3 stockpile areas and 2 orphan banks. The location and detailed information for each area are illustrated in Figure 3.14. While each type of dump truck can be arranged to do several different jobs, there are usually some physical constraints for job assignments since some jobs are equipment specified. In this case study, there are massive schists and limestone in Borrow Area 2 and Stockpile Area 3, which should be transported to the dam. These massive schists and limestone are both extremely large rockfill, and only a Perlini DP755 ($30m^3/65t$) can transport them. Thus a Perlini DP755 ($30m^3/65t$) is arranged to transport the massive schists and limestone from Borrow Area 2 to the dam as well as from Stockpile Area 3 to the dam. The job assignments of the other types of dump truck are similar. As a result, each type of dump truck is arranged to do 2 kinds of jobs as shown in Table 3.8 (i.e., $M_i$=2 for $i = 1, 2, \cdots, 6$). The initial number for each type of dump truck at the beginning of the whole period and the expected value of the equipment productivity for each type of dump truck are shown in Table 3.8. The fuzzy expected value for equipment productivity for each type of dump truck is estimated using Eq. (3.28).

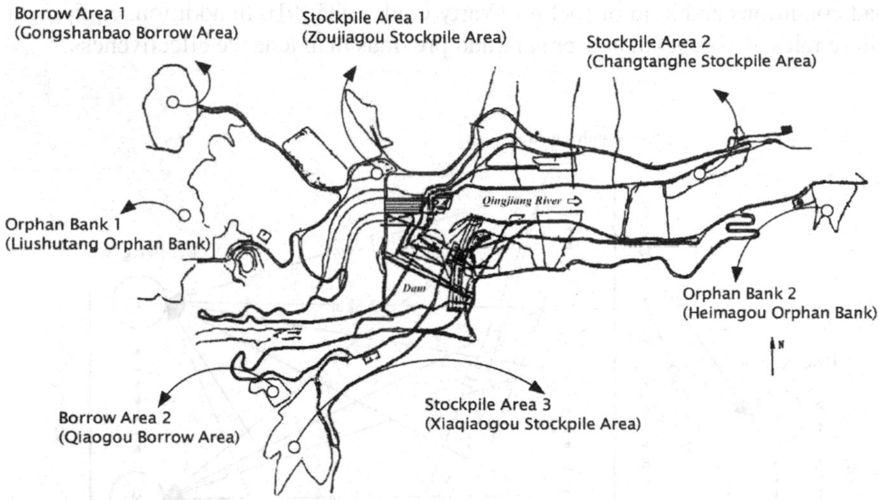


**Fig. 3.14** Layout of borrow areas, stockpile areas, and orphan banks in Shuibuya Hydropower Project

## 3.5.4 *Solution Methods*

This section presents the solution methods for the example problems at the Xiaolangdi Hydropower Project and the Shuibuya Hydropower Project.

### Xiaolangdi Hydropower Project

The solution method employed to solve the DMAP at the Xiaolangdi Hydropower Project is the theoretical algorithm. Based on the framework of the theoretical algorithm for the DMAP and the collected data from the Xiaolangdi Hydropower Project in the previous section, the solution can be computed using MATLAB 7.0.

**Shuibuya Hydropower Project**

In this case, the PSO algorithm is employed to solve the DMAP at the Shuibuya Hydropower Project. The PSO parameters were set based on the results of some preliminary experiments that were carried out to observe the behavior of the algorithm at different parameter settings. By comparing several sets of parameters, including population size, iteration number, acceleration constant, initial velocity, and inertia weight, the most reasonable parameters for each type of dump truck was chosen. Since more particles require more evaluation runs, leading to a higher optimization cost (Trelea 2003 [40]), 50 particles were selected as the population size and 100 times as the iteration number for all types of construction equipment. The inertia weight $w(\tau)$ was set to be varying with the iteration:

$$w(\tau) = w(T) + \frac{\tau - T}{1 - T}[w(1) - w(T)] \tag{3.75}$$

where $\tau$ = iteration index $= 1, 2, \cdots, T$; $T$ = iteration limit. Other parameters were selected by comparing the results and observing the dynamic search behavior of the swarm. Table 3.11 summarizes the values of all parameters selected for the 6 types of dump truck.

**Table 3.11** Parameter selection for the 6 types of dump trucks

| Type index | Equipment | Population size | Iteration number | Acceleration constant | | Initial velocity | | | | Inertia weight | |
|---|---|---|---|---|---|---|---|---|---|---|---|
| $i$ | *(Dump truck)* | $L$ | $T$ | $c_p$ | $c_g$ | $v_{i1}^j(0)$ | $v_{i2}^j(0)$ | $v_{i3}^j(0)$ | $v_{i4}^j(0)$ | $w(1)$ | $w(T)$ |
| 1 | Terex TA25 ($13.5m^3/23t$) | 50 | 100 | 2 | 3 | 2 | 2 | 2 | 2 | 0.9 | 0.1 |
| 2 | Terex TA28 ($17m^3/28t$) | 50 | 100 | 5 | 6 | 2 | 2 | 2 | 2 | 0.9 | 0.1 |
| 3 | K29N-6.4 ($11m^3/20t$) | 50 | 100 | 10 | 15 | 3 | 3 | 3 | 3 | 0.9 | 0.1 |
| 4 | K30N-8.4 ($16m^3/26t$) | 50 | 100 | 2.5 | 3 | 2 | 2 | 2 | 2 | 0.9 | 0.1 |
| 5 | Perlini DP366 ($20m^3/36t$) | 50 | 100 | 5 | 6 | 2 | 2 | 2 | 2 | 0.9 | 0.1 |
| 6 | Perlini DP755 ($30m^3/65t$) | 50 | 100 | 5 | 6 | 2 | 2 | 2 | 2 | 0.9 | 0.1 |

### 3.5.5 *Results and Analysis*

This section discusses the results and provides analysis for the example problems at the Xiaolangdi and Shuibuya Hydropower Projects.

**Xiaolangdi Hydropower Project**

Based on the theoretical algorithm discussed in previous sections, the results of the example problem at the Xiaolangdi Hydropower Project are presented in Table 3.12. Comparing the algorithm result (total construction or production throughput) with the actual data from each project in Table 3.12, the findings indicate that there are differences between the solutions generated by the theoretical algorithm developed in this chapter and the actual production throughput. The net increase in the total production throughput for each type of machine ranged from 9,357 to 36,373 $m^3$

for the Xiaolangdi Hydropower Project, and the rate of increase, was from 2.03% to 5.82% for the Xiaolangdi Hydropower Project. This improvement for the construction or production efficiency could bring a considerable economic benefit for each project, especially for a large scale one. While this finding is quite encouraging, at present it is not 100% certain that all these differences represent a net improvement because the mathematical model was formulated with some assumptions. Therefore, some portion of these differences could result from possible modeling errors. To some extent, it can be concluded from the resulting data that the theoretical optimized values for total construction or production throughput for the design variables are better than the actual values for each project. This means that the current practice for each project is not optimal. In fact, in the current practice, decision makers usually make their decisions on experience only. Thus, our results can be used to provide the decision makers with a theoretical optimal dynamic equipment allocation plan to guide current practice.

**Shuibuya Hydropower Project**

In order to show the practicality and efficiency of the optimization method for the DEAP presented in this chapter, the PSO algorithm was implemented to solve the above problem, and then compared with the actual data collected from the Shuibuya Hydropower Project. For the purpose of reducing potential statistical errors, the convergence iteration number and computing time of the algorithm for each type of dump trucks was set at a value of 10 implementations.

The results and performance of the algorithm, compared to the actual data from the project, are listed in Table 3.13. As shown, the solution to the DEAP for Equipment $i$ is expressed by $u^*_{i\theta}(k)$ and $x^*_i(k)$. Based on the resulting value of $u^*_{i\theta}(k)$ and

**Table 3.12** Results of theoretical algorithm for DMAP in Xiaolangdi Hydropower Project

| Machine | Type index $i$ | Machine type | Variable | Week index $k$: 0 | 1 | 2 | 3 | Total construction throughput: Algorithm result ($m^3$) | Actual data from the project ($m^3$) | Net increase ($m^3$) | Rate of increase |
|---|---|---|---|---|---|---|---|---|---|---|---|
| Dump truck | 1 | Perlini DP366 ($20m^3/36t$) | $x^*_1(k)$ | 34 | 33 | 29 | 22 | 446,220 | 421,670 | 24,550 | 5.82% |
| | | | $u^*_{11}(k)$ | 0 | 33 | 29 | 0 | | | | |
| | | | $u^*_{12}(k)$ | 34 | 0 | 0 | 22 | | | | |
| | 2 | Perlini DP755 ($30m^3/65t$) | $x^*_2(k)$ | 41 | 40 | 36 | 28 | 687,720 | 651,347 | 36,373 | 5.58% |
| | | | $u^*_{21}(k)$ | 0 | 40 | 36 | 0 | | | | |
| | | | $u^*_{22}(k)$ | 41 | 0 | 0 | 28 | | | | |
| Hydraulic excavator | 3 | Hitachi EX400 ($2.2m^3$) | $x^*_3(k)$ | 28 | 27 | 24 | 19 | 387,690 | 369,873 | 17,817 | 4.82% |
| | | | $u^*_{31}(k)$ | 28 | 27 | 0 | 19 | | | | |
| | | | $u^*_{32}(k)$ | 0 | 0 | 24 | 0 | | | | |
| | 4 | Hitachi EX1100 ($5.1m^3$) | $x^*_4(k)$ | 32 | 31 | 27 | 20 | 487,300 | 472,114 | 15,186 | 3.22% |
| | | | $u^*_{41}(k)$ | 32 | 0 | 0 | 20 | | | | |
| | | | $u^*_{42}(k)$ | 0 | 31 | 27 | 0 | | | | |
| | 5 | Hitachi EX1800 ($10.3m^3$) | $x^*_5(k)$ | 27 | 26 | 24 | 19 | 469,230 | 459,873 | 9,357 | 2.03% |
| | | | $u^*_{51}(k)$ | 0 | 26 | 24 | 0 | | | | |
| | | | $u^*_{52}(k)$ | 27 | 0 | 0 | 19 | | | | |
| | 6 | Komatsu PC220 ($1.0m^3$) | $x^*_6(k)$ | 31 | 30 | 27 | 20 | 393,030 | 373,036 | 19,994 | 5.36% |
| | | | $u^*_{61}(k)$ | 0 | 0 | 27 | 20 | | | | |
| | | | $u^*_{62}(k)$ | 31 | 30 | 0 | 0 | | | | |

**Table 3.13** Results of PSO algorithm for DEAP in Shuibuya Hydropower Project

| Type index $i$ | Equipment (Dump truck) | Variable | Week index $k$: 0 | 1 | 2 | 3 | Total construction throughput: Fitness value ($m^3$) | Actual data from the project ($m^3$) | Net increase ($m^3$) | Rate of increase | Convergence iteration number | Computing time (s) |
|---|---|---|---|---|---|---|---|---|---|---|---|---|
| 1 | Terex TA25 ($13.5m^3/23t$) | $x_1^*(k)$ | 50 | 48 | 43 | 33 | 622470 | 601029 | 21441 | 3.57% | 24 | 15.765 |
| | | $u_{11}^*(k)$ | 50 | 0 | 0 | 33 | | | | | | |
| | | $u_{12}^*(k)$ | 0 | 48 | 43 | 0 | | | | | | |
| 2 | Terex TA28 ($17m^3/28t$) | $x_2^*(k)$ | 40 | 39 | 35 | 26 | 545400 | 513897 | 31503 | 6.13% | 26 | 15.938 |
| | | $u_{21}^*(k)$ | 40 | 39 | 0 | 26 | | | | | | |
| | | $u_{22}^*(k)$ | 0 | 0 | 35 | 0 | | | | | | |
| 3 | K29N-6.4 ($11m^3/20t$) | $x_3^*(k)$ | 40 | 39 | 35 | 26 | 592090 | 552193 | 39897 | 7.23% | 98 | 15.954 |
| | | $u_{31}^*(k)$ | 0 | 39 | 35 | 0 | | | | | | |
| | | $u_{32}^*(k)$ | 40 | 0 | 0 | 26 | | | | | | |
| 4 | K30N-8.4 ($16m^3/26t$) | $x_4^*(k)$ | 50 | 49 | 45 | 35 | 726810 | 689454 | 37356 | 5.42% | 26 | 16.328 |
| | | $u_{41}^*(k)$ | 0 | 0 | 45 | 0 | | | | | | |
| | | $u_{42}^*(k)$ | 50 | 49 | 0 | 35 | | | | | | |
| 5 | Perlini DP366 ($20m^3/36t$) | $x_5^*(k)$ | 34 | 33 | 29 | 22 | 446220 | 421670 | 24550 | 5.82% | 25 | 15.484 |
| | | $u_{51}^*(k)$ | 0 | 33 | 29 | 0 | | | | | | |
| | | $u_{52}^*(k)$ | 34 | 0 | 0 | 22 | | | | | | |
| 6 | Perlini DP755 ($30m^3/65t$) | $x_6^*(k)$ | 41 | 40 | 36 | 28 | 687720 | 651347 | 36373 | 5.58% | 32 | 14.972 |
| | | $u_{61}^*(k)$ | 0 | 40 | 36 | 0 | | | | | | |
| | | $u_{62}^*(k)$ | 41 | 0 | 0 | 28 | | | | | | |

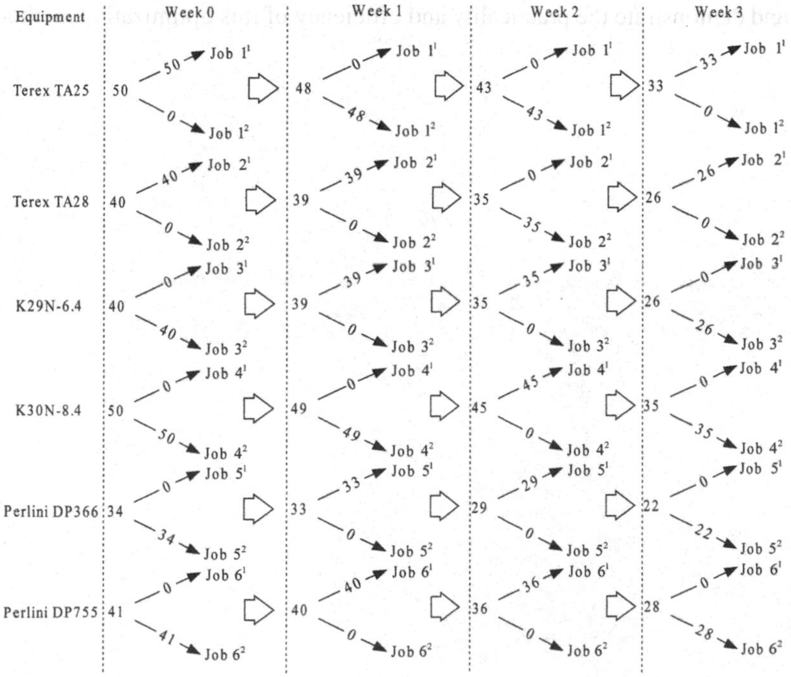


**Fig. 3.15** Dynamic allocation plan for the six types of dump trucks

$x_i^*(k)$ performed in Table 3.13, a detailed dynamic allocation plan for the 6 types of dump truck is determined as shown in Figure 3.15.

Comparing the fitness value (i.e., total construction throughput) with the actual data from the project in Table 3.13, the findings indicate that there are differences between the solutions generated by the algorithm and the actual construction throughput. The net increase in the total construction throughput for each type of equipment ranges from 21441 to 39897 $m^3$, and the rate of increase, from 3.57% to 7.23%. This improvement in construction efficiency could bring considerable economic benefit to the construction project, especially those of a larger scale. While this finding is quite encouraging, it is not 100% sure that all these differences represent a net improvement because the mathematical model was formulated with some assumptions and a portion of these differences could be a result of possible modeling errors. To some extent, it is possible to conclude from the resulting data that the theoretical optimized values for total construction throughput for the design variables are better than the actual values. This means that the current practice is not optimal. In fact, for current practice, the decision makers usually make their decisions on experience alone. Thus, these results could be used to provide decision makers with a theoretical optimal dynamic equipment allocation plan to guide current practice.

However, since the aim of this study is to provide an alternative and effective method to optimize equipment allocation in construction project using the PSO algorithm, this solution method and computational results are sufficient and significant, and demonstrate the practicality and efficiency of this optimization method.

# 4

# Fuzzy MOMSDM for Closed Multiclass Queuing Networks

Queuing decision problems are often encountered in many practical systems, such as flexible manufacturing systems (FMS), telecommunication systems, field service support systems, and flow-shop-type production systems (Gross and Harris 1998 [229]; Hillier and Lieberman 2001 [230]; Taha 2003 [240]; Balsamo et al. 2003 [212]; Lazowska et al. 1984 [237]; Kim 2009 [234]). Queuing decision models play an important role in queuing system designs that typically involve making one or a combination of decisions, such as the number of servers at a service facility, the efficiency of the servers, and the number of service facilities (Chen 2007 [215]). In fact, the construction system queuing network is quite different from the above systems (Govil and Fu 1999 [225]; Van and Vandaele 2007 [241]). In the concrete transportation system at the Jinping-I Hydropower Project, the queuing network has four transportation subsystems, with the arrival flow intensity depending significantly on the system itself in state, and with a closed transportation route network. The most suitable statement for describing this system is a closed multiclass queuing networks problem (CMQNP). As there are a limited number of vehicles and unloading equipment, the decision-makers need to determine a suitable schedule for dynamically allocating these amongst the different transportation subsystems so that the total operating cost and construction duration are minimized. This chapter mainly focuses on a discussion of multistage decision making for closed multiclass queuing network problems under a fuzzy environment.

## 4.1 Statement of CMQNP

There has been significant research about closed queuing networks including the development of methodologies, theoretical studies, and applications (Daduna and Szekli 2002 [216]; Kogan 2002 [235]; Abramov 2001 [210]; Gerasimov 2000 [224]; Casale 2011 [214]; George and Xia 2011 [223]). More specifically, Harrison and Coury (2002) [228] studied an expression for the generating function of the normalizing constant for a closed multiclass Markovian queuing network in terms of similar functions for networks with one fewer class. Berger et al. (1999) [213] also

J. Xu and Z. Zeng, *Fuzzy-Like Multiple Objective Multistage Decision Making*, Studies in Computational Intelligence 533,
DOI: 10.1007/978-3-319-03398-3_4, © Springer International Publishing Switzerland 2014

presented a bottleneck analysis in multiclass closed queuing networks, and the case of two bottlenecks was illustrated by its application to the problem of dimensioning bandwidth for different data sources in packet-switched communication networks. While these studies have significantly improved queuing decision theory and applications, they are incapable of reflecting the queuing features in construction systems. The statement for a CMQNP in a concrete transportation system is introduced in this section.

### *4.1.1 Concrete Construction in High Arch Dam Project*

In a concrete high arch dam construction project, the concrete needs to be transported by vehicles from the concrete production site to the concrete pouring site. The concrete construction process can be divided into three systems; concrete production, transportation, and pouring systems, as shown in Figure 4.1.

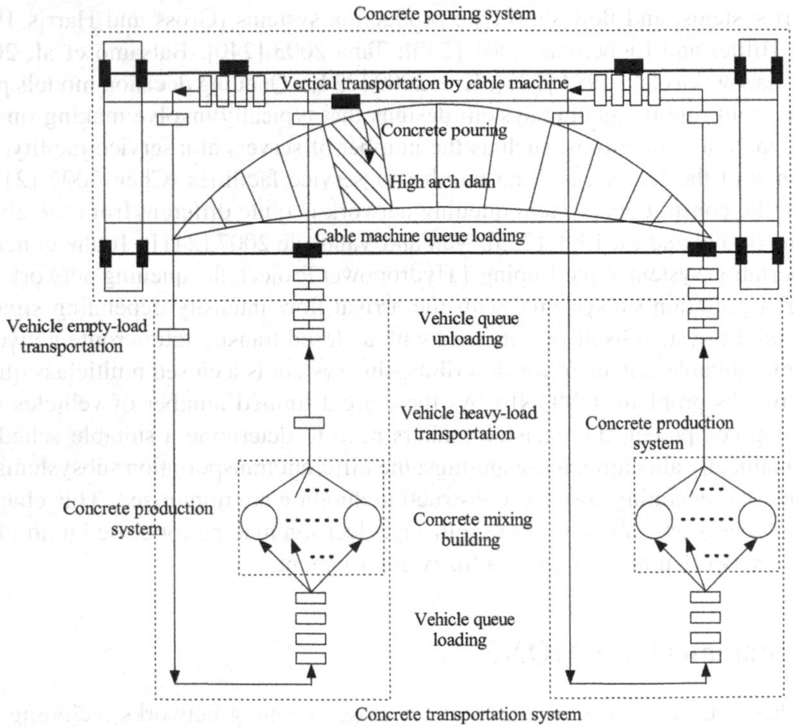


**Fig. 4.1** Concrete production, transportation, and pouring systems

In a large scale hydropower project, there may be one or multiple concrete production systems. Each concrete production system is composed of several concrete mixing buildings, which produce concrete by mixing cement, water, aggregate, and fly ash. Since different pouring areas need different kinds of concrete (i.e., concrete of different proportions) at different stages, the layout of the concrete production sites should be appropriately planned to provide adequate concrete in line with the pouring intensity of the corresponding pouring areas. Vehicles queue at the concrete mixing buildings to load the concrete, and then, under a heavy-load, transport it to unloading sites at the concrete pouring systems. To reduce the influence of external temperature, transportation time is usually controlled to within 20-30 minutes. When the vehicles arrive at the corresponding unloading site, they queue to unload the concrete into the waiting cable machines (i.e., unloading equipment), and then with an empty-load return to the concrete mixing building for the next load. This circulation is referred to as a closed queuing system for concrete transportation.

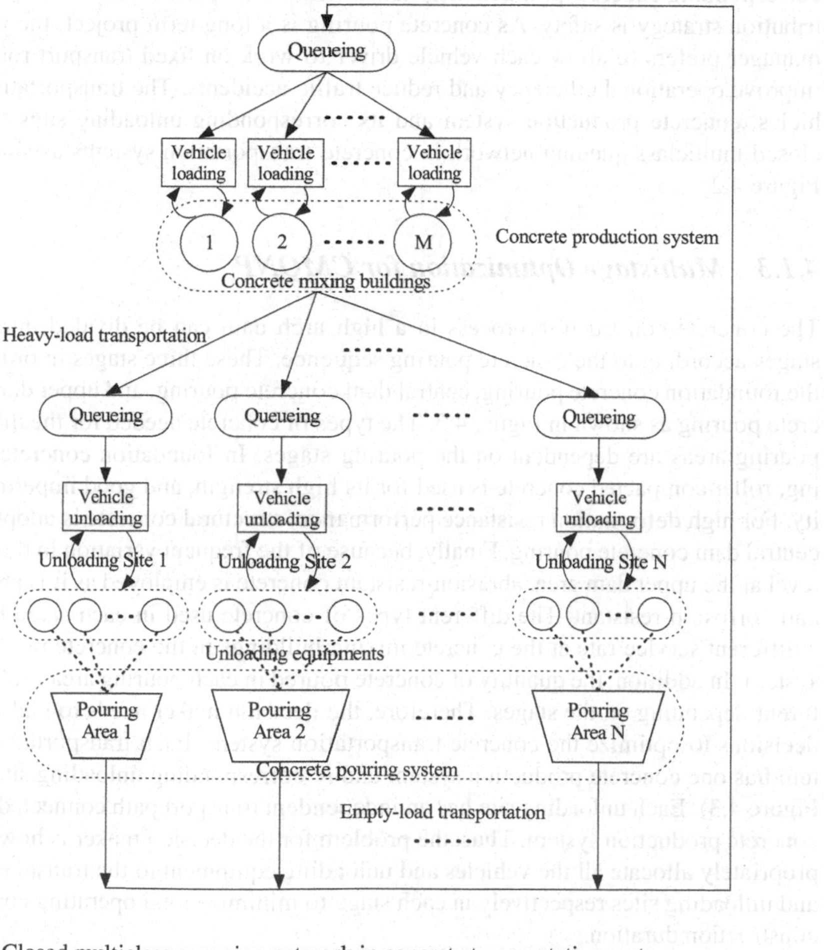


**Fig. 4.2** Closed multiclass queueing network in concrete transportation system

### *4.1.2 Closed Multiclass Queuing Networks in Concrete Transportation System*

It is usual that one concrete producing system is responsible for supplying concrete to several unloading sites, especially in large scale construction projects. Each unloading site is in charge of its corresponding pouring area and may have several pieces of unloading equipment (i.e., cable machines) working between them. The vehicles move between the concrete production system and its corresponding unloading site. Since several unloading sites may share the same concrete production system, and the transport paths between the concrete production system and different unloading sites are quite dissimilar, including transportation distance, road condition, and transportation intensity, the vehicles are usually divided into several teams to transport the concrete between the different unloading sites and the corresponding concrete production systems. Another important reason for this distribution strategy is safety. As concrete pouring is a long term project, the project manager prefers to allow each vehicle driver to work on fixed transport routes to improve operational efficiency and reduce traffic accidents. The transportation vehicles, concrete production system and its corresponding unloading sites form a closed multiclass queuing network in concrete transportation systems as shown in Figure 4.2.

### *4.1.3 Multistage Optimization for CMQNP*

The concrete construction process in a high arch dam can be divided into three stages according to the concrete pouring sequence. These three stages in order are; the foundation concrete pouring, central dam concrete pouring, and upper dam concrete pouring as shown in Figure 4.3. The types of concrete needed for the different pouring areas are dependent on the pouring stages. In foundation concrete pouring, roller compacted concrete is used for its high strength, and good impermeability. For high deformation resistance performance, structural concrete is adopted for central dam concrete pouring. Finally, because of the frequent variation in the water level at the upper dam area, abrasion-resistant concrete is employed as it is abrasion and corrosion-resistant. The different types of concrete used in each stage lead to a different service rate at the concrete mixing buildings in the concrete production system. In addition, the quantity of concrete poured in each pouring area is also different depending on the stages. Therefore, the decision maker needs to make stage decisions to optimize the concrete transportation system. Each transportation system has one concrete production system and $N$ corresponding unloading sites (see Figure 4.3). Each unloading site has an independent transport path connected to the concrete production system. Thus, the problem for the decision maker is how to appropriately allocate all the vehicles and unloading equipment to the transport paths and unloading sites respectively in each stage to minimize total operating costs and construction duration.

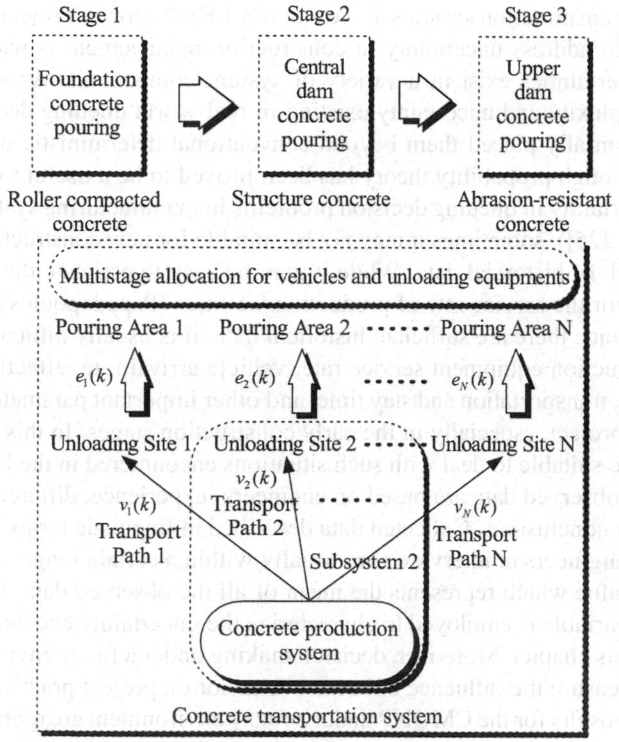


**Fig. 4.3** Multistage optimization for queueing decision problems

## 4.1.4 Motivation for Employing Fuzzy Variables in CMQNP

As is known, there is uncertain information in a CMQNP, most of which has been paid little attention in previous research. For example, the service rate and arrival rate are usually considered stochastic distributions (De Vuyst et al. 2002 [217]; Ivnitski 2001 [231]). However, in a new construction project, because activities tend to be unique and therefore lack historical data, it is difficult for a project manager to characterize these random variables correctly. For this reason, the fuzzy method is considered an effective approach for such situations. First proposed by Zadeh (1965) [22], and consequently developed by researchers such as Dubois and Prade (1988) [221], and Nahmias (1978) [238], fuzzy theory has been a useful tool in dealing with ambiguous information. For example, the unit cost of customer waiting time may vary within a certain time frame. Taha (2003) [240] pointed out that the main barrier to implementing cost models of this kind was that it may be difficult to obtain a reliable estimate of the unit wait cost, particularly when human behavior impacts the operation. In this chapter, a CMQNP under a fuzzy environment is considered and fuzzy theory is used to deal with the uncertain information in the modeling of the concrete transportation systems at the Jinping-I Hydropower Project.

The need to address uncertainty in construction management is widely recognized, as uncertainties exist in a variety of system components. As a result, the inherent complexity and uncertainty existing in real-world queuing decision problems has essentially placed them beyond conventional deterministic optimization methods. Although probability theory has been proved to be a useful tool for dealing with uncertainty in queuing decision problems in manufacturing systems (Govil and Fu 1999 [225]), sometimes it may not be suitable for new construction projects due to the lack of historical data. While it may be easy to estimate the probability distributions for the service rate of production facilities (Papadopoulos and Heavey 1996 [232]) since there are sufficient historical data, it is usually difficult to do this for the construction equipment service rate, vehicle arrival rate, effective monthly working days, transportation and stay time, and other important parameters in a new construction project, especially in the early construction stages. In this case, fuzzy theory is more suitable to deal with such situations encountered in the CMQNP. As the collected observed data are based on engineers' experience, different engineers have different conclusions. Collected data described in linguistic terms by different experienced engineers in interviews are usually within a certain range, and there is, therefore, a value which represents the mean of all the observed data. For this reason, a fuzzy variable is employed to characterize the uncertainty encountered in the CMQNP in this chapter. Moreover, decision making under a fuzzy environment can adequately measure the influence of human behavior on project practice. Thus, the optimization results for the CMQNP under a fuzzy environment are quite useful and may serve as a reference for decision makers in guiding current practice.

### 4.1.5 *Construction of Fuzzy Membership Functions*

In most cases, the methods for formulating fuzzy membership functions can be classified into three approaches: construction using the analyst's judgment, construction using experiments, or construction using a given numerical data set (Lee et al. 2006 [226]). Selecting a method to determine the membership functions depends on many conditions, including the characteristics of the study and the available data set associated with the study (Lee and Donnell 2006 [227]). In this study, a triangular membership function is considered the most appropriate type for characterizing the observed data. Since the triangular membership function can be expressed as a triangular fuzzy number (i.e., $\tilde{\gamma} = (r_1, r_2, r_3)$), it should be noted that the mean (i.e., $r_2$), the left border (i.e., $r_1$) and the right border (i.e., $r_3$) of the fuzzy number are determined from the engineers' experience. To collect the data, interview were done with differently experienced engineers (i.e., $\vartheta = 1, 2, \cdots, E$, where $\vartheta$ is the index of engineers). The observed data are described in linguistic terms such as "between 0.28 and 0.40, with the most likely value being 0.34." It should be noted that the engineers presented the linguistic terms based on their observed data over time. Generally, the view of each engineer can be denoted as a range in the observed data

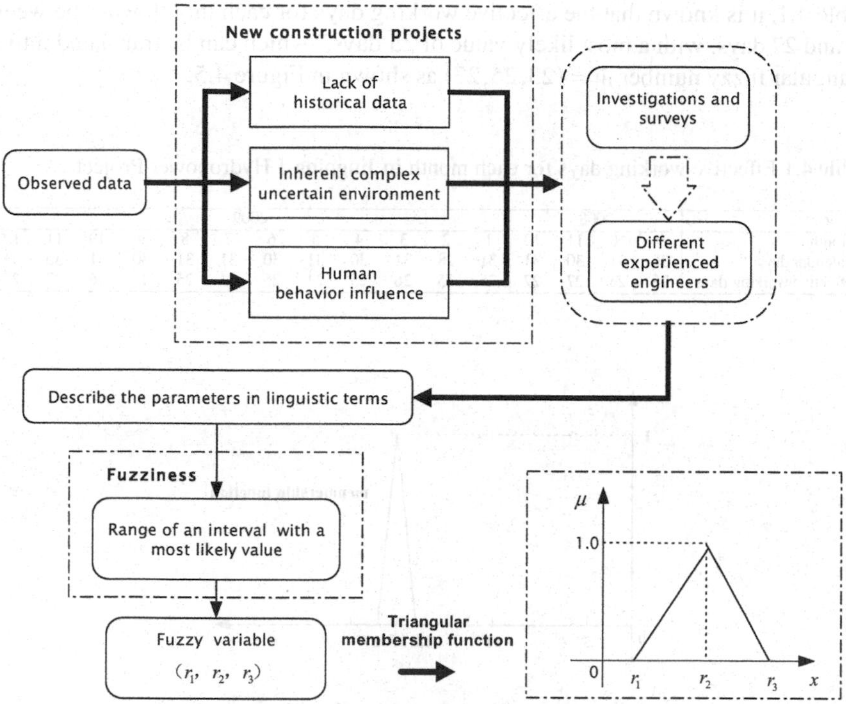


**Fig. 4.4** Flowchart of construction of fuzzy membership functions

(i.e., $[r_1^{\vartheta}, r_3^{\vartheta}]$) and a most likely value (i.e., $r_2^{\vartheta}$). Since different engineers have different views on the observed data, the minimal value of all $r_1^{\vartheta}$ and the maximal value of all $r_3^{\vartheta}$ (for $\vartheta = 1, 2, \cdots, E$) are selected as the left border (i.e., $r_1$) and the right border (i.e., $r_3$) of the triangular fuzzy number $\tilde{\gamma}$, respectively. The expected value of all the $r_2^{\vartheta}$ (for $\vartheta = 1, 2, \cdots, E$) is regarded as the mean (i.e., $r_2$) of the triangular fuzzy number $\tilde{\gamma}$. Figure 4.4 shows the flowchart for the construction of the fuzzy membership functions.

## *4.1.6 Typical Fuzzy Variables in CMQNP*

### Effective Working Days for Each Month

The effective working days for each month is a typical uncertain variable, which can fluctuate in a month because of lots of factors, such as the weather condition, weekends, festival and holiday events. While the uncertainty of weekends, festival and holiday events in each month are epistemic, the uncertainty of the weather condition is mostly aleatoric. Table 4.1 lists the effective working days for each month at the Jingping-I Hydropower Project from September 2008 to December 2009. Let $\tilde{w}$ be the effective working days for each month. Based on the data shown in

Table 4.1, it is known that the effective working days for each month was "between 23 and 27 days, with a most likely value of 25 days," which can be translated into a triangular fuzzy number $\tilde{w} = (23, 25, 27)$ as shown in Figure 4.5.

**Table 4.1** Effective working days for each month in Jingping-I Hydropower Project

| Year | 2008 | | | | 2009 | | | | | | | | | | | |
|---|---|---|---|---|---|---|---|---|---|---|---|---|---|---|---|---|
| Month | 9 | 10 | 11 | 12 | 1 | 2 | 3 | 4 | 5 | 6 | 7 | 8 | 9 | 10 | 11 | 12 |
| Calendar days | 30 | 31 | 30 | 31 | 31 | 28 | 31 | 30 | 31 | 30 | 31 | 31 | 30 | 31 | 30 | 31 |
| Effective working days | 25 | 26 | 27 | 27 | 25 | 25 | 26 | 25 | 24 | 25 | 23 | 25 | 25 | 26 | 27 | 27 |

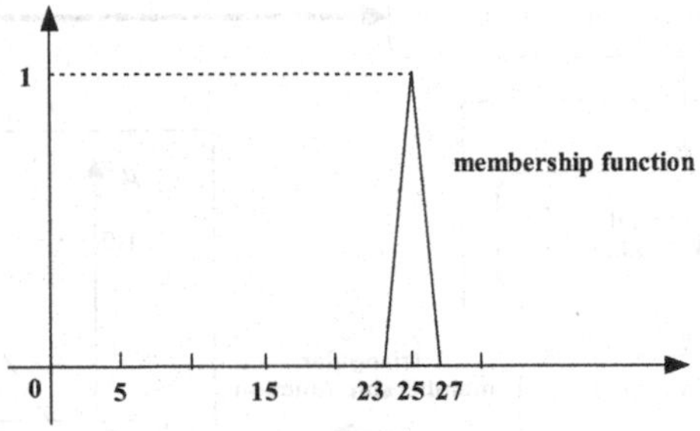


**Fig. 4.5** Membership function of the effective working days for each month

## Transportation and Stay Time

The transportation time for each vehicle is determined by many factors, such as travel distance, personal driving technique, and route condition, which vary from case to case. For each transport path, it may be difficult to obtain a reliable estimate of the transportation time, particularly when human behavior impacts the operation of the situation. Specifically, in many practical applications, statistical information may be obtained subjectively, i.e., the transportation patterns are more suitably described using linguistic terms such as "between 15 and 20 minutes, with the most likely value being 17 minutes," which can be translated into a triangular fuzzy number $(15, 17, 20)$. Fuzzy set theory deals with a set of objects characterized by a membership (characteristic) function that assigns to each object a grade of membership ranging between zero (no membership) and one (full membership) (Zadeh 1965 [22]). The concept of "unsharp boundaries" that fuzzy set theory represents mimics the human way of thinking, which works with shades of grey rather than black and white. The situation is similar with the stay time for each vehicle. The mathematical operations for the fuzzy numbers (i.e., addition, subtraction, multiplication, and division) can be processed using the definition given by Dubois and Prade (1978) [337].

Let $\tilde{T}_h^i(k)$ and $\tilde{T}_e^i(k)$ be the heavy-load and empty-load transportation time of Path $i$ in Stage $k$, respectively, $\tilde{T}_u^i(k)$ and $\tilde{T}_c(k)$, the stay time at Unloading Site $i$ and the concrete production system, respectively. So, the total time $\tilde{T}_i(k)$ for a vehicle when transporting between Unloading Site $i$ and the concrete production system in Stage $k$ is:

$$\tilde{T}_i(k) = \tilde{T}_h^i(k) + \tilde{T}_e^i(k) + \tilde{T}_u^i(k) + \tilde{T}_c(k), \quad \forall i,k. \tag{4.1}$$

### Arrival Rate of Vehicles

The arrival rate of each vehicle is determined by its total transportation time. Since the transportation time is fuzzy, the arrival rate is also fuzzy. Let $\tilde{\lambda}_i(k)$ be the arrival rate of a vehicle when arriving at Unloading Site $i$. Based on Eq. (4.1), the following equation is obtained

$$\tilde{\lambda}_i(k) = \frac{1}{\tilde{T}_i(k)} = \frac{1}{\tilde{T}_h^i(k) + \tilde{T}_e^i(k) + \tilde{T}_u^i(k) + \tilde{T}_c(k)}, \quad \forall i,k. \tag{4.2}$$

In closed transportation systems, the arrival rate of each subsystem (i.e., unloading site) is dependent on the state of the subsystem itself. As Figure 4.3 shows, Subsystem $i$ is made up of Unloading Site $i$, Transport Path $i$, and the concrete production system. Each subsystem is a closed transportation subsystem. If $v_i(k)$ is the number of vehicles that are allocated to Subsystem $i$ in Stage $k$, it can be argued that at any moment the vehicles in Subsystem $i$ that are not at Unloading Site $i$ are the arrival resource for Unloading Site $i$. Let $\tilde{\lambda}_i^j(k)$ be the arrival rate of Unloading Site $i$ in Stage $k$ when there are $j$ vehicles unloading and queuing, then:

$$\tilde{\lambda}_i^j(k) = (v_i(k) - j)\tilde{\lambda}_i(k), \quad \forall i,k. \tag{4.3}$$

### Service Rate of Facilities

The service rates for unloading equipment at the unloading sites can also be extracted as triangular fuzzy numbers from experts. Let $e_i(k)$ be the number of unloading equipment at Unloading Site $i$, and $\tilde{\mu}_i(k)$, the service rate of the unit unloading equipment at Unloading Site $i$ in Stage $k$. When there are $j$ vehicles at Unloading Site $i$, the service rate at Unloading Site $i$ in Stage $k$ (i.e., $\tilde{\mu}_i^j(k)$) can be expressed as:

$$\tilde{\mu}_i^j(k) = \begin{cases} j\tilde{\mu}_i(k) & j = 0,1,\cdots,e_i(k)-1 \\ e_i(k)\tilde{\mu}_i(k) & j = e_i(k), e_i(k)+1,\cdots,v_i(k) \end{cases} \tag{4.4}$$

### State Transition of Transportation System

Based on the above statement, the arrival rate of vehicles and the service rate of the facilities are the motivation for the transportation system state transition, especially for the pouring areas. Let $S_i^j(k)$ be the state of Unloading Site $i$ in Stage $k$ when there are $j$ vehicles unloading and queuing. All possible states are listed in Table 4.2. Each state can transfer to the neighboring state through the occurrence of an arrival event or service event. For example, the intensity of event flow that transfers state $S_i^0(k)$ to state $S_i^1(k)$ at Unloading Site $i$ in Stage $k$ is $\tilde{\lambda}_i^0(k) = v_i(k)\tilde{\lambda}_i(k)$, and the intensity of

**Table 4.2** The possible state of transportation system

| State | Vehicle number in Unloading Site $i$ | Arrival rate | Service rate |
|---|---|---|---|
| $S_i^0(k)$ | 0 | $\tilde{\lambda}_i^0(k) = v_i(k)\tilde{\lambda}_i(k)$ | $\tilde{\mu}_i^0(k) = (0,0,0)$ |
| $S_i^1(k)$ | 1 | $\tilde{\lambda}_i^1(k) = (v_i(k)-1)\tilde{\lambda}_i(k)$ | $\tilde{\mu}_i^1(k) = \tilde{\mu}_i(k)$ |
| $\cdots$ | $\cdots$ | $\cdots$ | $\cdots$ |
| $S_i^{e_i(k)-1}(k)$ | $e_i(k)-1$ | $\tilde{\lambda}_i^{e_i(k)-1}(k) = (v_i(k)-e_i(k)+1)\tilde{\lambda}_i(k)$ | $\tilde{\mu}_i^{e_i(k)-1}(k) = (e_i(k)-1)\tilde{\mu}_i(k)$ |
| $S_i^{e_i(k)}(k)$ | $e_i(k)$ | $\tilde{\lambda}_i^{e_i(k)}(k) = (v_i(k)-e_i(k))\tilde{\lambda}_i(k)$ | $\tilde{\mu}_i^{e_i(k)}(k) = e_i(k)\tilde{\mu}_i(k)$ |
| $S_i^{e_i(k)+1}(k)$ | $e_i(k)+1$ | $\tilde{\lambda}_i^{e_i(k)+1}(k) = (v_i(k)-e_i(k)-1)\tilde{\lambda}_i(k)$ | $\tilde{\mu}_i^{e_i(k)+1}(k) = e_i(k)\tilde{\mu}_i(k)$ |
| $\cdots$ | $\cdots$ | $\cdots$ | $\cdots$ |
| $S_i^{v_i(k)}(k)$ | $v_i(k)$ | $\tilde{\lambda}_i^{v_i(k)}(k) = (0,0,0)$ | $\tilde{\mu}_i^{v_i(k)}(k) = e_i(k)\tilde{\mu}_i(k)$ |

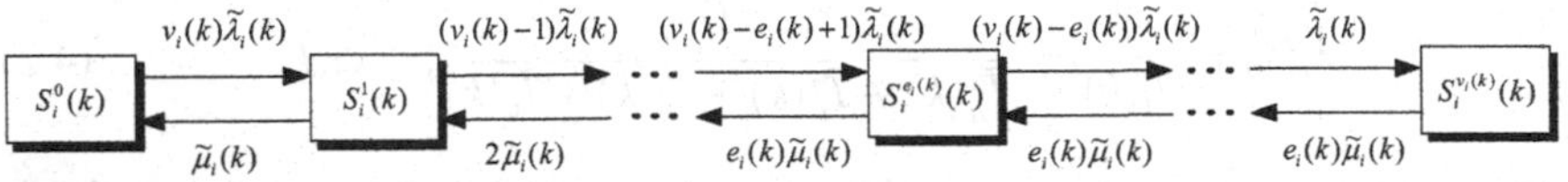


**Fig. 4.6** State transition of Unloading Site $i$

event flow that transfers state $S_i^1(k)$ to state $S_i^0(k)$ at Unloading Site $i$ in Stage $k$ is $\tilde{\mu}_i^1(k) = \tilde{\mu}_i(k)$. Figure 4.6 shows the state transition for Unloading Site $i$.

The above state transition has two characteristics: (1) the first is that each state may be transferred to either of the rest states; (2) the second is that the number of all possible states is finite. Let $p_i^j(k)$ be the limit probability of state $S_i^j(k)$ $(j = 0, 1, \cdots, v_i(k))$. Based on the properties of continuous-time Markov chains (Puterman 2005 [239]), the limit probability of each state should satisfy the Kolmogorov equations as follows:

$$\begin{cases} v_i(k)\tilde{\lambda}_i(k)p_i^0(k) = \tilde{\mu}_i(k)p_i^1(k) \\ (v_i(k)-1)\tilde{\lambda}_i(k)p_i^1(k) = 2\tilde{\mu}_i(k)p_i^2(k) \\ \cdots \\ (v_i(k)-e_i(k)+1)\tilde{\lambda}_i(k)p_i^{e_i(k)-1}(k) = e_i(k)\tilde{\mu}_i(k)p_i^{e_i(k)}(k) \\ \cdots \\ \tilde{\lambda}_i(k)p_i^{v_i(k)-1}(k) = e_i(k)\tilde{\mu}_i(k)p_i^{v_i(k)} \end{cases} \tag{4.5}$$

i.e.,

$$\begin{cases} p_i^1(k) = \frac{v_i(k)\tilde{\lambda}_i(k)}{\tilde{\mu}_i(k)}p_i^0(k) = v_i(k)\tilde{\rho}_i(k)p_i^0(k) \\ p_i^2(k) = \frac{v_i(k)(v_i(k)-1)(\tilde{\lambda}_i(k))^2}{(\tilde{\mu}_i(k))^2}p_i^0(k) = \frac{v_i(k)(v_i(k)-1)}{2!}(\tilde{\rho}_i(k))^2p_i^0(k) \\ \cdots \\ p_i^{e_i(k)}(k) = \frac{v_i(k)(v_i(k)-1)\cdots(v_i(k)-e_i(k)+1)}{e_i(k)!}(\tilde{\rho}_i(k))^{e_i(k)}p_i^0(k) \\ \cdots \\ p_i^{v_i(k)}(k) = \frac{v_i(k)!}{e_i(k)!e_i(k)^{v_i(k)-e_i(k)}}(\tilde{\rho}_i(k))^{v_i(k)}p_i^0(k) \end{cases} \tag{4.6}$$

where $\tilde{\rho}_i(k) = \frac{\tilde{\lambda}_i(k)}{\tilde{\mu}_i(k)}$. Since all states are possibly available, it is obvious that

$$\sum_{j=0}^{v_i(k)} p_i^j(k) = 1, \tag{4.7}$$

based on Eq. (4.6) and (4.7), then

$$v_i(k)\tilde{\rho}_i(k)p_i^0(k) + \frac{v_i(k)(v_i(k)-1)}{2!}(\tilde{\rho}_i(k))^2 p_i^0(k) + \cdots + \frac{v_i(k)(v_i(k)-1)\cdots(v_i(k)-e_i(k)+1)}{e_i(k)!} (\tilde{\rho}_i(k))^{e_i(k)} p_i^0(k) + \cdots + \frac{v_i(k)!}{e_i(k)!e_i(k)^{v_i(k)-e_i(k)}}(\tilde{\rho}_i(k))^{v_i(k)} p_i^0(k) = 1$$

i.e.,

$$p_i^0(k) = \left[\sum_{j=0}^{e_i(k)-1} \frac{v_i(k)!}{(v_i(k)-j)!j!}(\tilde{\rho}_i(k))^j + \sum_{j=e_i(k)}^{v_i(k)} \frac{v_i(k)!}{(v_i(k)-j)!e_i(k)!e_i(k)^{j-e_i(k)}}(\tilde{\rho}_i(k))^j\right]^{-1}. \tag{4.8}$$

Further more, from Eqs. (4.6) and (4.8), then

$$p_i^j(k) = \begin{cases} \frac{v_i(k)!}{(v_i(k)-j)!j!}(\tilde{\rho}_i(k))^j p_i^0(k), & 1 \le j < e_i(k) \\ \frac{v_i(k)!}{(v_i(k)-j)!e_i(k)!e_i(k)^{j-e_i(k)}}(\tilde{\rho}_i(k))^j p_i^0, & e_i(k) \le j \le v_i(k) \end{cases} \tag{4.9}$$

## 4.2 Modelling Process of CMQNP

To model the multistage optimization for queuing decision problems under a fuzzy environment in this chapter, the state equations, initial conditions, constraint conditions, and objective functions are presented.

### *4.2.1 State Equations*

Let $A_v(k)$ (or $A_e(k)$) be the number of available vehicles (or unloading equipment) for a transportation system in Stage $k$. Suppose there are $N$ subsystems in a transportation system, and $v_i(k)$ (or $e_i(k)$) is the number of vehicles (or unloading equipment) that are allocated to Subsystem (or Unloading Site) $i$ ($i = 1, 2, \cdots, N$) in Stage $k$. Since all available vehicles (or unloading equipment) at the beginning of every stage should be allocated to jobs, Eqs. (4.10) and (4.11) can be used to describe the relationship between the state variables (i.e., $A_v(k)$ and $A_e(k)$) and the decision variables (i.e., $v_i(k)$ and $e_i(k)$) as follows,

$$A_v(k) = \sum_{i=1}^{N} v_i(k), \quad \forall k. \tag{4.10}$$

and

$$A_e(k) = \sum_{i=1}^{N} e_i(k), \quad \forall k. \tag{4.11}$$

### 4.2.2 Initial Conditions

It should be noted that the number of available vehicles (or unloading equipment) allocated to a transportation system could be different at different stages, which is decided by the project manager according to the construction requirements for that stage. Let $\alpha_k$ (or $\beta_k$) be the number of available vehicles (unloading equipment) allocated to a transportation system in Stage $k$, thus the state variables $A_v(k)$ (or $A_e(k)$) can be initialized as follows,

$$A_v(k) = \alpha_k, \quad \forall k, \tag{4.12}$$

and

$$A_e(k) = \beta_k, \quad \forall k. \tag{4.13}$$

### 4.2.3 Constraint Conditions

Obviously, the number of vehicles (or unloading equipment) that are allocated to Subsystem (or Unloading Site) $i$ $(i = 1, 2, \cdots, N)$ in Stage $k$ (i.e., $v_i(k)$ (or $e_i(k)$)) must be more than zero and less than the total number of available vehicles (or unloading equipment) for a transportation system in Stage $k$ (i.e., $A_v(k)$ (or $A_e(k)$)). Therefore, the constraint conditions are as follows,

$$0 < v_i(k) < A_v(k), \quad \forall i, k. \tag{4.14}$$

and

$$0 < e_i(k) < A_e(k), \quad \forall i, k. \tag{4.15}$$

On the other hand, since the limit probability of each state (i.e., $p_i^j(k)$) is a function of the decision variables (i.e., $v_i(k)$ and $e_i(k)$), Eq. (4.8) and (4.9) are also considered constraint conditions.

### 4.2.4 Objective Functions

The objective of the decision maker is to appropriately allocate all vehicles to the transport paths in each stage while minimizing total construction duration and operating costs. This leads to the need to look at this as a multiple objective decision making problem. The details of the two objective functions are presented in this part.

#### Construction Duration

Let $t$ $(hour/day)$ be the vehicle work time, $H$ $(m^3/vehicle)$, the heaped capacity for a vehicle, $I_i^d(k)$ $(m^3/day)$, the pouring intensity of Pouring Area $i$ in Stage $k$, and $\xi_i(k)$ $(vehicle/hour)$, the service rate of Unloading Site $i$ in Stage $k$. Since $\xi_i(k)$ is a random variable, the value for which is $\tilde{\mu}_i^j(k)$ $(j = 0, 1, \cdots, v_i(k))$ as shown in Eq. (4.4), its expected value can be given based on the stochastic distribution expressed in Eqs. (4.8) and (4.9) as follows,

$$E(\xi_i(k)) = \sum_{j=0}^{v_i(k)} p_i^j(k)\tilde{\mu}_i^j(k), \quad \forall i,k. \tag{4.16}$$

Thus, the pouring intensity ($m^3/day$) of Pouring Area $i$ in Stage $k$ could be,

$$I_i^d(k) = tHE(\xi_i(k)) = tH\sum_{j=0}^{v_i(k)} p_i^j(k)\tilde{\mu}_i^j(k), \quad \forall i,k. \tag{4.17}$$

Let $I_i^m(k)$ ($m^3/month$) be the pouring intensity for Pouring Area $i$ in Stage $k$, $Q_i(k)$ ($m^3$), the required concrete pouring quantity of Pouring Area $i$ in Stage $k$, and $D_i(k)$ ($month$), the concrete construction duration of Pouring Area $i$ in Stage $k$. Since $\tilde{w}$ ($day/month$) is the effective working days for each month, then

$$D_i(k) = \frac{Q_i(k)}{I_i^m(k)} = \frac{Q_i(k)}{\tilde{w}I_i^d(k)} = \frac{Q_i(k)}{\tilde{w}tH\sum_{j=0}^{v_i(k)} p_i^j(k)\tilde{\mu}_i^j(k)}, \quad \forall i,k. \tag{4.18}$$

Note that, for each transportation system, only when the concrete construction of all pouring areas in the current stage has been finished can the next concrete construction stage be started. Let $f_d(\mathbf{x},\tilde{\theta})$ ($month$) be the objective function of construction duration, where $\mathbf{x} = (v_1(\cdot), v_2(\cdot), \cdots, v_N(\cdot), e_1(\cdot), e_2(\cdot), \cdots, e_N(\cdot))$, $\tilde{\theta} = (\tilde{w}, \tilde{\lambda}_1(\cdot), \tilde{\lambda}_2(\cdot), \cdots, \tilde{\lambda}_N(\cdot), \tilde{\mu}_1(\cdot), \tilde{\mu}_2(\cdot), \cdots, \tilde{\mu}_N(\cdot))$. Therefore, the total concrete construction duration for a transportation system is,

$$f_d(\mathbf{x},\tilde{\theta}) = \sum_{k=1}^{3} \max_i\{D_i(k)\} = \sum_{k=1}^{3} \max_i \left\{ \frac{Q_i(k)}{\tilde{w}tH\sum_{j=0}^{v_i(k)} p_i^j(k)\tilde{\mu}_i^j(k)} \right\}. \tag{4.19}$$

**Operation Cost**

Let $C_b$ ($CNY/day$) be the unit operating cost of a concrete mixing building in a construction production system, $C_v$ ($CNY/day$), the unit operating cost of a vehicle, $C_e$ ($CNY/day$), the unit operating cost of a piece of unloading equipment at a pouring area, and $f_c(\mathbf{x},\tilde{\theta})$ ($CNY$), the objective function for the operating cost. Suppose there are $M$ concrete mixing buildings in a construction production system, then the total operating cost for a transportation system is,

$$f_c(\mathbf{x},\tilde{\theta}) = \sum_{k=1}^{3} \tilde{w}(C_bM + C_vA_v(k) + C_eA_e(k)) \max_i\{D_i(k)\},$$

i.e.,

$$f_c(\mathbf{x}, \tilde{\theta}) = \sum_{k=1}^{3} \tilde{w}(C_b M + C_v A_v(k) + C_e A_e(k)) \max_i \left\{ \frac{Q_i(k)}{\tilde{w}tH \sum_{j=0}^{v_i(k)} p_i^j(k)\tilde{\mu}_i^j(k)} \right\}. \tag{4.20}$$

### 4.2.5 *Dealing with Fuzzy Coefficients*

It is difficult to handle a multistage multiple objective problem when it involves uncertain information. The coefficients involved in the queuing decision model are triangular fuzzy numbers, which means that the model solution is difficult to find. One strategy is to employ a transformation method to convert the current model into a deterministic equivalent. In this section, a crisp approach is used to integrate the decision maker's optimistic-pessimistic attitudes in adopting real world practice. The fuzzy coefficients in the model are defuzzified using an expected value operator with an optimistic-pessimistic index. To calculate the expected value of the triangular fuzzy numbers, a new measure with an optimistic-pessimistic adjustment index is introduced to characterize the real life problems. The definition for this measure can be found in Xu and Zhou (2011) [281].

Let $\tilde{\gamma} = (r_1, r_2, r_3)$ be any triangular fuzzy number in the model. In fact, in real world problems, especially in a queuing decision problem in large-scale construction projects, the case is often encountered when $r_1 > 0$, so the expected value of $\tilde{\gamma}$ is presented as below,

$$E^{Me}[\tilde{\gamma}] = \frac{(1-\eta)}{2} r_1 + \frac{1}{2} r_2 + \frac{\eta}{2} r_3, \tag{4.21}$$

where $\eta$ is the optimistic-pessimistic index to determine the combined attitude of a decision maker.

It should be noted that if $\tilde{\lambda}_i(k) = (a_1, a_2, a_3)$ and $\tilde{\mu}_i(k) = (b_1, b_2, b_3)$, then according to the algebraic operations for triangular fuzzy numbers (Xu and Zhou, 2011 [281]), $\frac{\tilde{\lambda}_i(k)}{\tilde{\mu}_i(k)} = (\frac{a_1}{b_3}, \frac{a_2}{b_2}, \frac{a_3}{b_1})$. Since $\tilde{\rho}_i(k) = \frac{\tilde{\lambda}_i(k)}{\tilde{\mu}_i(k)}$, based on Eq. (4.21), its expected value is,

$$E^{Me}[\tilde{\rho}_i(k)] = \frac{(1-\eta)}{2} \cdot \frac{a_1}{b_3} + \frac{1}{2} \cdot \frac{a_2}{b_2} + \frac{\eta}{2} \cdot \frac{a_3}{b_1}. \tag{4.22}$$

Similarly, the expected value for $\tilde{\mu}_i^j(k)$ is,

$$E^{Me}[\tilde{\mu}_i^j(k)] = \begin{cases} jE^{Me}[\tilde{\mu}_i(k)] = j(\frac{(1-\eta)}{2} b_1 + \frac{1}{2} b_2 + \frac{\eta}{2} b_3) & j = 0, 1, \cdots, e_i(k) - 1 \\ e_i(k)E^{Me}[\tilde{\mu}_i(k)] = e_i(k)(\frac{(1-\eta)}{2} b_1 + \frac{1}{2} b_2 + \frac{\eta}{2} b_3) & j = e_i(k), \\ e_i(k) + 1, \cdots, v_i(k) & \end{cases} \tag{4.23}$$

### 4.2.6 Equivalent Crisp Model

Based on the above approach that transformed the fuzzy coefficients in the objective functions and the constraints into expected values, the multistage multiple objective queuing decision problems can be formulated as following:

$$
\begin{array}{l}
\min f_d(\mathbf{x},\tilde{\theta}) = \sum_{k=1}^{3} \max_i \left\{ \frac{Q_i(k)}{E^{Me}[\tilde{w}]tH\sum_{j=0}^{v_i(k)} p_i^j(k)E^{Me}[\tilde{\mu}_i^j(k)]} \right\},\\
\min f_c(\mathbf{x},\tilde{\theta}) = \sum_{k=1}^{3} E^{Me}[\tilde{w}](C_bM + C_vA_v(k) + C_eA_e(k))\max_i \\
\qquad \left\{ \frac{Q_i(k)}{E^{Me}[\tilde{w}]tH\sum_{j=0}^{v_i(k)} p_i^j(k)E^{Me}[\tilde{\mu}_i^j(k)]} \right\},\\
\text{s.t.} \left\{ \begin{array}{l}
A_v(k) = \sum_{i=1}^{N} v_i(k), \quad \forall k,\\
A_e(k) = \sum_{i=1}^{N} e_i(k), \quad \forall k,\\
A_v(k) = \alpha_k, \forall k,\\
A_e(k) = \beta_k, \quad \forall k,\\
0 < v_i(k) < A_v(k), \quad \forall i,k,\\
0 < e_i(k) < A_e(k), \quad \forall i,k,\\
p_i^0(k) = \left[\sum_{j=0}^{e_i(k)-1} \frac{v_i(k)!}{(v_i(k)-j)!j!}(E^{Me}[\tilde{\rho}_i(k)])^j + \sum_{j=e_i(k)}^{v_i(k)} \frac{v_i(k)!}{(v_i(k)-j)!e_i(k)!e_i(k)^{j-e_i(k)}}\right.\\
\qquad \left.(E^{Me}[\tilde{\rho}_i(k)])^j\right]^{-1},\\
p_i^j(k) = \left\{ \begin{array}{ll}
\frac{v_i(k)!}{(v_i(k)-j)!j!}(E^{Me}[\tilde{\rho}_i(k)])^j p_i^0(k), & 1 \le j < e_i(k)\\
\frac{v_i(k)!}{(v_i(k)-j)!e_i(k)!e_i(k)^{j-e_i(k)}}(E^{Me}[\tilde{\rho}_i(k)])^j p_i^0(k), & e_i(k) \le j \le v_i(k)
\end{array}\right.
\end{array}\right.
\end{array}
\tag{4.24}
$$

It is difficult to handle multiple objective problems when each objective function has a different dimensionality, as it is not possible to simply transform the multiple objective model into a single objective model by introducing weight coefficients. Thus it is necessary to determine an equivalent crisp model. Suppose there is a most endurable construction duration for a transportation system (i.e., $D$) which is accepted by the decision makers, so the objective function for the construction duration $\min f_d(\mathbf{x},\tilde{\theta})$ can be converted into an inequality as below:

$$
\sum_{k=1}^{3} \max_i \left\{ \frac{Q_i(k)}{E^{Me}[\tilde{w}]tH\sum_{j=0}^{v_i(k)} p_i^j(k)E^{Me}[\tilde{\mu}_i^j(k)]} \right\} \le D. \tag{4.25}
$$

Following these introductions, Model (4.24) can now be transformed into an equivalent crisp model (4.26):

$$
\begin{aligned}
&\min f_c(\mathbf{x},\tilde{\theta}) = \textstyle\sum_{k=1}^{3} E^{Me}[\tilde{w}](C_bM + C_vA_v(k) + C_eA_e(k)) \max_i \\
&\qquad \left\{ \frac{Q_i(k)}{E^{Me}[\tilde{w}]tH\sum_{j=0}^{v_i(k)} p_i^j(k)E^{Me}[\tilde{\mu}_i^j(k)]} \right\}, \\
&s.t. \begin{cases}
A_v(k) = \sum_{i=1}^{N} v_i(k), \quad \forall k, \\
A_e(k) = \sum_{i=1}^{N} e_i(k), \quad \forall k, \\
A_v(k) = \alpha_k, \quad \forall k, \\
A_e(k) = \beta_k, \quad \forall k, \\
0 < v_i(k) < A_v(k), \quad \forall i,k, \\
0 < e_i(k) < A_e(k), \quad \forall i,k, \\
p_i^0(k) = \Big[\sum_{j=0}^{e_i(k)-1} \frac{v_i(k)!}{(v_i(k)-j)!j!}(E^{Me}[\tilde{\rho}_i(k)])^j + \sum_{j=e_i(k)}^{v_i(k)} \frac{v_i(k)!}{(v_i(k)-j)!e_i(k)!e_i(k)^{j-e_i(k)}} \\
\qquad (E^{Me}[\tilde{\rho}_i(k)])^j\Big]^{-1}, \\
p_i^j(k) = \begin{cases}
\frac{v_i(k)!}{(v_i(k)-j)!j!}(E^{Me}[\tilde{\rho}_i(k)])^j p_i^0(k), & 1 \le j < e_i(k) \\
\frac{v_i(k)!}{(v_i(k)-j)!e_i(k)!e_i(k)^{j-e_i(k)}}(E^{Me}[\tilde{\rho}_i(k)])^j p_i^0(k), & e_i(k) \le j \le v_i(k)
\end{cases} \\
\sum_{k=1}^{3} \max_i \left\{ \frac{Q_i(k)}{E^{Me}[\tilde{w}]tH\sum_{j=0}^{v_i(k)} p_i^j(k)E^{Me}[\tilde{\mu}_i^j(k)]} \right\} \le D.
\end{cases}
\end{aligned}
\tag{4.26}
$$

## 4.3 Antithetic Method-Based Particle Swarm Optimization

Since there is an objective function and constraints (i.e., $p_i^0(k)$ and $p_i^j(k)$) that are not linear in the model, and the number of summation items in the objective function and constraints changes depending on the decision variables, the above equivalent crisp model is in fact an NP-hard problem (Kunigahalli and Russell 1995 [236]), which cannot be solved using general linear programming method. To solve this multistage queuing decision problem (i.e., the equivalent crisp model (4.26)), an antithetic method-based particle swarm optimization (AM-based PSO) algorithm is developed. Particle swarm optimization (PSO) is an optimization technique based on swarm intelligence first proposed by Kennedy and Eberhart (1995) [23]. Particles are "flown" through an n-dimensional search space, with each particle being attracted towards the best solution found by the swarm and the best solution found by the particle (Eberhart and Shi 2001 [24]; Clerc and Kennedy 2002 [25]). PSO is a powerful optimization tool for large-scale, NP-hard problems that might have nonlinear, discontinuous, and nonconvex objective functions and constraints (Anghinolfi and Paolucci 2009 [211]; Yapicioglu et al. 2007 [244]; Zhang et al. 2006a [27]; Xu and Zeng 2011 [242]). It is important when using PSO to have a carefully designed computational experiment with a clear rationale for the solution representation, particle-updating mechanism, and parameters being used as well as a clear framework for assessing the search performance. These key features of the proposed AM-based PSO are explained in detail in the following sections.

### 4.3.1 Solution Representation of AM-Based PSO

In the AM-based PSO, each particle is encoded as an array with $2N \times 3$ (i.e., $H = 2N \times 3$) dimensions or elements, representing a possible solution in a multi-dimensional parameter space. For easier understanding, each particle is divided into 3 parts, representing the decisions variables (i.e., $v_i(k)$ and $e_i(k)$) made in the three stages respectively, as follows:

$$P_l(\tau) = [p_{l1}(\tau), p_{l2}(\tau), \cdots, p_{lH}(\tau)] = [Y_l^1(\tau), Y_l^2(\tau), Y_l^3(\tau)] \tag{4.27}$$

$$\begin{aligned} Y_l^k(\tau) &= [y_{l1}^k(\tau), y_{l2}^k(\tau), \cdots, y_{l(2N)}^k(\tau)] \\ &\Longleftrightarrow [v_1(k), v_2(k), \cdots, v_N(k), e_1(k), e_2(k), \cdots, e_N(k)], \end{aligned} \tag{4.28}$$

where $l$ (i.e., particle index) $= 1, 2, \cdots, L$; $L$ is the population size; $\tau$ (i.e., iteration index) $= 0, 1, \cdots, T$; $T$ is the iteration limit; $P_l(\tau) = [p_{l1}(\tau), p_{l2}(\tau), \cdots, p_{lH}(\tau)]$ denotes the $H$-dimension position for the $l$th particle in the $\tau$th iteration; $Y_l^k(\tau) =$ the $k$th part of the $l$th particle in the $\tau$th generation (i.e., the decision variables $v_i(k)$ and $e_i(k)$) made in Stage $k$. Note that every part of a particle is a $2N$-dimension vector, where $y_{lq}^k(\tau) =$ the $q$th dimension of $Y_l^k(\tau)$ for the $l$th particle in the $\tau$th generation; $q = 1, \cdots, 2N$. Since all available vehicles (or unloading equipment) at the beginning of every stage must be allocated to do jobs, reflecting the state constraints denoted by Eqs. (4.10) and (4.11), the elements of each particle should satisfy the following equations:

$$\sum_{q=1}^{N} y_{lq}^k(\tau) = A_v(k) = \alpha_k, \quad \forall k, \tag{4.29}$$

and

$$\sum_{q=N+1}^{2N} y_{lq}^k(\tau) = A_e(k) = \beta_k, \quad \forall k. \tag{4.30}$$

For example, a solution representation for a transportation system with $N$ subsystems, $\alpha$ vehicles, and $\beta$ unloading equipment can be represented as particles, as shown in Figure 4.7.

### 4.3.2 Antithetic Particle-Updating Mechanism

In the traditional PSO, all dimensions or elements of a multidimensional particle are independent of each other, thus, updating of the velocity and particle based on the traditional updating mechanism (Kennedy and Eberhart 1995 [23]) is performed independently for each element. The position, $P_l(\tau+1)$, of the $l$th particle is adjusted by stochastic velocity $V_l(\tau+1)$, which depends on the distance that the particle is from its own best solution and that of the swarm (Shi and Eberhart 1998a [34]), as shown below:

$$V_l(\tau+1) = w(\tau)V_l(\tau) + c_p r_p(PBest_l - P_l(\tau)) + c_g r_g(GBest - P_l(\tau)), \tag{4.31}$$

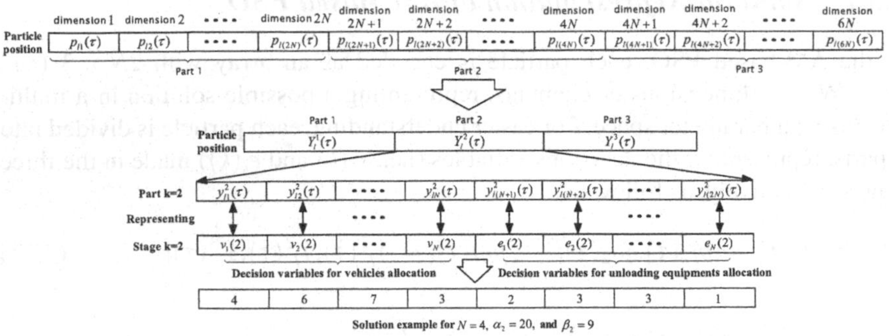

Fig. 4.7 Solution representation for a particle

$$P_l(\tau+1) = V_l(\tau+1) + P_l(\tau). \tag{4.32}$$

where $V_l(\tau) = [v_{l1}(\tau), v_{l2}(\tau), \cdots, v_{lH}(\tau)]$ denotes the $H$-dimension velocity (i.e., distance change) for the $l$th particle in the $\tau$th iteration; $PBest_l = [pBest_{l1}, pBest_{l2}, \cdots, pBest_{lH}]$, the personal best of the $l$th particle encountered after $\tau$ iterations; $GBest = [gBest_1, gBest_2, \cdots, gBest_H]$, the global best among all of the swarm of particles achieved so far; $c_p$ and $c_g$, the positive constants (namely, learning factors) that determine the relative weight that the global best has versus the personal best; $r_p$ and $r_g$, random numbers between 0 and 1; and $w(\tau)$, the inertia weight used to control the impact of the previous velocities on the current velocity that influences the trade-off between the global and the local exploration abilities during the search.

On the other hand, the elements of a feasible particle must be positive integers, but the traditional updating of the velocity may lead to non-positive and decimal elements in the particles. As a result, if the elements are simply rounded-off, the sum of the specific elements may no longer satisfy the state constraints denoted by Eqs. (4.29) and (4.30), which leads to an infeasible solution. Therefore, the updating mechanism for the traditional PSO must be modified so as to eliminate such conflicts when using the particle to represent the array-encoded solution for the CMQNP. To do this, the feasibility of the initialized particle position needs to be first guaranteed.

## Initializing Method for Generating Feasible Position

The element in the multidimensional particle position can be initialized as follows to avoid an infeasible position:

*Step 1:* Set $k = 1$. (Start to initialize the elements in Part $k$ of a particle.)

*Step 2:* Set $q = 1$. (Start to initialize Element $q$ in Part $k$ of a particle.)

*Step 3:* Initialize $y^k_{lq}(1)$ by generating a random positive integer number within $[1, \alpha_k - N + q]$, then let $q = q+1$. (Initialize Element 1 for the decision variable $v_1(k)$ in Part $k$ of a particle. In order to guarantee that the value of each element (for $q = 1, 2, \cdots, N$) is a positive integer, the upper limit for the value of Element 1 is set to be $\alpha_k - N + q$.)

*Step 4:* Initialize $y_{lq}^k(1)$ by generating a random positive integer number within $[1, \alpha_k - \sum_{j=1}^{q-1} y_{lj}^k(1) - N + q]$, then let $q = q+1$. (Initialize Element $q$ for the decision variable $v_i(k)$ in Part $k$ of a particle. In order to guarantee the value of each element (for $q = 1, 2, \cdots, N$) is a positive integer, the upper limit for the value of Element $q$ is set to be $\alpha_k - \sum_{j=1}^{q-1} y_{lj}^k(1) - N + q$.)

*Step 5:* If $q < N$, return to Step 4; else if $q = N$, initialize $y_{lq}^k(1) = \alpha_k - \sum_{j=1}^{q-1} y_{lj}^k(1)$, then let $q = q+1$.

*Step 6:* Initialize $y_{lq}^k(1)$ by generating a random positive integer number within $[1, \beta_k - 2N + q]$, then let $q = q+1$. (Initialize Element $N+1$ for the decision variable $e_1(k)$ in Part $k$ of a particle. In order to guarantee the value of each element (for $q = N+1, N+2, \cdots, 2N$) is a positive integer, the upper limit for the value of Element $N+1$ is set to be $\beta_k - 2N + q$.)

*Step 7:* Initialize $y_{lq}^k(1)$ by generating a random positive integer number within $[1, \beta_k - \sum_{j=N+1}^{q-1} y_{lj}^k(1) - 2N + q]$, then let $q = q+1$. (Initialize Element $q$ for the decision variable $e_i(k)$ in Part $k$ of a particle. In order to guarantee the value of each element (for $q = N+1, N+2, \cdots, 2N$) is a positive integer, the upper limit for the value of Element $q$ is set to be $\beta_k - \sum_{j=N+1}^{q-1} y_{lj}^k(1) - 2N + q$.)

*Step 8:* If $q < 2N$, return to Step 7; else if $q = 2N$, initialize $y_{lq}^k(1) = \beta_k - \sum_{j=N+1}^{q-1} y_{lj}^k(1)$.

*Step 9:* If the stopping criterion is met, i.e., $k = 3$, then the initialization for the $l$th particle is completed; otherwise, let $k = k+1$ and return to Step 2.

The above initialization method can efficiently guarantee the feasibility of the generated particle positions. To maintain the feasibility of the particle positions during the particle-updating process, the velocity needs to be initialized according to three empirical principles. For easier understanding, each velocity is also divided into 3 parts, respectively, as follows,

$$V_l(\tau) = [v_{l1}(\tau), v_{l2}(\tau), \cdots, v_{lH}(\tau)] = [Z_l^1(\tau), Z_l^2(\tau), Z_l^3(\tau)] \tag{4.33}$$

where $V_l(\tau) = [v_{l1}(\tau), v_{l2}(\tau), \cdots, v_{lH}(\tau)]$ denotes the $H$-dimension velocity (i.e., distance change) for the $l$th particle in the $\tau$th iteration; $Z_l^k(\tau) =$ the $k$th part of the $l$th velocity in the $\tau$th generation. Note that every part of a velocity is a $2N$-dimension vector, which can be denoted as:

$$Z_l^k(\tau) = [z_{l1}^k(\tau), z_{l2}^k(\tau), \cdots, z_{l(2N)}^k(\tau)], \tag{4.34}$$

where $z_{lq}^k(\tau) =$ the $q$th dimension of $Z_l^k(\tau)$ for the $l$th velocity in the $\tau$th generation; $q = 1, \cdots, 2N$. Then the three empirical principles are provided as follows:

(1) The value of each element in a velocity should be an integer;

(2) The sum of the specific elements should be zero, i.e., $\sum_{q=1}^{N} z_{lq}^k(\tau) = 0$, and $\sum_{q=N+1}^{2N} z_{lq}^k(\tau) = 0$, for $k = 1, 2, 3$;

(3) The absolute value of each element should not exceed $\frac{\alpha_k}{N}$ (for $q = 1, 2, \cdots, N$), or $\frac{\beta_k}{N}$ (for $q = N+1, N+2, \cdots, 2N$), in order to avoid infeasible positions after particle-updating.

**Velocity Clamping**

It has been observed that velocities may quickly explode to large values if each component of the velocity is not appropriately clamped (van den Bergh and Engelbrecht 2010 [122]). To avoid this erratic behavior, it is necessary to bind the maximum values of the components of the velocity vector. Similar to the initialization of velocity discussed above, the absolute value of each component of the velocity (i.e., $z_{lq}^k(\tau)$) is set to not exceed $\frac{\alpha_k}{N}$ (for $q = 1,2,\cdots,N$), or $\frac{\beta_k}{N}$ (for $q = N+1,N+2,\cdots,2N$), where $\alpha_k$ is the number of available vehicles allocated to a transportation system in Stage $k$, and $\beta_k$, the amount of available unloading equipment allocated to a transportation system in Stage $k$. If the absolute value of the velocity component (i.e., $z_{lq}^k(\tau)$) exceeds the maximal limitation during the updating process, i.e., $z_{lq}^k(\tau) > \frac{\alpha_k}{N}$ for $q = 1,2,\cdots,N$, or $z_{lq}^k(\tau) > \frac{\beta_k}{N}$ for $q = N+1,N+2,\cdots,2N$, then it is set that $z_{lq}^k(\tau) = \frac{\alpha_k}{N}$ for $q = 1,2,\cdots,N$, or $z_{lq}^k(\tau) = \frac{\beta_k}{N}$ for $q = N+1,N+2,\cdots,2N$.

**Original and Antithetic Elements**

It can be easily proved that if the initialized particle positions are all feasible, and the initialized velocities satisfy the above three empirical principles, then, based on Eq. (4.31), the updated velocities will also satisfy empirical principle (2). But the elements of these updated velocities may become decimals, the absolute value of which may be quite large, which would lead to an infeasible position if Eq. (4.32) were used to update particles.

In this study, the original elements and antithetic elements are introduced to develop the antithetic particle-updating mechanism, so that the particles are updated feasibly during the evolutionary process.

The velocity computed using Eq. (4.31) is a distance measure used to decide on a new position for the updated particle. A larger velocity means a more distant position or area is being explored. When used for an array-encoded particle representation, the $H$-dimension velocity (i.e., $V_l(\tau) = [v_{l1}(\tau), v_{l2}(\tau), \cdots, v_{lH}(\tau)]$) computed using Eq. (4.31) corresponds to the distance or gap between the current particle position (i.e., decision variables $v_i(k)$ and $e_i(k)$) and its own personal best and the global best positions (i.e., experiences) found so far. A larger gap means that such a particle position is more likely to be updated. Therefore, the absolute value of the velocity is used here to represent the probability that the particle position will change.

Since the elements in the $k$th part of a velocity (i.e., $z_{lq}^k(\tau)$) can be positive or nonpositive, they can be regarded as the original or antithetic elements, respectively. Each part of a velocity has two groups of original and antithetic elements. One group (i.e., $q = 1,2,\cdots,N$) is for the vehicle decision variables (i.e., $v_i(k)$), and the other (i.e., $q = N+1,N+2,\cdots,2N$) for the uploading equipment decision variables (i.e., $e_i(k)$). For each group of original and antithetic elements, if $z_{lq}^k(\tau) > 0$, so is regarded as an original element; otherwise, if $z_{lq}^k(\tau) \leq 0$, it is regarded as an antithetic element. It should be noted that if the number of original elements in a group is $O$, then the number of antithetic elements in that group must be $N - O$. Let $Or_g^k$ ($k = 1,2,3$; $g = 1,2$) be the set of original elements in Group $g$ of Part $k$ of a velocity (or particle), $An_g^k$, the set of antithetic elements in Group $g$ of Part $k$ of a velocity (or particle). It has also been defined that if $z_{lq}^k(\tau)$ is regarded as an original element (i.e., $z_{lq}^k(\tau) \in Or_g^k$) (or an antithetic element (i.e., $z_{lq}^k(\tau) \in An_g^k$)), then the corresponding element of the particle position (i.e., $y_{lq}^k(\tau)$) is regarded as an original element (i.e., $y_{lq}^k(\tau) \in Or_g^k$) (or an antithetic element (i.e., $y_{lq}^k(\tau) \in An_g^k$)) as well.

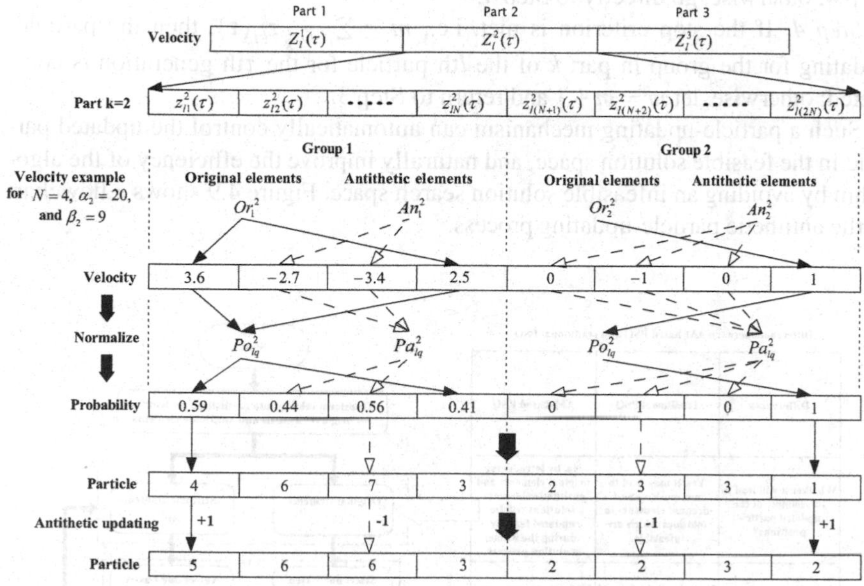


**Fig. 4.8** Antithetic particle-updating mechanism

## Procedure of Antithetic Particle-Updating Mechanism

Having outlined the approaches for dealing with particle-feasibility and state-constraint problems, the procedure for the antithetic particle-updating mechanism is presented in Figure 4.8 and described in detail as follows:

*Step 1:* For each iteration of the particle updating, the original and antithetic elements in each velocity group are transformed into an absolute value and then normalized to the range $[0,1]$ by

$$Po_{lq}^k = \frac{z_{lq}^k(\tau)}{\sum_{j \in Or_g^k} z_{lj}^k(\tau)}, \forall q \in Or_g^k, \quad Pa_{lq}^k = \frac{z_{lq}^k(\tau)}{\sum_{j \in An_g^k} z_{lj}^k(\tau)}, \forall q \in An_g^k,$$

respectively, where $Po_{lq}^k$ is the normalized probability of the original elements, and $Pa_{lq}^k$, the normalized probability of the antithetic elements.

*Step 2:* Set the updating number $m = 1$. (In order to sufficiently update the particles, the upper limit of the updating number is set to be $\sum_{j \in Or_g^k} z_{lj}^k(\tau)$.)

*Step 3:* For each particle group, one original element is selected based on the normalized probability $Po_{lq}^k$, and then one is added (i.e., $y_{lq}^k(\tau) = y_{lq}^k(\tau) + 1$, $y_{lq}^k(\tau) \in Or_g^k$). At the same time, for each particle group, one antithetic element is also selected based on the normalized probability $Pa_{lq}^k$, and then one is subtracted (i.e., $y_{lq}^k(\tau) = y_{lq}^k(\tau) - 1$, $y_{lq}^k(\tau) \in An_g^k$). For the above $y_{lq}^k(\tau) \in An_g^k$, if $y_{lq}^k(\tau) = 0$, the above operations are cancelled for the original and antithetic elements, and move to Step 4; otherwise, go directly to Step 4.

*Step 4:* If the stop criterion is met, i.e., $m \geq \sum_{j \in Or_g^k} z_{lj}^k(\tau)$, then the particle-updating for the group in part $k$ of the $l$th particle for the $\tau$th generation is completed; otherwise, let $m = m + 1$ and return to Step 3.

Such a particle-updating mechanism can automatically control the updated particle in the feasible solution space, and naturally improve the efficiency of the algorithm by avoiding an infeasible solution search space. Figure 4.9 shows a flowchart of the antithetic particle-updating process.

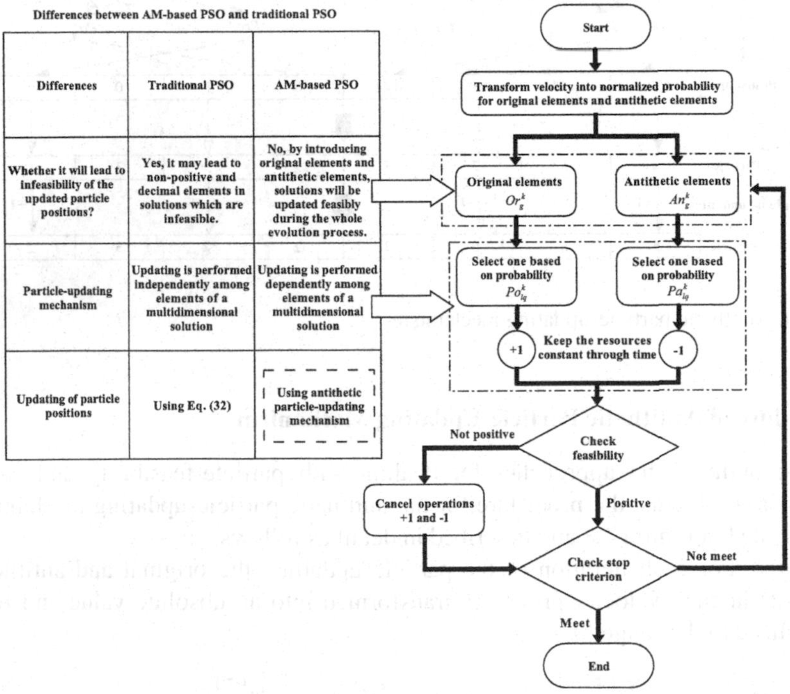


**Fig. 4.9** Flowchart of antithetic particle-updating process

The fitness value used to evaluate a particle is just the value of the objective function in Model (4.26), i.e.

$$
\begin{aligned}
&Fitness(P_l(\tau)) \\
&= \sum_{k=1}^{3} E^{Me}[\tilde{w}](C_b M + C_v A_v(k) + C_e A_e(k)) \max_i \left\{ \frac{Q_i(k)}{E^{Me}[\tilde{w}]tH \sum_{j=0}^{y_{li}^k(\tau)} p_i^j(k) E^{Me}[\tilde{\mu}_i^j(k)]} \right\}
\end{aligned}
\tag{4.35}
$$

### 4.3.3 The Framework of AM-Based PSO

From the above basic concepts (i.e., solution-representation, antithetic particle-updating mechanism, fitness value function), the framework for the implementation of an AM-based PSO for queuing decision optimization is developed.

The initial positions of the $L$ particles, i.e., $P_l(1) = [p_{l1}(1), p_{l2}(1), \cdots, p_{l(2N\times 3)}(1)]$ $(l = 1,2,\cdots,L)$, are randomly generated using the initializing method to avoid an infeasible position. The initial velocities of the $L$ particles, i.e., $V_l(1) = [v_{l1}(1), v_{l2}(1), \cdots, v_{l(2N\times 3)}(1)]$ $(l = 1,2,\cdots,L)$, are also randomly generated based on the three empirical principles. The velocity-updating mechanism represented by Eq. (4.31) is used to update the velocities (i.e., $V_l(\tau)$), and the antithetic particle-updating mechanism is employed instead of Eq. (4.32) in the traditional PSO to update the position (i.e., $P_l(\tau)$) of the particles to automatically avoid infeasible particle-updating until the optimal solution is found.

The initial and updated positions must be evaluated with respect to the fitness value in order to identify the personal best solution for each particle (i.e., $PBest_l = [pBest_{l1}, pBest_{l2}, \cdots, pBest_{l(2N\times 3)}]$) $(l = 1,2,\cdots,L)$ and the global best solution (i.e., $GBest = [gBest_1, gBest_2, \cdots, gBest_{2N\times 3}]$) in the swarm.

The objective function for construction duration is converted into an inequality as shown in Eq. (4.25) to transform the multiple objective model into a single objective equivalent. The solution feasibility of the construction duration constraint is checked when either one of the following two criteria is met: (1) the convergence iteration limit (i.e., maximum number of iterations since the last updating of the global best solution); (2) the total iteration limit (i.e., $T$). If the global best solution satisfies the construction duration constraint (i.e., Eq. (4.25)), then, the global best particle-represented sequence (i.e., $GBest = [gBest_1, gBest_2, \cdots, gBest_{2N\times 3}]$) can be transformed into the problem solution for the equivalent crisp model (4.26), which is the final solution to the multistage queuing decision problem, and the algorithm is ended. Otherwise, the algorithm needs to be restarted to find the feasible optimal solution.

## 4.4 Case Study: Multistage Queuing Decision Problem

Rapid economic growth and social advancement have created more pressing demands on energy all over the world. New and renewable sources of energy have become more important, of which hydropower resources are recognized as one of the

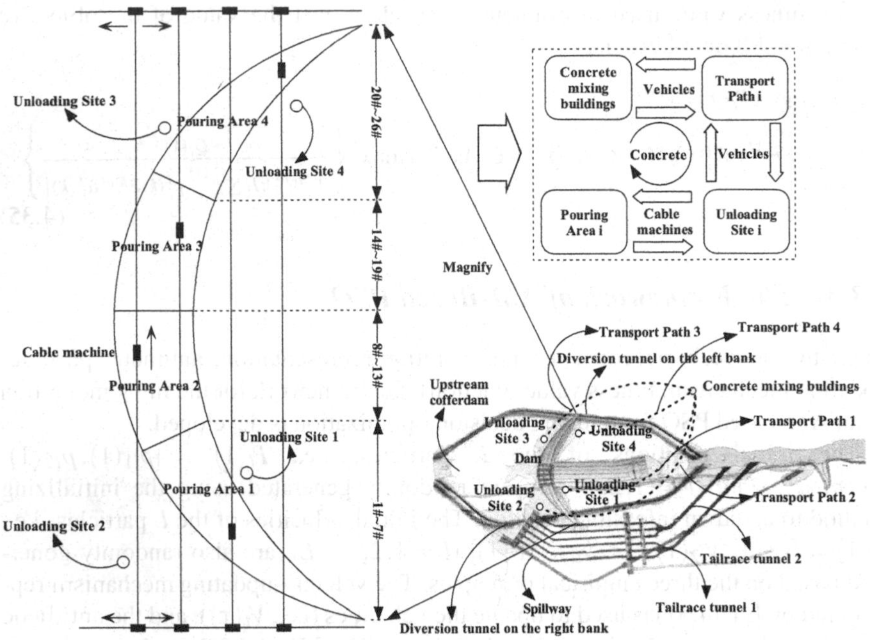


**Fig. 4.10** Construction Layout of Jinping-I Hydropower Project

best. Hydropower plays an important role in China, especially in Sichuan Province. The Chinese government has emphasized renewable energy development particularly in areas of water conservancy and hydropower. The Yalong River Hydropower Development Company, Ltd. (YRHDC) was appointed to supply clean and renewable energy to support the economic and social development of the Sichuan-Chongqing region through the development of the hydropower resources on the Yalong River in Sichuan. The Jinping-I Hydropower Project is one of YRHDC's projects currently under construction. This example focuses on the vehicle and facilities scheduling for the concrete transportation system queuing network optimization at the Jinping-I Hydropower Project to minimize total operating costs and construction duration.

### *4.4.1 Presentation of Case Problem*

An actual large-scale construction project, the Jinping-I Hydropower Project is used as a practical application example to demonstrate the practicality of the modeling approach and the efficacy of the AM-based PSO algorithm presented here. The Jinping-I Hydropower Project is located in the counties of Yanyuan and Muli, Liangshan Yi Autonomous Prefecture, Sichuan Province, China. The project consists of five permanent structures; water retention, spillway and dissipation, power tunnels, and the powerhouse complex. The 305$m$-high double curvature concrete arch dam is the world's highest dam. The construction layout of Jinping-I Hydropower Project is shown in Figure 4.10. One concrete production system which contains three concrete mixing buildings (i.e., $M = 3$) is located on the left bank of the dam. There are four (i.e., $N = 4$) transport paths that connect the concrete production system with the corresponding unloading sites. When the concrete is transported to the corresponding unloading site by the vehicles, cable machines unload the concrete from the vehicles and transport it to the corresponding pouring area. The working control area for the parallel mobile type of cable machines is a rectangle. Different cable machine groups are in charge of different pouring areas, and are installed at different elevations to avoid interference.

### 4.4.2 Data Collection

Detailed data for the Jinping-I Hydropower Project were obtained from the Yalong River Hydropower Development Company, Ltd. For modeling, the fuzzy data, including effective working days each month, transportation and stay time, were obtained based on previous data and expert experience. The details are shown in Table 4.3.

**Table 4.3** The data information of effective working days for each month, transportation and stay time

| Model parameters | Stage index $k=1$ | $k=2$ | $k=3$ | Measurement unit |
|---|---|---|---|---|
| $\tilde{T}_h^1(k)$ | (0.28,0.34,0.40) | (0.30,0.37,0.44) | (0.26,0.32,0.38) | *hour* |
| $\tilde{T}_e^1(k)$ | (0.20,0.25,0.27) | (0.23,0.26,0.29) | (0.18,0.22,0.25) | *hour* |
| $\tilde{T}_u^1(k)$ | (0.19,0.23,0.27) | (0.17,0.21,0.25) | (0.16,0.19,0.23) | *hour* |
| $\tilde{T}_h^2(k)$ | (0.38,0.42,0.47) | (0.40,0.44,0.48) | (0.36,0.40,0.45) | *hour* |
| $\tilde{T}_e^2(k)$ | (0.22,0.27,0.31) | (0.24,0.30,0.35) | (0.20,0.24,0.29) | *hour* |
| $\tilde{T}_u^2(k)$ | (0.17,0.21,0.26) | (0.18,0.23,0.27) | (0.16,0.20,0.25) | *hour* |
| $\tilde{T}_h^3(k)$ | (0.36,0.40,0.44) | (0.37,0.41,0.45) | (0.32,0.37,0.41) | *hour* |
| $\tilde{T}_e^3(k)$ | (0.28,0.33,0.37) | (0.26,0.34,0.39) | (0.22,0.29,0.35) | *hour* |
| $\tilde{T}_u^3(k)$ | (0.18,0.22,0.25) | (0.19,0.23,0.28) | (0.18,0.23,0.26) | *hour* |
| $\tilde{T}_h^4(k)$ | (0.23,0.27,0.34) | (0.25,0.29,0.33) | (0.22,0.27,0.31) | *hour* |
| $\tilde{T}_e^4(k)$ | (0.14,0.21,0.26) | (0.15,0.22,0.27) | (0.13,0.20,0.24) | *hour* |
| $\tilde{T}_u^4(k)$ | (0.19,0.22,0.26) | (0.18,0.23,0.27) | (0.18,0.22,0.25) | *hour* |
| $\tilde{T}_c(k)$ | (0.12,0.15,0.19) | (0.11,0.14,0.18) | (0.12,0.14,0.17) | *hour* |
| $\tilde{w}$ | (23,25,27) | | | *day/month* |

Based on Eqs. (4.1) and (4.2), the arrival rate of vehicles (i.e., $\tilde{\lambda}_i(k)$) can be easily computed and this is summarized with the cable machine service rates (i.e., $\tilde{\mu}_i(k)$) which were obtained from previous data and expert experience in Table 4.4. Since $\tilde{\rho}_i(k) = \frac{\tilde{\lambda}_i(k)}{\tilde{\mu}_i(k)}$, the results for $\tilde{\rho}_i(k)$ were also computed based on the algebraic operations for triangular fuzzy numbers (Xu and Zhou 2011 [281]) and are shown in Table 4.4.

Other important parameters including the number of available vehicles and unloading equipment in each stage (i.e., $\alpha_k$ and $\beta_k$), the work time (*hour/day*) for the

**Table 4.4** The data information for arrival rate of vehicles, service rate of cable machines, and $\tilde{\rho}_i(k)$

| Model parameters | Stage index $k=1$ | $k=2$ | $k=3$ | Measurement unit |
|---|---|---|---|---|
| $\tilde{\lambda}_1(k)$ | (0.8850,1.0309,1.2658) | (0.8621,1.0204,1.2346) | (0.9709,1.1494,1.3889) | *vehicle/hour* |
| $\tilde{\mu}_1(k)$ | (3.7037,4.3478,5.2632) | (4.0000,4.7619,5.8824) | (4.3478,5.2632,6.2500) | *vehicle/hour* |
| $\tilde{\rho}_1(k)$ | (0.1681,0.2371,0.3418) | (0.1466,0.2143,0.3086) | (0.1553,0.2184,0.3194) | - |
| $\tilde{\lambda}_2(k)$ | (0.8130,0.9524,1.1236) | (0.7813,0.9009,1.0753) | (0.8621,1.0204,1.1905) | *vehicle/hour* |
| $\tilde{\mu}_2(k)$ | (3.8462,4.7619,5.8824) | (3.7037,4.3478,5.5556) | (4.0000,5.0000,6.2500) | *vehicle/hour* |
| $\tilde{\rho}_2(k)$ | (0.1382,0.2000,0.2921) | (0.1406,0.2072,0.2903) | (0.1379,0.2041,0.2976) | - |
| $\tilde{\lambda}_3(k)$ | (0.8000,0.9091,1.0638) | (0.7692,0.8929,1.0753) | (0.8403,0.9709,1.1905) | *vehicle/hour* |
| $\tilde{\mu}_3(k)$ | (4.0000,4.5455,5.5556) | (3.5714,4.3478,5.2632) | (3.8462,4.3478,5.5556) | *vehicle/hour* |
| $\tilde{\rho}_3(k)$ | (0.1440,0.2000,0.2660) | (0.1462,0.2054,0.3011) | (0.1513,0.2233,0.3095) | - |
| $\tilde{\lambda}_4(k)$ | (0.9524,1.1765,1.4706) | (0.9524,1.1364,1.4493) | (1.0309,1.2048,1.5385) | *vehicle/hour* |
| $\tilde{\mu}_4(k)$ | (3.8462,4.5455,5.2632) | (3.7037,4.3478,5.5556) | (4.0000,4.5455,5.5556) | *vehicle/hour* |
| $\tilde{\rho}_4(k)$ | (0.1810,0.2588,0.3824) | (0.1714,0.2614,0.3913) | (0.1856,0.2651,0.3846) | - |

vehicles (i.e., $t$), the heaped capacity for a vehicle $(m^3/vehicle)$ (i.e., $H$), the required concrete pouring quantity at Pouring Area $i$ in Stage $k$ (i.e., $Q_i(k)$), the unit operation cost $(CNY/day)$ of a concrete mixing building (i.e., $C_b$), the unit operation cost $(CNY/day)$ of a vehicle (i.e., $C_v$), the unit operation cost $(CNY/day)$ of a cable machine (i.e., $C_e$), and the most endurable construction duration for a transportation system (i.e., $D$) are presented in Table 4.5.

**Table 4.5** The other important parameters

| Stage index k | $\alpha_k$ | $\beta_k$ | $Q_1(k)$ $(10^4m^3)$ | $Q_2(k)$ | $Q_3(k)$ | $Q_4(k)$ | $t$ $(hour/day)$ | $H$ $(m^3/vehicle)$ | $C_b$ $(CNY/day)$ | $C_v$ | $C_e$ | $D$ $(month)$ |
|---|---|---|---|---|---|---|---|---|---|---|---|---|
| 1 | 48 | 5 | 20.75 | 28.62 | 31.34 | 20.48 | 18 | 11 | 992 | 551 | 1226 | 62 |
| 2 | 56 | 6 | 65.76 | 86.92 | 83.17 | 64.82 | 18 | 11 | 992 | 551 | 1226 | 62 |
| 3 | 52 | 6 | 29.72 | 40.58 | 37.29 | 28.37 | 18 | 11 | 992 | 551 | 1226 | 62 |

## 4.4.3 Parameters Selection for AM-Based PSO

The PSO parameters were determined based on the results of some preliminary experiments that were carried out to observe the behavior of the algorithm at different parameter settings. From a comparison of several sets of parameters, including population size, iteration number, acceleration constant, initial velocity, and inertia weight, the most reasonable parameters were identified. Note that the population size (i.e., the number of particles) determines the evaluation runs, and thus impacts the optimization cost (Trelea 2003 [40]). After some testing of the solution algorithm, 50 particles were selected as the population size and 100 times as the iteration number. The inertia weight $w(\tau)$ was set to be varying with the iteration as follows:

$$w(\tau) = w(T) + \frac{\tau - T}{1 - T}[w(1) - w(T)] \tag{4.36}$$

where $\tau$ = iteration index $= 1, 2, \cdots, T$; $T$ = iteration limit. Through further experiments, $w(1) = 1.0$ and $w(T) = 0.1$ were found to be the most suitable in controlling the impact of the previous velocities on the current velocity, and to influence the trade-off between the global and local experiences. Other parameters were selected by comparing the results with the observations from the dynamic search behavior of the swarm. The selection of the acceleration coefficients $c_p$ and $c_g$ affects both the convergence speed and the capability of escaping from the local minima. Initial PSO studies often used $c_p = c_g = 2.0$ (van den Bergh and Engelbrecht 2010 [122]). Although good results were obtained, it was observed that the velocities quickly exploded to large values which lead to a slower convergence speed and an inadequate computing stability. To avoid being trapped in a local optimal solution and to improve the convergence speed, the acceleration coefficients $c_p$ and $c_g$ were selected based on the comparison of 9 parameter combinations. A sensitivity analysis was performed as shown in Table 4.6. It should be noted that the result values in Table 4.6 were obtained from 50 runs of the experiment for each combination, and the optimistic-pessimistic index was set as $\eta = 0.5$. It can be concluded that the performance of the algorithm can be slightly influenced by a change in the acceleration coefficients $c_p$ and $c_g$. While all combinations were able to obtain the same best fitness value (i.e., 52.72), it appeared that with an increase in the acceleration coefficients (both for $c_p$ and $c_g$), there was also an increasing trend for the worst and average fitness values, the variance of fitness value, and the average convergence number and computing time. The reason for this may be explained by the fact that a larger acceleration coefficient means that the velocities are more likely to explode to large values which will slow down the convergence speed and result in longer computing time. From the results shown in Table 4.6, Combination 1 has the lowest fitness value variance (i.e., 0.6073), the smallest average convergence iteration number (i.e., 23), and the shortest average computing time (i.e., 170.094 s). Thus, it is appropriate to select Combination 1 as the best acceleration coefficients setting (i.e., $c_p = 1.0$, $c_g = 1.0$) for the AM-based PSO. The initial velocity values were selected based on the empirical principles described in section "Antithetic particle-updating mechanism". Table 4.7 summarizes all the parameter values selected for the AM-based PSO in the computational experiments.

**Table 4.6** Sensitivity analysis for acceleration coefficients selection of AM-based PSO

| Parameter combination | Acceleration coefficient | | Fitness value ($10^6$ yuan) | | | Sample variance of fitness value | Average Convergence Iteration number | Average Computing Time (s) |
|---|---|---|---|---|---|---|---|---|
| | $c_p$ | $c_g$ | Best | Worst | Average | | | |
| Combination 1 | 1.0 | 1.0 | 52.72 | 54.48 | 53.17 | 0.6073 | 23 | 170.094 |
| Combination 2 | 1.0 | 1.5 | 52.72 | 54.96 | 53.77 | 0.6732 | 26 | 183.671 |
| Combination 3 | 1.0 | 2.0 | 52.72 | 55.24 | 54.03 | 0.8216 | 28 | 198.335 |
| Combination 4 | 1.5 | 1.0 | 52.72 | 55.03 | 53.86 | 0.7244 | 26 | 185.428 |
| Combination 5 | 1.5 | 1.5 | 52.72 | 55.18 | 53.92 | 0.8045 | 29 | 200.669 |
| Combination 6 | 1.5 | 2.0 | 52.72 | 55.68 | 54.59 | 0.9205 | 32 | 216.301 |
| Combination 7 | 2.0 | 1.0 | 52.72 | 55.26 | 54.17 | 0.8832 | 30 | 209.228 |
| Combination 8 | 2.0 | 1.5 | 52.72 | 55.53 | 54.41 | 0.9543 | 32 | 220.844 |
| Combination 9 | 2.0 | 2.0 | 52.72 | 55.79 | 54.73 | 0.9933 | 34 | 231.021 |

**Table 4.7** Parameter selection for AM-based PSO

| Population size | Iteration number | Acceleration coefficient | | Inertia weight | |
|---|---|---|---|---|---|
| $L$ | $T$ | $c_p$ | $c_g$ | $w(1)$ | $w(T)$ |
| 50 | 100 | 1.0 | 1.0 | 1.0 | 0.1 |

## 4.4.4 Results and Sensitivity Analysis

To show the practicality and efficiency of the optimization method for the multi-stage queuing decision problem at the Jinping-I Hydropower Project presented in this chapter, the AM-based PSO algorithm was implemented to solve the problem previously described, and the results compared with the actual data collected from the Jinping-I Hydropower Project. The results are shown in Table 4.8. It should be noted that the results were obtained based on the optimistic-pessimistic index, i.e., $\eta = 0.5$. When comparing the total operational cost fitness value and the algorithm value for the total construction duration with the actual data from the project (see Table 4.8), the findings indicated that there were differences between the solutions generated by the algorithm presented here and the actual data. The net reduction value for the total operating costs was $4.17 \times 10^6$ yuan, and the rate of reduction was 7.33%. The net reduction value for the total construction duration was 3.97 months, and the rate of reduction was 6.91%. This improvement in construction efficiency could bring considerable economic benefit to construction projects, especially those of a large-scale. Although this finding is quite encouraging, for now it is not 100% certain that all these differences represent a net improvement because the mathematical model was formulated with some assumptions, so a portion of these differences could be the result of possible modeling errors. To some extent, it can be concluded from the results data that the theoretical optimized values for the total operating costs and construction duration for the decision variables are better than the actual values. This means that current practice is not optimal. In fact, in current practice, decision makers usually make their decisions on experience alone. Therefore, these results could be used to provide decision makers with a theoretical optimal vehicle and unloading equipment allocation plan to guide current practice.

**Table 4.8** Results for the multistage queueing decision problem in Jinping-I Hydropower Project

| Stage index | Subsystem index | | | | | | | |
|---|---|---|---|---|---|---|---|---|
| | $i=1$ | | $i=2$ | | $i=3$ | | $i=4$ | |
| k | $v_1(k)$ | $e_1(k)$ | $v_2(k)$ | $e_2(k)$ | $v_3(k)$ | $e_3(k)$ | $v_4(k)$ | $e_4(k)$ |
| 1 | 11 | 1 | 13 | 1 | 14 | 2 | 10 | 1 |
| 2 | 12 | 1 | 16 | 2 | 15 | 2 | 13 | 1 |
| 3 | 11 | 1 | 15 | 2 | 14 | 2 | 12 | 1 |

| Total operational cost | | | | Total construction duration | | | |
|---|---|---|---|---|---|---|---|
| Fitness value ($10^6$ yuan) | Actual data from the project ($10^6$ yuan) | Net reduction value ($10^6$ yuan) | Rate of reduction | Algorithm value (month) | Actual data from the project (month) | Net reduction value (month) | Rate of reduction |
| 52.72 | 56.89 | 4.17 | 7.33% | 53.50 | 57.47 | 3.97 | 6.91% |

To gain insight into the selection principle for the optimistic-pessimistic index (i.e., $\eta$), a sensitivity analysis was conducted against this parameter (see Table 4.9). The decision makers are able to fine tune this parameter to obtain different solutions under different levels. The solutions reflect the different optimistic-pessimistic attitudes for uncertainty. It should be noted that the optimistic-pessimistic index (i.e., $\eta$) was used to determine the combined attitude of a decision maker, which is relevant to the total operating costs and construction duration. Table 4.9 summarizes the different fitness values (i.e., total operation cost) and total construction durations for the case problem with respect to the different values of $\eta$.

**Table 4.9** Sensitivity analysis on the optimistic-pessimistic index

| | Optimistic-pessimistic index | | | | | | | | | | |
|---|---|---|---|---|---|---|---|---|---|---|---|
| | $\eta=0$ | $\eta=0.1$ | $\eta=0.2$ | $\eta=0.3$ | $\eta=0.4$ | $\eta=0.5$ | $\eta=0.6$ | $\eta=0.7$ | $\eta=0.8$ | $\eta=0.9$ | $\eta=1.0$ |
| Fitness value ($10^6$ yuan) | 58.61 | 57.31 | 56.08 | 54.91 | 54.90 | 52.72 | 51.69 | 50.70 | 49.75 | 48.83 | 47.95 |
| Total construction duration (month) | 59.48 | 58.17 | 56.92 | 55.73 | 55.72 | 53.50 | 52.45 | 51.45 | 50.48 | 49.56 | 48.66 |

As shown in Table 4.9, the analytical results obtained from the case problem indicate that a change in the decision maker's optimistic-pessimistic index has a significant impact on the decision. If the parameter $\eta$ is larger, the fitness value and total construction duration decrease. For this case problem, since the objective was to minimize total operating costs and construction duration over the whole construction process, $\eta = 0$ is actually a pessimistic extreme, and $\eta = 1$, on the contrary, is actually an optimistic extreme. Thus, the following conclusion can be drawn: the result is more optimistic if the optimistic-pessimistic parameter goes up. Therefore, it is the decision-makers's attitudes which ultimately decide the optimal result.

### *4.4.5 Comparison with Results from Standard PSO*

The optimization results for the multistage queuing decision problem at the Jinping-I Hydropower Project using an AM-based PSO were compared with the standard PSO. To carry out the comparisons under a similar environment, the parameters stated in Table 4.7 were also adopted for the standard PSO; that is, a population size of 50, an iteration number of 100, the two acceleration coefficients $c_p$ and $c_g$ were set at values of 1.0 and 1.0 respectively, and the inertia weight was the same as in Eq. (4.36), and $w(1) = 1.0$, $w(T) = 0.1$. The other parameters and the updating mechanism for the standard PSO were selected as follows: (1) the initial velocities were taken as the same as for the AM-based PSO; and (2) Eq. (4.32) was used for the particle updating. If the final solution is infeasible, the standard PSO algorithm restarts until a feasible solution is found. The convergence histories (Figure 4.11) are based on the average of the optimal results from 50 runs of the experiment (for both the AM-based PSO and the Standard PSO, respectively) which are able to achieve a minimal fitness value, excluding locally trapped ones.

**Table 4.10** Comparison results between AM-based PSO and standard PSO

| Approach | Sample size | Fitness value | | | Convergence iteration number | | Computing time (s) | |
|---|---|---|---|---|---|---|---|---|
| | | Minimal | Average | Sample standard deviation | Average | Sample standard deviation | Average | Sample standard deviation |
| AM-based PSO | 50 | 52.72 | 53.17 | 0.7793 | 23 | 2.3212 | 170.094 | 12.5432 |
| Standard PSO | 50 | 54.84 | 58.96 | 1.0232 | 79 | 2.9872 | 2654.072 | 15.2342 |

Table 4.10 shows the comparison results obtained from 50 runs (i.e., sample size) for the AM-based PSO and the Standard PSO respectively, including the minimal, average, and sample standard deviation of the fitness value, the average and sample standard deviation of the convergence iteration number and the computing times. To detect if there were statistically significant differences in performance between the AM-based PSO and the Standard PSO, the t-test statistic (Livingston 2004 [41]) was employed here. Since the t-test can only be used if the variance in the two sample groups is also assumed to be equal, the F-test (Lomax 2007 [42]) was used first to ensure that the variance in the fitness values, the convergence iteration numbers, and the computing times for the AM-based PSO and the Standard PSO were equal. The test results are shown in Table 4.11, where the significance levels for $\alpha$ are considered to be 0.05 in all cases. The hypothesis for the F-test is as follows:

$$H_0^{\kappa} : \sigma_{A_\kappa}^2 = \sigma_{S_\kappa}^2; \quad H_1^{\kappa} : \sigma_{A_\kappa}^2 \neq \sigma_{S_\kappa}^2,$$

for $\kappa = 1,2,3$, where $H_0^{\kappa}$ and $H_1^{\kappa}$ are the null hypothesis and alternative hypothesis for the F-test respectively; $\sigma_{A_\kappa}^2$ and $\sigma_{S_\kappa}^2$ are the variance in the fitness values (i.e., $\kappa = 1$), the convergence iteration numbers (i.e., $\kappa = 2$), and the computing times (i.e, $\kappa = 3$) for the AM-based PSO and the Standard PSO. When $\alpha = 0.05$, the rejection region of $H_0^{\kappa}$ is $W_\kappa = \{F_\kappa > F_{1-\alpha/2}^{\kappa}(49,49)\}$. Since $F_\kappa \notin W_\kappa$ for $\kappa = 1,2,3$, thus the null hypothesis $H_0^{\kappa}$ (for $\kappa = 1,2,3$) is accepted, i.e., the assumption that the variance in the fitness values, the convergence iteration numbers, and the computing times for the AM-based PSO and the Standard PSO are equal can be considered to be true under the significance level $\alpha = 0.05$.

Similarly, the hypotheses for the t-test are as follows:

$$H_0^{*\kappa} : \gamma_{A_\kappa} = \gamma_{S_\kappa}; \quad H_1^{*\kappa} : \gamma_{A_\kappa} \neq \gamma_{S_\kappa},$$

for $\kappa = 1,2,3$, where $H_0^{*\kappa}$ and $H_1^{*\kappa}$ are null hypothesis and the alternative hypothesis for the t-test respectively; $\gamma_{A_\kappa}$ and $\gamma_{S_\kappa}$ are the fitness value means (i.e., $\kappa = 1$), the

**Table 4.11** Statistical significance test between AM-based PSO and standard PSO

| Performance item | Index | Significance level | Sample size | Statistical significance test | | | |
|---|---|---|---|---|---|---|---|
| | | | | F-test | | t-test | |
| | $\kappa$ | $\alpha$ | | $F_\kappa$ | $F_{1-\alpha/2}^{\kappa}(49,49)$ | $t_\kappa$ | $t_{1-\alpha/2}^{\kappa}(98)$ |
| Fitness value | 1 | 0.05 | 50 | 1.7239 | 1.81 | -31.8307 | 1.991 |
| Convergence Iteration number | 2 | 0.05 | 50 | 1.6562 | 1.81 | -104.6729 | 1.991 |
| Computing time | 3 | 0.05 | 50 | 1.4751 | 1.81 | -890.0914 | 1.991 |

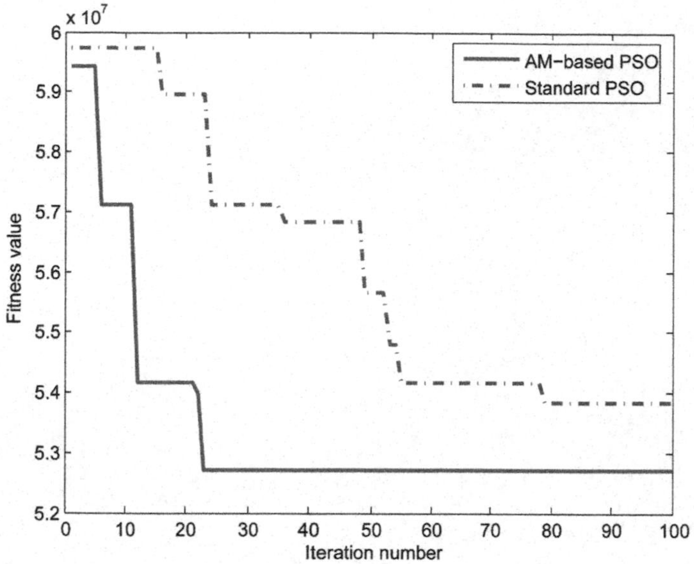


**Fig. 4.11** Convergence histories of AM-based PSO and Standard PSO

convergence iteration numbers (i.e., $\kappa = 2$), and the computing times (i.e, $\kappa = 3$) for the AM-based PSO and the Standard PSO. When $\alpha = 0.05$, the rejection region of $H_0^{*\kappa}$ is $W_\kappa^* = \{|t_\kappa| > t_{1-\alpha/2}^{\kappa}(98)\}$. Since $t_\kappa \in W_\kappa^*$ for $\kappa = 1,2,3$, thus the null hypothesis $H_0^{*\kappa}$ (for $\kappa = 1,2,3$) cannot be accepted, i.e., there are statistically significant differences between the fitness values means (i.e., $\kappa = 1$), the convergence iteration numbers (i.e., $\kappa = 2$), and the computing times (i.e, $\kappa = 3$) for the AM-based PSO and the Standard PSO under the significance level $\alpha = 0.05$.

From the comparison results shown in Table 4.10, it can be concluded that there is a statistically significant improved computational efficiency in the AM-based PSO compared with the standard PSO. The reason for these improvements can be explained as follows. On the one hand, by employing the antithetic particle-updating mechanism, the AM-based PSO can maintain the feasibility of the particle positions during the particle-updating process, which naturally improves the efficiency of the algorithm as it avoids the infeasible solution search space. The standard PSO algorithm, on the other hand, continually restarts when the final solution is infeasible until a feasible solution is found, which leads to a much slower convergence speed and a much longer computing time. Further, the updating is performed independently among the elements of a multidimensional particle in the standard PSO which largely reduces the probability of finding an optimal solution in the feasible solution space, thus, the computing stability of the standard PSO is also reduced and compared with that of the AM-based PSO.

# 5

# Fuzzy Random MOMSDM for Inventory Management

Inventory management for large scale construction projects is quite different than in other fields. Classical inventory models generally deal with a single-item (Naddor 1966 [43]), but in real world situations, especially in large scale construction systems, there is seldom a single-item inventory as multi-item inventory is much more common. To date, many studies have tackled the multi-item inventory problem, such as (Kiesmüller 2010 [44]; Feng et al. 2010 [33]; Suo et al. 2011 [45]; Rezaei and Davoodi 2008 [46]; Shah and Avittathur 2007 [47]; Bhattacharya 2005 [48]; Rahim and Ohta 2005 [50]; Xu and Liu 2008 [53]; Xu and Yao 2011 [282]). While these studies have significantly improved multi-item system inventory management, most have assumed that the different items are independent of one another, but in the real world, different items may have relationships, linear or nonlinear, with other items. According to the rules of international engineering bidding and Fédération Internationale Des Ingénieurs-Conseils (FIDIC) terms of the contract (Baker et al. 2009 [57]), important materials in large scale construction projects must be purchased by an invitation to bid, and there should be only one successful bidder for each item. Regulations and standard procedures for the invitation to bid can be found in the literature (Missbauer and Hauber 2006 [56]). Hence, for large scale construction projects, there is only one supplier for each inventory item. In addition, the successful bidder should dynamically supply the materials following a series of rules based on the FIDIC terms of the contract. Because of the relationships between the items, the associated optimization objectives may be in conflict with one another. Therefore, the inventory problem in large scale construction systems can be represented as a multi-objective multi-stage multi-item problem. In this chapter, a statement of an inventory management problem for large-scale construction projects (IMP-LCP) is presented. The modeling process for the IMP-LCP is explained in detail. Then a model analysis of the IMP-LCP is discussed in which a theoretical algorithm is designed for small scale problems. To solve large scale problems, a dynamic programming based particle swarm optimization algorithm is developed. Finally, a case study from the China Yangtze Three Gorges Project (TGP) is given as an application example.

J. Xu and Z. Zeng, *Fuzzy-Like Multiple Objective Multistage Decision Making*, Studies in Computational Intelligence 533,
DOI: 10.1007/978-3-319-03398-3_5, 

## 5.1 Statement of IMP-LCP

Inventory management problems for large-scale construction projects (IMP-LCP) are presented in this section. The purchasing behavior under rules of international bidding, purchasing strategies under rules of contract, and inventory conditions in IMP-LCP are explained in detail. The motivation for employing fuzzy random variables in IMP-LCP is also discussed.

### *5.1.1 Purchasing Behavior under Rules of International Bidding*

In general inventory problems there are usually several suppliers available for each of the construction material items, which is unique to large-scale construction projects. As noted earlier, based on international engineering bidding processes and the FIDIC terms of contract, important materials in large scale construction projects must be purchased by an invitation to bid, and there should be only one successful bidder for each item (Baker et al. 2009 [57]). This chapter tackles an "inventory management problem for large-scale construction projects," labeled the IMP-LCP, which is more complex than it appears as there are multiple relationships between the inventory items. For instance, in a real-life concrete dam construction project, the quantities purchased of the key item (e.g., reinforcing steel) and the auxiliary items (e.g., cement, fly ash, and aggregate) are in proportion with each other depending on the mix proportion design of the reinforced concrete project for the dam construction. Table 5.1 shows a part of the reinforced concrete mix proportion design used in building the China Yangze Three Gorges Project. Based on Table 5.1, the construction manager can decide on the purchasing proportion for each item.

**Table 5.1** Mix proportion design of reinforced concrete in Three Gorges Hydropower Project

| Construction Area | Kind of Reinforced Concrete | The volume of items (construction materials) | | | |
|---|---|---|---|---|---|
| | | Reinforcing steel ($kg/m^3$) | Cement ($kg/m^3$) | Fly ash content ($kg/m^3$) | Aggregate content ($kg/m^3$) |
| Dam Internal | Structure reinforced concrete | 279 | 224 | 56 | 948 |
| Dam External | Abrasion-resistant concrete | 340 | 360 | 90 | 1120 |
| Dam foundation | Roller compacted concrete | 315 | 306 | 77 | 1050 |

### *5.1.2 Purchasing Strategy under Rules of Contract*

For a large scale construction project, it is often not economical or practical to purchase all the construction materials at the beginning of a multi-year construction. Rather, the contractor purchases construction materials in multiple batches at different times to meet the operational needs over the entire construction duration. The span of each stage may range from one week to several months, depending on the construction duration and the type of construction project. The purchase of material items is similar to the process involved in options and futures trading. After signing the contract, the contractor pays an option premium for each item in each stage to obtain the right to buy the item in each stage.

In practice, most large scale construction projects have introduced a new way to determine the purchase price, which is quite different from the traditional pricing method which has a fixed price in the contract. In the new method, a pricing formula is defined for each item at each stage, usually prescribed in detail in the contract. The maximum and minimum purchase quantities of each item in each stage are also determined by the contractor and the successful bidder, and are written with the corresponding pricing formula in the contract. The purchasing strategy under the rules of contract in large scale construction projects is illustrated in Figure 5.1. As shown, the contractor purchases construction materials in multiple stages, and the purchase quantity of each item in each stage should be within a specified range. Since the item prices are time varying over the entire construction duration, the new method has the potential to maximize the benefits for both the contractor and the successful bidder. The key is to dynamically optimize the purchase quantity of each item in each stage according to inventory level and construction material demand, and minimize the total cost of each item over the entire construction duration. Because of the existence of the relationships between the inventory items, the inventory item optimization objectives may be in conflict with one another, which means this is considered a multi-objective multi-stage multi-item dynamic programming problem.

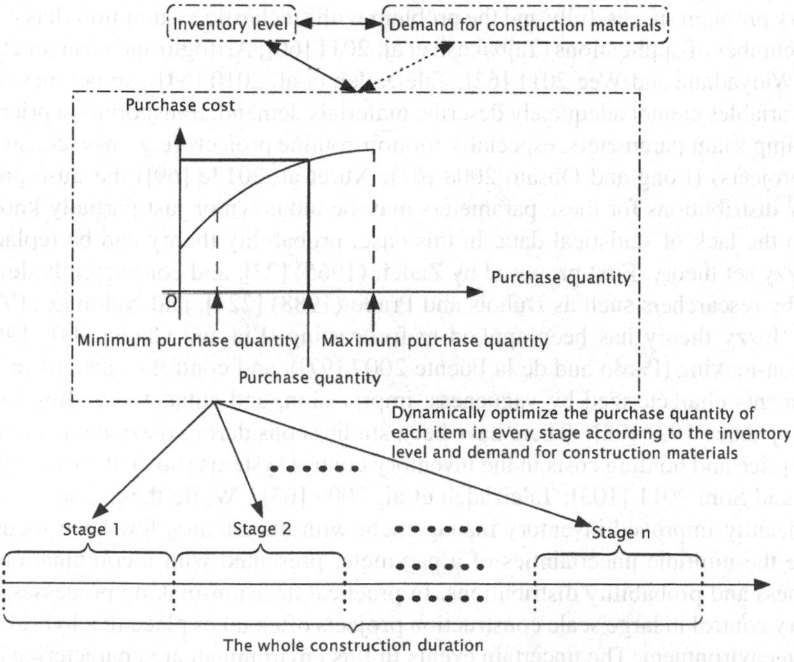


**Fig. 5.1** Purchasing strategy under rules of contract in large scale construction projects

### 5.1.3 Inventory Conditions in IMP-LCP

There are several constraints on the IMP-LCP inventory conditions as follows: (1) In this chapter, inventory items are considered to be always perfect throughout the construction process. Such an implicit assumption is reasonable in view of the fact that deterioration in the items is insignificant because of the advancements in construction material inventory technology. (2) Shortages are allowed and completely backlogged. The inventory shortages are penalized through proportional backorder costs (i.e., shortage costs) (Çetinkaya and Parlar 1998 [49]). (3) Inventories are reviewed periodically, and the replenishment for each item occurs at the beginning of each stage. Negative inventory levels represent outstanding backorders. (4) The warehouse has a fixed capacity for each inventory item. Additional explanations for these inventory conditions can be found in the following sections.

### 5.1.4 Motivation for Employing Fuzzy Random Variables

In practice, construction is often plagued by uncertainties (Tommelein 1998 [58]; Xu and Zeng 2012 [198]) because of unforeseen factors such as changing weather, equipment breakdown, labor inefficiency, and lack of coordination among stakeholders (Zhang et al. 2003 [59]). Though probability theory has been applied to inventory problem successfully and the problem with stochastic assumptions has got a great number of applications (Tajbakhsh et al. 2011 [60]; Arifoglu and Ozekici 2011 [61]; Widyadana and Wee 2011 [62]; Taleizadeh et al. 2010 [64]), sometimes random variables cannot adequately describe materials demand, transportation price or other important parameters, especially for non-routine projects (e.g., new construction projects) (Long and Ohsato 2008 [67]; Xu et al. 2013e [69]), because probability distributions for these parameters may be unknown or just partially known due to the lack of statistical data. In this case, probability theory can be replaced by fuzzy set theory. First proposed by Zadeh (1965) [22], and consequently developed by researchers such as Dubois and Prade (1988) [221], and Nahmias (1978) [238], fuzzy theory has been applied to forecasting (Lia and Cheng 2007 [96]), decision-making (Pardo and de la Fuente 2007 [97]), and control of actions in environments characterized by vagueness, imprecision, and subjectivity (Bojadziev and Bojadziev 1997 [98]). There have been studies considering fuzzy demand, purchase price and holding costs in the inventory control systems (Su et al. 2011 [101]; Shah and Soni 2011 [103]; Taleizadeh et al. 2009 [63]). While these studies have significantly improved inventory management with uncertainty, few have focused on the the multiple uncertainties of a parameter presented with a combination of fuzziness and probability distributions. In practical decision-making processes, inventory control in large scale construction projects often takes place in a hybrid uncertain environment. The uncertain events in this environment are characterized by both fuzziness and randomness, the so-called "twofold" uncertainty. In this case, the fuzzy random variables, introduced by Kwakernaak (1978, 1979) [94, 95], Puri and Ralescu (1986) [93], and Kruse and Meyer (1987) [479] in modeling and analyzing "imprecise values" associated with the sample space of a random experiment through the use of fuzzy-set functions (Gil et al. 2006 [105]), can be employed.

The need to address uncertainty in inventory management is widely recognized, as uncertainties exist in a variety of system components. As a result, the inherent complexity and stochastic uncertainty existing in real-world inventory control decision making processes have essentially placed them beyond conventional deterministic optimization methods. Although probability theory has been proved to be a useful tool for dealing with uncertainty in inventory problems under market mechanisms, sometimes it may not be suitable for new construction projects due to a lack of historical data. While it may be easy to estimate the probability distributions for the market demand (Schmitt et al. 2010 [104]; Rahim 2006 [51]; Chen and Simchi-Levi 2004 [52]) as there is sufficient historical data, it is usually difficult to do this for material demand, transportation price, and other important parameters in a new construction project, especially in the early construction stages. Some scholars have observed this uncertainty and imprecision and have dealt with it using fuzzy theory (Su et al. 2011 [101]; Handfield et al. 2009 [102]). However, a large-scale construction project is often faced with a hybrid uncertain environment where fuzziness and randomness co-exist in a decision making process. In fact, over the years, advancements in engineering technology have provided more alternatives for construction operations and, subsequently, decision-makers are now faced with more imprecise information than before. Therefore, fuzzy random variables that can take into account both fuzziness and randomness are favored by decision-makers to describe the uncertain and vague information encountered in reality.

To collect the data, investigations were conducted with experienced engineers (i.e., $q = 1,2,\cdots,E$, where $q$ is the index of engineers). The engineers are usually unable to give an exact expression of the required parameters. Instead, based on their experience, they describe the parameters in linguistic terms as an interval for each item (i.e., $[l_q, r_q]$) with a most possible value (i.e., $m_q$), such as "the monthly demand is between 3.60 and 5.78 million $kg$, and the most possible value being 4.91 million $kg$." Since different engineers have different views on the parameters, the minimum value of all $l_q$ and the maximal value of all $r_q$ (for $q = 1,2,\cdots,E$) are selected as the left border (i.e., $a$) and the right border (i.e., $b$) of the fuzzy random number, respectively. The fluctuation of the most possible values (i.e., $m_q$) can be characterized by a stochastic distribution. For example, the most possible value (i.e., $m_q$) for the monthly demand can be regarded as a random variable (i.e., $\varphi(\omega)$). By comparing the most possible values (i.e., $m_q$ for $q = 1,2,\cdots,E$) obtained from different engineers, it is found that $\varphi(\omega)$ approximately follows a normal distribution (i.e., $\mathscr{N}(\zeta, \nu^2)$), which can be estimated by using the maximum likelihood method and justified using a chi-square goodness-of-fit test. Thus, the triangular fuzzy random number for the monthly demand can be derived as $(a, \varphi(\omega), b)$, where $\varphi(\omega) \sim \mathscr{N}(\zeta, \nu^2)$. This modeling technique is recognized as fuzzy random optimization, which can benefit decision-makers more via personal experiences than simply use the fuzzy method. In brief, the fuzzy random description method is able to handle complex uncertainty through the leveraging of the valuable personal experience of the engineers. Figure 5.2 explains the reason why triangular fuzzy random

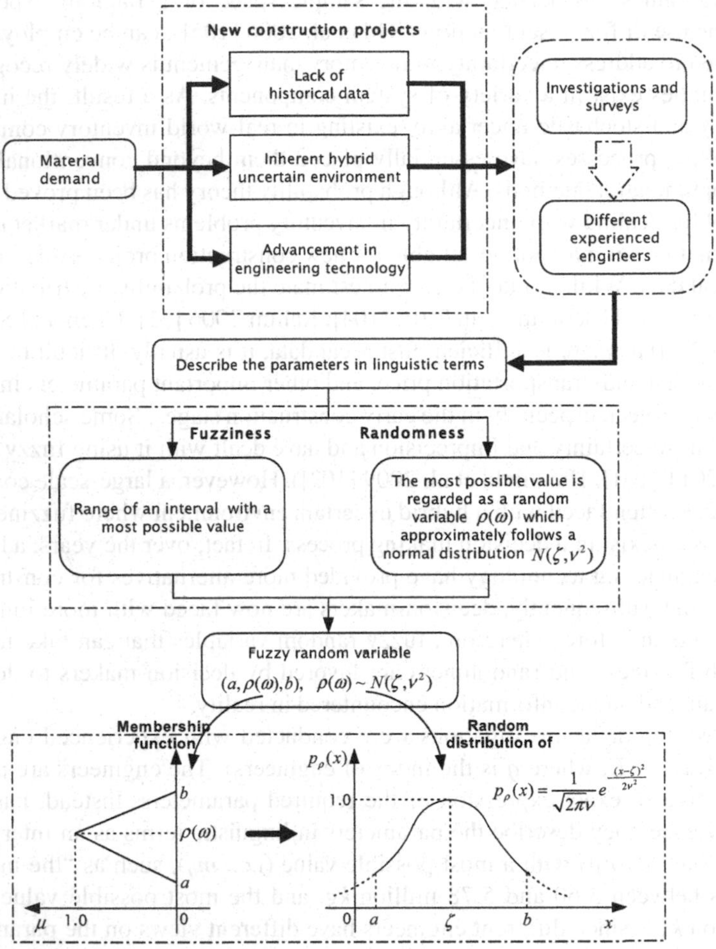


**Fig. 5.2** Flowchart of why material demand is a fuzzy random variable

variables are used to characterize the complex twofold uncertain environment encountered in the IMP-LCP.

While modeling the IMP-LCP under a fuzzy random environment, monthly demand, transportation price, and the conversion factor for the holding cost of each item in each stage are the most common factors, as these have a greater level of uncertainty containing both fuzziness and randomness. Therefore, these factors are represented as triangular fuzzy random numbers in this chapter.

## 5.2 Modelling Process of IMP-LCP

The objectives of the contractor are to minimize the total cost of each item in terms of the purchase, holding, transportation, and shortage costs. However, the decision-making is far more complex than it appears since there is a mutual relationship between the inventory items. It should be noted that the significance of each item in a large scale construction project is quite different. The costs of key items are usually very high whereas the costs of auxiliary items are much lower. Because of the relationships between the items, the inventory item optimization objectives may be in conflict with one another. Therefore, it is not appropriate to express total cost as a single-objective by aggregating over the item type, and it is necessary to separate the total cost into $n$ terms based on the item type and form a multi-objective model. The objective function for item $i$ is:

$$\begin{aligned} f_i(\mathbf{x}_i,\tilde{\bar{\xi}}_i) = {} & f_i^{PC}(Q_i(\cdot)) + f_i^{HC}(I_i(\cdot),Q_i(\cdot),\tilde{\bar{c}}_i(\cdot)) + f_i^{TC}(I_i(\cdot),Q_i(\cdot),\tilde{\bar{D}}_i(\cdot),\tilde{\bar{T}}_i^{sw},\tilde{\bar{T}}_i^{wc}) \\ & + f_i^{SC}(I_i(\cdot),Q_i(\cdot),\tilde{\bar{D}}_i(\cdot)), \end{aligned}$$

for $i = 1,2,\cdots,n$, where $\mathbf{x}_i = (I_i(\cdot),Q_i(\cdot))$, $\tilde{\bar{\xi}}_i = (\tilde{\bar{c}}_i(\cdot),\tilde{\bar{D}}_i(\cdot),\tilde{\bar{T}}_i^{sw},\tilde{\bar{T}}_i^{wc})$, and $f_i^{PC}$, $f_i^{HC}$, $f_i^{TC}$, $f_i^{SC}$ are the purchase, holding, transportation, and shortage cost respectively. Further details about these cost components are explained in following section.

Taking into account all inventory items, their relationships, and the purchasing strategy under the international bidding rules, the research problem IMP-LCP can be formulated as in the following multiple-objective multiple-stage, multiple-item mathematical model:

$$\begin{cases} \min\{f_1(\mathbf{x}_1,\tilde{\bar{\xi}}_1), f_2(\mathbf{x}_2,\tilde{\bar{\xi}}_2),\cdots,f_n(\mathbf{x}_n,\tilde{\bar{\xi}}_n)\} \\ \text{s.t.} \begin{cases} I_i(k+1) = I_i(k) + Q_i(k) - \tilde{\bar{D}}_i(k), & \forall i\in\Psi, \quad k\in\Phi, \\ I_i(0) = a_i, \quad \forall i\in\Psi, \\ I_i(m) = b_i, \quad \forall i\in\Psi, \\ Q_j^L Q_1(k) \leqslant Q_j(k) \leqslant Q_j^U Q_1(k), \quad \forall j\in\Omega, \quad k\in\Phi, \\ Q_{i,k}^{min} \leqslant Q_i(k) \leqslant Q_{i,k}^{max} \quad or \quad Q_i(k) = 0, \quad \forall i\in\Psi, \quad k\in\Phi, \\ I_i(k) \leqslant I_i^{max}, \quad \forall i\in\Psi, \quad k\in\Phi, \end{cases} \end{cases} \tag{5.1}$$

where $\Psi = \{1,2,\cdots,n\}$, $\Phi = \{0,1,\cdots,m-1\}$, $\Omega = \{2,3,\cdots,n\}$. Detailed explanations of the mathematical model are as follows.

### 5.2.1 *State Equation with Fuzzy Random Variables*

The state equation describes the relationship among the inventory level $I_i(k)$, $I_i(k+1)$, purchase quantity $Q_i(k)$, and demand $\tilde{\bar{D}}_i(k)$. Since $Q_i(k)$ is the purchase quantity of Item $i$ at the beginning of $(k+1)^{th}$ stage, $I_i(k)$ is the inventory level of Item $i$ in the corresponding warehouse at the beginning of the $(k+1)^{th}$ stage before the replenishment (i.e., $Q_i(k)$) occurs, and $\tilde{\bar{D}}_i(k)$ is the demand for Item $i$ in the $(k+1)^{th}$ stage, so the relationship between the inventory level, purchase quantity, and demand for construction materials is:

$$I_i(k+1) = I_i(k) + Q_i(k) - \tilde{\bar{D}}_i(k), \quad \forall i \in \Psi, \quad k \in \Phi. \tag{5.2}$$

It should be mentioned that negative inventory levels represent unfilled demand. If $I_i(k+1) \geqslant 0$, this indicates that no inventory shortage occurs. Otherwise, if $I_i(k+1) < 0$, this indicates there is a stockout for Item $i$. In this case, the unsatisfied demand (i.e., $-I_i(k+1)$) of Item $i$ in the $(k+1)^{th}$ stage is considered to be completely backlogged, and this penalty for the inventory shortage is described as proportional shortage costs which is explained later in the objective function. Note that Eq. (5.2) is the state equation for Item $i$ in the $(k+1)^{th}$ stage.

### 5.2.2 *Initial and Terminal Conditions*

Initial conditions describe the inventory level at the beginning of the construction duration. Let $a_i$ be the initial inventory level of Item $i$ in the corresponding warehouse at the beginning of the first stage, so the initial conditions are:
when $k = 0$,

$$I_i(0) = a_i, \quad \forall i \in \Psi \tag{5.3}$$

Similarly, terminal conditions describe the inventory level at the end of the entire construction duration. Let $b_i$ be the terminal inventory level of Item $i$ in the corresponding warehouse at the end of the construction duration, so the terminal conditions are:
when $k = m$,

$$I_i(m) = b_i, \quad \forall i \in \Psi \tag{5.4}$$

### 5.2.3 *Constraint Conditions*

As stated in the previous section, the purchase quantity of every item is in proportion to each other, so let $Q_j^U$ be the upper limit of the proportionality coefficient for the purchase quantity of Item $j$, and $Q_j^L$, the lower limit of the proportionality coefficient for the purchase quantity of Item $j$. Then the relationships between the key item (i.e., $Q_1(k)$) and the auxiliary items (i.e., $Q_j(k)$, $j \in \Omega$) can be expressed using the following inequalities:

$$Q_j^L Q_1(k) \leqslant Q_j(k) \leqslant Q_j^U Q_1(k), \quad \forall k \in \Phi, \quad j \in \Omega \tag{5.5}$$

On the other hand, the contractor may use the purchasing strategy described in the previous section to determine the quantity of Item $i$ in the $(k+1)^{th}$ stage, so then the purchase quantity of Item $i$ in the very stage should be within the specified range between the maximum and minimum purchase quantity. If the contractor decides not to buy Item $i$ in the $(k+1)^{th}$ stage, then the purchase quantity of Item $i$ in the $(k+1)^{th}$ stage, $Q_i(k)$, should be zero. Let $Q_{i,k}^{min}$ be the minimum purchase quantity of Item $i$ in the $(k+1)^{th}$ stage, and $Q_{i,k}^{max}$ be the maximum purchase quantity of Item $i$ in the $(k+1)^{th}$ stage, then $Q_i(k)$ should satisfy the following constraints:

$$Q_{i,k}^{min} \leqslant Q_i(k) \leqslant Q_{i,k}^{max} \quad or \quad Q_i(k) = 0, \quad \forall i \in \Psi, \quad k \in \Phi \tag{5.6}$$

Additionally, in this section, a periodic review and replenishment to level inventory control policy are considered. The inventory level of every item in every stage should not exceed the maximum inventory level for each item. Let $I_i^{max}$ be the maximum inventory level of Item $i$, then the following inequalities hold:

$$I_i(k) \leqslant I_i^{max}, \quad \forall i \in \Psi, \quad k \in \Phi \tag{5.7}$$

### 5.2.4 Objective Function with Fuzzy Random Coefficients

The objective function defines the total costs for Item $i$ over the whole construction duration. The aim of the contractor is to minimize total costs for each item, which is made up of four components: the purchase cost, the holding cost, the transportation cost, and the shortage cost. The details of each component are given below.

**Purchase Cost:** As stated in previous section, most large scale construction projects have adopted a new purchasing protocol to purchase construction materials in stages, which is similar to the process employed in options and futures trading,. In fact, there have been several pricing formula versions that are usually used in practice. In this research, the most popular pricing formula is used.

Let $r_i$ be the rise and fall rate for the price of Item $i$. If $r_i > 0$, then it means the contractor and supplier reach an agreement that the price of Item $i$ will rise in the whole construction duration. If $r_i < 0$, means they think the price of Item $i$ will fall in the whole construction duration. Let $\delta_i$ be the spot price of Item $i$ in the first stage, then the price of Item $i$ in the $(k+1)^{th}$ stage should be $\delta_i(1+r_i)^k$, and the purchase cost of Item $i$ in the $(k+1)^{th}$ stage is naturally $\delta_i(1+r_i)^k Q_i(k)$. However, there are usually extremely large purchase quantities in large scale construction projects, so it is not fair to use the linear purchase cost $\delta_i(1+r_i)^k Q_i(k)$ if the contractors desire quantity discounts from suppliers. To maximize the benefits of both contractors and suppliers while minimizing the risk, the following rules are usually observed to define the purchase cost:

**Rule** (1) The purchase cost of Item $i$ in the $(k+1)^{th}$ stage must be equal to $\delta_i(1+r_i)^k Q_{i,k}^{min}$, if the purchase quantity of Item $i$ in the $(k+1)^{th}$ stage $Q_i(k) = Q_{i,k}^{min}$;

**Rule** (2) Let $\gamma_i$ be the discount percentage of maximum purchase quantity of Item $i$, then the purchase cost of Item $i$ in the $(k+1)^{th}$ stage must be equal to $\gamma_i\delta_i(1+r_i)^k Q_{i,k}^{max}$, if the purchase quantity of Item $i$ in the $(k+1)^{th}$ stage $Q_i(k) = Q_{i,k}^{max}$;

**Rule** (3) If $Q_{i,k}^{min} < Q_i(k) < Q_{i,k}^{max}$, then the purchase cost of Item $i$ in the $(k+1)^{th}$ stage should be less than $\delta_i(1+r_i)^k Q_i(k)$.

Based on the above rules, the most popular purchase cost formula is in quadratic form. Let $f_i^{PC}(Q_i(\cdot))$ be the purchase cost of Item $i$ over the whole construction duration, then:

$$f_i^{PC}(Q_i(\cdot)) = \sum_{k=0}^{m-1} [\alpha_i(k) + \beta_i(k)\delta_i(1+r_i)^k Q_i^2(k)], \quad \forall i \in \Psi \tag{5.8}$$

In Eq. (5.8), $\alpha_i(k)$ is called the option premium of Item $i$ in the $(k+1)^{th}$ stage, and $\beta_i(k)$, the conversion coefficient of Item $i$ in the $(k+1)^{th}$ stage. According to Rules (1) and (2), $\alpha_i(k)$ and $\beta_i(k)$ can be calculated by the following equations:

$$\begin{cases} \alpha_i(k) + \beta_i(k)\delta_i(1+r_i)^k (Q_{i,k}^{min})^2 = \delta_i(1+r_i)^k Q_{i,k}^{min} \\ \alpha_i(k) + \beta_i(k)\delta_i(1+r_i)^k (Q_{i,k}^{max})^2 = \gamma_i\delta_i(1+r_i)^k Q_{i,k}^{max} \end{cases} \tag{5.9}$$

in which $r_i$, $\gamma_i$, $Q_{i,k}^{min}$, and $Q_{i,k}^{max}$ are prescribed in the contract by the contractor and supplier, $\forall i \in \Psi = \{1,2,\cdots,n\}, k \in \Phi = \{0,1,\cdots,m-1\}$.

It is not difficult to prove that if the discount percentage (i.e., $\gamma_i$) of the maximum purchase quantity of Item $i$ decreases, then the option premium of Item $i$ in the $(k+1)^{th}$ stage $\alpha_i(k)$ increases accordingly. The relationship between the purchase cost and purchase quantity for Item $i$ in a stage is illustrated in Figure 5.3.

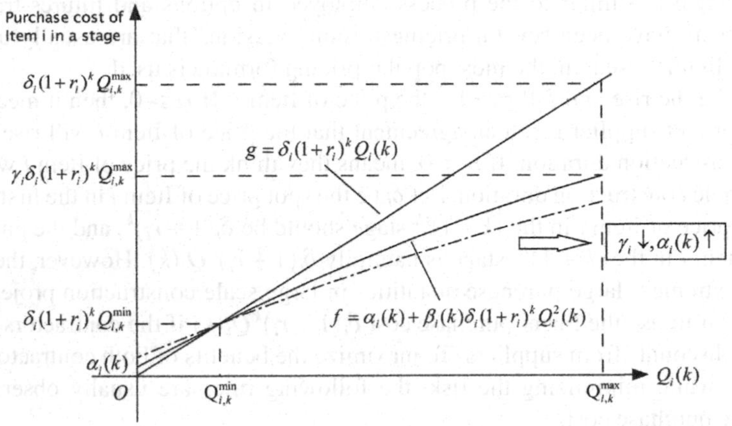


**Fig. 5.3** The relationship between the purchase cost and purchase quantity of Item $i$ in a stage

**Holding Cost:** In this section, it is considered that the replenishment $(Q_i(k))$ for each item occurs instantly at the beginning of each stage, thus the inventory level of Item $i$ in the corresponding warehouse at the beginning of the $(k+1)^{th}$ stage after the replenishment (i.e., $Q_i(k)$) should be $I_i(k) + Q_i(k)$. Let $h_i$ $(\$/(unit \cdot stage))$ be the holding price of Item $i$, if $I_i(k) + Q_i(k) > 0$, and since not all units of Item $i$ are usually stored in the warehouse for the entire stage, the holding cost of Item $i$ in the $(k+1)^{th}$ stage should be less than $h_i(I_i(k) + Q_i(k))$. As a result, a conversion factor is introduced to balance the difference between $h_i(I_i(k) + Q_i(k))$ and the actual holding cost of Item $i$ in the $(k+1)^{th}$ stage. Otherwise, if $I_i(k) + Q_i(k) \leq 0$, the inventory for Item $i$ is exhausted in the $(k+1)^{th}$ stage. In this case, the conversion factor is set to be 0. Let $\tilde{\bar{c}}_i(k)$ be the conversion factor for Item $i$ in the $(k+1)^{th}$ stage, $I_i^C(t)$ be the function of current inventory for Item $i$ in the entire construction duration, in which the unit of $t$ is one stage and $I_i^C(k) = I_i(k) + Q_i(k)$, then $\tilde{\bar{c}}_i(k)$ can be defined using the following equation:

$$\begin{cases} \tilde{\bar{c}}_i(k)h_i(I_i(k)+Q_i(k))[(k+1)-k] = \int_k^{k+1} h_i \max\{I_i^C(t),0\}dt, & \text{if} I_i(k)+Q_i(k) > 0, \\ \tilde{\bar{c}}_i(k) = 0, \quad \text{if} \quad I_i(k)+Q_i(k) \leq 0, \end{cases} \tag{5.10}$$

$\forall i \in \Psi, k \in \Phi$,
i.e.,

$$\tilde{\bar{c}}_i(k) = \begin{cases} \frac{\int_k^{k+1} \max\{I_i^C(t),0\}dt}{I_i(k)+Q_i(k)}, & \text{if} \quad I_i(k)+Q_i(k) > 0, \\ 0, \quad \text{if} \quad I_i(k)+Q_i(k) \leq 0, \end{cases} \quad \forall i \in \Psi, k \in \Phi \tag{5.11}$$

According to Eq. (5.11), it can be proved that $0 \leq \tilde{\bar{c}}_i(k) < 1$. However, because of incomplete information, the value of $\tilde{\bar{c}}_i(k)$ is usually difficult to determine. For example, in practice, engineers can usually estimate that the value of $\tilde{\bar{c}}_i(k)$ is between 0.34 and 0.71, and its most possible value follows a stochastic distribution, which can be justified using a chi-square goodness-of-fit test. Take the normal distribution $\mathcal{N}(\zeta, \nu^2)$ as an example. The parameters of the distribution can be estimated using a maximum likelihood estimation. Thus, the engineers can calculate that the expected value of the most possible value $\zeta$ is 0.51. Then, $\tilde{\bar{c}}_i(k)$ can be represented by a triangular fuzzy random variable $(0.34, \varphi(\omega), 0.7)$, in which $\varphi(\omega)$ follows a normal distribution $\mathcal{N}(0.51, \nu^2)$.

Let $f_i^{HC}(I_i(\cdot), Q_i(\cdot), \tilde{\bar{c}}_i(\cdot))$ be the total holding cost of Item $i$ over the whole construction duration, then the holding cost is:

$$f_i^{HC}(I_i(\cdot), Q_i(\cdot), \tilde{\bar{c}}_i(\cdot)) = \sum_{k=0}^{m-1} \tilde{\bar{c}}_i(k)h_i(I_i(k)+Q_i(k)), \quad \forall i \in \Psi \tag{5.12}$$

**Transportation Cost:** Let $\tilde{\bar{T}}_i^{sw}$ be the transportation price of Item $i$ from the supplier to the corresponding warehouse, $\tilde{\bar{T}}_i^{wc}$ be the transportation price of Item $i$ from the warehouse to the construction sites, and $f_i^{TC}(I_i(\cdot), Q_i(\cdot), \tilde{\bar{D}}_i(\cdot), \tilde{\bar{T}}_i^{sw}, \tilde{\bar{T}}_i^{wc})$ be the total transportation cost of Item $i$ over the entire construction duration, since $Q_i(k)$ is the transportation quantity of Item $i$ from the supplier to the corresponding warehouse in the $(k+1)^{th}$ stage, and $\tilde{\bar{D}}_i(k)$, the demand for Item $i$ in the $(k+1)^{th}$ stage, the transportation quantity of Item $i$ from the warehouse to the construction sites in the $(k+1)^{th}$ stage (i.e., $\varpi_i(k)$) should be discussed in five kinds of cases, which are outlined below (See Figure 5.4):

(1) If $I_i(k) + Q_i(k) \leq 0$, the inventory level of Item $i$ in the corresponding warehouse at the beginning of the $(k+1)^{th}$ stage after the replenishment (i.e., $Q_i(k)$) is sill in a stockout situation, thus all the replenishment (i.e., $Q_i(k)$) is transported from the warehouse to the construction sites in the $(k+1)^{th}$ stage to alleviate the unsatisfied demand for Item $i$ in the last stage, i.e., $\varpi_i(k) = Q_i(k)$.

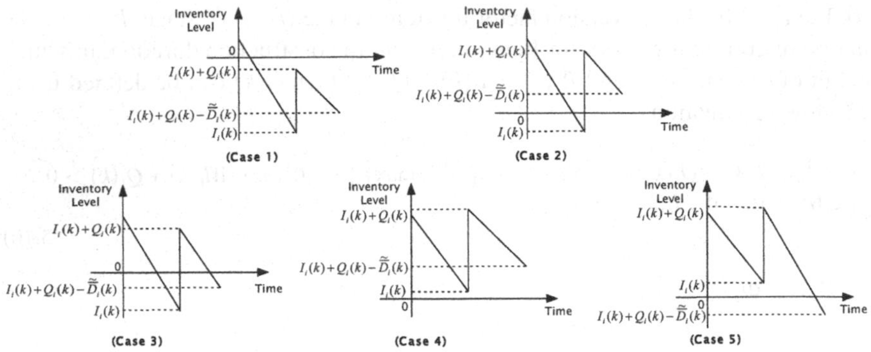


**Fig. 5.4** Graphical representations of inventory systems for five kinds of cases

(2) If $I_i(k)+Q_i(k)>0$, $I_i(k)<0$, and $I_i(k)+Q_i(k)-\tilde{\tilde{D}}_i(k)\geq 0$, all the demand for Item $i$ in the $(k+1)^{th}$ stage (i.e., $\tilde{\tilde{D}}_i(k)$) and the unfilled demand for Item $i$ in the last stage (i.e., $-I_i(k)$) were both completely satisfied, thus $\varpi_i(k)=\tilde{\tilde{D}}_i(k)-I_i(k)$.

(3) If $I_i(k)+Q_i(k)>0$, $I_i(k)<0$, and $I_i(k)+Q_i(k)-\tilde{\tilde{D}}_i(k)<0$, although the replenishment (i.e., $Q_i(k)$) is significant, the demand for Item $i$ in the $(k+1)^{th}$ stage is still not completely satisfied, thus all the replenishment (i.e., $Q_i(k)$) will be transported, i.e., $\varpi_i(k)=Q_i(k)$.

(4) If $I_i(k)+Q_i(k)>0$, $I_i(k)\geq 0$, and $I_i(k)+Q_i(k)-\tilde{\tilde{D}}_i(k)\geq 0$, there is no unfilled demand for Item $i$ in the last stage, and all demand for Item $i$ in the $(k+1)^{th}$ stage (i.e., $\tilde{\tilde{D}}_i(k)$) is completely satisfied, thus $\varpi_i(k)=\tilde{\tilde{D}}_i(k)$.

(5) If $I_i(k)+Q_i(k)>0$, $I_i(k)\geq 0$, and $I_i(k)+Q_i(k)-\tilde{\tilde{D}}_i(k)<0$, while no inventory shortage occurred in the last stage, the demand for Item $i$ in the $(k+1)^{th}$ stage (i.e., $\tilde{\tilde{D}}_i(k)$) is not completely satisfied, thus all the available inventory of Item $i$ will be transported, i.e., $\varpi_i(k)=I_i(k)+Q_i(k)$.

Finally, the expression of $\varpi_i(k)$ can be summarized as:

$$\varpi_i(k)=\min\{\tilde{\tilde{D}}_i(k)-\min\{I_i(k),0\},\max\{I_i(k),0\}+Q_i(k)\},$$

then the transportation cost is:

$$f_i^{TC}(I_i(\cdot),Q_i(\cdot),\tilde{\tilde{D}}_i(\cdot),\tilde{\tilde{T}}_i^{sw},\tilde{\tilde{T}}_i^{wc})=\sum_{k=0}^{m-1}[\tilde{\tilde{T}}_i^{sw}Q_i(k)+\tilde{\tilde{T}}_i^{wc}\varpi_i(k)],\quad \forall i\in\Psi \quad (5.13)$$

**Shortage Cost:** The shortage cost describes the penalty if the demand for Item $i$ cannot be met. Let $f_i^{SC}(I_i(\cdot),Q_i(\cdot),\tilde{\tilde{D}}_i(\cdot))$ be the shortage cost of Item $i$ over the construction duration, $s_i$ (\$/*unit*) be the penalty price if the demand for Item $i$ cannot be met in the $(k+1)^{th}$ stage, let

$$[\tilde{\tilde{D}}_i(k)-I_i(k)-Q_i(k)]^+=\max\{\tilde{\tilde{D}}_i(k)-I_i(k)-Q_i(k),0\}$$

then the shortage cost is:

$$f_i^{SC}(I_i(\cdot),Q_i(\cdot),\tilde{\bar{D}}_i(\cdot)) = \sum_{k=0}^{m-1} s_i[\tilde{\bar{D}}_i(k) - I_i(k) - Q_i(k)]^+, \quad \forall i \in \Psi \tag{5.14}$$

Since the total costs for each item are comprised of the purchase cost, holding cost, transportation cost, and shortage cost, let $\mathbf{x}_i = (I_i(\cdot),Q_i(\cdot))$, $\tilde{\bar{\xi}}_i = (\tilde{\bar{c}}_i(\cdot),\tilde{\bar{D}}_i(\cdot),\tilde{\bar{T}}_i^{sw},\tilde{\bar{T}}_i^{wc})$, and $f_i(\mathbf{x}_i,\tilde{\bar{\xi}}_i)$ be the total costs of Item $i$ over the entire construction duration, then the objective function for item $i$ is:

$$\begin{aligned} f_i(\mathbf{x}_i,\tilde{\bar{\xi}}_i) = {} & f_i^{PC}(Q_i(\cdot)) + f_i^{HC}(I_i(\cdot),Q_i(\cdot),\tilde{\bar{c}}_i(\cdot)) + f_i^{TC}(I_i(\cdot),Q_i(\cdot),\tilde{\bar{D}}_i(\cdot),\tilde{\bar{T}}_i^{sw},\tilde{\bar{T}}_i^{wc}) + \\ & f_i^{SC}(I_i(\cdot),Q_i(\cdot),\tilde{\bar{D}}_i(\cdot)) \end{aligned} \tag{5.15}$$

### 5.2.5 *Dealing with Fuzzy Random Variables*

It is difficult to handle multi-objective problems when there is uncertain information. Because the coefficients in the IMP-LCP model are triangular fuzzy random numbers, Model (5.1) is difficult to solve. One strategy is to employ a transformation method to convert the current model into a deterministic equivalent. In this section, a hybrid crisp approach is used to integrate the decision maker's optimistic-pessimistic attitudes in adopting real world practice. This method first transforms fuzzy random parameters into $(r,\sigma)$-level trapezoidal fuzzy variables, which are subsequently defuzzified using an expected value operator with an optimistic-pessimistic index. The following describes the transformation process of converting the fuzzy random variables into the objective function and the constraints into $(r,\sigma)$-level trapezoidal fuzzy variables.

With no loss of generality, the fuzzy random coefficient is denoted by $\tilde{\bar{\kappa}} = ([\kappa]_L, \varphi(\omega), [\kappa]_R)$, where $\varphi(\omega)$ is a random variable with a probability density function $p_\varphi(x)$. Furthermore, it is assumed that $\varphi(\omega)$ follows a Normal distribution $\mathcal{N}(\zeta, v^2)$, then $p_\varphi(x) = \frac{1}{\sqrt{2\pi}v} e^{-\frac{(x-\zeta)^2}{2v^2}}$. Let $\sigma$ be any given probability level of a random variable, $r$ be any given possibility level of a fuzzy variable. The parameters $\sigma \in [0, \sup p_\varphi(x)]$ and $r \in [\frac{[\kappa]_R-[\kappa]_L}{[\kappa]_R-[\kappa]_L+\varphi_\sigma^R-\varphi_\sigma^L}, 1]$ reflect the decision-maker's degree of optimism. For an easier description, $\sigma$ is called the *probability* level and $r$ , the *possibility* level. A stepwise transformation method is presented below:

***Step.1.*** Estimate the parameters $[\kappa]_L$, $[\kappa]_R$, $\zeta$, and $v$ from collected data and professional experience using statistical methods.

***Step.2.*** Obtain the decision-maker's degree of optimism, i.e., the values of a *probability* level $\sigma \in [0, \sup p_\varphi(x)]$ and *possibility* level $r \in [\frac{[\kappa]_R-[\kappa]_L}{[\kappa]_R-[\kappa]_L+\varphi_\sigma^R-\varphi_\sigma^L}, 1]$, which often can be determined through a group decision making approach (Herrera et al. 1996 [66]; Wu and Xu 2012 [65]).

***Step.3.*** Let $\varphi_\sigma$ be the $\sigma$-level sets (or $\sigma$-cuts) of the random variable $\varphi(\omega)$, i.e. $\varphi_\sigma = [\varphi_\sigma^L, \varphi_\sigma^R] = \{x \in R | p_\varphi(x) \geqslant \sigma\}$, then compute the values of $\varphi_\sigma^L$ and $\varphi_\sigma^R$. They are:

$$\varphi_\sigma^L = \inf\{x \in R | p_\varphi(x) \geqslant \sigma\} = \inf p_\varphi^{-1}(\sigma) = \zeta - \sqrt{-2\nu^2 \ln(\sqrt{2\pi}\nu\sigma)}, \quad (5.16)$$

$$\varphi_\sigma^R = \sup\{x \in R | p_\varphi(x) \geqslant \sigma\} = \sup p_\varphi^{-1}(\sigma) = \zeta + \sqrt{-2\nu^2 \ln(\sqrt{2\pi}\nu\sigma)}. \quad (5.17)$$

Both Eqs. (5.16) and (5.17) are depicted in Figure 5.5.

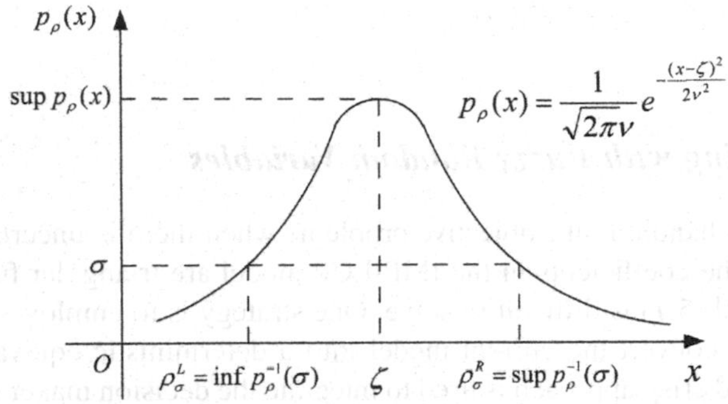


**Fig. 5.5** The probability density function of random variable $\varphi$

***Step.4.*** Transform the fuzzy random variable $\tilde{\bar{\kappa}} = ([\kappa]_L, \varphi(\omega), [\kappa]_R)$ into the $(r, \sigma)$-level trapezoidal fuzzy variable $\tilde{\kappa}_{(r,\sigma)}$ using the following equation:

$$\tilde{\bar{\kappa}} \to \tilde{\kappa}_{(r,\sigma)} = ([\kappa]_L, \underline{\kappa}, \overline{\kappa}, [\kappa]_R),$$

where

$$\underline{\kappa} = [\kappa]_R - r([\kappa]_R - \varphi_\sigma^L) = [\kappa]_R - r([\kappa]_R - \zeta + \sqrt{-2\nu^2 \ln(\sqrt{2\pi}\nu\sigma)}), \quad (5.18)$$

$$\overline{\kappa} = [\kappa]_L + r(\varphi_\sigma^R - [\kappa]_L) = [\kappa]_L + r(\zeta + \sqrt{-2\nu^2 \ln(\sqrt{2\pi}\nu\sigma)} - [\kappa]_L). \quad (5.19)$$

The membership function of $\tilde{\kappa}_{(r,\sigma)}$ is as below:

$$\mu_{\tilde{\kappa}_{(r,\sigma)}}(x) = \begin{cases} 0 & if \quad x < [\kappa]_L, x > [\kappa]_R, \\ \frac{x-[\kappa]_L}{\underline{\kappa}-[\kappa]_L} & if \quad [\kappa]_L \leqslant x < \underline{\kappa}, \\ 1 & if \quad \underline{\kappa} \leqslant x \leqslant \overline{\kappa}, \\ \frac{[\kappa]_R-x}{[\kappa]_R-\overline{\kappa}} & if \quad \overline{\kappa} < x \leqslant [\kappa]_R. \end{cases}$$

The above transformation process, from the fuzzy random variable $\tilde{\bar{\kappa}}$ into the $(r, \sigma)$-level trapezoidal fuzzy variable $\tilde{\kappa}_{(r,\sigma)}$, is illustrated in Figure 5.6. As shown, the

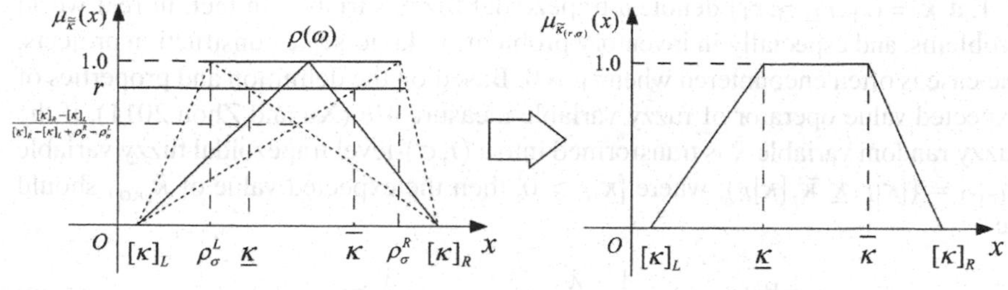

**Fig. 5.6** The transformation process from the fuzzy random variable $\tilde{\bar{\kappa}}$ to the $(r,\sigma)$-level trapezoidal fuzzy variable $\tilde{\kappa}_{(r,\sigma)}$

*possibility* level (i.e., $r$) should be within $[\frac{[\kappa]_R-[\kappa]_L}{[\kappa]_R-[\kappa]_L+\varphi_\sigma^R-\varphi_\sigma^L}, 1]$ in order to deal with the real world problems. The decision maker can choose the value of $r$ accordingly. When the *possibility* level increases, it enlarges the distance between $\underline{\kappa}$ and $\overline{\kappa}$ in the $(r,\sigma)$-level trapezoidal fuzzy variable, which means the decision maker holds a pessimistic attitude about the accuracy of data. Contrarily, when the *possibility* level decreases, it reduces the distance between $\underline{\kappa}$ and $\overline{\kappa}$ in the $(r,\sigma)$-level trapezoidal fuzzy variable, which means the decision maker holds an optimistic attitude about the accuracy of data.

Based on the above method, the fuzzy random variables $\tilde{\bar{c}}_i(k)$, $\tilde{\bar{D}}_i(k)$, $\tilde{\bar{T}}_i^{sw}$, and $\tilde{\bar{T}}_i^{wc}$ in our mathematical model (see Model (5.1)) can be transformed into the $(r,\sigma)$-level trapezoidal fuzzy variables as follows:

$$\tilde{\bar{c}}_i(k) \to \tilde{c}_i(k)_{(r,\sigma)} = ([c_i(k)]_L, \underline{c_i(k)}, \overline{c_i(k)}, [c_i(k)]_R),$$

$$\tilde{\bar{D}}_i(k) \to \tilde{D}_i(k)_{(r,\sigma)} = ([D_i(k)]_L, \underline{D_i(k)}, \overline{D_i(k)}, [D_i(k)]_R),$$

$$\tilde{\bar{T}}_i^{sw} \to \tilde{T}_{i(r,\sigma)}^{sw} = ([T_i^{sw}]_L, \underline{T_i^{sw}}, \overline{T_i^{sw}}, [T_i^{sw}]_R),$$

$$\tilde{\bar{T}}_i^{wc} \to \tilde{T}_{i(r,\sigma)}^{wc} = ([T_i^{wc}]_L, \underline{T_i^{wc}}, \overline{T_i^{wc}}, [T_i^{wc}]_R).$$

### 5.2.6 FEVM of IMP-LCP

To calculate the expected value of the above $(r,\sigma)$-level trapezoidal fuzzy variables, a new fuzzy measure with an optimistic-pessimistic adjustment index is introduced to characterize the real life problems. The definition of this fuzzy measure *Me*, which is a convex combination of *Pos* and *Nec*, can be found in Xu and Zhou (2011) [281], and the basic knowledge for the measures *Pos* and *Nec* are in Dubois and Prade (1988) [221].

Let $\tilde{\kappa} = (r_1, r_2, r_3, r_4)$ denote a trapezoidal fuzzy variable. In fact, in real world problems, and especially in inventory problems in large-scale construction projects, the case is often encountered when $r_1 > 0$. Based on the definition and properties of expected value operator of fuzzy variable measure $Me$ (Xu and Zhou 2011), if the fuzzy random variable $\tilde{\bar{\kappa}}$ is transformed into a $(r, \sigma)$-level trapezoidal fuzzy variable $\tilde{\kappa}_{(r,\sigma)} = ([\kappa]_L, \underline{\kappa}, \overline{\kappa}, [\kappa]_R)$, where $[\kappa]_L > 0$, then the expected value of $\tilde{\kappa}_{(r,\sigma)}$ should be

$$E^{Me}[\tilde{\kappa}_{(r,\sigma)}] = \frac{1-\lambda}{2}([\kappa]_L + \underline{\kappa}) + \frac{\lambda}{2}(\overline{\kappa} + [\kappa]_R). \tag{5.20}$$

The linearity of $E^{Me}$ was proved in Xu and Zhou (2011) [281], thus it is suitable to employ this expected value operator to deal with the linear objective functions and constraints that contain the $(r, \sigma)$-level trapezoidal fuzzy variable.

Based on the above methods, the expected value of the $(r, \sigma)$-level trapezoidal fuzzy variables in each objective function and state equation can be calculated using Eq. (5.20). Therefore, the research problem IMP-LCP (see Model (5.1)) can be transformed into the following multi-objective expected value model:

$$\begin{cases} \min\{E^{Me}[f_1(\mathbf{x}_1, \tilde{\bar{\xi}}_{1(r_1,\sigma_1)})], E^{Me}[f_2(\mathbf{x}_2, \tilde{\bar{\xi}}_{2(r_2,\sigma_2)})], \cdots, E^{Me}[f_n(\mathbf{x}_n, \tilde{\bar{\xi}}_{n(r_n,\sigma_n)})]\} \\ s.t. \begin{cases} I_i(k+1) = I_i(k) + Q_i(k) - E^{Me}[\tilde{D}_i(k)_{(r_i,\sigma_i)}], & \forall i \in \Psi, \quad k \in \Phi, \\ I_i(0) = a_i, \quad \forall i \in \Psi, \\ I_i(m) = b_i, \quad \forall i \in \Psi, \\ Q_j^L Q_1(k) \leqslant Q_j(k) \leqslant Q_j^U Q_1(k), \quad \forall j \in \Omega, \quad k \in \Phi, \\ Q_{i,k}^{min} \leqslant Q_i(k) \leqslant Q_{i,k}^{max} \quad or \quad Q_i(k) = 0, \quad \forall i \in \Psi, \quad k \in \Phi, \\ I_i(k) \leqslant I_i^{max}, \quad \forall i \in \Psi, \quad k \in \Phi, \end{cases} \end{cases} \tag{5.21}$$

where $\mathbf{x}_i = (I_i(\cdot), Q_i(\cdot))$, $\tilde{\bar{\xi}}_{i(r_i,\sigma_i)} = (\tilde{c}_i(\cdot)_{(r_i,\sigma_i)}, \tilde{D}_i(\cdot)_{(r_i,\sigma_i)}, T_i^{\tilde{sw}}{}_{(r_i,\sigma_i)}, T_i^{\tilde{wc}}{}_{(r_i,\sigma_i)})$.

However, it is not easy to find an optimal solution to the above model as it is a multi-objective optimization. In this section, the weighted sum method (Marler and Arora 2010 [68]) is adopted to deal with the model. It should be noted that the weighted sum method is applicable only when all the objective function values are expressed in the same unit. Fortunately, since each objective function is described as total cost in the same unit for each item, Model (5.21) can be transformed into a single-objective equivalent using the following process. Suppose $E^{Me}[f(\mathbf{x}, \tilde{\bar{\xi}}_{(r,\sigma)})] = \sum_{i=1}^{n} \omega_i E^{Me}[f_i(\mathbf{x}_i, \tilde{\bar{\xi}}_{i(r_i,\sigma_i)})]$, in which the weight coefficient $\omega_i$ expresses the importance of $E^{Me}[f_i(\mathbf{x}_i, \tilde{\bar{\xi}}_{i(r_i,\sigma_i)})]$ by the decision-maker where $\mathbf{x} = (\mathbf{x}_1, \mathbf{x}_2, \cdots, \mathbf{x}_n)$ and $\tilde{\bar{\xi}}_{(r,\sigma)} = (\tilde{\bar{\xi}}_{1(r_1,\sigma_1)}, \tilde{\bar{\xi}}_{2(r_2,\sigma_2)}, \cdots, \tilde{\bar{\xi}}_{n(r_n,\sigma_n)})$. Hence, Model (5.21) can be converted into the following:

$$\begin{cases} \min\{\sum_{i=1}^{n} \omega_i E^{Me}[f_i(\mathbf{x}_i, \tilde{\xi}_{i(r_i,\sigma_i)})]\} \\ s.t. \begin{cases} I_i(k+1) = I_i(k) + Q_i(k) - E^{Me}[\tilde{D}_i(k)_{(r_i,\sigma_i)}], \quad \forall i \in \Psi, \quad k \in \Phi, \\ I_i(0) = a_i, \quad \forall i \in \Psi, \\ I_i(m) = b_i, \quad \forall i \in \Psi, \\ Q_j^L Q_1(k) \leqslant Q_j(k) \leqslant Q_j^U Q_1(k), \quad \forall j \in \Omega, \quad k \in \Phi, \\ Q_{i,k}^{min} \leqslant Q_i(k) \leqslant Q_{i,k}^{max} \quad or \quad Q_i(k) = 0, \quad \forall i \in \Psi, \quad k \in \Phi, \\ I_i(k) \leqslant I_i^{max}, \forall i \in \Psi, \quad k \in \Phi, \\ \sum_{i=1}^{n} \omega_i = 1, \end{cases} \end{cases} \tag{5.22}$$

where $\omega_i$ is the weight coefficient that expresses the importance of $E^{Me}[f_i(\mathbf{x}_i, \tilde{\xi}_{i(r_i,\sigma_i)})]$ by the decision-maker. Model (5.22) is in effect a single-objective expected value model that can be pragmatically solved.

To summarize the transformation methods and the model transformation process addressed from previous sections, Figure 5.7 is provided as an easy-to-follow roadmap to understand the model transformation process from the IMP-LCP (i.e., Model (5.1)) to Model (5.22). A very effective and relatively efficient solution algorithm is developed for Model (5.22), as detailed in the following section.

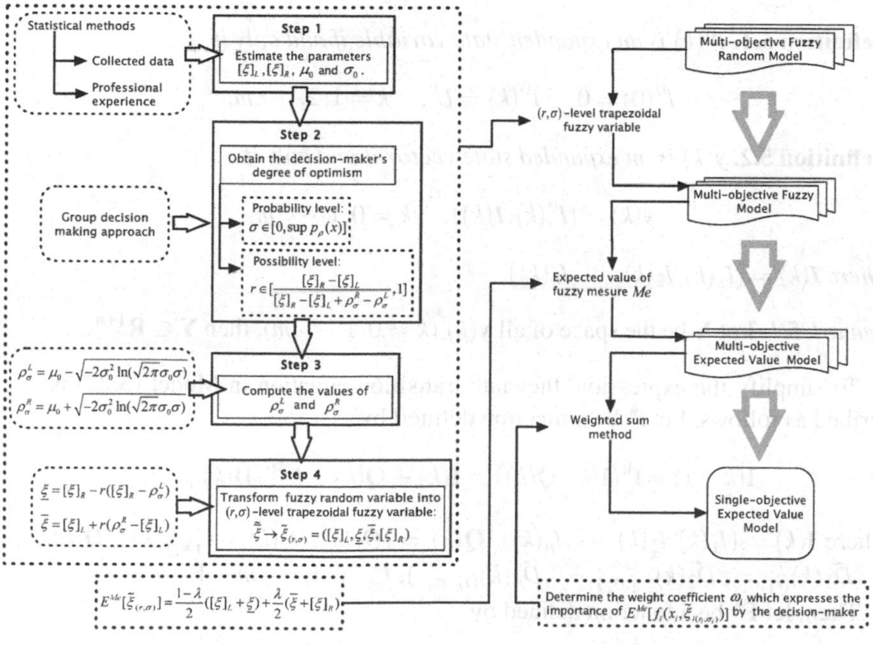


**Fig. 5.7** Flowchart of model transformation process

## 5.3 Model Analysis of IMP-LCP

In order to solve the expected value model above, the support function and the Hamilton function are introduced to derive the theorem necessary to find the optimal solution to the problem posed here. Before the theorem can be proved, a series of mathematical statements and definitions are introduced.

### 5.3.1 Expanded Continuous State Attainable Set

Firstly, the expected value for the objective function in Model (5.22) is examined. To prove the crucial theorem, a function which is similar to (5.22) and useful in the following proof process needs to be defined. This function is described below.

Let $L^k$ be a function defined by

$$L^k = \Sigma_{i=1}^n \omega_i\{\Sigma_{p=0}^{k-1}[\alpha_i(p)+\beta_i(p)\delta_i(1+r_i)^p Q_i^2(p)] + \Sigma_{p=0}^{k-1} E^{Me}[\tilde{\bar{c}}_i(p)]h_i(I_i(p)+ Q_i(p)) + \Sigma_{p=0}^{k-1}[E^{Me}[\tilde{\bar{T}}_i^{sw}]Q_i(p) + E^{Me}[\tilde{\bar{T}}_i^{wc}]\varpi_i(p)] + \Sigma_{p=0}^{k-1} s_i[E^{Me}[\tilde{\bar{D}}_i(p)] - I_i(p) - Q_i(p)]^+\},$$

where $k = 1,2,\cdots,m$.

Then, the expanded state variable and expanded state vector need to be defined, so, in this case, it is convenient to discuss the problem in Euclidean space.

**Definition 5.1.** *$I^0(k)$ is an expanded state variable if and only if*

$$I^0(0) = 0, \quad I^0(k) = L^k, \quad k = 1,2,\cdots,m.$$

**Definition 5.2.** *$y(k)$ is an expanded state vector if and only if*

$$y(k) = (I^0(k), \boldsymbol{I}(k)), \quad k = 0,1,\cdots,m.$$

*where $\boldsymbol{I}(k) = (I_1(k), I_2(k),\cdots,I_n(k))$.*

*Remark 5.1.* Let $\mathbf{Y}$ be the space of all $\mathbf{y}(k)$ ($k = 0,1,\cdots,m$), then $\mathbf{Y} \subset \mathbf{R}^{1+n}$.

To simplify the expression, the state transition equation in Model (5.22) is described as follows. Let $\mathbf{f^k}$ be a function defined by

$$\mathbf{I}(k+1) = \mathbf{f^k}(\mathbf{I}(k), \mathbf{Q}(k)) = \mathbf{I}(k) + \mathbf{Q}(k) - E^{Me}[\tilde{\mathbf{D}}(k)_{(\mathbf{r},\sigma)}],$$

where $\mathbf{I}(k) = (I_1(k), I_2(k),\cdots,I_n(k))$, $\mathbf{Q}(k) = (Q_1(k), Q_2(k),\cdots,Q_n(k))$, $\tilde{\mathbf{D}}(k)_{(\mathbf{r},\sigma)} = (\tilde{D_1}(k)_{(r_1,\sigma_1)}, \tilde{D_2}(k)_{(r_2,\sigma_2)},\cdots,\tilde{D_n}(k)_{(r_n,\sigma_n)})$, $k = 0,1,\cdots,m-1$.

Then, let $\mathbf{F^k}$ be a function defined by

$$\mathbf{F^k} = (L^{k+1}, \mathbf{f^k}), \quad k = 0,1,\cdots,m-1.$$

Therefore, we have $\mathbf{y}(0) = (0, \mathbf{I}(0)) = (0, \mathbf{a})$, $\mathbf{y}(k+1) = \mathbf{F^k}(\mathbf{y}(k), \mathbf{Q}(k))$, where $\mathbf{a} = (a_1, a_2, \cdots, a_n)$, $k = 0, 1, \cdots, m-1$.

Base on the above, now the state attainable set and the expanded state attainable set are presented as follows.

The state attainable set of the $2^{nd}$ stage $\mathbf{X^1}(\mathbf{I}(0))$, which is derived from the initial condition $\mathbf{I}(0) = \mathbf{a}$ is defined by

$$\mathbf{X^1}(\mathbf{I}(0)) = \left\{ \mathbf{I}(1) \,\middle|\, \begin{array}{l} \mathbf{I}(1) = \mathbf{f^0}(\mathbf{I}(0), \mathbf{Q}(0)), Q_j^L Q_1(0) \leqslant Q_j(0) \leqslant Q_j^U Q_1(0), \\ Q_{i,0}^{min} \leqslant Q_i(0) \leqslant Q_{i,0}^{max} \quad or \quad Q_i(0) = 0, \forall j \in \Omega, i \in \Psi \end{array} \right\}.$$

**Definition 5.3.** *$X^k(I(k-1))$ is the state attainable set of the $(k+1)$th stage, which is derived from $I(k-1)$ if and only if*

$$\begin{aligned} &X^k(I(k-1)) \\ &= \left\{ I(k) \,\middle|\, \begin{array}{l} I(k) = f^{k\text{-}1}(I(k-1), Q(k-1)), Q_j^L Q_1(k-1) \leqslant Q_j(k-1) \leqslant Q_j^U Q_1(k-1), \\ Q_{i,k-1}^{min} \leqslant Q_i(k-1) \leqslant Q_{i,k-1}^{max} \quad or \quad Q_i(k-1) = 0, I_i(k-1) \leqslant I_i^{max}, \forall j \in \Omega, i \in \Psi \end{array} \right\}, \end{aligned}$$

$\forall I(k-1) \in X^{k\text{-}1}$, $k = 2, 3, \cdots, m$.

The expanded state attainable set of the $2^{nd}$ stage $\mathbf{Y^1}(\mathbf{y}(0))$, which is in the corresponding expanded space $\mathbf{Y}$, is defined by

$$\mathbf{Y^1}(\mathbf{y}(0)) = \left\{ \mathbf{y}(1) \,\middle|\, \begin{array}{l} \mathbf{y}(1) = \mathbf{F^0}(\mathbf{y}(0), \mathbf{Q}(0)), Q_j^L Q_1(0) \leqslant Q_j(0) \leqslant Q_j^U Q_1(0), \\ Q_{i,0}^{min} \leqslant Q_i(0) \leqslant Q_{i,0}^{max} \quad or \quad Q_i(0) = 0, \ \forall j \in \Omega, i \in \Psi \end{array} \right\}.$$

**Definition 5.4.** *$Y^k(y(k-1))$ is the expanded state attainable set of the $(k+1)$th stage, which is derived from $y(k-1)$ if and only if*

$$\begin{aligned} &Y^k(y(k-1)) \\ &= \left\{ y(k) \,\middle|\, \begin{array}{l} y(k) = F^{k\text{-}1}(y(k-1), Q(k-1)), Q_j^L Q_1(k-1) \leqslant Q_j(k-1) \leqslant Q_j^U Q_1(k-1), \\ Q_{i,k-1}^{min} \leqslant Q_i(k-1) \leqslant Q_{i,k-1}^{max} \quad or \quad Q_i(k-1) = 0, I_i(k-1) \leqslant I_i^{max}, \forall j \in \Omega, i \in \Psi \end{array} \right\}, \end{aligned}$$

$\forall y(k-1) \in Y^{k\text{-}1}, k = 2, 3, \cdots, m$.

All of the state attainable set of the $(k+1)$th stage $\mathbf{X^k}$ is defined by

$$\mathbf{X^k} = \bigcup \{\mathbf{X^k}(\mathbf{I}(k-1)) | \mathbf{I}(k-1) \in \mathbf{X^{k\text{-}1}}\}, \quad k = 2, 3, \cdots, m.$$

All of the expanded state attainable set of the $(k+1)$th stage $\mathbf{Y^k}$ is defined by

$$\mathbf{Y^k} = \bigcup \{\mathbf{Y^k}(\mathbf{y}(k-1)) | \mathbf{y}(k-1) \in \mathbf{Y^{k\text{-}1}}\}, \quad k = 2, 3, \cdots, m.$$

Up to now, a series of ideas about the solution space have been listed. The conditional function of the optimal value and the function of optimal value need to be defined at this point to allow a further preparation of the proof.

### 5.3.2 Conditional Function of Optimal Value

The conditional function of the optimal value of the $2^{nd}$ stage $fov^1(\mathbf{I}(1)|\mathbf{y}(0))$ is defined by

$$fov^1(\mathbf{I}(1)|\mathbf{y}(0)) = \min\left\{L^1(\mathbf{Q}(0))\left|\begin{array}{l} Q_j^L Q_1(0) \leqslant Q_j(0) \leqslant Q_j^U Q_1(0), Q_{i,0}^{min} \leqslant Q_i(0) \leqslant Q_{i,0}^{max} \\ or \quad Q_i(0) = 0, \mathbf{f}^0(\mathbf{I}(0), \mathbf{Q}(0)) = \mathbf{I}(1) \quad \forall j \in \Omega, i \in \Psi \end{array}\right.\right\}.$$

The function of the optimal value of the $2^{nd}$ stage $FOV^1(\mathbf{I}(1))$ is defined by

$$FOV^1(\mathbf{I}(1)) = \min\left\{L^1(\mathbf{Q}(0))\left|\begin{array}{l} Q_j^L Q_1(0) \leqslant Q_j(0) \leqslant Q_j^U Q_1(0), Q_{i,0}^{min} \leqslant Q_i(0) \leqslant Q_{i,0}^{max} \\ or \quad Q_i(0) = 0, \mathbf{f}^0(\mathbf{I}(0), \mathbf{Q}(0)) = \mathbf{I}(1) \quad \forall j \in \Omega, i \in \Psi \end{array}\right.\right\}.$$

Here, the value of the conditional function of the optimal value $fov^1(\mathbf{I}(1)|\mathbf{y}(0))$ is calculated under the condition that the value of the expanded state vector $\mathbf{y}(0)$ is fixed. Since $\mathbf{y}(0) = (0, \mathbf{I}(0)) = (0, \mathbf{a})$ is the only fixed value in the $2^{nd}$ stage, the conditional function of the optimal value of the $2^{nd}$ stage $fov^1(\mathbf{I}(1)|\mathbf{y}(0))$ is the same as the function of the optimal value of the $2^{nd}$ stage $FOV^1(\mathbf{I}(1))$ in this special case, i.e. $fov^1(\mathbf{I}(1)|\mathbf{y}(0)) = FOV^1(\mathbf{I}(1))$. But there are differences between the two functions when $k = 2, 3, \cdots, m$.

**Definition 5.5.** *$fov^k(\boldsymbol{I}(k)|\boldsymbol{y}(k-1))$ is the conditional function of the optimal value of the $(k+1)$th stage if and only if*

$$fov^k(\boldsymbol{I}(k)|\boldsymbol{y}(k-1))$$
$$= \min\left\{L^k(\boldsymbol{Q}(k-1))\left|\begin{array}{l} Q_j^L Q_1(k-1) \leqslant Q_j(k-1) \leqslant Q_j^U Q_1(k-1), \\ Q_{i,k-1}^{min} \leqslant Q_i(k-1) \leqslant Q_{i,k-1}^{max} \quad or \\ Q_i(k-1) = 0, I_i(k-1) \leqslant I_i^{max}, \forall j \in \Omega, i \in \Psi \\ L^k = I^0(k-1) + \sum_{i=1}^n \omega_i\{[\alpha_i(k-1) + \beta_i(k-1)\delta_i(1+r_i)^{k-1} \\ Q_i^2(k-1)] + E^{Me}[\tilde{\bar{c}}_i(k-1)]h_i(I_i(k-1) + Q_i(k-1)) \\ +[E^{Me}[\tilde{\bar{T}}_j^{sw}]Q_i(k-1) + E^{Me}[\tilde{\bar{T}}_i^{wc}]\varpi_i(k-1)]+ \\ s_i[E^{Me}[\tilde{\bar{D}}_i(k-1)] - I_i(k-1) - Q_i(k-1)]^+\}, \\ \boldsymbol{f}^{k-1}(\boldsymbol{I}(k-1), \boldsymbol{Q}(k-1)) = \boldsymbol{I}(k) \end{array}\right.\right\},$$

$\forall i \in \Psi, k = 2, 3, \cdots, m.$

**Lemma 5.1.** *The conditional function of the optimal value of the $(k+1)$th stage $fov^k(\boldsymbol{I}(k)|\boldsymbol{y}(k-1))$ is differentiable of its argument $\boldsymbol{I}(k)$, $k = 1, 2, \cdots, m$.*

*Proof.* According to Def. 5.5 and the statement of $fov^1(\mathbf{I}(1)|\mathbf{y}(0))$, $k=1,2,\cdots,m$, then

$$fov^k(\mathbf{I}(k)|\mathbf{y}(k-1)) = \min\left\{ L^k(\mathbf{Q}(k-1)) \left| \begin{array}{l} Q_j^L Q_1(k-1) \leqslant Q_j(k-1) \leqslant Q_j^U Q_1(k-1), \\ Q_{i,k-1}^{min} \leqslant Q_i(k-1) \leqslant Q_{i,k-1}^{max} \quad or \\ Q_i(k-1)=0, I_i(k-1) \leqslant I_i^{max}, \forall j \in \Omega, i \in \Psi \\ L^k = I^0(k-1) + \sum_{i=1}^n \omega_i\{[\alpha_i(k-1)+\beta_i(k-1)\delta_i(1+r_i)^{k-1} \\ Q_i^2(k-1)] + E^{Me}[\tilde{\bar{c}}_i(k-1)]h_i(I_i(k-1)+Q_i(k-1)) \\ +[E^{Me}[\tilde{\bar{T}}_i^{sw}]Q_i(k-1) + E^{Me}[\tilde{\bar{T}}_i^{wc}]\varpi_i(k-1)]+ \\ s_i[E^{Me}[\tilde{\bar{D}}_i(k-1)] - I_i(k-1) - Q_i(k-1)]^+\}, \\ \mathbf{f^{k\text{-}1}}(\mathbf{I}(k-1),\mathbf{Q}(k-1)) = \mathbf{I}(k) \end{array} \right. \right\}.$$

Since

$$\begin{aligned} & L^k(\mathbf{Q}(k-1)) \\ =\ & \textstyle\sum_{i=1}^n \omega_i\{\sum_{p=0}^{k-1}[\alpha_i(p)+\beta_i(p)\delta_i(1+r_i)^p Q_i^2(p)] + \sum_{p=0}^{k-1} E^{Me}[\tilde{\bar{c}}_i(p)]h_i(I_i(p)+Q_i(p))+ \\ & \textstyle\sum_{p=0}^{k-1}[E^{Me}[\tilde{\bar{T}}_i^{sw}]Q_i(p) + E^{Me}[\tilde{\bar{T}}_i^{wc}]\varpi_i(p)] + \sum_{p=0}^{k-1} s_i[E^{Me}[\tilde{\bar{D}}_i(p)] - I_i(p) - Q_i(p)]^+\} \\ =\ & \textstyle\sum_{i=1}^n \omega_i\{\sum_{p=0}^{k-2}[\alpha_i(p)+\beta_i(p)\delta_i(1+r_i)^p Q_i^2(p)] + \sum_{p=0}^{k-2} E^{Me}[\tilde{\bar{c}}_i(p)]h_i(I_i(p)+Q_i(p))+ \\ & \textstyle\sum_{p=0}^{k-2}[E^{Me}[\tilde{\bar{T}}_i^{sw}]Q_i(p) + E^{Me}[\tilde{\bar{T}}_i^{wc}]\varpi_i(p)] + \sum_{p=0}^{k-2} s_i[E^{Me}[\tilde{\bar{D}}_i(p)] - I_i(p) - Q_i(p)]^+\} \\ & \textstyle +\sum_{i=1}^n \omega_i\{[\alpha_i(k-1)+\beta_i(k-1)\delta_i(1+r_i)^{k-1}Q_i^2(k-1)] + E^{Me}[\tilde{\bar{c}}_i(k-1)]h_i(I_i(k-1) \\ & +Q_i(k-1)) + [E^{Me}[\tilde{\bar{T}}_i^{sw}]Q_i(k-1) + E^{Me}[\tilde{\bar{T}}_i^{wc}]\varpi_i(k-1)] + s_i[E^{Me}[\tilde{\bar{D}}_i(k-1)]- \\ & I_i(k-1) - Q_i(k-1)]^+\}, \end{aligned}$$

and $\mathbf{I}(k-1)$ is fixed by the value of $\mathbf{y}(k-1) = (I^0(k-1), \mathbf{I}(k-1))$, so $\mathbf{Q}(k-1)$ and $\mathbf{I}(k)$ have one to one correspondence as a result of the equation $\mathbf{f^{k\text{-}1}}(\mathbf{I}(k-1),\mathbf{Q}(k-1)) = \mathbf{I}(k)$. Since the transportation quantity of Item $i$ from the warehouse to the construction sites in the $(k+1)^{th}$ stage (i.e., $\varpi_i(k)$) can be divided into five kinds of cases. Here, each case is discussed as follows.

(1) If $I_i(p)+Q_i(p) \leq 0$, this means the inventory level of Item $i$ in the corresponding warehouse at the beginning of the $(p+1)^{th}$ stage after the replenishment (i.e., $Q_i(p)$) is sill under a stockout situation, thus all the replenishment (i.e., $Q_i(p)$) will be transported from the warehouse to the construction sites in the $(p+1)^{th}$ stage for alleviating the unsatisfied demand for Item $i$ in the last stage, i.e., $\varpi_i(p) = Q_i(p)$, and it is obvious that

$$[E^{Me}[\tilde{\bar{D}}_i(p)] - I_i(p) - Q_i(p)]^+ = E^{Me}[\tilde{\bar{D}}_i(p)] - I_i(p) - Q_i(p),$$

where $p = 0,1,\cdots,k-1$. Thus,

$$
\begin{aligned}
& fov^k(\mathbf{I}(k)|\mathbf{y}(k-1)) \\
&= \min\left\{ L^k(\mathbf{Q}(k-1)) \left| \begin{array}{l} Q_j^L Q_1(k-1) \leqslant Q_j(k-1) \leqslant Q_j^U Q_1(k-1), \\ Q_{i,k-1}^{min} \leqslant Q_i(k-1) \leqslant Q_{i,k-1}^{max} \quad or \\ Q_i(k-1)=0, I_i(k-1) \leqslant I_i^{max}, \forall j \in \Omega, i \in \Psi \\ L^k = I^0(k-1) + \sum_{i=1}^n \omega_i\{[\alpha_i(k-1)+\beta_i(k-1)\delta_i(1+r_i)^{k-1} \\ Q_i^2(k-1)] + E^{Me}[\tilde{\bar{c}}_i(k-1)]h_i(I_i(k-1)+Q_i(k-1)) \\ +[E^{Me}[\tilde{\bar{T}}_i^{sw}]Q_i(k-1) + E^{Me}[\tilde{\bar{T}}_i^{wc}]\varpi_i(k-1)] + \\ s_i[E^{Me}[\tilde{\bar{D}}_i(k-1)] - I_i(k-1) - Q_i(k-1)]^+\}, \\ \mathbf{f}^{\mathbf{k-1}}(\mathbf{I}(k-1),\mathbf{Q}(k-1)) = \mathbf{I}(k) \end{array} \right. \right\} \\
&= \min\left\{ \begin{array}{l} \sum_{i=1}^n \omega_i\{\sum_{p=0}^{k-2}[\alpha_i(p)+\beta_i(p)\delta_i(1+r_i)^p Q_i^2(p)] \\ +\sum_{p=0}^{k-2} E^{Me}[\tilde{\bar{c}}_i(p)]h_i(I_i(p)+Q_i(p)) + \\ \sum_{p=0}^{k-2}[E^{Me}[\tilde{\bar{T}}_i^{sw}]Q_i(p) + E^{Me}[\tilde{\bar{T}}_i^{wc}]\varpi_i(p)] + \\ \sum_{p=0}^{k-2} s_i[E^{Me}[\tilde{\bar{D}}_i(p)] - I_i(p) - Q_i(p)]^+\} \end{array} \left| \begin{array}{l} Q_j^L Q_1(p) \leqslant Q_j(p) \leqslant Q_j^U Q_1(p), \\ Q_{i,p}^{min} \leqslant Q_i(p) \leqslant Q_{i,p}^{max} \quad or \\ Q_i(p)=0, I_i(p) \leqslant I_i^{max}, \forall j \in \Omega, \\ i \in \Psi, p=0,\cdots,k-2 \end{array} \right. \right\} \\
&\quad + \sum_{i=1}^n \omega_i\{[\alpha_i(k-1)+\beta_i(k-1)\delta_i(1+r_i)^{k-1}Q_i^2(k-1)] + E^{Me}[\tilde{\bar{c}}_i(k-1)]h_i(I_i(k-1) \\
&\quad +Q_i(k-1)) + [E^{Me}[\tilde{\bar{T}}_i^{sw}]Q_i(k-1) + E^{Me}[\tilde{\bar{T}}_i^{wc}]\varpi_i(k-1)] + s_i[E^{Me}[\tilde{\bar{D}}_i(k-1)] - I_i(k-1) \\
&\quad -Q_i(k-1)]^+\} \\
&= \min\left\{ \begin{array}{l} \sum_{i=1}^n \omega_i\{\sum_{p=0}^{k-2}[\alpha_i(p)+\beta_i(p)\delta_i(1+r_i)^p Q_i^2(p)] \\ +\sum_{p=0}^{k-2} E^{Me}[\tilde{\bar{c}}_i(p)]h_i(I_i(p)+Q_i(p)) + \\ \sum_{p=0}^{k-2}[E^{Me}[\tilde{\bar{T}}_i^{sw}]Q_i(p) + E^{Me}[\tilde{\bar{T}}_i^{wc}]Q_i(p)] + \\ \sum_{p=0}^{k-2} s_i(E^{Me}[\tilde{\bar{D}}_i(p)] - I_i(p) - Q_i(p))\} \end{array} \left| \begin{array}{l} Q_j^L Q_1(p) \leqslant Q_j(p) \leqslant Q_j^U Q_1(p), \\ Q_{i,p}^{min} \leqslant Q_i(p) \leqslant Q_{i,p}^{max} \quad or \\ Q_i(p)=0, I_i(p) \leqslant I_i^{max}, \forall j \in \Omega, \\ i \in \Psi, p=0,\cdots,k-2 \end{array} \right. \right\} \\
&\quad + \sum_{i=1}^n \omega_i\{[\alpha_i(k-1)+\beta_i(k-1)\delta_i(1+r_i)^{k-1}Q_i^2(k-1)] + E^{Me}[\tilde{\bar{c}}_i(k-1)]h_i(I_i(k-1) \\
&\quad +Q_i(k-1)) + [E^{Me}[\tilde{\bar{T}}_i^{sw}]Q_i(k-1) + E^{Me}[\tilde{\bar{T}}_i^{wc}]Q_i(k-1)] + s_i(E^{Me}[\tilde{\bar{D}}_i(k-1)] - I_i(k-1) \\
&\quad -Q_i(k-1))\}
\end{aligned}
$$

After simply combine like terms, the following equation can be obtained

$$
\begin{aligned}
& fov^k(\mathbf{I}(k)|\mathbf{y}(k-1)) \\
&= \min\left\{ \begin{array}{l} \sum_{i=1}^n \omega_i\{\sum_{p=0}^{k-2}(\beta_i(p)\delta_i(1+r_i)^p Q_i^2(p) + \\ (E^{Me}[\tilde{\bar{c}}_i(p)]h_i + E^{Me}[\tilde{\bar{T}}_i^{sw}] + E^{Me}[\tilde{\bar{T}}_i^{wc}] \\ -s_i)Q_i(p) + (E^{Me}[\tilde{\bar{c}}_i(p)]h_i - s_i)I_i(p) + \\ s_i E^{Me}[\tilde{\bar{D}}_i(p)] + \alpha_i(p))\} \end{array} \left| \begin{array}{l} Q_j^L Q_1(p) \leqslant Q_j(p) \leqslant Q_j^U Q_1(p), \\ Q_{i,p}^{min} \leqslant Q_i(p) \leqslant Q_{i,p}^{max} \quad or \\ Q_i(p)=0, I_i(p) \leqslant I_i^{max}, \forall j \in \Omega, \\ i \in \Psi, p=0,\cdots,k-2 \end{array} \right. \right\} \\
&\quad + \sum_{i=1}^n \omega_i\{\beta_i(k-1)\delta_i(1+r_i)^{k-1}Q_i^2(k-1) + (E^{Me}[\tilde{\bar{c}}_i(k-1)]h_i + E^{Me}[\tilde{\bar{T}}_i^{sw}] + E^{Me}[\tilde{\bar{T}}_i^{wc}] \\
&\quad -s_i)Q_i(k-1) + (E^{Me}[\tilde{\bar{c}}_i(k-1)]h_i - s_i)I_i(k-1) + s_i E^{Me}[\tilde{\bar{D}}_i(k-1)] + \alpha_i(k-1)\}
\end{aligned}
$$

Let $Q_i^*(p)$ $(p = 0, 1, \cdots, k-2)$ be the value that can minimize $I^0(k-1)$, while meeting the constraints as follows:

$$Q_j^L Q_1(p) \leqslant Q_j(p) \leqslant Q_j^U Q_1(p), \quad \forall j \in \Omega,$$

$$Q_{i,p}^{min} \leqslant Q_i(p) \leqslant Q_{i,p}^{max} \quad or \quad Q_i(p) = 0, \quad \forall i \in \Psi.$$

Here,

$$I^0(k-1) = \sum_{i=1}^n \omega_i\{\sum_{p=0}^{k-2}(G_{1i}^p(Q_i(p)) + C_{1i}^p I_i(p))\}$$

where

$$
\begin{aligned}
G_{1i}^p(Q_i(p)) = \beta_i(p)\delta_i(1+r_i)^p Q_i^2(p) + (E^{Me}[\tilde{\bar{c}}_i(p)]h_i + E^{Me}[\tilde{\bar{T}}_i^{sw}] + E^{Me}[\tilde{\bar{T}}_i^{wc}] - s_i) \\
Q_i(p) + s_i E^{Me}[\tilde{\bar{D}}_i(p)] + \alpha_i(p),
\end{aligned}
$$

is quadratic polynomial of its argument $Q_i(p)$ and

$$C_{1i}^{p} = E^{Me}[\tilde{\bar{c}}_i(p)]h_i - s_i.$$

Thus, it can be obtained that

$$fov^k(\mathbf{I}(k)|\mathbf{y}(k-1)) = \Sigma_{i=1}^{n}\omega_i\{\Sigma_{p=0}^{k-2}(G_{1i}^{p}(Q_i^*(p)) + C_{1i}^{p}I_i(p))\} + \Sigma_{i=1}^{n}\omega_i\{G_{1i}^{k-1}(Q_i(k-1)) + C_{1i}^{k-1}I_i(k-1)\}$$

when $I_i(p) + Q_i(p) \le 0$, where $p = 0, 1, \cdots, k-1$.

(2) If $I_i(p)+Q_i(p) > 0$, $I_i(p) < 0$, and $I_i(p)+Q_i(p) - E^{Me}[\tilde{\bar{D}}_i(p)] \ge 0$, this means all the demand for Item $i$ in the $(p+1)^{th}$ stage (i.e., $\tilde{\bar{D}}_i(p)$) and the unfilled demand for Item $i$ in the last stage (i.e., $-I_i(p)$) could be both completely satisfied, thus $\varpi_i(p) = E^{Me}[\tilde{\bar{D}}_i(p)] - I_i(p)$, and in this case it is obvious that

$$[E^{Me}[\tilde{\bar{D}}_i(p)] - I_i(p) - Q_i(p)]^+ = 0,$$

where $p = 0, 1, \cdots, k-1$. Thus,

$$\begin{aligned}
& fov^k(\mathbf{I}(k)|\mathbf{y}(k-1)) \\
&= \min\left\{\begin{array}{l|l}
\Sigma_{i=1}^{n}\omega_i\{\Sigma_{p=0}^{k-2}[\alpha_i(p) + \beta_i(p)\delta_i(1+r_i)^p Q_i^2(p)] & Q_j^L Q_1(p) \leqslant Q_j(p) \leqslant Q_j^U Q_1(p), \\
+\Sigma_{p=0}^{k-2}E^{Me}[\tilde{\bar{c}}_i(p)]h_i(I_i(p) + Q_i(p)) + & Q_{i,p}^{min} \leqslant Q_i(p) \leqslant Q_{i,p}^{max} \quad or \\
\Sigma_{p=0}^{k-2}[E^{Me}[\tilde{\bar{T}}_i^{sw}]Q_i(p) + E^{Me}[\tilde{\bar{T}}_i^{wc}]\varpi_i(p)] + & Q_i(p) = 0, I_i(p) \leqslant I_i^{max}, \forall j \in \Omega, \\
\Sigma_{p=0}^{k-2}s_i[E^{Me}[\tilde{\bar{D}}_i(p)] - I_i(p) - Q_i(p)]^+\} & i \in \Psi, p = 0, \cdots, k-2
\end{array}\right\} \\
&\quad + \Sigma_{i=1}^{n}\omega_i\{[\alpha_i(k-1) + \beta_i(k-1)\delta_i(1+r_i)^{k-1}Q_i^2(k-1)] + E^{Me}[\tilde{\bar{c}}_i(k-1)]h_i(I_i(k-1) \\
&\quad + Q_i(k-1)) + [E^{Me}[\tilde{\bar{T}}_i^{sw}]Q_i(k-1) + E^{Me}[\tilde{\bar{T}}_i^{wc}]\varpi_i(k-1)] + s_i[E^{Me}[\tilde{\bar{D}}_i(k-1)] - I_i(k-1) \\
&\quad - Q_i(k-1)]^+\} \\
&= \min\left\{\begin{array}{l|l}
\Sigma_{i=1}^{n}\omega_i\{\Sigma_{p=0}^{k-2}[\alpha_i(p) + \beta_i(p)\delta_i(1+r_i)^p Q_i^2(p)] & Q_j^L Q_1(p) \leqslant Q_j(p) \leqslant Q_j^U Q_1(p), \\
+\Sigma_{p=0}^{k-2}E^{Me}[\tilde{\bar{c}}_i(p)]h_i(I_i(p) + Q_i(p)) + & Q_{i,p}^{min} \leqslant Q_i(p) \leqslant Q_{i,p}^{max} \quad or \\
\Sigma_{p=0}^{k-2}[E^{Me}[\tilde{\bar{T}}_i^{sw}]Q_i(p) + E^{Me}[\tilde{\bar{T}}_i^{wc}] & Q_i(p) = 0, I_i(p) \leqslant I_i^{max}, \forall j \in \Omega, \\
(E^{Me}[\tilde{\bar{D}}_i(p)] - I_i(p))]\} & i \in \Psi, p = 0, \cdots, k-2
\end{array}\right\} \\
&\quad + \Sigma_{i=1}^{n}\omega_i\{[\alpha_i(k-1) + \beta_i(k-1)\delta_i(1+r_i)^{k-1}Q_i^2(k-1)] + E^{Me}[\tilde{\bar{c}}_i(k-1)]h_i(I_i(k-1) \\
&\quad + Q_i(k-1)) + [E^{Me}[\tilde{\bar{T}}_i^{sw}]Q_i(k-1) + E^{Me}[\tilde{\bar{T}}_i^{wc}](E^{Me}[\tilde{\bar{D}}_i(k-1)] - I_i(k-1))]\}
\end{aligned}$$

After simply combine like terms, it can be obtained that

$$\begin{aligned}
& fov^k(\mathbf{I}(k)|\mathbf{y}(k-1)) \\
&= \min\left\{\begin{array}{l|l}
\Sigma_{i=1}^{n}\omega_i\{\Sigma_{p=0}^{k-2}(\beta_i(p)\delta_i(1+r_i)^p Q_i^2(p) & Q_j^L Q_1(p) \leqslant Q_j(p) \leqslant Q_j^U Q_1(p), \\
+(E^{Me}[\tilde{\bar{c}}_i(p)]h_i + E^{Me}[\tilde{\bar{T}}_i^{sw}])Q_i(p) & Q_{i,p}^{min} \leqslant Q_i(p) \leqslant Q_{i,p}^{max} \quad or \\
+(E^{Me}[\tilde{\bar{c}}_i(p)]h_i - E^{Me}[\tilde{\bar{T}}_i^{wc}])I_i(p) + & Q_i(p) = 0, I_i(p) \leqslant I_i^{max}, \forall j \in \Omega, \\
E^{Me}[\tilde{\bar{T}}_i^{wc}]E^{Me}[\tilde{\bar{D}}_i(p)] + \alpha_i(p))\} & i \in \Psi, p = 0, \cdots, k-2
\end{array}\right\} \\
&\quad + \Sigma_{i=1}^{n}\omega_i\{\beta_i(k-1)\delta_i(1+r_i)^{k-1}Q_i^2(k-1) + (E^{Me}[\tilde{\bar{c}}_i(k-1)]h_i + E^{Me}[\tilde{\bar{T}}_i^{sw}])Q_i(k-1) + \\
&\quad (E^{Me}[\tilde{\bar{c}}_i(k-1)]h_i - E^{Me}[\tilde{\bar{T}}_i^{wc}])I_i(k-1) + E^{Me}[\tilde{\bar{T}}_i^{wc}]E^{Me}[\tilde{\bar{D}}_i(k-1)] + \alpha_i(k-1)\}
\end{aligned}$$

Let $Q_i^*(p)$ $(p = 0, 1, \cdots, k-2)$ be the value that can minimize $I^0(k-1)$, while meeting the constraints as follows:

$$Q_j^L Q_1(p) \leqslant Q_j(p) \leqslant Q_j^U Q_1(p), \quad \forall j \in \Omega,$$

$$Q_{i,p}^{min} \leqslant Q_i(p) \leqslant Q_{i,p}^{max} \quad or \quad Q_i(p) = 0, \quad \forall i \in \Psi.$$

Here,

$$I^0(k-1) = \sum_{i=1}^{n} \omega_i \{\sum_{p=0}^{k-2} (G_{2i}^p(Q_i(p)) + C_{2i}^p I_i(p))\}$$

where

$$G_{2i}^p(Q_i(p)) = \beta_i(p)\delta_i(1+r_i)^p Q_i^2(p) + (E^{Me}[\tilde{\bar{c}}_i(p)]h_i + E^{Me}[\tilde{\bar{T}}_i^{sw}])Q_i(p) \\ +E^{Me}[\tilde{\bar{T}}_i^{wc}]E^{Me}[\tilde{\bar{D}}_i(p)] + \alpha_i(p)$$

is quadratic polynomial of its argument $Q_i(p)$ and

$$C_{2i}^p = E^{Me}[\tilde{\bar{c}}_i(p)]h_i - E^{Me}[\tilde{\bar{T}}_i^{wc}].$$

Thus, we can have

$$fov^k(\mathbf{I}(k)|\mathbf{y}(k-1)) = \sum_{i=1}^{n} \omega_i \{\sum_{p=0}^{k-2} (G_{2i}^p(Q_i^*(p)) + C_{2i}^p I_i(p))\} + \sum_{i=1}^{n} \omega_i \{G_{2i}^{k-1}(Q_i(k-1)) \\ +C_{2i}^{k-1} I_i(k-1)\}$$

when $I_i(p) + Q_i(p) > 0$, $I_i(p) < 0$, and $I_i(p) + Q_i(p) - E^{Me}[\tilde{\bar{D}}_i(p)] \geq 0$, where $p = 0, 1, \cdots, k-1$.

(3) If $I_i(p) + Q_i(p) > 0$, $I_i(p) < 0$, and $I_i(p) + Q_i(p) - E^{Me}[\tilde{\bar{D}}_i(p)] < 0$, this means although the replenishment (i.e., $Q_i(p)$) is significant, the demand for Item $i$ in the $(p+1)^{th}$ stage is still not satisfied completely, thus all the replenishment (i.e., $Q_i(p)$) will be transported, i.e., $\varpi_i(p) = Q_i(p)$, and in this case it is obvious that

$$[E^{Me}[\tilde{\bar{D}}_i(p)] - I_i(p) - Q_i(p)]^+ = E^{Me}[\tilde{\bar{D}}_i(p)] - I_i(p) - Q_i(p),$$

where $p = 0, 1, \cdots, k-1$. Therefore, the situation is the same with that in Case (1). Then, we can have

$$fov^k(\mathbf{I}(k)|\mathbf{y}(k-1)) = \sum_{i=1}^{n} \omega_i \{\sum_{p=0}^{k-2} (G_{1i}^p(Q_i^*(p)) + C_{1i}^p I_i(p))\} + \sum_{i=1}^{n} \omega_i \{G_{1i}^{k-1}(Q_i(k-1)) \\ +C_{1i}^{k-1} I_i(k-1)\}$$

when $I_i(p) + Q_i(p) > 0$, $I_i(p) < 0$, and $I_i(p) + Q_i(p) - E^{Me}[\tilde{\bar{D}}_i(p)] < 0$, where $p = 0, 1, \cdots, k-1$.

(4) If $I_i(p) + Q_i(p) > 0$, $I_i(p) \geq 0$, and $I_i(p) + Q_i(p) - E^{Me}[\tilde{\bar{D}}_i(p)] \geq 0$, this means there exists no unfilled demand for Item $i$ in the last stage, and all the demand for Item $i$ in the $(p+1)^{th}$ stage (i.e., $E^{Me}[\tilde{\bar{D}}_i(p)]$) are completely satisfied, thus $\varpi_i(p) = E^{Me}[\tilde{\bar{D}}_i(p)]$, and in this case it is obvious that

$$[E^{Me}[\tilde{\bar{D}}_i(p)] - I_i(p) - Q_i(p)]^+ = 0,$$

where $p = 0, 1, \cdots, k-1$. Thus,

$$
\begin{aligned}
& fov^k(\mathbf{I}(k)|\mathbf{y}(k-1)) \\
&= \min\left\{\begin{array}{l|l}
\Sigma_{i=1}^n \omega_i\{\Sigma_{p=0}^{k-2}[\alpha_i(p)+\beta_i(p)\delta_i(1+r_i)^p Q_i^2(p)] & Q_j^L Q_1(p) \leqslant Q_j(p) \leqslant Q_j^U Q_1(p), \\
+\Sigma_{p=0}^{k-2} E^{Me}[\tilde{\bar{c}}_i(p)]h_i(I_i(p)+Q_i(p))+ & Q_{i,p}^{min} \leqslant Q_i(p) \leqslant Q_{i,p}^{max} \quad or \\
\Sigma_{p=0}^{k-2}[E^{Me}[\tilde{\bar{T}}_i^{sw}]Q_i(p)+E^{Me}[\tilde{\bar{T}}_i^{wc}]\varpi_i(p)]+ & Q_i(p)=0, I_i(p)\leqslant I_i^{max}, \forall j\in\Omega, \\
\Sigma_{p=0}^{k-2} s_i[E^{Me}[\tilde{\bar{D}}_i(p)]-I_i(p)-Q_i(p)]^+\} & i\in\Psi, p=0,\cdots,k-2
\end{array}\right\} \\
&+\Sigma_{i=1}^n \omega_i\{[\alpha_i(k-1)+\beta_i(k-1)\delta_i(1+r_i)^{k-1}Q_i^2(k-1)]+E^{Me}[\tilde{\bar{c}}_i(k-1)]h_i(I_i(k-1) \\
&+Q_i(k-1))+[E^{Me}[\tilde{\bar{T}}_i^{sw}]Q_i(k-1)+E^{Me}[\tilde{\bar{T}}_i^{wc}]\varpi_i(k-1)]+s_i[E^{Me}[\tilde{\bar{D}}_i(k-1)]-I_i(k-1) \\
&-Q_i(k-1)]^+\} \\
&= \min\left\{\begin{array}{l|l}
\Sigma_{i=1}^n \omega_i\{\Sigma_{p=0}^{k-2}[\alpha_i(p)+\beta_i(p)\delta_i(1+r_i)^p Q_i^2(p)] & Q_j^L Q_1(p) \leqslant Q_j(p) \leqslant Q_j^U Q_1(p), \\
+\Sigma_{p=0}^{k-2} E^{Me}[\tilde{\bar{c}}_i(p)]h_i(I_i(p)+Q_i(p))+\Sigma_{p=0}^{k-2} & Q_{i,p}^{min} \leqslant Q_i(p) \leqslant Q_{i,p}^{max} \quad or \\
(E^{Me}[\tilde{\bar{T}}_i^{sw}]Q_i(p)+E^{Me}[\tilde{\bar{T}}_i^{wc}]E^{Me}[\tilde{\bar{D}}_i(p)])\} & Q_i(p)=0, I_i(p)\leqslant I_i^{max}, \forall j\in\Omega, \\
 & i\in\Psi, p=0,\cdots,k-2
\end{array}\right\} \\
&+\Sigma_{i=1}^n \omega_i\{[\alpha_i(k-1)+\beta_i(k-1)\delta_i(1+r_i)^{k-1}Q_i^2(k-1)]+E^{Me}[\tilde{\bar{c}}_i(k-1)]h_i(I_i(k-1) \\
&+Q_i(k-1))+(E^{Me}[\tilde{\bar{T}}_i^{sw}]Q_i(k-1)+E^{Me}[\tilde{\bar{T}}_i^{wc}]E^{Me}[\tilde{\bar{D}}_i(k-1)])\}
\end{aligned}
$$

After simply combine like terms, it can be obtained that

$$
\begin{aligned}
& fov^k(\mathbf{I}(k)|\mathbf{y}(k-1)) \\
&= \min\left\{\begin{array}{l|l}
\Sigma_{i=1}^n \omega_i\{\Sigma_{p=0}^{k-2}(\beta_i(p)\delta_i(1+r_i)^p Q_i^2(p) & Q_j^L Q_1(p) \leqslant Q_j(p) \leqslant Q_j^U Q_1(p), \\
+(E^{Me}[\tilde{\bar{c}}_i(p)]h_i+E^{Me}[\tilde{\bar{T}}_i^{sw}])Q_i(p) & Q_{i,p}^{min} \leqslant Q_i(p) \leqslant Q_{i,p}^{max} \quad or \\
+E^{Me}[\tilde{\bar{c}}_i(p)]h_i I_i(p)+E^{Me}[\tilde{\bar{T}}_i^{wc}] & Q_i(p)=0, I_i(p)\leqslant I_i^{max}, \forall j\in\Omega, \\
E^{Me}[\tilde{\bar{D}}_i(p)]+\alpha_i(p))\} & i\in\Psi, p=0,\cdots,k-2
\end{array}\right\} \\
&+\Sigma_{i=1}^n \omega_i\{\beta_i(k-1)\delta_i(1+r_i)^{k-1}Q_i^2(k-1)+(E^{Me}[\tilde{\bar{c}}_i(k-1)]h_i+E^{Me}[\tilde{\bar{T}}_i^{sw}])Q_i(k-1)+ \\
&E^{Me}[\tilde{\bar{c}}_i(k-1)]h_i I_i(k-1)+E^{Me}[\tilde{\bar{T}}_i^{wc}]E^{Me}[\tilde{\bar{D}}_i(k-1)]+\alpha_i(k-1)\}
\end{aligned}
$$

Let $Q_i^*(p)$ $(p=0,1,\cdots,k-2)$ be the value that can minimize $I^0(k-1)$, while meeting the constraints as follows:

$$
\begin{aligned}
&Q_j^L Q_1(p) \leqslant Q_j(p) \leqslant Q_j^U Q_1(p), \quad \forall j\in\Omega, \\
&Q_{i,p}^{min} \leqslant Q_i(p) \leqslant Q_{i,p}^{max} \quad or \quad Q_i(p)=0, \quad \forall i\in\Psi.
\end{aligned}
$$

Here,

$$I^0(k-1)=\Sigma_{i=1}^n \omega_i\{\Sigma_{p=0}^{k-2}(G_{2i}^p(Q_i(p))+C_{2i}^p I_i(p))\}$$

where

$$
\begin{aligned}
G_{2i}^p(Q_i(p)) = {} & \beta_i(p)\delta_i(1+r_i)^p Q_i^2(p)+(E^{Me}[\tilde{\bar{c}}_i(p)]h_i+E^{Me}[\tilde{\bar{T}}_i^{sw}])Q_i(p) \\
&+E^{Me}[\tilde{\bar{T}}_i^{wc}]E^{Me}[\tilde{\bar{D}}_i(p)]+\alpha_i(p)
\end{aligned}
$$

is quadratic polynomial of its argument $Q_i(p)$ and

$$C_{3i}^p=E^{Me}[\tilde{\bar{c}}_i(p)]h_i.$$

Thus, the following equation can be obtained

$$
\begin{aligned}
fov^k(\mathbf{I}(k)|\mathbf{y}(k-1)) = {} & \Sigma_{i=1}^n \omega_i\{\Sigma_{p=0}^{k-2}(G_{2i}^p(Q_i^*(p))+C_{3i}^p I_i(p))\}+\Sigma_{i=1}^n \omega_i\{G_{2i}^{k-1}(Q_i(k-1)) \\
&+C_{3i}^{k-1}I_i(k-1)\}
\end{aligned}
$$

when $I_i(p)+Q_i(p)>0$, $I_i(p)\geq 0$, and $I_i(p)+Q_i(p)-E^{Me}[\tilde{\bar{D}}_i(p)]\geq 0$, where $p=0,1,\cdots,k-1$.

(5) If $I_i(p)+Q_i(p)>0$, $I_i(p)\geq 0$, and $I_i(p)+Q_i(p)-E^{Me}[\tilde{\bar{D}}_i(p)]<0$, this means while no inventory shortage occurred in the last stage, the demand for Item $i$ in the $(p+1)^{th}$ stage (i.e., $E^{Me}[\tilde{\bar{D}}_i(p)]$) are not completely satisfied, thus all the available inventory of Item $i$ will be transported, i.e., $\varpi_i(p)=I_i(p)+Q_i(p)$, and in this case it is obvious that

$$[E^{Me}[\tilde{\bar{D}}_i(p)]-I_i(p)-Q_i(p)]^+=E^{Me}[\tilde{\bar{D}}_i(p)]-I_i(p)-Q_i(p),$$

where $p=0,1,\cdots,k-1$. Thus,

$$\begin{aligned}
&fov^k(\mathbf{I}(k)|\mathbf{y}(k-1))\\
&=\min\left\{\begin{array}{l|l}
\sum_{i=1}^n\omega_i\{\sum_{p=0}^{k-2}[\alpha_i(p)+\beta_i(p)\delta_i(1+r_i)^pQ_i^2(p)] & Q_j^LQ_1(p)\leqslant Q_j(p)\leqslant Q_j^UQ_1(p),\\
+\sum_{p=0}^{k-2}E^{Me}[\tilde{\bar{c}}_i(p)]h_i(I_i(p)+Q_i(p))+\sum_{p=0}^{k-2} & Q_{i,p}^{min}\leqslant Q_i(p)\leqslant Q_{i,p}^{max}\quad or\\
(E^{Me}[\tilde{\bar{T}}_i^{sw}]Q_i(p)+E^{Me}[\tilde{\bar{T}}_i^{wc}](I_i(p)+Q_i(p))) & Q_i(p)=0,I_i(p)\leqslant I_i^{max},\forall j\in\Omega,\\
+\sum_{p=0}^{k-2}s_i(E^{Me}[\tilde{\bar{D}}_i(p)]-I_i(p)-Q_i(p))\} & i\in\Psi,p=0,\cdots,k-2
\end{array}\right\}\\
&+\sum_{i=1}^n\omega_i\{[\alpha_i(k-1)+\beta_i(k-1)\delta_i(1+r_i)^{k-1}Q_i^2(k-1)]+E^{Me}[\tilde{\bar{c}}_i(k-1)]h_i(I_i(k-1)+\\
&Q_i(k-1))+(E^{Me}[\tilde{\bar{T}}_i^{sw}]Q_i(k-1)+E^{Me}[\tilde{\bar{T}}_i^{wc}](I_i(k-1)+Q_i(k-1)))+s_i(E^{Me}[\tilde{\bar{D}}_i(k-1)]\\
&-I_i(k-1)-Q_i(k-1))\}
\end{aligned}$$

After simply combine like terms, it can be obtained that

$$\begin{aligned}
&fov^k(\mathbf{I}(k)|\mathbf{y}(k-1))\\
&=\min\left\{\begin{array}{l|l}
\sum_{i=1}^n\omega_i\{\sum_{p=0}^{k-2}(\beta_i(p)\delta_i(1+r_i)^pQ_i^2(p) & Q_j^LQ_1(p)\leqslant Q_j(p)\leqslant Q_j^UQ_1(p),\\
+(E^{Me}[\tilde{\bar{c}}_i(p)]h_i+E^{Me}[\tilde{\bar{T}}_i^{sw}]+E^{Me}[\tilde{\bar{T}}_i^{wc}] & Q_{i,p}^{min}\leqslant Q_i(p)\leqslant Q_{i,p}^{max}\quad or\\
-s_i)Q_i(p)+(E^{Me}[\tilde{\bar{c}}_i(p)]h_i+E^{Me}[\tilde{\bar{T}}_i^{wc}] & Q_i(p)=0,I_i(p)\leqslant I_i^{max},\forall j\in\Omega,\\
-s_i)I_i(p)+s_iE^{Me}[\tilde{\bar{D}}_i(p)]+\alpha_i(p))\} & i\in\Psi,p=0,\cdots,k-2
\end{array}\right\}\\
&+\sum_{i=1}^n\omega_i\{\beta_i(k-1)\delta_i(1+r_i)^{k-1}Q_i^2(k-1)+(E^{Me}[\tilde{\bar{c}}_i(k-1)]h_i+E^{Me}[\tilde{\bar{T}}_i^{sw}]+\\
&E^{Me}[\tilde{\bar{T}}_i^{wc}]-s_i)Q_i(k-1)+(E^{Me}[\tilde{\bar{c}}_i(k-1)]h_i+E^{Me}[\tilde{\bar{T}}_i^{wc}]-s_i)I_i(k-1)+s_i\\
&E^{Me}[\tilde{\bar{D}}_i(k-1)]+\alpha_i(k-1)\}
\end{aligned}$$

Let $Q_i^*(p)$ $(p=0,1,\cdots,k-2)$ be the value that can minimize $I^0(k-1)$, while meeting the constraints as follows:

$$Q_j^LQ_1(p)\leqslant Q_j(p)\leqslant Q_j^UQ_1(p),\quad\forall j\in\Omega,$$

$$Q_{i,p}^{min}\leqslant Q_i(p)\leqslant Q_{i,p}^{max}\quad or\quad Q_i(p)=0,\quad\forall i\in\Psi.$$

Here,

$$I^0(k-1)=\sum_{i=1}^n\omega_i\{\sum_{p=0}^{k-2}(G_{1i}^p(Q_i(p))+C_{4i}^pI_i(p))\}$$

where

$$C_{4i}^p=E^{Me}[\tilde{\bar{c}}_i(k-1)]h_i+E^{Me}[\tilde{\bar{T}}_i^{wc}]-s_i.$$

Thus, we can have

$$fov^k(\mathbf{I}(k)|\mathbf{y}(k-1)) = \Sigma_{i=1}^n \omega_i\{\Sigma_{p=0}^{k-2}(G_{1i}^p(Q_i^*(p)) + C_{4i}^p I_i(p))\} + \Sigma_{i=1}^n \omega_i\{G_{1i}^{k-1}(Q_i(k-1)) + C_{4i}^{k-1} I_i(k-1)\}$$

when $I_i(p) + Q_i(p) > 0$, $I_i(p) \geq 0$, and $I_i(p) + Q_i(p) - E^{Me}[\tilde{\bar{D}}_i(p)] < 0$, where $p = 0, 1, \cdots, k-1$.

To sum up, it can be obtained that

$$fov^k(\mathbf{I}(k)|\mathbf{y}(k-1)) = \begin{cases} \Sigma_{i=1}^n \omega_i\{\Sigma_{p=0}^{k-2}(G_{1i}^p(Q_i^*(p)) + C_{1i}^p I_i(p))\} + \Sigma_{i=1}^n \omega_i\{G_{1i}^{k-1}(Q_i(k-1)) + C_{1i}^{k-1} I_i(k-1)\}, \\ I_i(p) + Q_i(p) \leq 0, or I_i(p) + Q_i(p) > 0, I_i(p) < 0, and I_i(p) + Q_i(p) - E^{Me}[\tilde{\bar{D}}_i(p)] < 0; \\ \Sigma_{i=1}^n \omega_i\{\Sigma_{p=0}^{k-2}(G_{2i}^p(Q_i^*(p)) + C_{2i}^p I_i(p))\} + \Sigma_{i=1}^n \omega_i\{G_{2i}^{k-1}(Q_i(k-1)) + C_{2i}^{k-1} I_i(k-1)\}, \\ I_i(p) + Q_i(p) > 0, I_i(p) < 0, and I_i(p) + Q_i(p) - E^{Me}[\tilde{\bar{D}}_i(p)] \geq 0; \\ \Sigma_{i=1}^n \omega_i\{\Sigma_{p=0}^{k-2}(G_{2i}^p(Q_i^*(p)) + C_{3i}^p I_i(p))\} + \Sigma_{i=1}^n \omega_i\{G_{2i}^{k-1}(Q_i(k-1)) + C_{3i}^{k-1} I_i(k-1)\}, \\ I_i(p) + Q_i(p) > 0, I_i(p) \geq 0, and I_i(p) + Q_i(p) - E^{Me}[\tilde{\bar{D}}_i(p)] \geq 0; \\ \Sigma_{i=1}^n \omega_i\{\Sigma_{p=0}^{k-2}(G_{1i}^p(Q_i^*(p)) + C_{4i}^p I_i(p))\} + \Sigma_{i=1}^n \omega_i\{G_{1i}^{k-1}(Q_i(k-1)) + C_{4i}^{k-1} I_i(k-1)\}, \\ I_i(p) + Q_i(p) > 0, I_i(p) \geq 0, and I_i(p) + Q_i(p) - E^{Me}[\tilde{\bar{D}}_i(p)] < 0. \end{cases} \quad (5.23)$$

From the above expression, then $fov^k(\mathbf{I}(k)|\mathbf{y}(k-1))$ is a linear function and differentiable of its argument $\mathbf{I}(k-1)$. Let

$$g^1(\mathbf{I}(k)) = -\Sigma_{i=1}^n \omega_i\{G_{1i}^k(Q_i(k)) + C_{1i}^k I_i(k)\},$$

$$g^2(\mathbf{I}(k)) = -\Sigma_{i=1}^n \omega_i\{G_{2i}^k(Q_i(k)) + C_{2i}^k I_i(k)\},$$

$$g^3(\mathbf{I}(k)) = -\Sigma_{i=1}^n \omega_i\{G_{2i}^k(Q_i(k)) + C_{3i}^k I_i(k)\},$$

$$g^4(\mathbf{I}(k)) = -\Sigma_{i=1}^n \omega_i\{G_{1i}^k(Q_i(k)) + C_{4i}^k I_i(k)\}.$$

According to Eq. (5.23), it can be obtained that

$$fov^k(\mathbf{I}(k)|\mathbf{y}(k-1)) = \begin{cases} fov^{k+1}(\mathbf{I}(k+1)|\mathbf{y}(k)) + g^1(\mathbf{I}(k)), \quad I_i(p) + Q_i(p) \leq 0, \\ or I_i(p) + Q_i(p) > 0, I_i(p) < 0, and I_i(p) + Q_i(p) - E^{Me}[\tilde{\bar{D}}_i(p)] < 0; \\ fov^{k+1}(\mathbf{I}(k+1)|\mathbf{y}(k)) + g^2(\mathbf{I}(k)), \quad I_i(p) + Q_i(p) > 0, I_i(p) < 0, \\ and I_i(p) + Q_i(p) - E^{Me}[\tilde{\bar{D}}_i(p)] \geq 0; \\ fov^{k+1}(\mathbf{I}(k+1)|\mathbf{y}(k)) + g^3(\mathbf{I}(k)), \quad I_i(p) + Q_i(p) > 0, I_i(p) \geq 0, \\ and I_i(p) + Q_i(p) - E^{Me}[\tilde{\bar{D}}_i(p)] \geq 0; \\ fov^{k+1}(\mathbf{I}(k+1)|\mathbf{y}(k)) + g^4(\mathbf{I}(k)), \quad I_i(p) + Q_i(p) > 0, I_i(p) \geq 0, \\ and I_i(p) + Q_i(p) - E^{Me}[\tilde{\bar{D}}_i(p)] < 0. \end{cases}$$

It has been proved that $fov^{k+1}(\mathbf{I}(k+1)|\mathbf{y}(k))$ is a linear function and is differentiable of its argument $\mathbf{I}(k)$. On the other hand, it is obvious that $g^1(\mathbf{I}(k))$, $g^2(\mathbf{I}(k))$, $g^3(\mathbf{I}(k))$, and $g^4(\mathbf{I}(k))$ are all linear function and differentiable of their argument $\mathbf{I}(k)$. Therefore, the linear combinations

$$fov^{k+1}(\mathbf{I}(k+1)|\mathbf{y}(k)) + g^1(\mathbf{I}(k)),$$

$$fov^{k+1}(\mathbf{I}(k+1)|\mathbf{y}(k)) + g^2(\mathbf{I}(k)),$$
$$fov^{k+1}(\mathbf{I}(k+1)|\mathbf{y}(k)) + g^3(\mathbf{I}(k)),$$
$$fov^{k+1}(\mathbf{I}(k+1)|\mathbf{y}(k)) + g^4(\mathbf{I}(k))$$

are all differentiable of their argument $\mathbf{I}(k)$. That means $fov^k(\mathbf{I}(k)|\mathbf{y}(k-1))$ is differentiable of its argument $\mathbf{I}(k)$. The proof is completed. □

**Definition 5.6.** *$FOV^k(\boldsymbol{I}(k))$ is the function of the optimal value of the $(k+1)$th stage if and only if*

$$FOV^k(\boldsymbol{I}(k)) = \min\left\{ L^k(\cdot) \left| \begin{array}{l} Q_j^L Q_1(p) \leqslant Q_j(p) \leqslant Q_j^U Q_1(p), \\ Q_{i,p}^{min} \leqslant Q_i(p) \leqslant Q_{i,p}^{max} \quad or \quad Q_i(p) = 0, \\ I_i(p) \leqslant I_i^{max}, \forall j \in \Omega, i \in \Psi, p = 0, 1, \cdots, k-1 \\ \boldsymbol{f}^0(\boldsymbol{I}(0), \boldsymbol{Q}(0)) = \boldsymbol{I}(1), \boldsymbol{f}^1(\boldsymbol{I}(1), \boldsymbol{Q}(1)) = \boldsymbol{I}(2), \\ \cdots, \boldsymbol{f}^{k-1}(\boldsymbol{I}(k-1), \boldsymbol{Q}(k-1)) = \boldsymbol{I}(k) \end{array} \right. \right\},$$

$\forall i \in \Psi, k = 2, 3, \cdots, m.$

**Lemma 5.2.** *The function of the optimal value of the $(k+1)$th stage $FOV^k(\boldsymbol{I}(k))$ is differentiable of its argument $\boldsymbol{I}(k)$, $\forall i \in \Psi$, $k = 1, 2, \cdots, m$.*

*Proof.* According to Def. 5.6 and the statement of $FOV^1(\mathbf{I}(1))$, $\forall i \in \Psi$, $k = 1, 2, \cdots, m$, then

$$FOV^k(\mathbf{I}(k)) = \min\left\{ L^k(\cdot) \left| \begin{array}{l} Q_j^L Q_1(p) \leqslant Q_j(p) \leqslant Q_j^U Q_1(p), \\ Q_{i,p}^{min} \leqslant Q_i(p) \leqslant Q_{i,p}^{max} \quad or \quad Q_i(p) = 0, \\ I_i(p) \leqslant I_i^{max}, \forall j \in \Omega, i \in \Psi, p = 0, 1, \cdots, k-1 \\ \mathbf{f}^0(\mathbf{I}(0), \mathbf{Q}(0)) = \mathbf{I}(1), \mathbf{f}^1(\mathbf{I}(1), \mathbf{Q}(1)) = \mathbf{I}(2), \\ \cdots, \mathbf{f}^{k-1}(\mathbf{I}(k-1), \mathbf{Q}(k-1)) = \mathbf{I}(k) \end{array} \right. \right\}.$$

Since

$$L^k = \Sigma_{i=1}^n \omega_i \{\Sigma_{p=0}^{k-1}[\alpha_i(p) + \beta_i(p)\delta_i(1+r_i)^p Q_i^2(p)] + \Sigma_{p=0}^{k-1} E^{Me}[\tilde{\bar{c}}_i(p)] h_i(I_i(p) + Q_i(p)) + \Sigma_{p=0}^{k-1}[E^{Me}[\tilde{\bar{T}}_i^{sw}]Q_i(p) + E^{Me}[\tilde{\bar{T}}_i^{wc}]\varpi_i(p)] + \Sigma_{p=0}^{k-1} s_i [E^{Me}[\tilde{\bar{D}}_i(p)] - I_i(p) - Q_i(p)]^+\},$$

if $I_i(p) + Q_i(p) \leq 0$, or $I_i(p) + Q_i(p) > 0$, $I_i(p) < 0$, and $I_i(p) + Q_i(p) - E^{Me}[\tilde{\bar{D}}_i(p)] < 0$, then $\varpi_i(p) = Q_i(p)$, and it is obvious that

$$[E^{Me}[\tilde{\bar{D}}_i(p)] - I_i(p) - Q_i(p)]^+ = E^{Me}[\tilde{\bar{D}}_i(p)] - I_i(p) - Q_i(p),$$

where $p = 0, 1, \cdots, k-1$. Thus, the minimum value of $L^k(\cdot)$ can be arrived when $Q_i(p) = Q_i^*(p)$, where $p = 0, \cdots, k-1$, thus

$$FOV^k(\mathbf{I}(k)) = \sum_{i=1}^n \omega_i \{\sum_{p=0}^{k-1} G_{1i}^p(Q_i^*(p)) + \sum_{p=0}^{k-2} C_{1i}^p I_i(p) + C_{1i}^{k-1} I_i(k-1)\};$$

if $I_i(p)+Q_i(p)>0$, $I_i(p)<0$, and $I_i(p)+Q_i(p)-E^{Me}[\tilde{\bar{D}}_i(p)]\geq 0$, then $\varpi_i(p)=E^{Me}[\tilde{\bar{D}}_i(p)]-I_i(p)$, and in this case it is obvious that

$$[E^{Me}[\tilde{\bar{D}}_i(p)]-I_i(p)-Q_i(p)]^+=0,$$

where $p=0,1,\cdots,k-1$. Thus,

$$FOV^k(\mathbf{I}(k))=\sum_{i=1}^{n}\omega_i\{\sum_{p=0}^{k-1}G_{2i}^p(Q_i^*(p))+\sum_{p=0}^{k-2}C_{2i}^pI_i(p)+C_{2i}^{k-1}I_i(k-1)\};$$

if $I_i(p)+Q_i(p)>0$, $I_i(p)\geq 0$, and $I_i(p)+Q_i(p)-E^{Me}[\tilde{\bar{D}}_i(p)]\geq 0$, then $\varpi_i(p)=E^{Me}[\tilde{\bar{D}}_i(p)]$, and in this case it is obvious that

$$[E^{Me}[\tilde{\bar{D}}_i(p)]-I_i(p)-Q_i(p)]^+=0,$$

where $p=0,1,\cdots,k-1$. Thus,

$$FOV^k(\mathbf{I}(k))=\sum_{i=1}^{n}\omega_i\{\sum_{p=0}^{k-1}G_{2i}^p(Q_i^*(p))+\sum_{p=0}^{k-2}C_{3i}^pI_i(p)+C_{3i}^{k-1}I_i(k-1)\};$$

if $I_i(p)+Q_i(p)>0$, $I_i(p)\geq 0$, and $I_i(p)+Q_i(p)-E^{Me}[\tilde{\bar{D}}_i(p)]<0$, then $\varpi_i(p)=I_i(p)+Q_i(p)$, and in this case it is obvious that

$$[E^{Me}[\tilde{\bar{D}}_i(p)]-I_i(p)-Q_i(p)]^+=E^{Me}[\tilde{\bar{D}}_i(p)]-I_i(p)-Q_i(p),$$

where $p=0,1,\cdots,k-1$. Thus,

$$FOV^k(\mathbf{I}(k))=\sum_{i=1}^{n}\omega_i\{\sum_{p=0}^{k-1}G_{1i}^p(Q_i^*(p))+\sum_{p=0}^{k-2}C_{4i}^pI_i(p)+C_{4i}^{k-1}I_i(k-1)\};$$

To sum up, it can be obtained that

$$FOV^k(\mathbf{I}(k))$$
$$=\begin{cases}\sum_{i=1}^n\omega_i\{\sum_{p=0}^{k-1}G_{1i}^p(Q_i^*(p))+\sum_{p=0}^{k-2}C_{1i}^pI_i(p)+C_{1i}^{k-1}I_i(k-1)\}, & I_i(p)+Q_i(p)\leq 0,\\ or I_i(p)+Q_i(p)>0, I_i(p)<0, and I_i(p)+Q_i(p)-E^{Me}[\tilde{\bar{D}}_i(p)]<0;\\ \sum_{i=1}^n\omega_i\{\sum_{p=0}^{k-1}G_{2i}^p(Q_i^*(p))+\sum_{p=0}^{k-2}C_{2i}^pI_i(p)+C_{2i}^{k-1}I_i(k-1)\}, & I_i(p)+Q_i(p)>0,\\ I_i(p)<0, and I_i(p)+Q_i(p)-E^{Me}[\tilde{\bar{D}}_i(p)]\geq 0;\\ \sum_{i=1}^n\omega_i\{\sum_{p=0}^{k-1}G_{2i}^p(Q_i^*(p))+\sum_{p=0}^{k-2}C_{3i}^pI_i(p)+C_{3i}^{k-1}I_i(k-1)\}, & I_i(p)+Q_i(p)>0,\\ I_i(p)\geq 0, and I_i(p)+Q_i(p)-E^{Me}[\tilde{\bar{D}}_i(p)]\geq 0;\\ \sum_{i=1}^n\omega_i\{\sum_{p=0}^{k-1}G_{1i}^p(Q_i^*(p))+\sum_{p=0}^{k-2}C_{4i}^pI_i(p)+C_{4i}^{k-1}I_i(k-1)\}, & I_i(p)+Q_i(p)>0,\\ I_i(p)\geq 0, and I_i(p)+Q_i(p)-E^{Me}[\tilde{\bar{D}}_i(p)]<0.\end{cases} \tag{5.24}$$

From the above expression, it is obviously to know that $FOV^k(\mathbf{I}(k))$ is a linear function and is differentiable of its argument $\mathbf{I}(k-1)$. So it is obvious that

$FOV^{k+1}(\mathbf{I}(k+1))$ is a linear function and is differentiable of its argument $\mathbf{I}(k)$. According to Eq. (5.24), then

$$FOV^k(\mathbf{I}(k)) = \begin{cases} FOV^{k+1}(\mathbf{I}(k+1)) - \sum_{i=1}^n \omega_i\{G_{1i}^k(Q_i^*(k)) + C_{1i}^k I_i(k)\}, I_i(p) + Q_i(p) \le 0, \\ or I_i(p) + Q_i(p) > 0, I_i(p) < 0, and I_i(p) + Q_i(p) - E^{Me}[\tilde{\bar{D}}_i(p)] < 0; \\ FOV^{k+1}(\mathbf{I}(k+1)) - \sum_{i=1}^n \omega_i\{G_{2i}^k(Q_i^*(k)) + C_{2i}^k I_i(k)\}, I_i(p) + Q_i(p) > 0, \\ I_i(p) < 0, and I_i(p) + Q_i(p) - E^{Me}[\tilde{\bar{D}}_i(p)] \ge 0; \\ FOV^{k+1}(\mathbf{I}(k+1)) - \sum_{i=1}^n \omega_i\{G_{2i}^k(Q_i^*(k)) + C_{3i}^k I_i(k)\}, I_i(p) + Q_i(p) > 0, \\ I_i(p) \ge 0, and I_i(p) + Q_i(p) - E^{Me}[\tilde{\bar{D}}_i(p)] \ge 0; \\ FOV^{k+1}(\mathbf{I}(k+1)) - \sum_{i=1}^n \omega_i\{G_{1i}^k(Q_i^*(k)) + C_{4i}^k I_i(k)\}, I_i(p) + Q_i(p) > 0, \\ I_i(p) \ge 0, and I_i(p) + Q_i(p) - E^{Me}[\tilde{\bar{D}}_i(p)] < 0. \end{cases}$$

Since $\sum_{i=1}^n \omega_i\{G_{1i}^k(Q_i^*(k)) + C_{1i}^k I_i(k)\}$, $\sum_{i=1}^n \omega_i\{G_{2i}^k(Q_i^*(k)) + C_{2i}^k I_i(k)\}$, $\sum_{i=1}^n \omega_i\{G_{2i}^k(Q_i^*(k)) + C_{3i}^k I_i(k)\}$ and $\sum_{i=1}^n \omega_i\{G_{1i}^k(Q_i^*(k)) + C_{4i}^k I_i(k)\}$ are all linear functions and differentiable of their argument $\mathbf{I}(k)$. Therefore, the linear combinations

$$FOV^{k+1}(\mathbf{I}(k+1)) - \sum_{i=1}^n \omega_i\{G_{1i}^k(Q_i^*(k)) + C_{1i}^k I_i(k)\},$$

$$FOV^{k+1}(\mathbf{I}(k+1)) - \sum_{i=1}^n \omega_i\{G_{2i}^k(Q_i^*(k)) + C_{2i}^k I_i(k)\},$$

$$FOV^{k+1}(\mathbf{I}(k+1)) - \sum_{i=1}^n \omega_i\{G_{2i}^k(Q_i^*(k)) + C_{3i}^k I_i(k)\},$$

and

$$FOV^{k+1}(\mathbf{I}(k+1)) - \sum_{i=1}^n \omega_i\{G_{1i}^k(Q_i^*(k)) + C_{4i}^k I_i(k)\}$$

are all differentiable of their argument $\mathbf{I}(k)$. That means $FOV^k(\mathbf{I}(k))$ is differentiable of its argument $\mathbf{I}(k)$. The proof is completed. □

Base on above, the conditional lower bound surface and the lower bound surface can be defined.

The conditional lower bound surface of the $2^{nd}$ stage is defined as follows:

Let $\Pi^1(\mathbf{y}(0))$ be the conditional lower bound surface of the $2^{nd}$ stage which is defined by

$$\Pi^1(\mathbf{y}(0)) = \{\mathbf{y}(1)|I^0(1) = fov^1(\mathbf{I}(1)|\mathbf{y}(0)), \mathbf{I}(1) \in \mathbf{X^1}(\mathbf{I}(0))\}.$$

Let $\Pi^1$ be the lower bound surface of the $2^{nd}$ stage which is defined by

$$\Pi^1 = \{\mathbf{y}(1)|I^0(1) = FOV^1(\mathbf{I}(1)), \mathbf{I}(1) \in \mathbf{X^1}\}.$$

Since $fov^1(\mathbf{I}(1)|\mathbf{y}(0)) = FOV^1(\mathbf{I}(1))$ is known, it is obvious that $\Pi^1(\mathbf{y}(0)) = \Pi^2$ in this special case. However, the situations are different when $k = 2, 3, \cdots, m$, as seen in the following definitions.

**Definition 5.7.** *$\Pi^k(y(k-1))$ is the conditional lower bound surface of the $(k+1)$th stage if and only if*

$$\Pi^k(y(k-1)) = \{y(k)|I^0(k) = fov^k(\boldsymbol{I}(k)|y(k-1)), \boldsymbol{I}(k) \in \boldsymbol{X}^k(\boldsymbol{I}(k-1))\},$$

$\forall y(k-1) \in \boldsymbol{Y}^{k\text{-}1}, k = 2,3,\cdots,m.$

**Definition 5.8.** *$\Pi^k$ is the lower bound surface of the $(k+1)$th stage if and only if*

$$\Pi^k = \{y(k)|I^0(k) = FOV^k(\boldsymbol{I}(k)), \boldsymbol{I}(k) \in \boldsymbol{X}^k\}, \quad k = 2,3,\cdots,m.$$

According to Def. 5.5, 5.6 and the above statement of $fov^1(\mathbf{I}(1)|\mathbf{y}(0))$ and $FOV^1(\mathbf{I}(1))$, it is obvious that

$$FOV^k(\mathbf{I}(k)) \leqslant fov^k(\mathbf{I}(k)|\mathbf{y}(k-1)), \quad \forall \mathbf{y}(k-1) \in \mathbf{Y^{k\text{-}1}}, \quad k = 1,2,\cdots,m. \quad (5.25)$$

Assume that $\{\mathbf{Q}^*(0),\cdots,\mathbf{Q}^*(m-1)\}$ is the optimal solution sequence for the purchasing quantity at the beginning of every stage; $\{\mathbf{I}^*(0) = \mathbf{a}, \mathbf{I}^*(1),\cdots,\mathbf{I}^*(m)\}$ is the optimal solution sequence for the inventory level at the beginning of each stage; $\{\mathbf{y}^*(0),\cdots,\mathbf{y}^*(m)\}$ is the optimal solution sequence for the corresponding expanded vectors, so then the following lemmas are applied.

**Lemma 5.3.** *The point $\mathbf{y}^*(k)$ is on the lower bound surface of the $(k+1)$ stage $\Pi^k$, $k = 1,2,\cdots,m$.*

*Proof.* When $k = m$, assume that $\mathbf{y}^*(m) \notin \Pi^m$, then there exists sequence $\{\bar{\mathbf{Q}}(0),\cdots,\bar{\mathbf{Q}}(m-1)\}$ such that the corresponding expanded vector $\bar{\mathbf{y}}(m) = (\bar{L}^m, \bar{\mathbf{I}}(m))$ satisfies

$$\bar{\mathbf{I}}(m) = \mathbf{I}^*(m),$$

$$\bar{L}^m = \Sigma_{i=1}^n \omega_i\{\Sigma_{p=0}^{m-1}[\alpha_i(p) + \beta_i(p)\delta_i(1+r_i)^p\bar{Q}_i^2(p)] + \Sigma_{p=0}^{m-1} E^{Me}[\tilde{\bar{c}}_i(p)]h_i(I_i(p)+\bar{Q}_i(p)) + \Sigma_{p=0}^{m-1}[E^{Me}[\tilde{\bar{T}}_i^{sw}]\bar{Q}_i(p) + E^{Me}[\tilde{\bar{T}}_i^{wc}]\bar{\varpi}_i(p)] + \Sigma_{p=0}^{m-1} s_i[E^{Me}[\tilde{\bar{D}}_i(p)] - I_i(p) - \bar{Q}_i(p)]^+\},$$

$$\bar{L}^m < FOV^m(\bar{\mathbf{I}}(m)), \quad \bar{\mathbf{I}}(m) = \mathbf{I}^*(m).$$

Thus,

$$\bar{L}^m < FOV^m(\mathbf{I}^*(m)) = \min\{\sum_{i=1}^n \omega_i E^{Me}[f_i(\mathbf{x}_i, \xi_{i(r_i,\sigma_i)})]\} = C^{m*}],$$

where $C^{m*}$ is the optimal weight sum total cost of all kinds of materials in a whole period. That means the sequence $\{\bar{\mathbf{Q}}(0),\cdots,\bar{\mathbf{Q}}(m-1)\}$ is better than the optimal solution sequence $\{\mathbf{Q}^*(0),\cdots,\mathbf{Q}^*(m-1)\}$, but it is impossible.

When $1 \leqslant k \leqslant m-1$, assume that $\mathbf{y}^*(k) \notin \Pi^k$, then there exists sequence $\{\bar{\mathbf{Q}}(0),\cdots,\bar{\mathbf{Q}}(k-1)\}$ such that the corresponding expanded vector $\bar{\mathbf{y}}(k) = (\bar{L}^k, \bar{\mathbf{I}}(k))$ satisfies

$$\bar{\mathbf{I}}(k) = \mathbf{I}^*(k),$$

$$\bar{L}^k = \sum_{i=1}^n \omega_i\{\sum_{p=0}^{k-1}[\alpha_i(p)+\beta_i(p)\delta_i(1+r_i)^p\bar{Q}_i^2(p)]+\sum_{p=0}^{k-1}E^{Me}[\tilde{\bar{c}}_i(p)]h_i(I_i(p)+\bar{Q}_i(p))+ \sum_{p=0}^{k-1}[E^{Me}[\tilde{\bar{T}}_i^{sw}]\bar{Q}_i(p)+E^{Me}[\tilde{\bar{T}}_i^{wc}]\varpi_i(p)]+\sum_{p=0}^{k-1}s_i[E^{Me}[\tilde{\bar{D}}_i(p)]-I_i(p)-\bar{Q}_i(p)]^+\},$$

$$\bar{L}^k < FOV^k(\bar{\mathbf{I}}(k)) = FOV^k(\mathbf{I}^*(k)) = C^{k*} = \sum_{i=1}^n \omega_i\{\sum_{p=0}^{k-1}[\alpha_i(p)+\beta_i(p)\delta_i(1+r_i)^pQ_i^{2*}(p)] +\sum_{p=0}^{k-1}E^{Me}[\tilde{\bar{c}}_i(p)]h_i(I_i(p)+Q_i^*(p))+\sum_{p=0}^{k-1}[E^{Me}[\tilde{\bar{T}}_i^{sw}]Q_i^*(p)+E^{Me}[\tilde{\bar{T}}_i^{wc}]\varpi_i(p)] +\sum_{p=0}^{k-1}s_i[E^{Me}[\tilde{\bar{D}}_i(p)]-I_i(p)-Q_i^*(p)]^+\}. \tag{5.26}$$

Now consider the sequence $\{\bar{\mathbf{Q}}(0),\cdots,\bar{\mathbf{Q}}(k-1),\mathbf{Q}^*(k),\cdots,\mathbf{Q}^*(m-1)\}$, let

$$\hat{L}^m = \bar{L}^k + \sum_{i=1}^n \omega_i\{\sum_{p=k}^{m-1}[\alpha_i(p)+\beta_i(p)\delta_i(1+r_i)^pQ_i^{2*}(p)]+\sum_{p=k}^{m-1}E^{Me}[\tilde{\bar{c}}_i(p)]h_i(I_i(p) +Q_i^*(p))+\sum_{p=k}^{m-1}[E^{Me}[\tilde{\bar{T}}_i^{sw}]Q_i^*(p)+E^{Me}[\tilde{\bar{T}}_i^{wc}]\varpi_i(p)]+\sum_{p=k}^{m-1}s_i[E^{Me}[\tilde{\bar{D}}_i(p)]-I_i(p) -Q_i^*(p)]^+\}. \tag{5.27}$$

From Eq. (5.26) and (5.27), then

$$\hat{L}^m < \sum_{i=1}^n \omega_i\{\sum_{p=0}^{m-1}[\alpha_i(p)+\beta_i(p)\delta_i(1+r_i)^pQ_i^{2*}(p)]+\sum_{p=0}^{m-1}E^{Me}[\tilde{\bar{c}}_i(p)]h_i(I_i(p)+Q_i^*(p))+ \sum_{p=0}^{m-1}[E^{Me}[\tilde{\bar{T}}_i^{sw}]Q_i^*(p)+E^{Me}[\tilde{\bar{T}}_i^{wc}]\varpi_i(p)]+\sum_{p=0}^{m-1}s_i[E^{Me}[\tilde{\bar{D}}_i(p)]-I_i(p)-Q_i^*(p)]^+\} = C^{m*}.$$

That means the sequence $\{\bar{\mathbf{Q}}(0),\cdots,\bar{\mathbf{Q}}(k-1),\mathbf{Q}^*(k),\cdots,\mathbf{Q}^*(m-1)\}$ is better than the optimal solution sequence $\{\mathbf{Q}^*(0),\cdots,\mathbf{Q}^*(m-1)\}$, but it is impossible. The proof is completed. □

From Lemma 5.3, it is not hard to prove the following lemma.

**Lemma 5.4.** *The point $\mathbf{y}^*(k)$ is on the conditional lower bound surface of the $(k+1)$th stage $\Pi^k(\mathbf{y}(k-1))$, $k=1,\cdots,m$.*

*Proof.* Let $\mathbf{y}^*(k) = (L^{k*},\mathbf{I}^*(k))$, where

$$L^{k*} = \sum_{i=1}^n \omega_i\{\sum_{p=0}^{k-1}[\alpha_i(p)+\beta_i(p)\delta_i(1+r_i)^pQ_i^{2*}(p)]+\sum_{p=0}^{k-1}E^{Me}[\tilde{\bar{c}}_i(p)]h_i(I_i(p)+Q_i^*(p))+ \sum_{p=0}^{k-1}[E^{Me}[\tilde{\bar{T}}_i^{sw}]Q_i^*(p)+E^{Me}[\tilde{\bar{T}}_i^{wc}]\varpi_i(p)]+\sum_{p=0}^{k-1}s_i[E^{Me}[\tilde{\bar{D}}_i(p)]-I_i(p)-Q_i^*(p)]^+\},$$

$k=1,2,\cdots,m$. According to Def. 5.5 and the statement of $fov^1(\mathbf{I}(1)|\mathbf{y}(0))$, it can be obtained that

$$L^{k*} \geqslant fov^k(\mathbf{I}^*(k)|\mathbf{y}^*(k-1)),$$

but according to Lemma 5.3, it is obvious that

$$L^{k*} = FOV^k(\mathbf{I}^*(k)). \tag{5.28}$$

On the other hand, from Eq. (5.25), it can be obtained that

$$fov^k(\mathbf{I}^*(k)|\mathbf{y}^*(k-1)) \geqslant FOV^k(\mathbf{I}^*(k)).$$

Hence,

$$fov^k(\mathbf{I}^*(k)|\mathbf{y}^*(k-1)) = L^{k*}. \tag{5.29}$$

The above equation means the point $\mathbf{y}^*(k)$ is on the conditional lower bound surface of the $(k+1)$th stage $\Pi^k(\mathbf{y}(k-1))$, $k=1,2,\cdots,m$. The proof is completed. □

### 5.3.3 Support Function of Surface

Now the support function of surface is introduced in the following definition.

**Definition 5.9.** *Let $S^k(\boldsymbol{I}(k)|\boldsymbol{y}(k-1))$ be a finite function which is continuous for all real values $\boldsymbol{I}(k)$ and is defined on the state attainable set of the $(k+1)$th stage $\boldsymbol{X^k}(\boldsymbol{I}(k-1))$ which is derived from $\boldsymbol{I}(k-1)$, so then $S^k(\boldsymbol{I}(k)|\boldsymbol{y}(k-1))$ is called the support function of the conditional lower bound surface of the $(k+1)$th stage $\Pi^k(\boldsymbol{y}(k-1))$ at point $\hat{\boldsymbol{I}}(k)$, if and only if there is a real number $\alpha^k$ such that:*

$$S^k(\boldsymbol{I}(k)|\boldsymbol{y}(k-1)) - \alpha^k \leqslant fov^k(\boldsymbol{I}(k)|\boldsymbol{y}(k-1)), \quad \forall \boldsymbol{I}(k) \in \boldsymbol{X^k}(\boldsymbol{I}(k-1)), \tag{5.30}$$

$$S^k(\hat{\boldsymbol{I}}(k)|\boldsymbol{y}(k-1)) - \alpha^k = fov^k(\hat{\boldsymbol{I}}(k)|\boldsymbol{y}(k-1)), \tag{5.31}$$

$k = 1, 2, \cdots, m.$

**Lemma 5.5.** *Let $S^k(\boldsymbol{I}(k)|\boldsymbol{y}^*(k-1))$ be the support function of the conditional lower bound surface of the $(k+1)$th stage $\Pi^k(\boldsymbol{y}^*(k-1))$ at point $\boldsymbol{I}^*(k)$, if $S^k(\boldsymbol{I}(k)|\boldsymbol{y}^*(k-1))$ is differentiable of its argument $I_i(k)$, and $\boldsymbol{I}^*(k)$ are the interior points of $\boldsymbol{X^k}(\boldsymbol{I}^*(k-1))$ then*

$$\frac{\partial S^k(\boldsymbol{I}^*(k)|\boldsymbol{y}^*(k-1))}{\partial I_i(k)} = \frac{\partial fov^k(\boldsymbol{I}^*(k)|\boldsymbol{y}^*(k-1))}{\partial I_i(k)} = \frac{\partial FOV^k(\boldsymbol{I}^*(k))}{\partial I_i(k)}, \quad \forall i \in \Psi, \quad k = 1, 2, \cdots, m.$$

*Proof.* From Eq. (5.30) and (5.31), it can be obtained that

$$S^k(\mathbf{I}(k)|\mathbf{y}^*(k-1)) - fov^k(\mathbf{I}(k)|\mathbf{y}^*(k-1)) \leqslant \alpha^k,$$
$$S^k(\mathbf{I}^*(k)|\mathbf{y}^*(k-1)) - fov^k(\mathbf{I}^*(k)|\mathbf{y}^*(k-1)) = \alpha^k,$$

i.e. $S^k(\mathbf{I}(k)|\mathbf{y}^*(k-1)) - fov^k(\mathbf{I}(k)|\mathbf{y}^*(k-1))$ arrives the maximum value $\alpha^k$ at point $\mathbf{I}^*(k)$, $k = 1, 2, \cdots, m$. Since $S^k(\mathbf{I}(k)|\mathbf{y}^*(k-1))$ is differentiable of its argument $I_i(k)$, and from Lemma 5.1, it is obvious that the conditional function of optimal value of the $(k+1)$th stage $fov^k(\mathbf{I}(k)|\mathbf{y}^*(k-1))$ is differentiable of its argument $I_i(k)$, then

$$\frac{\partial S^k(\mathbf{I}^*(k)|\mathbf{y}^*(k-1))}{\partial I_i(k)} - \frac{\partial fov^k(\mathbf{I}^*(k)|\mathbf{y}^*(k-1))}{\partial I_i(k)} = 0, \quad \forall i \in \Psi, \quad k = 1, 2, \cdots, m,$$

i.e.

$$\frac{\partial S^k(\mathbf{I}^*(k)|\mathbf{y}^*(k-1))}{\partial I_i(k)} = \frac{\partial fov^k(\mathbf{I}^*(k)|\mathbf{y}^*(k-1))}{\partial I_i(k)}, \quad \forall i \in \Psi, \quad k = 1, 2, \cdots, m.$$

From Eq. (5.28) and (5.29), it can be obtained that

$$FOV^k(\mathbf{I}^*(k)) - fov^k(\mathbf{I}^*(k)|\mathbf{y}^*(k-1)) = 0,$$

from Eq. (5.25), then

$$FOV^k(\mathbf{I}(k)) - fov^k(\mathbf{I}(k)|\mathbf{y}^*(k-1)) \leqslant 0,$$

i.e. $FOV^k(\mathbf{I}(k)) - fov^k(\mathbf{I}(k)|\mathbf{y}^*(k-1))$ arrives the maximum value 0 at point $\mathbf{I}^*(k)$, $k = 1, 2, \cdots, m$. From Lemma 5.2, it is obvious that the function of optimal value of the $(k+1)$th stage $FOV^k(\mathbf{I}(k))$ is differentiable of their argument $I_i(k)$. Then

$$\frac{\partial fov^k(\mathbf{I}^*(k)|\mathbf{y}^*(k-1))}{\partial I_i(k)} - \frac{\partial FOV^k(\mathbf{I}^*(k))}{\partial I_i(k)} = 0, \quad \forall i \in \Psi, \quad k = 1, 2, \cdots, m,$$

i.e.

$$\frac{\partial fov^k(\mathbf{I}^*(k)|\mathbf{y}^*(k-1))}{\partial I_i(k)} = \frac{\partial FOV^k(\mathbf{I}^*(k))}{\partial I_i(k)}, \quad \forall i \in \Psi, \quad k = 1, 2, \cdots, m.$$

The proof is completed. □

Then, the generalized Hamilton function is introduced in the following definition.

**Definition 5.10.** *Let $S^k(\boldsymbol{I}(k)|\boldsymbol{y}(k-1))$ be the support function for the conditional lower bound surface of the $(k+1)$th stage $\Pi^k(\boldsymbol{y}(k-1))$ at point $\boldsymbol{I}(k)$, then $H^k(\boldsymbol{I}(k), \boldsymbol{Q}(k), S^{k+1})$ is the generalized Hamilton function if and only if*

$$\begin{aligned} H^k(\boldsymbol{I}(k), \boldsymbol{Q}(k), S^{k+1}) = &-\textstyle\sum_{i=1}^n \omega_i\{[\alpha_i(k) + \beta_i(k)\delta_i(1+r_i)^k Q_i^2(k)] + E^{Me}[\tilde{\bar{c}}_i(k)]h_i(I_i(k) \\ &+ Q_i(k)) + [E^{Me}[\tilde{\bar{T}}_i^{sw}]Q_i(k) + E^{Me}[\tilde{\bar{T}}_i^{wc}]\varpi_i(k)] + s_i[E^{Me}[\tilde{\bar{D}}_i(k)] \\ &- I_i(k) - Q_i(k)]^+\} + S^{k+1}(\boldsymbol{I}(k+1)|\boldsymbol{y}(k)), \end{aligned} \tag{5.32}$$

$k = 0, 1, \cdots, m-1$.

Based on Lemma 5.4, the above statements and definitions, the most important lemma can be proved as below.

**Lemma 5.6.** *$S^{k+1}(\boldsymbol{I}(k+1)|\boldsymbol{y}^*(k))$ is the support function of the conditional lower bound surface of the $(k+1)$th stage $\Pi^{k+1}(\boldsymbol{y}^*(k))$ at point $\boldsymbol{I}^*(k+1)$, if and only if $H^k(\boldsymbol{I}^*(k), \boldsymbol{Q}(k), S^{k+1})$ arrives at a maximum value at point $\boldsymbol{Q}^*(k)$, where $Q_j^L Q_1^*(k) \leqslant Q_j^*(k) \leqslant Q_j^U Q_1^*(k)$, $Q_{i,k}^{min} \leqslant Q_i^*(k) \leqslant Q_{i,k}^{max}$ or $Q_i^*(k) = 0$, $I_i(k) \leqslant I_i^{max}$, $\forall j \in \Omega$, $i \in \Psi$, $k = 0, 1, \cdots, m-1$.*

*Proof.* Let $S^{k+1}(\mathbf{I}(k+1)|\mathbf{y}^*(k))$ be the support function of the conditional lower bound surface of the $(k+1)$th stage $\Pi^{k+1}(\mathbf{y}^*(k))$ at point $\mathbf{I}^*(k+1)$, then

$$-fov^{k+1}(\mathbf{I}(k+1)|\mathbf{y}^*(k)) + S^{k+1}(\mathbf{I}(k+1)|\mathbf{y}^*(k)) \leqslant \alpha^{k+1}, \quad \forall \mathbf{I}(k+1) \in \mathbf{X^{k+1}}(\mathbf{I}^*(k)). \tag{5.33}$$

Let $Q_j^L \bar{Q}_1(k) \leqslant \bar{Q}_j(k) \leqslant Q_j^U \bar{Q}_1(k)$, $Q_{i,k}^{min} \leqslant \bar{Q}_i(k) \leqslant Q_{i,k}^{max}$ or $\bar{Q}_i(k) = 0$, $I_i(k) \leqslant I_i^{max}$, $\forall j \in \Omega, i \in \Psi$, and $\bar{\mathbf{I}}(k+1) = \mathbf{f^k}(\mathbf{I}^*(k), \bar{\mathbf{Q}}(k)) = \mathbf{I}^*(k) + \bar{\mathbf{Q}}(k) - E^{Me}[\tilde{\mathbf{D}}(k)_{(\mathbf{r},\sigma)}]$, from Def. 5.5 and the statement of $fov^1(\mathbf{I}(1)|\mathbf{y}(0))$, it can be obtained that

$$-I^{0*}(k) - \Sigma_{i=1}^n \omega_i\{[\alpha_i(k) + \beta_i(k)\delta_i(1+r_i)^k \bar{Q}_i^2(k)] + E^{Me}[\tilde{\bar{c}}_i(k)]h_i(I_i(k) + \bar{Q}_i(k)) + [E^{Me}[\tilde{\bar{T}}_i^{sw}] \bar{Q}_i(k) + E^{Me}[\tilde{\bar{T}}_i^{wc}]\varpi_i(k)] + s_i[E^{Me}[\tilde{\bar{D}}_i(k)] - I_i^*(k) - \bar{Q}_i(k)]^+\} \leqslant -fov^{k+1}(\bar{\mathbf{I}}(k+1)|\mathbf{y}^*(k)).$$

With Eq. (5.33), and by adding $S^{k+1}(\bar{\mathbf{I}}(k+1)|\mathbf{y}^*(k))$ to both sides of the above inequality, it can be obtained that

$$-I^{0*}(k) - \Sigma_{i=1}^n \omega_i\{[\alpha_i(k) + \beta_i(k)\delta_i(1+r_i)^k \bar{Q}_i^2(k)] + E^{Me}[\tilde{\bar{c}}_i(k)]h_i(I_i(k) + \bar{Q}_i(k)) + [E^{Me}[\tilde{\bar{T}}_i^{sw}] \bar{Q}_i(k) + E^{Me}[\tilde{\bar{T}}_i^{wc}]\varpi_i(k)] + s_i[E^{Me}[\tilde{\bar{D}}_i(k)] - I_i^*(k) - \bar{Q}_i(k)]^+\} + S^{k+1}(\bar{\mathbf{I}}(k+1)|\mathbf{y}^*(k)) \leqslant \alpha^{k+1},$$

i.e.

$$H^k(\mathbf{I}^*(k), \mathbf{Q}(k), S^{k+1}(\mathbf{I}(k+1)|\mathbf{y}^*(k))) \leqslant \alpha^{k+1} + I^{0*}(k),$$

where $Q_j^L Q_1(k) \leqslant Q_j(k) \leqslant Q_j^U Q_1(k)$, $Q_{i,k}^{min} \leqslant Q_i(k) \leqslant Q_{i,k}^{max}$ or $Q_i(k) = 0$, $I_i^*(k) \leqslant I_i^{max}$, $\forall j \in \Omega$, $i \in \Psi$. On the other hand, from Eq. (5.29) in Lemma 5.4, it can be obtained that

$$\begin{aligned} & fov^{k+1}(\mathbf{I}^*(k+1)|\mathbf{y}^*(k)) \\ & = I^{0*}(k) + \Sigma_{i=1}^n \omega_i\{[\alpha_i(k) + \beta_i(k)\delta_i(1+r_i)^k Q_i^{2*}(k)] + E^{Me}[\tilde{\bar{c}}_i(k)]h_i(I_i^*(k) + Q_i^*(k)) \\ & \quad + [E^{Me}[\tilde{\bar{T}}_i^{sw}]Q_i^*(k) + E^{Me}[\tilde{\bar{T}}_i^{wc}]\varpi_i(k)] + s_i[E^{Me}[\tilde{\bar{D}}_i(k)] - I_i^*(k) - Q_i^*(k)]^+\}. \end{aligned} \tag{5.34}$$

From Def. 5.9, it can be obtained that

$$-fov^{k+1}(\mathbf{I}^*(k+1)|\mathbf{y}^*(k)) + S^{k+1}(\mathbf{I}^*(k+1)|\mathbf{y}^*(k)) \leqslant \alpha^{k+1},$$

i.e.

$$H^k(\mathbf{I}^*(k), \mathbf{Q}^*(k), S^{k+1}(\mathbf{I}^*(k+1)|\mathbf{y}^*(k))) = \alpha^{k+1} + I^{0*}(k).$$

This means $H^k(\mathbf{I}^*(k), \mathbf{Q}(k), S^{k+1}(\mathbf{I}(k+1)|\mathbf{y}^*(k)))$ arrives the maximum value $\alpha^{k+1} + I^{0*}(k)$ at point $\mathbf{Q}^*(k)$, $k = 0, 1, \cdots, m-1$.

Contrarily, if $H^k(\mathbf{I}^*(k), \mathbf{Q}(k), S^{k+1}(\mathbf{I}(k+1)|\mathbf{y}^*(k)))$ arrives the maximum value at point $\mathbf{Q}^*(k)$, then there is a $\beta^k$ such that

$$\begin{aligned} & -\Sigma_{i=1}^n \omega_i\{[\alpha_i(k) + \beta_i(k)\delta_i(1+r_i)^k Q_i^2(k)] + E^{Me}[\tilde{\bar{c}}_i(k)]h_i(I_i^*(k) + Q_i(k)) + [E^{Me}[\tilde{\bar{T}}_i^{sw}]Q_i(k) \\ & + E^{Me}[\tilde{\bar{T}}_i^{wc}]\varpi_i(k)] + s_i[E^{Me}[\tilde{\bar{D}}_i(k)] - I_i^*(k) - Q_i(k)]^+\} + S^{k+1}(\mathbf{I}(k+1)|\mathbf{y}^*(k)) \leqslant \beta^k, \end{aligned} \tag{5.35}$$

where $Q_j^L Q_1(k) \leqslant Q_j(k) \leqslant Q_j^U Q_1(k)$, $Q_{i,k}^{min} \leqslant Q_i(k) \leqslant Q_{i,k}^{max}$ or $Q_i(k) = 0$, $I_i^*(k) \leqslant I_i^{max}$, $\forall j \in \Omega, i \in \Psi$,

$$\begin{aligned} & -\Sigma_{i=1}^n \omega_i\{[\alpha_i(k) + \beta_i(k)\delta_i(1+r_i)^k Q_i^{2*}(k)] + E^{Me}[\tilde{\bar{c}}_i(k)]h_i(I_i^*(k) + Q_i^*(k)) + [E^{Me}[\tilde{\bar{T}}_i^{sw}]Q_i^*(k) \\ & + E^{Me}[\tilde{\bar{T}}_i^{wc}]\varpi_i(k)] + s_i[E^{Me}[\tilde{\bar{D}}_i(k)] - I_i^*(k) - Q_i^*(k)]^+\} + S^{k+1}(\mathbf{I}^*(k+1)|\mathbf{y}^*(k)) = \beta^k. \end{aligned} \tag{5.36}$$

Let $\alpha^{k+1} = \beta^k - I^{0*}(k)$, from Eq. (5.34) and (5.36), it can be obtained that

$$-fov^{k+1}(\mathbf{I}^*(k+1)|\mathbf{y}^*(k)) + S^{k+1}(\mathbf{I}^*(k+1)|\mathbf{y}^*(k)) = \beta^k - I^{0*}(k) = \alpha^{k+1}.$$

On the other hand, from Eq. (5.35), it can be obtained that

$$-I^{0*}(k) - \Sigma_{i=1}^n \omega_i\{[\alpha_i(k) + \beta_i(k)\delta_i(1+r_i)^k Q_i^2(k)] + E^{Me}[\tilde{\bar{c}}_i(k)]h_i(I_i^*(k) + Q_i(k)) + [E^{Me}[\tilde{\bar{T}}_i^{sw}] Q_i(k) + E^{Me}[\tilde{\bar{T}}_i^{wc}]\varpi_i(k)] + s_i[E^{Me}[\tilde{\bar{D}}_i(k)] - I_i^*(k) - Q_i(k)]^+\} + S^{k+1}(\mathbf{I}(k+1)|\mathbf{y}^*(k)) \leqslant \alpha^{k+1},$$

where $Q_j^L Q_1(k) \leqslant Q_j(k) \leqslant Q_j^U Q_1(k)$, $Q_{i,k}^{min} \leqslant Q_i(k) \leqslant Q_{i,k}^{max}$ or $Q_i(k) = 0$, $I_i^*(k) \leqslant I_i^{max}, \forall j \in \Omega, i \in \Psi$.

From Def. 5.5 and the statement of $fov^1(\mathbf{I}(1)|\mathbf{y}(0))$ and Eq. (5.29), it is obvious that

$$-fov^{k+1}(\mathbf{I}(k+1)|\mathbf{y}^*(k)) + S^{k+1}(\mathbf{I}(k+1)|\mathbf{y}^*(k)) \leqslant \alpha^{k+1}, \quad \forall \mathbf{I}(k+1) \in \mathbf{X^{k+1}}(\mathbf{I}^*(k)).$$

According to Def 5.9, it is obvious that $S^{k+1}(\mathbf{I}(k+1)|\mathbf{y}^*(k))$ is the support function of the conditional lower bound surface of the $(k+1)$th stage $\Pi^{k+1}(\mathbf{y}^*(k))$ at point $\mathbf{I}^*(k+1)$, $k = 0, 1, \cdots, m-1$. The proof is completed. □

**Theorem 5.1.** *Let $\{\boldsymbol{Q}^*(0), \cdots, \boldsymbol{Q}^*(m-1)\}$ be the optimal solution sequence of the purchase quantity at the beginning of each stage, and $\{\boldsymbol{I}^*(0) = \boldsymbol{a}, \boldsymbol{I}^*(1), \cdots, \boldsymbol{I}^*(m)\}$ be the optimal solution sequence for the inventory level at the beginning of each stage. Assume that $\boldsymbol{I}^*(k)$ is an interior point of $\mathbf{X^k}(\boldsymbol{I}^*(k-1))$, where $k = 0, 1, \cdots, m$. Then function $S^k(\boldsymbol{I}(k)|\boldsymbol{y}^*(k-1))$ is introduced, which is the support function of the conditional lower bound surface of the $(k+1)$th stage $\Pi^k(\boldsymbol{y}^*(k-1))$ at point $\boldsymbol{I}^*(k)$, $k = 1, 2, \cdots, m$, such that,*
*(i)*

$$\frac{\partial S^k(\boldsymbol{I}^*(k)|\boldsymbol{y}^*(k-1))}{\partial I_i(k)} = \frac{\partial H^k(\boldsymbol{I}^*(k), \boldsymbol{Q}^*(k), S^{k+1}(\boldsymbol{I}^*(k+1)|\boldsymbol{y}^*(k)))}{\partial I_i(k)}, \quad \forall i \in \Psi, \tag{5.37}$$

*where $k = 1, 2, \cdots, m-1$;*
*(ii)*

$$H^k(\boldsymbol{I}^*(k), \boldsymbol{Q}^*(k), S^{k+1*}) = \max\left[ H^k(\boldsymbol{I}^*(k), \boldsymbol{Q}(k), S^{k+1*}) \,\middle|\, \begin{array}{l} Q_j^L Q_1(k) \leqslant Q_j(k) \leqslant Q_j^U Q_1(k), \\ Q_{i,k}^{min} \leqslant Q_i(k) \leqslant Q_{i,k}^{max} \quad or \\ Q_i(k) = 0, I_i(k) \leqslant I_i^{max}, \forall j \in \Omega, i \in \Psi \end{array} \right], \tag{5.38}$$

*where $S^{k+1*} = S^{k+1}(\boldsymbol{I}^*(k+1)|\boldsymbol{y}^*(k))$, $S^{k+1} = S^{k+1}(\boldsymbol{I}(k+1)|\boldsymbol{y}^*(k))$, $k = 0, 1, \cdots, m-1$;*
*(iii)*

$$\frac{\partial S^m(\boldsymbol{I}^*(m)|\boldsymbol{y}^*(m-1))}{\partial I_i(m)} = 0, \quad \forall i \in \Psi. \tag{5.39}$$

*Proof.* From Eq. (5.25), (5.28) and (5.29), it can be obtained that

$$FOV^k(\mathbf{I}(k)) \leqslant fov^k(\mathbf{I}(k)|\mathbf{y}^*(k-1)), \quad k=1,2,\cdots,m,$$
$$FOV^k(\mathbf{I}^*(k)) = fov^k(\mathbf{I}^*(k)|\mathbf{y}^*(k-1)), \quad k=1,2,\cdots,m.$$

From Def. 5.9, it is obvious that $FOV^k(\mathbf{I}(k))$ is the support function of the conditional lower bound surface of the $(k+1)$th stage $\Pi^k(\mathbf{y}^*(k-1))$ at point $\mathbf{I}^*(k)$ (here $\alpha^k=0$). Now it is only needed to prove the function $FOV^k(\mathbf{I}^*(k))$ satisfies (i), (ii), (iii), so it is just the very support function $S^k(\mathbf{I}(k)|\mathbf{y}^*(k-1))$ which needs to be found.

Let $\hat{\mathbf{X}}^{\mathbf{k}}$ be a subset of $\mathbf{X}^{\mathbf{k}}$, $\hat{\mathbf{X}}^{\mathbf{k}}$ is defined as follows,

$$\hat{\mathbf{X}}^{\mathbf{k}} = \{\mathbf{I}(k)|\mathbf{I}(k)=\mathbf{f}^{\mathbf{k-1}}(\mathbf{I}(k-1),\mathbf{Q}^*(k-1)),\mathbf{Q}(k-1)\in\mathbf{X}^{\mathbf{k-1}}\}.$$

Let $\hat{F}^k(\mathbf{I}(k))$ be a function defined on the subset $\hat{\mathbf{X}}^{\mathbf{k}}$ as follows,

$$\begin{aligned}&\hat{F}^k(\mathbf{I}(k))\\&=\min\{FOV^{k-1}(\mathbf{I}(k-1))+\textstyle\sum_{i=1}^n \omega_i\{[\alpha_i(k-1)+\beta_i(k-1)\delta_i(1+r_i)^{k-1}Q_i^{2*}(k-1)]+\\&E^{Me}[\tilde{\bar{c}}_i(k-1)]h_i(I_i^*(k-1)+Q_i^*(k-1))+[E^{Me}[\tilde{\bar{T}}_i^{sw}]Q_i^*(k-1)+E^{Me}[\tilde{\bar{T}}_i^{wc}]\varpi_i(k-1)]+s_i\\&[E^{Me}[\tilde{\bar{D}}_i(k-1)]-I_i(k-1)-Q_i^*(k-1)]^+\}|\mathbf{f}^{k-1}(\mathbf{I}(k-1),\mathbf{Q}^*(k-1))=\mathbf{I}(k),\\&\mathbf{I}(k-1)\in\mathbf{X}^{\mathbf{k-1}}\}.\end{aligned}$$

From the above definition, it is obvious that

$$FOV^k(\mathbf{I}(k)) \leqslant \hat{F}^k(\mathbf{I}(k)), \quad \forall \mathbf{I}(k)\in\hat{\mathbf{X}}^{\mathbf{k}}, \tag{5.40}$$
$$FOV^k(\mathbf{I}^*(k)) = \hat{F}^k(\mathbf{I}^*(k)).$$

From Eq. (5.28), then

$$\begin{aligned}&FOV^{k+1}(\mathbf{I}^*(k+1))\\&=FOV^k(\mathbf{I}^*(k))+\textstyle\sum_{i=1}^n \omega_i\{[\alpha_i(k)+\beta_i(k)\delta_i(1+r_i)^kQ_i^{2*}(k)]+E^{Me}[\tilde{\bar{c}}_i(k)]h_i(I_i^*(k)+Q_i^*(k))\\&+[E^{Me}[\tilde{\bar{T}}_i^{sw}]Q_i^*(k)+E^{Me}[\tilde{\bar{T}}_i^{wc}]\varpi_i(k)]+s_i[E^{Me}[\tilde{\bar{D}}_i(k)]-I_i(k)-Q_i^*(k)]^+\}.\end{aligned} \tag{5.41}$$

According to Eq. (5.40) and the definition of $\hat{F}^k(\mathbf{I}(k))$, then

$$\begin{aligned}&FOV^{k+1}(\mathbf{I}(k+1))\leqslant\hat{F}^{k+1}(\mathbf{I}(k+1))\leqslant FOV^k(\mathbf{I}(k))+\textstyle\sum_{i=1}^n \omega_i\{[\alpha_i(k)+\beta_i(k)\delta_i(1+r_i)^k\\&Q_i^{2*}(k)]+E^{Me}[\tilde{\bar{c}}_i(k)]h_i(I_i^*(k)+Q_i^*(k))+[E^{Me}[\tilde{\bar{T}}_i^{sw}]Q_i^*(k)+E^{Me}[\tilde{\bar{T}}_i^{wc}]\varpi_i(k)]+s_i[E^{Me}[\tilde{\bar{D}}_i(k)]\\&-I_i(k)-Q_i^*(k)]^+\},\end{aligned} \tag{5.42}$$

$\forall\mathbf{I}(k)\in\mathbf{X}^{\mathbf{k}},\mathbf{I}(k+1)\in\hat{\mathbf{X}}^{\mathbf{k+1}}$.

Since, $\mathbf{I}(k+1)=\mathbf{f}^{\mathbf{k}}(\mathbf{I}(k),\mathbf{Q}(k))$, Eqs. (5.41) and (5.42) are transformed as follows

$$\begin{aligned}&-\textstyle\sum_{i=1}^n \omega_i\{[\alpha_i(k)+\beta_i(k)\delta_i(1+r_i)^kQ_i^{2*}(k)]+E^{Me}[\tilde{\bar{c}}_i(k)]h_i(I_i^*(k)+Q_i^*(k))+[E^{Me}[\tilde{\bar{T}}_i^{sw}]Q_i^*(k)\\&+E^{Me}[\tilde{\bar{T}}_i^{wc}]\varpi_i(k)]+s_i[E^{Me}[\tilde{\bar{D}}_i(k)]-I_i(k)-Q_i^*(k)]^+\}+FOV^{k+1}(\mathbf{f}^{\mathbf{k}}(\mathbf{I}^*(k),\mathbf{Q}^*(k)))-\\&FOV^k(\mathbf{I}^*(k))=0,\end{aligned} \tag{5.43}$$

$$-\Sigma_{i=1}^{n}\omega_i\{[\alpha_i(k)+\beta_i(k)\delta_i(1+r_i)^k Q_i^{2*}(k)]+E^{Me}[\tilde{\bar{c}}_i(k)]h_i(I_i^*(k)+Q_i^*(k))+[E^{Me}[\tilde{\bar{T}}_i^{sw}]Q_i^*(k) +E^{Me}[\tilde{\bar{T}}_i^{wc}]\varpi_i(k)]+s_i[E^{Me}[\tilde{\bar{D}}_i(k)]-I_i(k)-Q_i^*(k)]^+\}+FOV^{k+1}(\mathbf{f^k}(\mathbf{I}(k),\mathbf{Q}^*(k)))- FOV^k(\mathbf{I}(k))\leqslant 0, \tag{5.44}$$

$\forall \mathbf{I}(k)\in\mathbf{X^k}$.

Let

$$\Upsilon(\mathbf{I}(k))=-\Sigma_{i=1}^{n}\omega_i\{[\alpha_i(k)+\beta_i(k)\delta_i(1+r_i)^k Q_i^{2*}(k)]+E^{Me}[\tilde{\bar{c}}_i(k)]h_i(I_i^*(k)+Q_i^*(k))+ [E^{Me}[\tilde{\bar{T}}_i^{sw}]Q_i^*(k)+E^{Me}[\tilde{\bar{T}}_i^{wc}]\varpi_i(k)]+s_i[E^{Me}[\tilde{\bar{D}}_i(k)]-I_i(k)-Q_i^*(k)]^+\}+ FOV^{k+1}(\mathbf{f^k}(\mathbf{I}(k),\mathbf{Q}^*(k)))-FOV^k(\mathbf{I}(k)).$$

From Eq. (5.43) and (5.44), it is easy to know that $\Upsilon(\mathbf{I}(k))$ arrives the maximum value 0 at point $\mathbf{I}^*(k)$. It is obvious that

$$-\Sigma_{i=1}^{n}\omega_i\{[\alpha_i(k)+\beta_i(k)\delta_i(1+r_i)^k Q_i^{2*}(k)]+E^{Me}[\tilde{\bar{c}}_i(k)]h_i(I_i^*(k)+Q_i^*(k))+[E^{Me}[\tilde{\bar{T}}_i^{sw}]Q_i^*(k) +E^{Me}[\tilde{\bar{T}}_i^{wc}]\varpi_i(k)]+s_i[E^{Me}[\tilde{\bar{D}}_i(k)]-I_i(k)-Q_i^*(k)]^+\}$$

and

$$\mathbf{f^k}(\mathbf{I}(k),\mathbf{Q}^*(k))=\mathbf{I}(k)+\mathbf{Q}(k)-E^{Me}[\tilde{\bar{\mathbf{D}}}(k)_{(\mathbf{r},\sigma)}]$$

are both differentiable of their argument $I_i(k)$. From Lemma 5.2, it is obvious that $FOV^k(\mathbf{I}(k))$ is differentiable of its argument $I_i(k)$, so $\Upsilon(\mathbf{I}(k))$ is differentiable of its argument $I_i(k)$ as a result of the fact that it is a linear combination of $-\Sigma_{i=1}^{n}\omega_i\{[\alpha_i(k)+\beta_i(k)\delta_i(1+r_i)^k Q_i^{2*}(k)]+E^{Me}[\tilde{\bar{c}}_i(k)]h_i(I_i^*(k)+Q_i^*(k))+[E^{Me}[\tilde{\bar{T}}_i^{sw}]Q_i^*(k)+E^{Me}[\tilde{\bar{T}}_i^{wc}]\varpi_i(k)]+s_i[E^{Me}[\tilde{\bar{D}}_i(k)]-I_i(k)-Q_i^*(k)]^+\}$, $FOV^{k+1}(\mathbf{I}(k+1))$ and $FOV^k(\mathbf{I}(k))$. Then

$$\frac{\partial\Upsilon(\mathbf{I}^*(k))}{\partial I_i(k)}=0,\quad \forall i\in\Psi. \tag{5.45}$$

From Def. 5.10, it can be obtained that

$$\Upsilon(\mathbf{I}(k))=H^k(\mathbf{I}(k),\mathbf{Q}^*(k),FOV^{k+1}(\mathbf{f^k}(\mathbf{I}(k),\mathbf{Q}^*(k))))-FOV^k(\mathbf{I}(k)).$$

From Eq. (5.45), then

$$\frac{\partial FOV^k(\mathbf{I}^*(k))}{\partial I_i(k)}=\frac{\partial H^k(\mathbf{I}^*(k),\mathbf{Q}^*(k),FOV^{k+1}(\mathbf{I}^*(k+1)))}{\partial I_i(k)},\quad k=1,2,\cdots,m-1. \tag{5.46}$$

This means $FOV^k(\mathbf{I}(k))$ satisfies (i), $k=1,2,\cdots,m-1$.

Since it has been proved previously that $FOV^k(\mathbf{I}(k))$ is the support function of the conditional lower bound surface of the $(k+1)$th stage $\Pi^k(\mathbf{y}^*(k-1))$ on the point $\mathbf{I}^*(k)$, according to Lemma 5.6, it can be obtained that $FOV^{k+1}(\mathbf{I}(k+1))$ satisfies (ii), $k=0,1,\cdots,m-1$.

From Def. 5.6, Lemma 5.3 and Eq. (5.28), it is obvious that

$$FOV^n(\mathbf{I}(n))\geqslant \min[FOV^n(\mathbf{I}(n))|\mathbf{I}(n)\in\mathbf{X^n}]=L^{k*}=FOV^n(\mathbf{I}^*(n)),$$

i.e. $FOV^n(\mathbf{I}(n))$ arrives at a minimum value at point $\mathbf{I}^*(n)$, then

$$\frac{\partial FOV^n(\mathbf{I}^*(n))}{\partial I_i(n)} = 0.$$

Hence, $FOV^n(\mathbf{I}(n))$ satisfies (iii). The proof is completed. □

### 5.3.4 Existence of Linear Support Function

However, the Hamilton function in the Theorem 5.1 may not be linear. To solve the expected value model previously discussed, a linear form of the Hamilton function is defined as follows:

$$\begin{aligned} H^k(\mathbf{I}(k),\mathbf{Q}(k),\lambda(k+1)) = & -\Sigma_{i=1}^n \omega_i\{[\alpha_i(k)+\beta_i(k)\delta_i(1+r_i)^k Q_i^2(k)] + E^{Me}[\tilde{\bar{c}}_i(k)]h_i(I_i(k) \\ & +Q_i(k)) + [E^{Me}[\tilde{\bar{T}}_i^{sw}]Q_i(k) + E^{Me}[\tilde{\bar{T}}_i^{wc}]\varpi_i(k)] + s_i[E^{Me}[\tilde{\bar{D}}_i(k)] \\ & -I_i(k) - Q_i(k)]^+\} + \lambda^T(k+1)\{\mathbf{I}(k)+\mathbf{Q}(k) - E^{Me}[\tilde{\mathbf{D}}(k)_{(\mathbf{r},\sigma)}]\}, \end{aligned} \tag{5.47}$$

$k = 0, 1, \cdots, m-1$.
i.e.

$$\begin{aligned} H^k(\mathbf{I}(k),\mathbf{Q}(k),\lambda(k+1)) = & -\Sigma_{i=1}^n \omega_i\{[\alpha_i(k)+\beta_i(k)\delta_i(1+r_i)^k Q_i^2(k)] + E^{Me}[\tilde{\bar{c}}_i(k)]h_i(I_i(k) \\ & +Q_i(k)) + [E^{Me}[\tilde{\bar{T}}_i^{sw}]Q_i(k) + E^{Me}[\tilde{\bar{T}}_i^{wc}]\varpi_i(k)] + s_i[E^{Me}[\tilde{\bar{D}}_i(k)] \\ & -I_i(k) - Q_i(k)]^+\} + \lambda^T(k+1)\mathbf{f^k}(\mathbf{I}(k),\mathbf{Q}(k)), \end{aligned}$$

where

$$\mathbf{f^k}(\mathbf{I}(k),\mathbf{Q}(k)) = \mathbf{I}(k+1) = \mathbf{I}(k) + \mathbf{Q}(k) - E^{Me}[\tilde{\mathbf{D}}(k)_{(\mathbf{r},\sigma)}],$$

$k = 0, 1, \cdots, m-1$.

Compared with Def. 5.10, this indicates that the support function $S^{k+1}(\mathbf{I}(k+1)|\mathbf{y}(k))$ is of linear form as follows:

$$S^{k+1}(\mathbf{I}(k+1)) = \lambda^T(k+1)\mathbf{I}(k+1), \quad k = 0, 1, \cdots, m-1,$$

i.e.

$$S^{k+1}(\mathbf{f^k}(\mathbf{I}(k),\mathbf{Q}(k))) = \lambda^T(k+1)\mathbf{f^k}(\mathbf{I}(k),\mathbf{Q}(k)), \quad k = 0, 1, \cdots, m-1.$$

Now, the problem is whether or not there is a linear support function $S^{k+1}(\mathbf{I}(k+1)) = \lambda^T(k+1)\mathbf{I}(k+1)$. That indicates that the existence of the linear support function needs to be proved. Thus, the following lemmas are needed.

**Lemma 5.7.** *The state attainable set of the* $(k+1)$*th stage* $\mathbf{X^k}(\mathbf{I}(k-1))$ *which is derived from* $\mathbf{I}(k-1)$ *is a convex set,* $k = 1, 2, \cdots, m$.

*Proof.* Let $\delta \in [0,1]$, $\forall \mathbf{I}^1(k), \mathbf{I}^2(k) \in \mathbf{X^k}(\mathbf{I}(k-1))$, it is necessary to prove that $\delta\mathbf{I}^1(k) + (1-\delta)\mathbf{I}^2(k) \in \mathbf{X^k}(\mathbf{I}(k-1))$, $k = 1, 2, \cdots, m$. In fact, from Def. 5.3 and the statement of $\mathbf{X^2}(\mathbf{I}(1))$, it can be obtained that

$$\begin{aligned}
&\delta\mathbf{I}^1(k)+(1-\delta)\mathbf{I}^2(k)\\
&=\delta\mathbf{f}^{\mathbf{k}\text{-}\mathbf{1}}(\mathbf{I}(k-1),\mathbf{Q}^1(k-1))+(1-\delta)\mathbf{f}^{\mathbf{k}\text{-}\mathbf{1}}(\mathbf{I}(k-1),\mathbf{Q}^2(k-1))\\
&=\delta\{\mathbf{I}(k-1)+\mathbf{Q}^1(k-1)-E^{Me}[\tilde{\mathbf{D}}(k-1)_{(\mathbf{r},\sigma)}]\}+(1-\delta)\{\mathbf{I}(k-1)+\mathbf{Q}^2(k-1)-\\
&\quad E^{Me}[\tilde{\mathbf{D}}(k-1)_{(\mathbf{r},\sigma)}]\}\\
&=\mathbf{I}(k-1)+\delta\mathbf{Q}^1(k-1)+(1-\delta)\mathbf{Q}^2(k-1)-E^{Me}[\tilde{\mathbf{D}}(k-1)_{(\mathbf{r},\sigma)}]\\
&=\mathbf{f}^{\mathbf{k}\text{-}\mathbf{1}}(\mathbf{I}(k-1),\delta\mathbf{Q}^1(k-1)+(1-\delta)\mathbf{Q}^2(k-1)),
\end{aligned} \tag{5.48}$$

and

$$\delta Q_j^L Q_1(p)\leqslant\delta Q_j^1(p)\leqslant\delta Q_j^U Q_1(p),\quad (1-\delta)Q_j^L Q_1(p)\leqslant(1-\delta)Q_j^2(p)\leqslant(1-\delta)Q_j^U Q_1(p).$$

Hence,

$$Q_j^L Q_1(p)\leqslant\delta Q_j^1(p)+(1-\delta)Q_j^2(p)\leqslant Q_j^U Q_1(p). \tag{5.49}$$

On the other hand, since

$$\delta Q_{i,p}^{min}\leqslant\delta Q_i^1(p)\leqslant\delta Q_{i,p}^{max},\quad (1-\delta)Q_{i,p}^{min}\leqslant(1-\delta)Q_i^2(p)\leqslant(1-\delta)Q_{i,p}^{max}.$$

Therefore,

$$Q_{i,p}^{min}\leqslant\delta Q_i^1(p)+(1-\delta)Q_i^2(p)\leqslant Q_{i,p}^{max}. \tag{5.50}$$

From Def. 5.3 and the statement of $\mathbf{X}^2(\mathbf{I}(1))$, Eqs. (5.48) and (5.50), it is easy to know that

$$\delta\mathbf{I}^1(k)+(1-\delta)\mathbf{I}^2(k)\in\mathbf{X}^{\mathbf{k}}(\mathbf{I}(k-1)),k=1,2,\cdots,m.$$

This means $\mathbf{X}^{\mathbf{k}}(\mathbf{I}(k-1))$ is a convex set, $k=1,2,\cdots,m$. The proof is completed. □

**Definition 5.11.** *Let $z\in\boldsymbol{R}^2$, if $\forall\delta\in[0,1],\forall\boldsymbol{y}^1(k),\boldsymbol{y}^2(k)\in\boldsymbol{Y}^k(\boldsymbol{y}(k-1))$, then the set $\boldsymbol{Y}^k(\boldsymbol{y}(k-1))\subset\boldsymbol{R}^2$ is called convex in direction z, if and only if there exists $\gamma\geqslant 0$, such that $\delta\boldsymbol{y}^1(k)+(1-\delta)\boldsymbol{y}^2(k)-\gamma\boldsymbol{z}\in\boldsymbol{Y}^k(\boldsymbol{y}(k-1))$.*

**Lemma 5.8.** *The expanded state attainable set of the $(k+1)$th stage $\boldsymbol{Y}^k(\boldsymbol{y}(k-1))$ which is derived from $\mathbf{y}(k-1)$ is convex in direction $\boldsymbol{e}=(1,0)$, $k=1,2,\cdots,m$.*

*Proof.* Let $\delta\in[0,1]$, $\forall\mathbf{y}^1(k),\mathbf{y}^2(k)\in\mathbf{Y}^{\mathbf{k}}(\mathbf{y}(k-1))$, from Def. 5.11, it is necessary to prove that there exists $\gamma\geqslant 0$, such that $\delta\mathbf{y}^1(k)+(1-\delta)\mathbf{y}^2(k)-\gamma\mathbf{e}\in\mathbf{Y}^{\mathbf{k}}(\mathbf{y}(k-1))$, $k=1,2,\cdots,m$. In fact, from Def. 5.4 and the statement of $\mathbf{Y}^2(\mathbf{y}(1))$, it can be obtained that

$$\begin{aligned}
&\delta\mathbf{y}^1(k)+(1-\delta)\mathbf{y}^2(k)-\gamma\mathbf{e}\\
&=\delta\mathbf{F}^{\mathbf{k}\text{-}\mathbf{1}}(\mathbf{y}(k-1),\mathbf{Q}^1(k-1))+(1-\delta)\mathbf{F}^{\mathbf{k}\text{-}\mathbf{1}}(\mathbf{y}(k-1),\mathbf{Q}^2(k-1))-(\gamma,0)\\
&=\delta(L^k(\mathbf{Q}^1(k-1)),\mathbf{f}^{\mathbf{k}\text{-}\mathbf{1}}(\mathbf{I}(k-1),\mathbf{Q}^1(k-1)))+(1-\delta)(L^k(\mathbf{Q}^2(k-1)),\\
&\quad\mathbf{f}^{\mathbf{k}\text{-}\mathbf{1}}(\mathbf{I}(k-1),\mathbf{Q}^2(k-1)))-(\gamma,0)\\
&=(L^k(\delta\mathbf{Q}^1(k-1)+(1-\delta)\mathbf{Q}^2(k-1))-\gamma,\mathbf{f}^{\mathbf{k}\text{-}\mathbf{1}}(\mathbf{I}(k-1),\delta\mathbf{Q}^1(k-1)+(1-\delta)\mathbf{Q}^2(k-1))).
\end{aligned} \tag{5.51}$$

Let $\gamma = 0$, then

$$\begin{aligned}&(L^k(\delta \mathbf{Q}^1(k-1)+(1-\delta)\mathbf{Q}^2(k-1))-\gamma, \mathbf{f}^{\mathbf{k-1}}(\mathbf{I}(k-1),\delta \mathbf{Q}^1(k-1)+(1-\delta)\mathbf{Q}^2(k-1)))\\ &=(L^k(\delta \mathbf{Q}^1(k-1)+(1-\delta)\mathbf{Q}^2(k-1)),\mathbf{f}^{\mathbf{k-1}}(\mathbf{I}(k-1),\delta \mathbf{Q}^1(k-1)+(1-\delta)\mathbf{Q}^2(k-1)))\\ &=\mathbf{F}^{\mathbf{k-1}}(\mathbf{y}(k-1),\delta \mathbf{Q}^1(k-1)+(1-\delta)\mathbf{Q}^2(k-1)).\end{aligned}$$

From Def. 5.4 and the statement of $\mathbf{Y^2}(\mathbf{y}(1))$ and Eq. (5.50), it can be obtained that

$$\delta \mathbf{y^1}(k)+(1-\delta)\mathbf{y^2}(k)-\gamma \mathbf{e}=\mathbf{F}^{\mathbf{k-1}}(\mathbf{y}(k-1),\delta \mathbf{Q}^1(k-1)+(1-\delta)\mathbf{Q}^2(k-1))\in \mathbf{Y^k}(\mathbf{y}(k-1)).$$

This means $\mathbf{Y^k}(\mathbf{y}(k-1))$ $\mathbf{y}(k-1)$ is convex in direction $\mathbf{e}=(1,0)$, $k=1,2,\cdots,m$. The proof is completed. □

**Lemma 5.9.** *If the expanded state attainable set of the $(k+1)$th stage $\mathbf{Y^k}(\mathbf{y}(k-1))$ which is derived from $\mathbf{y}(k-1)$ is convex in direction $\mathbf{e}=(1,0)$, then the conditional function of the optimal value of the $(k+1)$th stage $fov^k(\mathbf{I}(k)|\mathbf{y}(k-1))$ is a convex function, $k=1,2,\cdots,m$.*

*Proof.* Assumed that $\mathbf{Y^k}(\mathbf{y}(k-1))$ is convex in direction $\mathbf{e}=(1,0)$. $k=1,2,\cdots,m$, let $\mathbf{I}^1(k),\mathbf{I}^2(k)\in \mathbf{X^k}(\mathbf{I}(k-1))$, chose $I^{0,1}(k),I^{0,2}(k)$, such that $\mathbf{y}^1(k)=(I^{0,1}(k),\mathbf{I}^1(k)),\mathbf{y}^2(k)=(I^{0,2}(k),\mathbf{I}^2(k))$, $\mathbf{y}^1(k)$, $\mathbf{y}^2(k)\in \mathbf{Y^k}(\mathbf{y}(k-1))$. From Def. 5.11, $\forall \delta \in [0,1]$, there exists $\gamma \geqslant 0$, such that

$$\delta \mathbf{y^1}(k)+(1-\delta)\mathbf{y^2}(k)-\gamma \mathbf{e}\in \mathbf{Y^k}(\mathbf{y}(k-1)),$$

i.e.

$$\delta \mathbf{I}^1(k)+(1-\delta)\mathbf{I}^2(k)\in \mathbf{X^k}(\mathbf{I}(k-1)).$$

Hence, $\mathbf{X^k}(\mathbf{I}(k-1))$ is a convex set. Now let $\overline{\mathbf{I}^1(k)},\overline{\mathbf{I}^2(k)}\in \mathbf{X^k}(\mathbf{I}(k-1))$, let

$$\overline{I^{0,1}(k)}=fov^k(\overline{\mathbf{I}^1(k)}|\mathbf{y}(k-1)),\quad \overline{I^{0,2}(k)}=fov^k(\overline{\mathbf{I}^2(k)}|\mathbf{y}(k-1)),$$

$$\overline{\mathbf{y}^1(k)}=(\overline{I^{0,1}(k)},\overline{\mathbf{I}^1(k)}),\quad \overline{\mathbf{y}^2(k)}=(\overline{I^{0,2}(k)},\overline{\mathbf{I}^2(k)}),$$

then $\overline{\mathbf{y}^1(k)},\overline{\mathbf{y}^2(k)}\in \mathbf{Y^k}(\mathbf{y}(k-1))$. Since $\mathbf{Y^k}(\mathbf{y}(k-1))$ is convex in direction $\mathbf{e}=(1,0)$, $\forall \delta \in [0,1]$, there exists $\bar{\gamma}\geqslant 0$, such that

$$\delta \overline{\mathbf{y}^1(k)}+(1-\delta)\overline{\mathbf{y}^2(k)}-\bar{\gamma}\mathbf{e}\in \mathbf{Y^k}(\mathbf{y}(k-1)),$$

i.e.

$$\begin{aligned}&\delta fov^k(\overline{\mathbf{I}^1(k)}|\mathbf{y}(k-1))+(1-\delta)fov^k(\overline{\mathbf{I}^2(k)}|\mathbf{y}(k-1))-\bar{\gamma}\\ &\geqslant fov^k(\delta \overline{\mathbf{I}^1(k)}+(1-\delta)\overline{\mathbf{I}^2(k)}|\mathbf{y}(k-1)).\end{aligned}$$

Since $\bar{\gamma}\geqslant 0$, then

$$\begin{aligned}&\delta fov^k(\overline{\mathbf{I}^1(k)}|\mathbf{y}(k-1))+(1-\delta)fov^k(\overline{\mathbf{I}^2(k)}|\mathbf{y}(k-1))\\ &\geqslant fov^k(\delta \overline{\mathbf{I}^1(k)}+(1-\delta)\overline{\mathbf{I}^2(k)}|\mathbf{y}(k-1)).\end{aligned}$$

This means $fov^k(\mathbf{I}(k)|\mathbf{y}(k-1))$ is a convex function, $k=1,2,\cdots,m$. The proof is completed. □

Based on Lemma 5.7, the following lemma can be proved.

**Lemma 5.10.** *If the conditional function of the optimal value of the $(k+1)$th stage $fov^k(\boldsymbol{I}(k)|\boldsymbol{y}(k-1))$ is convex on $\mathbf{X^k}(\boldsymbol{I}(k-1))$, and $\hat{\boldsymbol{I}}(k)$ is any interior point of $\mathbf{X^k}(\boldsymbol{I}(k-1))$, then there is a linear function $S^k(\boldsymbol{I}(k)) = \lambda^T(k)\boldsymbol{I}(k)$, which is the support function for the conditional lower bound surface of the $(k+1)$th stage $\Pi^k(\boldsymbol{y}(k-1))$ at point $\hat{\boldsymbol{I}}(k)$ $k = 1,2,\cdots,m$.*

*Proof.* Let $\overline{\mathbf{Y^k}}$ be a subset of $\mathbf{Y^k}(\mathbf{y}(k-1))$, $k = 1,2,\cdots,m$, it is defined as below

$$\overline{\mathbf{Y^k}} = \left\{\mathbf{y}(k) \,\middle|\, \begin{array}{l} fov^k(\mathbf{I}(k)|\mathbf{y}(k-1)) \leqslant I^0(k) \leqslant \sup[fov^k(\mathbf{I}(k)|\mathbf{y}(k-1))| \\ \mathbf{I}(k) \in \mathbf{X^k}(\mathbf{I}(k-1))], \mathbf{I}(k) \in \mathbf{X^k}(\mathbf{I}(k-1)) \end{array}\right\}.$$

Firstly, it is necessary to prove $\overline{\mathbf{Y^k}}$ is a convex set. Let $\mathbf{y}^1(k) = (I^{0,1}(k), \mathbf{I}^1(k))$, $\mathbf{y}^2(k) = (I^{0,2}(k), \mathbf{I}^2(k))$, $\mathbf{y}^1(k), \mathbf{y}^2(k) \in \overline{\mathbf{Y^k}}$, it is defined that

$$\mathbf{y}^k(\delta) = \delta\mathbf{y}^1(k) + (1-\delta)\mathbf{y}^2(k), \quad 0 \leqslant \delta \leqslant 1.$$

According to Lemma 5.7, $\mathbf{X^k}(\mathbf{I}(k-1))$ is a convex set. Let $\mathbf{I}^1(k), \mathbf{I}^2(k) \in \mathbf{X^k}(\mathbf{I}(k-1))$, then

$$\delta\mathbf{I}^1(k) + (1-\delta)\mathbf{I}^2(k) \in \mathbf{X^k}(\mathbf{I}(k-1)).$$

On the other hand, it is obvious that

$$I^{0,1}(k) \geqslant fov^k(\mathbf{I}^1(k)|\mathbf{y}(k-1)), \quad I^{0,2}(k) \geqslant fov^k(\mathbf{I}^2(k)|\mathbf{y}(k-1)),$$

since $fov^k(\mathbf{I}(k)|\mathbf{y}(k-1))$ is convex, then

$$\delta I^{0,1}(k) + (1-\delta)I^{0,2} \geqslant fov^k(\delta\mathbf{I}^1(k) + (1-\delta)\mathbf{I}^2(k)|\mathbf{y}(k-1)).$$

Since

$$I^{0,1}(k) \leqslant \sup[fov^k(\mathbf{I}(k)|\mathbf{y}(k-1))|\mathbf{I}(k) \in \mathbf{X^k}(\mathbf{I}(k-1))],$$
$$I^{0,2}(k) \leqslant \sup[fov^k(\mathbf{I}(k)|\mathbf{y}(k-1))|\mathbf{I}(k) \in \mathbf{X^k}(\mathbf{I}(k-1))],$$

it is obvious that

$$\delta I^{0,1}(k) + (1-\delta)I^{0,2} \leqslant \sup[fov^k(\mathbf{I}(k)|\mathbf{y}(k-1))|\mathbf{I}(k) \in \mathbf{X^k}(\mathbf{I}(k-1))].$$

This means $\mathbf{y}^k(\delta) \in \overline{\mathbf{Y^k}}$, so $\overline{\mathbf{Y^k}}$ is a convex set.

According to the theorems of convex set, $\forall\hat{\mathbf{y}}(k) = (\hat{I}^0(k), \hat{\mathbf{I}}(k))$ which is a boundary point of $\overline{\mathbf{Y^k}}$, there exists a support plane $\mathbf{P^k}$ which is defined as below

$$\mathbf{P^k} = \{\mathbf{y}(k)|\mathbf{y}(k) = (I^0(k), \mathbf{I}(k)), a^0(k)I^0(k) + \mathbf{a}^T(k)\mathbf{I}(k) = b(k)\},$$

where $a^0(k), \mathbf{a}(k)$ and $b(k)$ are all nonnegative real values. Then

$$a^0(k)\hat{I}^0(k) + \mathbf{a}^T(k)\hat{\mathbf{I}}(k) = b(k),$$

$$a^0(k)I^0(k) + \mathbf{a}^T(k)\mathbf{I}(k) \geqslant b(k), \quad \forall \mathbf{y}(k) \in \overline{\mathbf{Y^k}}.$$

Now let $\hat{\mathbf{I}}(k)$ be an interior point of $\mathbf{X^k}(\mathbf{I}(k-1))$, let $\hat{\mathbf{y}}(k) = (fov^k(\hat{\mathbf{I}}(k)|\mathbf{y}(k-1)), \hat{\mathbf{I}}(k))$, then by the definition of $\overline{\mathbf{Y^k}}$, it can be seen that $\hat{\mathbf{y}}(k)$ is a boundary point of $\overline{\mathbf{Y^k}}$, so

$$a^0(k)fov^k(\hat{\mathbf{I}}(k)|\mathbf{y}(k-1)) + \mathbf{a}^T(k)\hat{\mathbf{I}}(k) = b(k),$$

$$a^0(k)fov^k(\mathbf{I}(k)|\mathbf{y}(k-1)) + \mathbf{a}^T(k)\mathbf{I}(k) \geqslant b(k), \quad \forall \mathbf{I}(k) \in \mathbf{X^k}(\mathbf{I}(k-1)).$$

Here, it must be that $a^0(k) \neq 0$, otherwise

$$\mathbf{a}^T(k)\hat{\mathbf{I}}(k) = b(k),$$

$$\mathbf{a}^T(k)\mathbf{I}(k) \geqslant b(k), \quad \forall \mathbf{I}(k) \in \mathbf{X^k}(\mathbf{I}(k-1)).$$

This means $\mathbf{a}^T(k)\mathbf{I}(k) - b(k) = 0$ is just the support plane of $\mathbf{X^k}(\mathbf{I}(k-1))$ on the point $\hat{\mathbf{I}}(k)$, i.e. $\hat{\mathbf{I}}(k)$ is a boundary point of $\mathbf{X^k}(\mathbf{I}(k-1))$, but it is impossible since we have assumed that $\hat{\mathbf{I}}(k)$ is an interior point of $\mathbf{X^k}(\mathbf{I}(k-1))$.

As $a^0(k) \neq 0$, let $\alpha^k = -\frac{b(k)}{a^0(k)}$, $\lambda(k) = -\frac{\mathbf{a}(k)}{a^0(k)}$, then

$$\lambda^T(k)\hat{\mathbf{I}}(k) - \alpha^k = fov^k(\hat{\mathbf{I}}(k)|\mathbf{y}(k-1)),$$

$$\lambda^T(k)\mathbf{I}(k) - \alpha^k \leqslant fov^k(\mathbf{I}(k)|\mathbf{y}(k-1)), \quad \forall \mathbf{I}(k) \in \mathbf{X^k}(\mathbf{I}(k-1)).$$

From Def. 5.9, it can be obtained that

$$S^k(\mathbf{I}(k)) = \lambda^T(k)\mathbf{I}(k), \quad k = 1, 2, \cdots, m.$$

This is just the linear support function of the conditional lower bound surface of the $(k+1)$th stage $\Pi^k(\mathbf{y}(k-1))$ on the point $\hat{\mathbf{I}}(k)$. The proof is completed. □

Now the theorem for the existence of the linear support function can be proved.

**Theorem 5.2.** *If $\hat{\boldsymbol{I}}(k)$ is an interior point of $\mathbf{X^k}(\boldsymbol{I}(k-1))$, then there is a $\lambda(k)$ such that $S^k(\boldsymbol{I}(k)) = \lambda^T(k)\boldsymbol{I}(k)$ is a linear support function of the conditional lower bound surface of the $(k+1)$th stage $\Pi^k(\boldsymbol{y}(k-1))$ at point $\hat{\boldsymbol{I}}(k)$ $k = 1, 2, \cdots, m$.*

*Proof.* According to Lemmas 5.8, 5.9 and 5.10, the theorem for the existence of the linear support function can be proved as below.

From Lemma 5.8, it can be known that the expanded state attainable set of the $(k+1)$th stage $\mathbf{Y^k}(\mathbf{y}(k-1))$ which is derived from $\mathbf{y}(k-1)$ is convex in direction $\mathbf{e} = (1,0)$, $k = 1, 2, \cdots, m$. Hence, the conditional function of optimal value of the $(k+1)$th stage $fov^k(\mathbf{I}(k)|\mathbf{y}(k-1))$ is a convex function according to Lemma 5.9. From Lemma 5.10, it can be known that there is a $\lambda(k)$ such that $S^k(\mathbf{I}(k)) = \lambda^T(k)\mathbf{I}(k)$ is a linear support function of the conditional lower bound surface of the $(k+1)$th stage $\Pi^k(\mathbf{y}(k-1))$ at point $\hat{\mathbf{I}}(k)$ $k = 1, 2, \cdots, m$. The proof is completed. □

Since the existence of the linear support function is proved, now the linear form of the Hamilton function is defined as below.

**Definition 5.12.** *Let* $S^{k+1}(\boldsymbol{I}(k+1)) = \boldsymbol{\lambda}^T(k+1)\boldsymbol{I}(k+1)$ *be the linear support function of the conditional lower bound surface of the* $(k+1)$*th stage* $\Pi^{k+1}(\boldsymbol{y}(k))$ *at point* $\boldsymbol{I}(k+1)$*, then* $H^k(\boldsymbol{I}(k),\boldsymbol{Q}(k),\boldsymbol{\lambda}(k+1))$ *is the linear Hamilton function if and only if*

$$\begin{aligned}H^k(\boldsymbol{I}(k),\boldsymbol{Q}(k),\boldsymbol{\lambda}(k+1)) = &-\Sigma_{i=1}^n \omega_i\{[\alpha_i(k)+\beta_i(k)\delta_i(1+r_i)^k Q_i^2(k)]+E^{Me}[\tilde{\bar{c}}_i(k)]h_i(I_i(k)\\&+Q_i(k))+[E^{Me}[\tilde{\bar{T}}_i^{sw}]Q_i(k)+E^{Me}[\tilde{\bar{T}}_i^{wc}]\varpi_i(k)]+s_i[E^{Me}[\tilde{\bar{D}}_i(k)]\\&-I_i(k)-Q_i(k)]^+\}+\boldsymbol{\lambda}^T(k+1)\boldsymbol{f}^k(\boldsymbol{I}(k),\boldsymbol{Q}(k)),\end{aligned} \tag{5.52}$$

*where*

$$\boldsymbol{f}^k(\boldsymbol{I}(k),\boldsymbol{Q}(k)) = \boldsymbol{I}(k+1) = \boldsymbol{I}(k)+\boldsymbol{Q}(k)-E^{Me}[\tilde{\boldsymbol{D}}(k)_{(\boldsymbol{r},\sigma)}],$$

$k = 0,1,\cdots,m-1.$

### 5.3.5 The Maximal Theory

Now, the theorem crucial for finding the optimal solution to the problem based on Theorems 5.1 and 5.2 can be proved.

**Theorem 5.3.** *Let* $\{\boldsymbol{Q}^*(0),\cdots,\boldsymbol{Q}^*(m-1)\}$ *be the optimal solution sequence for purchasing quantities at the beginning of each stage, and* $\{\boldsymbol{I}^*(0)=\boldsymbol{a},\boldsymbol{I}^*(1),\cdots,\boldsymbol{I}^*(m)\}$ *be the optimal solution sequence of the inventory levels at the beginning of each stage. Assume that* $\boldsymbol{I}^*(k)$ *is an interior point of* $\boldsymbol{X}^k(\boldsymbol{I}^*(k-1))$*, where* $k=1,2,\cdots,m$*. Then there is a* $\boldsymbol{\lambda}^*(k) = (\lambda_1^*(k),\lambda_2^*(k),\cdots,\lambda_n^*(k))$*,* $k=1,2,\cdots,m$*, such that,*
*(i)*

$$\lambda_i^*(k) = \frac{\partial H^k(\boldsymbol{I}^*(k),\boldsymbol{Q}^*(k),\boldsymbol{\lambda}^*(k+1))}{\partial I_i(k)} = \lambda_i^*(k+1)+E^{Me}[\tilde{\bar{c}}_i(k)]h_i - s_i, \tag{5.53}$$

$\forall i \in \Psi, k = 1,2,\cdots,m-1;$
*(ii)*

$$\begin{aligned}&H^k(\boldsymbol{I}^*(k),\boldsymbol{Q}^*(k),\boldsymbol{\lambda}^*(k+1))\\&=\max\left[H^k(\boldsymbol{I}^*(k),\boldsymbol{Q}(k),\boldsymbol{\lambda}^*(k+1))\,\middle|\,\begin{array}{l}Q_j^L Q_1(k)\leqslant Q_j(k)\leqslant Q_j^U Q_1(k),\\ Q_{i,k}^{min}\leqslant Q_i(k)\leqslant Q_{i,k}^{max}\quad or\\ Q_i(k)=0, I_i(k)\leqslant I_i^{max}, \forall j\in\Omega, i\in\Psi\end{array}\right],\end{aligned} \tag{5.54}$$

$k = 0,1,\cdots,m-1;$
*(iii)*

$$\lambda_i^*(m) = 0, \quad \forall i \in \Psi. \tag{5.55}$$

*Proof.* From Theorem 5.2, since $\mathbf{I}^*(k)$ is an interior point of $\mathbf{X^k}(\mathbf{I}^*(k-1))$, it can be known that there exists a $\lambda^*(k)$ such that $\hat{S}^k(\mathbf{I}^*(k)) = \lambda^{T*}(k)\mathbf{I}^*(k)$ is a linear support function of the conditional lower bound surface of the $(k+1)$th stage $\Pi^k(\mathbf{y}^*(k-1))$ at point $\mathbf{I}^*(k)$, $k = 1,2,\cdots,m$. From Def. 5.12, it can be obtained that

$$\frac{\partial H^k(\mathbf{I}^*(k),\mathbf{Q}^*(k),\lambda^*(k+1))}{\partial I_i(k)} = E^{Me}[\tilde{\bar{c}}_i(k)]h_i - s_i + \frac{\partial \hat{S}^{k+1}(\mathbf{I}^*(k+1))}{\partial I_i(k+1)} \times \frac{\partial \mathbf{f^k}(\mathbf{I}^*(k),\mathbf{Q}^*(k))}{\partial I_i(k)}.$$

According to Lemma 5.5, it can be obtained that

$$\frac{\partial \hat{S}^{k+1}(\mathbf{I}^*(k+1))}{\partial I_i(k+1)} = \frac{\partial FOV_i^{k+1}(\mathbf{I}^*(k+1))}{\partial I_i(k+1)}. \tag{5.56}$$

This means

$$\frac{\partial H^k(\mathbf{I}^*(k),\mathbf{Q}^*(k),\lambda^*(k+1))}{\partial I_i(k)} = \frac{\partial H^k(\mathbf{I}^*(k),\mathbf{Q}^*(k),FOV^{k+1}(\mathbf{I}^*(k+1)))}{\partial I_i(k)}. \tag{5.57}$$

From Eq. (5.46) in Theorem 5.1, it can be obtained that

$$\frac{\partial FOV^k(\mathbf{I}^*(k))}{\partial I_i(k)} = \frac{\partial H^k(\mathbf{I}^*(k),\mathbf{Q}^*(k),FOV^{k+1}(\mathbf{I}^*(k+1)))}{\partial I_i(k)}. \tag{5.58}$$

From Eqs. (5.56), (5.57) and (5.58), it is obvious that

$$\frac{\partial H^k(\mathbf{I}^*(k),\mathbf{Q}^*(k),\lambda^*(k+1))}{\partial I_i(k)} = \frac{\partial \hat{S}^k(\mathbf{I}^*(k))}{\partial I_i(k)}. \tag{5.59}$$

Since

$$\frac{\partial \hat{S}^k(\mathbf{I}^*(k))}{\partial I_i(k)} = \lambda_i^*(k),$$

so

$$\lambda_i^*(k) = \frac{\partial H^k(\mathbf{I}^*(k),\mathbf{Q}^*(k),\lambda^*(k+1))}{\partial I_i(k)} = \lambda_i^*(k+1) + E^{Me}[\tilde{\bar{c}}_i(k)]h_i - s_i,$$

$\forall i \in \Psi$, $k = 1,2,\cdots,m-1$. This means $\lambda_i^*(k)$ satisfies (i).

Since

$$\hat{S}^k(\mathbf{I}^*(k)) = \lambda^{T*}(k)\mathbf{I}^*(k)$$

is a linear support function of the conditional lower bound surface of the $(k+1)$th stage $\Pi^k(\mathbf{y}^*(k-1))$ on the point $\mathbf{I}^*(k)$, according to Lemma 5.6, it can be known that

$$\hat{S}^{k+1}(\mathbf{I}^*(k+1)) = \lambda^{T*}(k+1)\mathbf{I}^*(k+1)$$

satisfies (ii), $k = 0,1,\cdots,m-1$.

According to Theorem 5.1, it has been proved that

$$\frac{\partial FOV^m(\mathbf{I}^*(m))}{\partial I_i(m)} = 0, \quad \forall i \in \Psi.$$

From Eq. (5.56), then

$$\frac{\partial \hat{S}^m(\mathbf{I}^*(m))}{\partial I_i(m)} = \frac{\partial FOV^m(\mathbf{I}^*(m))}{\partial I_i(m)}, \quad \forall i \in \Psi.$$

Since

$$\frac{\partial \hat{S}^m(\mathbf{I}^*(m))}{\partial I_i(m)} = \lambda_i^*(m), \quad \forall i \in \Psi,$$

then

$$\lambda_i^*(m) = 0, \quad \forall i \in \Psi.$$

This means $\lambda_i^*(m)$ satisfies (iii). The proof is completed. □

### 5.3.6 Theoretical Algorithm for IMP-LCP

Since the crucial theorem for finding the optimal solution to our problem has been proved, it is not difficult to develop a theoretical algorithm (TA) to solve the expected value model (EVM) for inventory problems to minimize the total cost of each item within an entire period. Based on Theorem 5.3, the steps for the development of the theoretical algorithm are shown as follows.

- ***Step.1***: From Eq. (5.53) $\lambda_i^*(k) = \lambda_i^*(k+1) + E^{Me}[\tilde{\bar{c}}_i(k)]h_i - s_i$ and Eq. (5.55) $\lambda_i^*(m) = 0$ in Theorem 5.3, it is possible to calculate $\lambda_i^*(k)$, $\forall i \in \Psi$, $k = 1, 2, \cdots, m-1$.

Then, to find the optimal solution for the expected value model, the Hamilton function is considered in Theorem 5.3.

- ***Step.2***: Eq. (5.54) in Theorem 5.3 is transformed as follows:
(1) If $I_i^*(k) + Q_i(k) \leq 0$, then

$$\begin{aligned}
&H^k(\mathbf{I}^*(k), \mathbf{Q}^*(k), \boldsymbol{\lambda}^*(k+1)) \\
&= \max \left[ H^k(\mathbf{I}^*(k), \mathbf{Q}(k), \boldsymbol{\lambda}^*(k+1)) \;\middle|\; \begin{array}{l} Q_j^L Q_1(k) \leqslant Q_j(k) \leqslant Q_j^U Q_1(k), \\ Q_{i,k}^{min} \leqslant Q_i(k) \leqslant Q_{i,k}^{max} \quad or \\ Q_i(k) = 0, I_i^*(k) \leqslant I_i^{max}, \forall j \in \Omega, i \in \Psi \end{array} \right] \\
&= \Sigma_{i=1}^n - \{\omega_i E^{Me}[\tilde{\bar{c}}_i(k)] h_i I_i^*(k) + \omega_i s_i [E^{Me}[\tilde{\bar{D}}_i(k)] - I_i^*(k)] - \lambda_i^*(k+1)(I_i^*(k) - E^{Me}[\tilde{\bar{D}}_i(k)])\} \\
&\quad + \max \left[ \begin{array}{l} -\Sigma_{i=1}^n \{\omega_i \beta_i(k) \delta_i (1+r_i)^k Q_i^2(k) \\ +(\omega_i E^{Me}[\tilde{\bar{c}}_i(k)] h_i + \omega_i E^{Me}[\tilde{\bar{T}}_i^{sw}] \\ +\omega_i E^{Me}[\tilde{\bar{T}}_i^{wc}] - \lambda_i^*(k+1) - \omega_i s_i) \\ Q_i(k) + \omega_i \alpha_i(k)\} \end{array} \;\middle|\; \begin{array}{l} Q_j^L Q_1(k) \leqslant Q_j(k) \leqslant Q_j^U Q_1(k), \\ Q_{i,k}^{min} \leqslant Q_i(k) \leqslant Q_{i,k}^{max} \quad or \\ Q_i(k) = 0, I_i^*(k) \leqslant I_i^{max}, \forall j \in \Omega, \\ i \in \Psi \end{array} \right]
\end{aligned} \tag{5.60}$$

$k = 0, 1, \cdots, m-1$.
(2) If $I_i^*(k) + Q_i(k) > 0$, $I_i^*(k) < 0$, and $I_i^*(k) + Q_i(k) - E^{Me}[\tilde{\bar{D}}_i(k)] \geq 0$, then

$$
\begin{aligned}
&H^k(\mathbf{I}^*(k),\mathbf{Q}^*(k),\lambda^*(k+1))\\
&=\max\left[H^k(\mathbf{I}^*(k),\mathbf{Q}(k),\lambda^*(k+1))\left|\begin{array}{l}Q_j^L Q_1(k)\leqslant Q_j(k)\leqslant Q_j^U Q_1(k),\\ Q_{i,k}^{min}\leqslant Q_i(k)\leqslant Q_{i,k}^{max}\quad or\\ Q_i(k)=0, I_i^*(k)\leqslant I_i^{max},\forall j\in\Omega, i\in\Psi\end{array}\right.\right]\\
&=\Sigma_{i=1}^n-\{\omega_i E^{Me}[\tilde{\tilde{c}}_i(k)]h_i I_i^*(k)+\omega_i E^{Me}[\tilde{\tilde{T}}_i^{wc}][E^{Me}[\tilde{\tilde{D}}_i(k)]-I_i^*(k)]-\lambda_i^*(k+1)(I_i^*(k)\\
&\quad -E^{Me}[\tilde{\tilde{D}}_i(k)])\}\\
&\quad+\max\left[\begin{array}{l}-\Sigma_{i=1}^n\{\omega_i\beta_i(k)\delta_i(1+r_i)^k Q_i^2(k)\\ +(\omega_i E^{Me}[\tilde{\tilde{c}}_i(k)]h_i+\omega_i E^{Me}[\tilde{\tilde{T}}_i^{sw}]\\ -\lambda_i^*(k+1))Q_i(k)+\omega_i\alpha_i(k)\}\end{array}\left|\begin{array}{l}Q_j^L Q_1(k)\leqslant Q_j(k)\leqslant Q_j^U Q_1(k),\\ Q_{i,k}^{min}\leqslant Q_i(k)\leqslant Q_{i,k}^{max}\quad or\\ Q_i(k)=0, I_i(k)\leqslant I_i^{max},\forall j\in\Omega,\\ i\in\Psi\end{array}\right.\right]
\end{aligned}
\tag{5.61}
$$

$k=0,1,\cdots,m-1$.

(3) If $I_i^*(k)+Q_i(k)>0$, $I_i^*(k)<0$, and $I_i^*(k)+Q_i(k)-E^{Me}[\tilde{\tilde{D}}_i(k)]<0$, then

$$
\begin{aligned}
&H^k(\mathbf{I}^*(k),\mathbf{Q}^*(k),\lambda^*(k+1))\\
&=\max\left[H^k(\mathbf{I}^*(k),\mathbf{Q}(k),\lambda^*(k+1))\left|\begin{array}{l}Q_j^L Q_1(k)\leqslant Q_j(k)\leqslant Q_j^U Q_1(k),\\ Q_{i,k}^{min}\leqslant Q_i(k)\leqslant Q_{i,k}^{max}\quad or\\ Q_i(k)=0, I_i^*(k)\leqslant I_i^{max},\forall j\in\Omega, i\in\Psi\end{array}\right.\right]\\
&=\Sigma_{i=1}^n-\{\omega_i E^{Me}[\tilde{\tilde{c}}_i(k)]h_i I_i^*(k)+\omega_i s_i[E^{Me}[\tilde{\tilde{D}}_i(k)]-I_i^*(k)]-\lambda_i^*(k+1)(I_i^*(k)\\
&\quad -E^{Me}[\tilde{\tilde{D}}_i(k)])\}\\
&\quad+\max\left[\begin{array}{l}-\Sigma_{i=1}^n\{\omega_i\beta_i(k)\delta_i(1+r_i)^k Q_i^2(k)\\ +(\omega_i E^{Me}[\tilde{\tilde{c}}_i(k)]h_i+\omega_i E^{Me}[\tilde{\tilde{T}}_i^{sw}]\\ +\omega_i E^{Me}[\tilde{\tilde{T}}_i^{wc}]-\lambda_i^*(k+1)-\omega_i s_i)\\ Q_i(k)+\omega_i\alpha_i(k)\}\end{array}\left|\begin{array}{l}Q_j^L Q_1(k)\leqslant Q_j(k)\leqslant Q_j^U Q_1(k),\\ Q_{i,k}^{min}\leqslant Q_i(k)\leqslant Q_{i,k}^{max}\quad or\\ Q_i(k)=0, I_i^*(k)\leqslant I_i^{max},\forall j\in\Omega,\\ i\in\Psi\end{array}\right.\right]
\end{aligned}
\tag{5.62}
$$

$k=0,1,\cdots,m-1$.

(4) If $I_i^*(k)+Q_i(k)>0$, $I_i^*(k)\geq 0$, and $I_i^*(k)+Q_i(k)-E^{Me}[\tilde{\tilde{D}}_i(k)]\geq 0$, then

$$
\begin{aligned}
&H^k(\mathbf{I}^*(k),\mathbf{Q}^*(k),\lambda^*(k+1))\\
&=\max\left[H^k(\mathbf{I}^*(k),\mathbf{Q}(k),\lambda^*(k+1))\left|\begin{array}{l}Q_j^L Q_1(k)\leqslant Q_j(k)\leqslant Q_j^U Q_1(k),\\ Q_{i,k}^{min}\leqslant Q_i(k)\leqslant Q_{i,k}^{max}\quad or\\ Q_i(k)=0, I_i^*(k)\leqslant I_i^{max},\forall j\in\Omega, i\in\Psi\end{array}\right.\right]\\
&=\Sigma_{i=1}^n-\{\omega_i E^{Me}[\tilde{\tilde{c}}_i(k)]h_i I_i^*(k)+\omega_i E^{Me}[\tilde{\tilde{T}}_i^{wc}]E^{Me}[\tilde{\tilde{D}}_i(k)]-\lambda_i^*(k+1)(I_i^*(k)\\
&\quad -E^{Me}[\tilde{\tilde{D}}_i(k)])\}\\
&\quad+\max\left[\begin{array}{l}-\Sigma_{i=1}^n\{\omega_i\beta_i(k)\delta_i(1+r_i)^k Q_i^2(k)\\ +(\omega_i E^{Me}[\tilde{\tilde{c}}_i(k)]h_i+\omega_i E^{Me}[\tilde{\tilde{T}}_i^{sw}]\\ -\lambda_i^*(k+1))Q_i(k)+\omega_i\alpha_i(k)\}\end{array}\left|\begin{array}{l}Q_j^L Q_1(k)\leqslant Q_j(k)\leqslant Q_j^U Q_1(k),\\ Q_{i,k}^{min}\leqslant Q_i(k)\leqslant Q_{i,k}^{max}\quad or\\ Q_i(k)=0, I_i(k)\leqslant I_i^{max},\forall j\in\Omega,\\ i\in\Psi\end{array}\right.\right]
\end{aligned}
\tag{5.63}
$$

$k=0,1,\cdots,m-1$.

(5) If $I_i^*(k)+Q_i(k)>0$, $I_i^*(k)\geq 0$, and $I_i^*(k)+Q_i(k)-E^{Me}[\tilde{\bar{D}}_i(k)]<0$, then

$$\begin{aligned}
&H^k(\mathbf{I}^*(k),\mathbf{Q}^*(k),\boldsymbol{\lambda}^*(k+1))\\
&=\max\left[H^k(\mathbf{I}^*(k),\mathbf{Q}(k),\boldsymbol{\lambda}^*(k+1))\,\middle|\,\begin{array}{l}Q_j^L Q_1(k)\leqslant Q_j(k)\leqslant Q_j^U Q_1(k),\\ Q_{i,k}^{min}\leqslant Q_i(k)\leqslant Q_{i,k}^{max}\quad or\\ Q_i(k)=0, I_i^*(k)\leqslant I_i^{max}, \forall j\in\Omega, i\in\Psi\end{array}\right]\\
&=\Sigma_{i=1}^n-\{\omega_i E^{Me}[\tilde{\bar{c}}_i(k)]h_i I_i^*(k)+\omega_i s_i[E^{Me}[\tilde{\bar{D}}_i(k)]-I_i^*(k)]+\omega_i E^{Me}[\tilde{\bar{T}}_i^{wc}]I_i^*(k)-\\
&\quad\lambda_i^*(k+1)(I_i^*(k)-E^{Me}[\tilde{\bar{D}}_i(k)])\}\\
&\quad+\max\left[\begin{array}{l}-\Sigma_{i=1}^n\{\omega_i\beta_i(k)\delta_i(1+r_i)^k Q_i^2(k)\\ +(\omega_i E^{Me}[\tilde{\bar{c}}_i(k)]h_i+\omega_i E^{Me}[\tilde{\bar{T}}_i^{sw}]\\ +\omega_i E^{Me}[\tilde{\bar{T}}_i^{wc}]-\lambda_i^*(k+1)-\omega_i s_i)\\ Q_i(k)+\omega_i\alpha_i(k)\}\end{array}\,\middle|\,\begin{array}{l}Q_j^L Q_1(k)\leqslant Q_j(k)\leqslant Q_j^U Q_1(k),\\ Q_{i,k}^{min}\leqslant Q_i(k)\leqslant Q_{i,k}^{max}\quad or\\ Q_i(k)=0, I_i^*(k)\leqslant I_i^{max}, \forall j\in\Omega,\\ i\in\Psi\end{array}\right]
\end{aligned} \tag{5.64}$$

$k=0,1,\cdots,m-1$.

Based on the deductions in Step. 2, information can be obtained by the simple analysis below.

- ***Step.3***: From Eq. (5.60), $\forall i\in\Psi, k=0,1,\cdots,m-1$, the value of $Q_i^*(k)$ can be determined by easy analysis. To maximize $H^k(\mathbf{I}^*(k),\mathbf{Q}(k),\boldsymbol{\lambda}^*(k+1))$, we only need to consider the last item of the above 5 cases. Let

$$\theta_i^l(k)=\max\{Q_i^L Q_1(k),Q_{i,k}^{min}\},\quad \theta_i^r(k)=\min\{Q_i^U Q_1(k),Q_{i,k}^{max}\},\quad \forall i\in\Psi$$

where $Q_1^L=1$, and $Q_1^U=1$. For the above 5 cases, the last item can be described as a sum of a series of quadratic polynomials, and each quadratic polynomial can be expressed as following:

$$a_i(k)Q_i^{2*}(k)+b_i(k)Q_i^*(k)+c_i(k)$$

where $Q_i^*(k)\in[\theta_i^l(k),\theta_i^r(k)]$. Since $\beta_i(k)<0$, it is not hard to know that $a_i(k)=-\omega_i\beta_i(k)\delta_i(1+r_i)^k>0$. According to the properties of quadratic polynomials, the symmetry axis $\phi_i(k)$ of each quadratic polynomial can be expressed based on the above 5 cases as follows:

(1) If $I_i^*(k)+Q_i(k)\leq 0$, then the symmetry axis $\phi_i(k)$ could be

$$\phi_i(k)=-\frac{b_i(k)}{2}=\frac{1}{2}(\omega_i E^{Me}[\tilde{\bar{c}}_i(k)]h_i+\omega_i E^{Me}[\tilde{\bar{T}}_i^{sw}]+\omega_i E^{Me}[\tilde{\bar{T}}_i^{wc}]-\lambda_i^*(k+1)-\omega_i s_i).$$

(2) If $I_i^*(k)+Q_i(k)>0$, $I_i^*(k)<0$, and $I_i^*(k)+Q_i(k)-E^{Me}[\tilde{\bar{D}}_i(k)]\geq 0$, then

$$\phi_i(k)=-\frac{b_i(k)}{2}=\frac{1}{2}(\omega_i E^{Me}[\tilde{\bar{c}}_i(k)]h_i+\omega_i E^{Me}[\tilde{\bar{T}}_i^{sw}]-\lambda_i^*(k+1)).$$

(3) If $I_i^*(k)+Q_i(k)>0$, $I_i^*(k)<0$, and $I_i^*(k)+Q_i(k)-E^{Me}[\tilde{\bar{D}}_i(k)]<0$, then

$$\phi_i(k)=-\frac{b_i(k)}{2}=\frac{1}{2}(\omega_i E^{Me}[\tilde{\bar{c}}_i(k)]h_i+\omega_i E^{Me}[\tilde{\bar{T}}_i^{sw}]+\omega_i E^{Me}[\tilde{\bar{T}}_i^{wc}]-\lambda_i^*(k+1)-\omega_i s_i).$$

(4) If $I_i^*(k)+Q_i(k)>0$, $I_i^*(k)\geq 0$, and $I_i^*(k)+Q_i(k)-E^{Me}[\tilde{\tilde{D}}_i(k)]\geq 0$, then

$$\phi_i(k)=-\frac{b_i(k)}{2}=\frac{1}{2}(\omega_i E^{Me}[\tilde{\tilde{c}}_i(k)]h_i+\omega_i E^{Me}[\tilde{\tilde{T}}_i^{sw}]-\lambda_i^*(k+1)).$$

(5) If $I_i^*(k)+Q_i(k)>0$, $I_i^*(k)\geq 0$, and $I_i^*(k)+Q_i(k)-E^{Me}[\tilde{\tilde{D}}_i(k)]<0$, then

$$\phi_i(k)=-\frac{b_i(k)}{2}=\frac{1}{2}(\omega_i E^{Me}[\tilde{\tilde{c}}_i(k)]h_i+\omega_i E^{Me}[\tilde{\tilde{T}}_i^{sw}]+\omega_i E^{Me}[\tilde{\tilde{T}}_i^{wc}]-\lambda_i^*(k+1)-\omega_i s_i).$$

***Case.1***: It is not hard to deduce that if $\phi_i(k)<\frac{\theta_i^l(k)+\theta_i^r(k)}{2}$, since $f(Q_i(k))=a_i(k)Q_i^2(k)+b_i(k)Q_i(k)+c_i(k)$ is a convex function, the quadratic polynomial $f(Q_i(k))$ can arrive its maximal value with in the interval $[\theta_i^l(k),\theta_i^r(k)]$ if and only if $Q_i^*(k)=\theta_i^r(k)$ as shown in Figure 5.8.

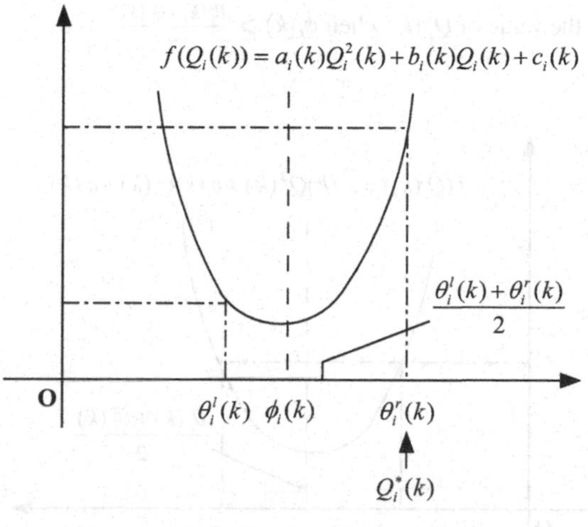


**Fig. 5.8** Determine the value of $Q_i^*(k)$ when $\phi_i(k)<\frac{\theta_i^l(k)+\theta_i^r(k)}{2}$

***Case.2***: Similarly, if $\phi_i(k)>\frac{\theta_i^l(k)+\theta_i^r(k)}{2}$, since $f(Q_i(k))=a_i(k)Q_i^2(k)+b_i(k)Q_i(k)+c_i(k)$ is a convex function, the quadratic polynomial $f(Q_i(k))$ can arrive its maximal value with in the interval $[\theta_i^l(k),\theta_i^r(k)]$ if and only if $Q_i^*(k)=\theta_i^l(k)$ as shown in Figure 5.9.
***Case.3***: If $\phi_i(k)=\frac{\theta_i^l(k)+\theta_i^r(k)}{2}$, then the quadratic polynomial $f(Q_i(k))$ can arrive its maximal value at the boundary points of the interval $[\theta_i^l(k),\theta_i^r(k)]$. So $Q_i^*(k)=\theta_i^l(k)$ or $Q_i^*(k)=\theta_i^r(k)$ as shown in Figure 5.10.
Finally, the optimal solution of our problem by Step. 4 can be obtained as shown below.

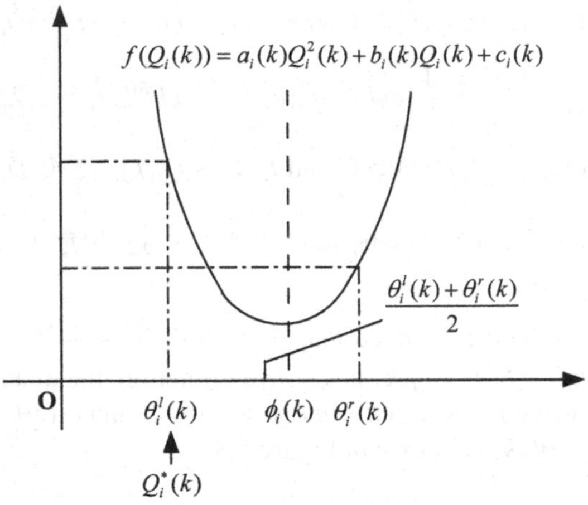


**Fig. 5.9** Determine the value of $Q_i^*(k)$ when $\phi_i(k) > \frac{\theta_i^l(k)+\theta_i^r(k)}{2}$

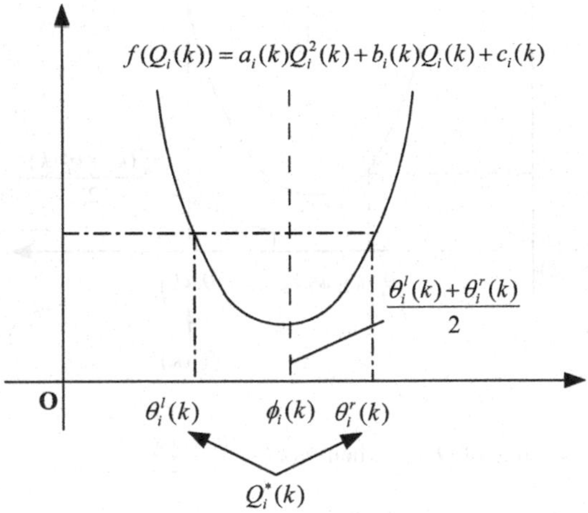


**Fig. 5.10** Determine the value of $Q_i^*(k)$ when $\phi_i(k) = \frac{\theta_i^l(k)+\theta_i^r(k)}{2}$

- ***Step.4***: Since the value of $\mathbf{Q}^*(k)$ $(k = 0, 1, \cdots, m-1)$ has been determined in Step. 3, the value of $\mathbf{I}^*(k)$ $(k = \{1, 2, \cdots, m-1\})$ can be calculated by solving Eq. (5.2).

## 5.4 Dynamic Programming-Based Particle Swarm Optimization

After a proper model transformation, the IMP-LCP becomes solvable and a dynamic programming-based particle swarm optimization (DP-based PSO) algorithm is developed to solve Model (5.22). The advantage of using the PSO algorithm over other techniques is that it is computationally inexpensive, easy-to-implement, and requires only the objective function's values, but not the gradient information. Therefore, a DP-based PSO algorithm is employed to solve the above problem. The key features of the DP-based PSO algorithm are explained in detail below, including the general mechanism, solution representation, the initialization method, the adjustment and particle decoding, the fitness value function, and the PSO framework of the algorithm.

### *5.4.1 General Mechanism of DP-Based PSO*

Particle swarm optimization (PSO), an optimization technique based on swarm intelligence, was proposed by Kennedy and Eberhart (1995) [23]. This technique simulates social behavior, such as birds flocking to a promising position, for certain objectives in a multidimensional space (Eberhart and Shi 2001 [24]; Clerc and Kennedy 2002 [25]). While there have been many variants (Gao et al. 2011 [435]; Cagnina et al. 2011 [100]; Anghinolfi and Paolucci 2009 [211]; Yapicioglu et al. 2007 [244]; Xu and Zeng 2011 [242]) of the PSO algorithm since its debut, in this study, a DP-based PSO algorithm is developed to solve the expected value model (5.22).

Unlike the standard PSO (Bratton and Kennedy 2007 [99]), the DP-based PSO can reduce the dimensions of a solution representation (i.e., a particle) using the state equation in the crisp dynamic programming model (5.22). The essential differences between the DP-based PSO and the standard PSO are in the solution representations. In the standard PSO, the algorithm is initialized with a population (called swarm) of $L$ random individuals (called particles). Each particle is represented by its position in an $H$-dimensional space, where $H$ is the problem dimension. For the IMP-LCP, the problem dimensions need to include the decision variables (i.e., $Q_i(k)$) and state variables (i.e., $I_i(k)$), which both appear in the objectives and constraints of Model (5.22) (refer to Eq. (5.15) for details). Note that if the decision variables are known, then the state variables can be determined based on the state equation (See Eq. (5.2)). Thus, for the IMP-LCP, the dimensions of the solution representation as well as the search space can be reduced using the state equation when developing the iterative solution algorithm for Model (5.22). The basic PSO formula is shown below:

$$V_l(\tau+1) = w(\tau)V_l(\tau) + c_p r_p(PBest_l - P_l(\tau)) + c_g r_g(GBest - P_l(\tau)), \qquad (5.65)$$

where $l$ (i.e., particle index) $= 1, 2, \cdots, L$; $L$ is the population size; $\tau$ (i.e., iteration index) $= 0, 1, \cdots, T$; $T$ is the iteration limit; $V_l(\tau) = [v_{l1}(\tau), v_{l2}(\tau), \cdots, v_{lH}(\tau)]$ denotes the $H$-dimension velocity (i.e., distance change) for the $l$th particle in the

$\tau$th iteration; $P_l(\tau) = [p_{l1}(\tau), p_{l2}(\tau), \cdots, p_{lH}(\tau)]$ denotes the $H$-dimension position for the $l$th particle in the $\tau$th iteration; $PBest_l = [pBest_{l1}, pBest_{l2}, \cdots, pBest_{lH}]$, the personal best of the $l$th particle encountered after $\tau$ iterations; and $GBest = [gBest_1, gBest_2, \cdots, gBest_H]$, the global best among all the swarm of particles achieved so far. The initial or updated positions (i.e., $P_l(\tau)$) must be evaluated with respect to a fitness value (i.e., Eq. (5.69)) in order to identify the personal best of each particle (i.e., $PBest_l$) and the global best (i.e., $GBest$) in the swarm. $c_p$ and $c_g$ are positive constants (namely, learning factors) that determine the relative weight that the global best has versus the personal best. Note that $r_p$ and $r_g$ are random numbers between 0 and 1, and $w(\tau)$ is the inertia weight used to control the impact of the previous velocities on the current velocity which influences the trade-off between the global and local exploration abilities during the search. The particle position is updated using the new velocity:

$$P_l(\tau+1) = V_l(\tau+1) + P_l(\tau). \tag{5.66}$$

Eq. (5.65) is used to calculate the particle's new velocity according to its previous velocity and the distances of its current position from its own best experience and the group's best experience. Eq. (5.66) is traditionally used to update a particle that flies toward a new position (Shi and Eberhart 1998a [34]). Figure 5.11 illustrates the differences in the solution representations between a DP-based PSO and a standard PSO.

### 5.4.2 Solution Representation of DP-Based PSO

As noted earlier, the essential differences between the DP-based PSO and the standard PSO is that the DP-basd PSO takes advantage of the iterative mechanism of Model (5.22) by using the state equation (Eq. (5.2)) to reduce the dimensions of the particles as well as the solution search space. To highlight the differences, a solution representation for the DP-based PSO for a dynamic programming model (5.22) is detailed below. (Note: the dimensions of a DP-based PSO is $n \times m$, which is one half of the size $2n \times m$ required by the standard PSO.)

In this study, the $n \times m$ dimensions (i.e., $H = n \times m$) of each particle is divided into $n$ parts, that is:

$$P_l(\tau) = [p_{l1}(\tau), p_{l2}(\tau), \cdots, p_{lH}(\tau)] = [Y_l^1(\tau), Y_l^2(\tau), \cdots, Y_l^n(\tau)] \tag{5.67}$$

where $Y_l^i(\tau) =$ the $i^{th}$ part of the $l^{th}$ particle in the $\tau^{th}$ generation. Note that every part of a particle is a $m$-dimension vector, which can be denoted as:

$$Y_l^i(\tau) = [y_{l1}^i(\tau), y_{l2}^i(\tau), \cdots, y_{lm}^i(\tau)] \tag{5.68}$$

where $y_{l(k+1)}^i(\tau) =$ the $(k+1)^{th}$ dimension of $Y_l^i(\tau)$ for the $l^{th}$ particle in the $\tau^{th}$ generation; $k = 0, 1, \cdots, m-1$. In order to be in line with the expression $P_l(\tau) = [p_{l1}(\tau), p_{l2}(\tau), \cdots, p_{lH}(\tau)]$, $y_{l(k+1)}^i(\tau) = p_{l(m\times(i-1)+k+1)}(\tau)$.

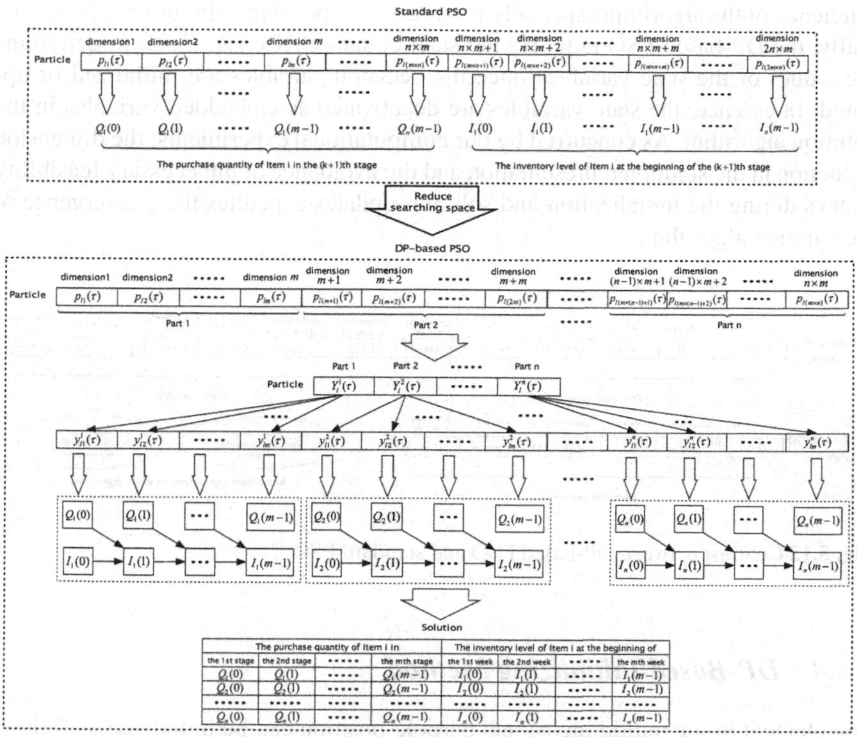


Fig. 5.11 The differences of solution representation between a DP-based PSO and a standard PSO

Note that the $(k+1)^{th}$ dimension of $Y_l^i(\tau)$ (i.e., $y_{l(k+1)}^i(\tau)$) represents the purchasing quantity of Item $i$ in the $(k+1)^{th}$ stage (i.e., $Q_i(k)$). Based on the state equation (i.e., Eq. (5.2)) and constraint conditions (i.e., Eqs. (5.5) and (5.6)), the $(k+1)^{th}$ dimension of $Y_l^i(\tau)$ (i.e., $y_{l(k+1)}^i(\tau)$), either at initialization or during updates, is limited to $[Q_{i,k}^{min}, Q_{i,k}^{max}]$ (i.e., $[p^{min}, p^{max}]$) while at the same time satisfying Eqs. (5.2) and (5.5) to ensure a feasible particle position. This dimension reduction in the solution representation expedites the PSO search.

### 5.4.3 Contrast between DP-Based PSO and Standard PSO

In contrast to the standard PSO which contains all the decision variables and state variables in each particle, the DP-based PSO contains only the decision variables in each particle (Figure 5.11). Thus, in a DP-based PSO, the particle dimension is halved, and the checking procedures required for the state variables in terms of their relationship with the decision variables during the initializing and updating processes can be eliminated. This special treatment significantly improves the efficiency of the algorithm especially for large scale problems (Figure 5.12). Specifically, the DP-based PSO employs the state equation (See Eq. (5.2)) to determine the values of the state variables once the decision variables are initialized or updated. In essence, the state variables are deactivated as embedded variables in the solution algorithm. As concurred by our computational experiments, the dimension reduction in the solution representation and the avoidance of unnecessary feasibility checks during the initialization and solution updates expedites the convergence of the solution algorithm.

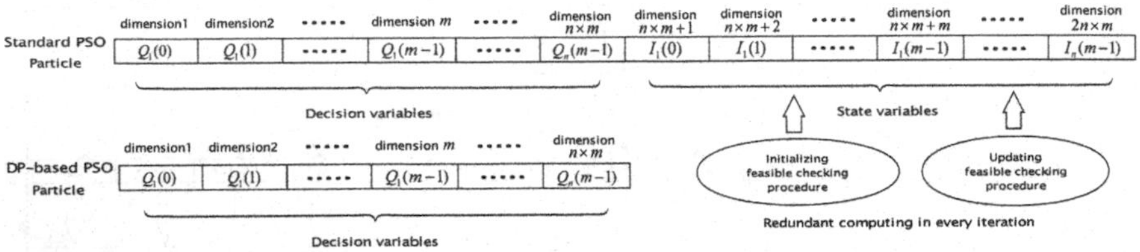


**Fig. 5.12** Contrast between DP-based PSO and standard PSO

## *5.4.4 DP-Based Initializing Method*

The element in the multidimensional particle position can be initialized as follows to avoid an infeasible position:

*Step 1:* Set $i = 1$.

*Step 2:* Set $k = 0$.

*Step 3:* Initialize $y^i_{l(k+1)}(\tau)$ by generating a random real number within $[Q^{min}_{i,k}, Q^{max}_{i,k}]$. Then, based on the first constraint in Eq. (5.22), $I_i(k+1) = I_i(k) + y^i_{l(k+1)}(\tau) - E^{Me}[\tilde{\bar{D}}_i(k)_{(r_i,\sigma_i)}]$ (note that $I_i(0) = a_i$, where $a_i$ denotes the initial inventory level of Item $i$ in the corresponding warehouse at the beginning of the first stage).

*Step 4:* If $I_i(k+1) \leqslant I^{max}_i$, then go to Step 5. Otherwise, return to Step 3.

*Step 5:* If $k = m-1$, then $i = i+1$ and go to Step 6. Otherwise, $k = k+1$ and return to Step 3.

*Step 6:* Set $k = 0$.

*Step 7:* Let $\underline{V} = \max\{Q^{min}_{i,k}, Q^L_i y^1_{l(k+1)}(\tau)\}$, and $\overline{V} = \min\{Q^{max}_{i,k}, Q^U_i y^1_{l(k+1)}(\tau)\}$. Initialize $y^i_{l(k+1)}(\tau)$ by generating a random real number within $[\underline{V}, \overline{V}]$.

*Step 8:* Based on the first constraint in Eq. (5.22), $I_i(k+1) = I_i(k) + y^i_{l(k+1)}(\tau) - E^{Me}[\tilde{\bar{D}}_i(k)_{(r_i,\sigma_i)}]$.

*Step 9:* If $I_i(k+1) \leqslant I^{max}_i$, then go to Step 10. Otherwise, return to Step 7.

*Step 10:* If the stopping criterion is met, i.e., $i = n$ and $k = m-1$, then the initialization for the $l^{th}$ particle is completed. Otherwise, if $k = m-1$, then $i = i+1$ and return to Step 6. If $k < m-1$, then $k = k+1$ and return to Step 7.

### 5.4.5 DP-Based Adjusting Method

The element in the multidimensional particle position can be adjusted as follows to avoid an infeasible position:

*Step 1:* Set $i = 1$.

*Step 2:* Set $k = 0$.

*Step 3:* If $y^i_{l(k+1)}(\tau) < Q^{min}_{i,k}$, then $y^i_{l(k+1)}(\tau) = Q^{min}_{i,k}$. If $y^i_{l(k+1)}(\tau) > Q^{max}_{i,k}$, then $y^i_{l(k+1)}(\tau) = Q^{max}_{i,k}$.

*Step 4:* Based on the first constraint in Eq. (5.22), $I_i(k+1) = I_i(k) + y^i_{l(k+1)}(\tau) - E^{Me}[\tilde{D}_i(k)_{(r_i,\sigma_i)}]$.

*Step 5:* If $I_i(k+1) \leqslant I^{max}_i$, then go to Step 6. Otherwise, let $U = I_i(k+1) - I^{max}_i$, $y^i_{l(k+1)}(\tau) = y^i_{l(k+1)}(\tau) - U$, and return to Step 4.

*Step 6:* If $k = m-1$, then $i = i+1$ and go to Step 7. Otherwise, $k = k+1$ and return to Step 3.

*Step 7:* Set $k = 0$.

*Step 8:* Let $\underline{V} = \max\{Q^{min}_{i,k}, \underline{R}_i y^1_{l(k+1)}(\tau)\}$, and $\overline{V} = \min\{Q^{max}_{i,k}, \overline{R}_i y^1_{l(k+1)}(\tau)\}$. If $y^i_{l(k+1)}(\tau) < \underline{V}$, then $y^i_{l(k+1)}(\tau) = \underline{V}$. If $y^i_{l(k+1)}(\tau) > \overline{V}$, then $y^i_{l(k+1)}(\tau) = \overline{V}$.

*Step 9:* Based on the first constraint in Eq. (5.22), $I_i(k+1) = I_i(k) + y^i_{l(k+1)}(\tau) - E^{Me}[\tilde{D}_i(k)_{(r_i,\sigma_i)}]$.

*Step 10:* If $I_i(k+1) \leqslant I^{max}_i$, then go to Step 11. Otherwise, let $U = I_i(k+1) - I^{max}_i$, $y^i_{l(k+1)}(\tau) = y^i_{l(k+1)}(\tau) - U$, and return to Step 8.

*Step 11:* If the stopping criterion is met, i.e., $i = n$ and $k = m-1$, then the adjustment for the $l^{th}$ particle is completed. Otherwise, if $k = m-1$, then $i = i+1$ and return to Step 7. If $k < m-1$, then $k = k+1$ and return to Step 8.

### 5.4.6 DP-Based Decoding Method

A DP-based decoding method for this representation in the problem solution starts by transforming a particle into a corresponding purchase quantity for Item $i$ at the beginning of each stage. The overall procedure for this decoding is detailed below:

*Step 1:* Decode the $(k+1)^{th}$ dimension of $Y^i_l(\tau)$ (i.e., $y^i_{l(k+1)}(\tau)$) into the purchase quantity of Item $i$ at the beginning of the $(k+1)^{th}$ week, i.e., $Q_i(k) = y^i_{l(k+1)}(\tau)$, $i = 1, 2, \cdots, n$, $k = 0, 1, \cdots, m-1$.

*Step 2:* Set $i = 0$.

*Step 3:* Set $k = 0$.

*Step 4:* Based on the first constraint in Eq. (5.22), $I_i(k+1) = I_i(k) + y^i_{l(k+1)}(\tau) - E^{Me}[\tilde{D}_i(k)_{(r_i,\sigma_i)}]$.

*Step 5:* If the stopping criterion is met, i.e., $i = n$ and $k = m-1$, then integrate $Q_i(k)$ and $I_i(k)$ (for $i = 1, 2, \cdots, n$, $k = 1, 2, \cdots, m-1$) to determine a solution to the IMP-LCP. Otherwise, if $k = m-1$, then $i = i+1$ and return to Step 3. If $k < m-1$, then $k = k+1$ and return to Step 4.

### 5.4.7 Fitness Value

The fitness value used to evaluate a particle is just the value of the objective function in Model (5.22), i.e.

$$Fitness(P_l(\tau)) = E^{Me}[\sum_{i=1}^{n} \omega_i f_i(\mathbf{x}_i, \tilde{\xi}_{i(r_i,\sigma_i)})] \tag{5.69}$$

### 5.4.8 DP-Based PSO Framework for IMP-LCP

Based on the above basic concepts (i.e., solution-representation, initialization method, adjustment method, decoding method, fitness value function), the framework to implement a DP-based PSO for the IMP-LCP optimization is developed.

The initial positions of the $L$ particles, i.e., $P_l(0) = [p_{l1}(0), p_{l2}(0), \cdots, p_{l(n\times m)}(0)]$ $(l = 1,2,\cdots,L)$, are randomly generated using a DP-based initialization method as described in the previous section to avoid an infeasible position. The initial velocities of the $L$ particles, i.e., $V_l(0) = [v_{l1}(0), v_{l2}(0), \cdots, v_{l(n\times m)}(0)]$ $(l = 1,2,\cdots,L)$, are also randomly generated within $[V^{min}, V^{max}]$, where $V^{min} = -\frac{1}{5}(Q_{i,k}^{max} - Q_{i,k}^{min})$ and $V^{max} = \frac{1}{5}(Q_{i,k}^{max} - Q_{i,k}^{min})$. In addition, the value $-\frac{1}{5}$ and $\frac{1}{5}$ are selected for the $V^{min}$ and $V^{max}$ to avoid the particles flying outside the feasible search space. The particle-updating mechanism represented by Eqs. (5.65) and (5.66) is used to update the velocities and positions (i.e., $V_l(\tau)$ and $P_l(\tau)$) of the particles until the optimal solution is found. During this DP-based PSO search, each element in the multidimensional particle position (i.e., $p_{l(k+1)}(\tau)$) is updated using the adjustment method as described in the previous section when it is beyond $[p^{min}, p^{max}]$.

The initial or updated positions must be evaluated with respect to a fitness value (i.e., Eq. (5.69)) in order to identify the personal best of each particle (i.e., $PBest_l = [pBest_{l1}, pBest_{l2}, \cdots, pBest_{l(n\times m)}]$) $(l = 1,2,\cdots,L)$ and the global best (i.e., $GBest = [gBest_1, gBest_2, \cdots, gBest_{n\times m}]$) of the swarm.

The above DP-based PSO search is terminated when either one of the following two stopping criteria is met: (1) convergence iteration limit (i.e., maximum number of iterations since last updating of the global best); (2) total iteration limit (i.e., $T$). Then, based on the DP-based decoding method as described in the previous section, the global best particle-represented solution (i.e., $GBest = [gBest_1, gBest_2, \cdots, gBest_{n\times m}]$) can be transformed into a problem solution for the IMP-LCP, which is the final solution to the IMP-LCP.

## 5.5 Case Study: Construction Materials Inventory

This section introduces an actual construction project, the China Yangtze Three Gorges Project (TGP), as an example for the IMP-LCP, and employs a DP-based PSO to solve the problem. The project description and data collection are explained in detail, and the specific features of each case are thoroughly interpreted. Results and analysis are also provided to allow for a full understanding of the solution methods.

### 5.5.1 Project Description

As one of the biggest complex hydropower projects in the world, TGP ranks as a key project for the improvement and development of the Yangtze River. The dam is located in the Xilingxia gorge, one of the three gorges of the river, and controls a drainage area of 1 million $km^2$ , with an average annual runoff of 451 billion $m^3$. The construction of the main body of the Three Gorges Water Conservancy Complex includes the following work: a rock-and-earth excavation of 102.83 million $m^3$, concrete placement of 27.94 million $m^3$, rock-and-earth refill of 31.98 million $m^3$, steel frame installation of 463,000 tons, metal frame installation of 256,500 tons, installation of twenty-six 700 $MW$ turbine-generator units (not including the six 700 $MW$ units in the Right Bank Underground Powerhouse under construction), and others. Except for the rock-and-earth refill index, all the preceding indices are the largest of all water conservancy projects either already completed or under construction.

### 5.5.2 Data Collection

As an actual large-scale construction project, the TGP is used as an application example to demonstrate the practicality of the modeling approach and the efficacy of the DP-based PSO algorithm.

**Table 5.2** Purchasing and inventory information of construction materials - I

| Item index $i$ | Item name | Initial inventory level $a_i$ ($10^4kg$) | Terminal inventory level $b_i$ ($10^4kg$) | Rise and fall rate $r_i$ % | Spot price $\delta_i$ ($yuan/kg$) | Discount percentage $\gamma_i$ % | Maximum inventory level $I_i^{max}$ ($10^4kg$) | Penalty price $s_i$ ($yuan/kg$) |
|---|---|---|---|---|---|---|---|---|
| 1 | Reinforcing steel | 165.31 | 77.03 | 1.3 | 2.68 | 85 | 8000 | 0.1 |
| 2 | Cement | 289.44 | 34.12 | 0.8 | 0.32 | 78 | 9500 | 0.08 |
| 3 | Fly ash | 78.13 | 19.20 | 1.2 | 0.08 | 90 | 3000 | 0.04 |
| 4 | Aggregate | 459.56 | 392.42 | 0.9 | 0.02 | 92 | 28000 | 0.01 |

The detailed data for the TGP were obtained from the China Three Gorges Corporation. In this project, all permanent equipment and construction materials are purchased through bidding (Dai et al. 2006 [26]). Specifically, the inventory management problem for the concrete construction of works contract Nos. *I&IIB* in the TGP was researched in-depth for this practical application. The works contract Nos. *I&IIB* mainly includes the spillway section and blocks Nos.10 – 14 of the left powerhouse section. It was started in 1999 and completed in 2001. For modeling, purchasing and inventory data for the four types of construction materials (i.e., reinforcing steel, cement, fly ash, and aggregate) were collected over 6 months during construction. The reinforcing steel is considered the key item, and the cement, fly ash, and aggregate are auxiliary items. Detailed information on these construction materials is shown in Tables 5.2-5.5.

It should be noted that the fuzzy random numbers (i.e., $(a, \varphi(\omega), b)$) in Tables 5.3-5.5 were obtained from the following steps: (1) To collect the data, interviews were conducted with experienced engineers (i.e., $q = 1, 2, \cdots, E$, where $q$ is the index of engineers). The data for each item was described in linguistic terms such as "between 48 and 57, with a most likely value of 53.". The engineers presented the linguistic terms based on their experience and the related data. Generally, the view of each engineer is able to be denoted as a range of the parameter (i.e., $[l_q, r_q]$) and a most possible value (i.e., $m_q$). (2) Since different engineers have different views of the parameters, the minimum value of all $l_q$ and the maximal value of all $r_q$ (for $q = 1, 2, \cdots, E$) were selected as the left border (i.e., $a$) and the right border (i.e., $b$) of the fuzzy random number, respectively. (3) Assume $m_q$ to be a random variable (i.e., $\varphi(\omega)$). By comparing the most possible values (i.e., $m_q$ for $q = 1, 2, \cdots, E$) obtained from different engineers, it was found that $\varphi(\omega)$ approximately follows a normal distribution (i.e., $\mathscr{N}(\zeta, \nu^2)$), which could be estimated using the maximum likelihood method and justified using a chi-square goodness-of-fit test. Table 5.6 provides a summary of the parameter values and test results produced for Item 1. As shown, the test results indicate that Item 1 has been properly modeled in terms of its parameter measures with a proof that they follow a normal distribution. (4) Finally, derive the fuzzy random number, $(a, \varphi(\omega), b)$ was derived, where $\varphi(\omega) \sim \mathscr{N}(\zeta, \nu^2)$.

**Table 5.3** Purchasing and inventory information of construction materials - II

| Item index $i$ | Item name | Lower limit $Q_j^L$ % | Upper limit $Q_j^U$ % | Holding price $h_i$ $(yuan/(kg \cdot month))$ | Transportation price from supplier to warehouse $\tilde{\bar{T}}_i^{sw}$ $(yuan/kg)$ | Transportation price from warehouse to construction site $\tilde{\bar{T}}_i^{wc}$ $(yuan/kg)$ |
|---|---|---|---|---|---|---|
| 1 | Reinforcing steel | - | - | 0.28 | $(0.12, \varphi(\omega), 0.23)$ $\varphi(\omega) \sim \mathscr{N}(0.18, 1.2)$ | $(0.02, \varphi(\omega), 0.06)$ $\varphi(\omega) \sim \mathscr{N}(0.04, 2.6)$ |
| 2 | Cement | 80.3 | 105.9 | 0.04 | $(0.073, \varphi(\omega), 0.098)$ $\varphi(\omega) \sim \mathscr{N}(0.086, 4.7)$ | $(0.014, \varphi(\omega), 0.022)$ $\varphi(\omega) \sim \mathscr{N}(0.017, 3.8)$ |
| 3 | Fly ash | 20.1 | 26.5 | 0.01 | $(0.017, \varphi(\omega), 0.025)$ $\varphi(\omega) \sim \mathscr{N}(0.022, 1.1)$ | $(0.003, \varphi(\omega), 0.008)$ $\varphi(\omega) \sim \mathscr{N}(0.006, 3.2)$ |
| 4 | Aggregate | 329.4 | 339.8 | 0.003 | $(0.002, \varphi(\omega), 0.007)$ $\varphi(\omega) \sim \mathscr{N}(0.004, 8.4)$ | $(0.0007, \varphi(\omega), 0.0012)$ $\varphi(\omega) \sim \mathscr{N}(0.001, 10.6)$ |

Based on the data from $r_i$, $\gamma_i$, $\delta_i$, $Q_{i,k}^{min}$, $Q_{i,k}^{max}$ in Tables 5.2-5.5, the option premium (i.e., $\alpha_i(k)$) and conversion coefficient (i.e., $\beta_i(k)$) can be easily computed and these are summarized in Table 5.7.

### 5.5.3 Parameters Selection for DP-Based PSO

The PSO parameters were determined based on the results of some preliminary experiments that were carried out to observe the behavior of the algorithm in different parameter settings. Through a comparison of several sets of parameters, including

**Table 5.4** Purchasing and inventory information of construction materials - III

| Variable | Month index $k=0$ | $k=1$ | $k=2$ |
|---|---|---|---|
| $Q_{1,k}^{min}$ $(10^4 kg)$ | 2500 | 2700 | 2400 |
| $Q_{1,k}^{max}$ $(10^4 kg)$ | 6400 | 7000 | 6800 |
| $\tilde{\bar{D}}_1(k)$ $(10^6 kg)$ | $(45,\varphi(\omega),54)$ $\varphi(\omega)\sim\mathcal{N}(51,12)$ | $(42,\varphi(\omega),56)$ $\varphi(\omega)\sim\mathcal{N}(50,14)$ | $(51,\varphi(\omega),59)$ $\varphi(\omega)\sim\mathcal{N}(54,22)$ |
| $\tilde{\bar{c}}_1(k)$ | $(0.54,\varphi(\omega),0.78)$ $\varphi(\omega)\sim\mathcal{N}(0.67,8)$ | $(0.62,\varphi(\omega),0.81)$ $\varphi(\omega)\sim\mathcal{N}(0.72,13)$ | $(0.56,\varphi(\omega),0.88)$ $\varphi(\omega)\sim\mathcal{N}(0.67,31)$ |
| $Q_{2,k}^{min}$ $(10^4 kg)$ | 1800 | 2100 | 1600 |
| $Q_{2,k}^{max}$ $(10^4 kg)$ | 7300 | 6800 | 7000 |
| $\tilde{\bar{D}}_2(k)$ $(10^6 kg)$ | $(36,\varphi(\omega),57)$ $\varphi(\omega)\sim\mathcal{N}(49,42)$ | $(40,\varphi(\omega),65)$ $\varphi(\omega)\sim\mathcal{N}(56,35)$ | $(38,\varphi(\omega),56)$ $\varphi(\omega)\sim\mathcal{N}(51,46)$ |
| $\tilde{\bar{c}}_2(k)$ | $(0.47,\varphi(\omega),0.83)$ $\varphi(\omega)\sim\mathcal{N}(0.72,11)$ | $(0.58,\varphi(\omega),0.89)$ $\varphi(\omega)\sim\mathcal{N}(0.75,9)$ | $(0.51,\varphi(\omega),0.79)$ $\varphi(\omega)\sim\mathcal{N}(0.69,23)$ |
| $Q_{3,k}^{min}$ $(10^4 kg)$ | 550 | 480 | 520 |
| $Q_{3,k}^{max}$ $(10^4 kg)$ | 1800 | 2000 | 1800 |
| $\tilde{\bar{D}}_3(k)$ $(10^6 kg)$ | $(8.9,\varphi(\omega),14.4)$ $\varphi(\omega)\sim\mathcal{N}(12.3,17)$ | $(9.2,\varphi(\omega),14.9)$ $\varphi(\omega)\sim\mathcal{N}(12.7,15)$ | $(10.2,\varphi(\omega),16.4)$ $\varphi(\omega)\sim\mathcal{N}(14.1,13)$ |
| $\tilde{\bar{c}}_3(k)$ | $(0.49,\varphi(\omega),0.75)$ $\varphi(\omega)\sim\mathcal{N}(0.69,4)$ | $(0.66,\varphi(\omega),0.89)$ $\varphi(\omega)\sim\mathcal{N}(0.75,3)$ | $(0.52,\varphi(\omega),0.91)$ $\varphi(\omega)\sim\mathcal{N}(0.73,11)$ |
| $Q_{4,k}^{min}$ $(10^4 kg)$ | 12000 | 13000 | 12000 |
| $Q_{4,k}^{max}$ $(10^4 kg)$ | 23000 | 22000 | 25000 |
| $\tilde{\bar{D}}_4(k)$ $(10^7 kg)$ | $(15.2,\varphi(\omega),17.9)$ $\varphi(\omega)\sim\mathcal{N}(16.8,18)$ | $(17.3,\varphi(\omega),20.5)$ $\varphi(\omega)\sim\mathcal{N}(19.2,11)$ | $(14.8,\varphi(\omega),16.7)$ $\varphi(\omega)\sim\mathcal{N}(15.4,13)$ |
| $\tilde{\bar{c}}_4(k)$ | $(0.51,\varphi(\omega),0.85)$ $\varphi(\omega)\sim\mathcal{N}(0.71,14)$ | $(0.59,\varphi(\omega),0.85)$ $\varphi(\omega)\sim\mathcal{N}(0.76,8)$ | $(0.61,\varphi(\omega),0.92)$ $\varphi(\omega)\sim\mathcal{N}(0.83,11)$ |

**Table 5.5** Purchasing and inventory information of construction materials - IV

| Variable | Month index $k=3$ | $k=4$ | $k=5$ |
|---|---|---|---|
| $Q_{1,k}^{min}$ $(10^4 kg)$ | 2400 | 2550 | 3000 |
| $Q_{1,k}^{max}$ $(10^4 kg)$ | 6600 | 6600 | 6400 |
| $\tilde{\bar{D}}_1(k)$ $(10^6 kg)$ | $(48,\varphi(\omega),57)$ $\varphi(\omega)\sim\mathcal{N}(53,24)$ | $(52,\varphi(\omega),60)$ $\varphi(\omega)\sim\mathcal{N}(57,32)$ | $(47,\varphi(\omega),55)$ $\varphi(\omega)\sim\mathcal{N}(52,42)$ |
| $\tilde{\bar{c}}_1(k)$ | $(0.64,\varphi(\omega),0.83)$ $\varphi(\omega)\sim\mathcal{N}(0.74,15)$ | $(0.66,\varphi(\omega),0.78)$ $\varphi(\omega)\sim\mathcal{N}(0.71,25)$ | $(0.61,\varphi(\omega),0.83)$ $\varphi(\omega)\sim\mathcal{N}(0.77,19)$ |
| $Q_{2,k}^{min}$ $(10^4 kg)$ | 2000 | 1800 | 1900 |
| $Q_{2,k}^{max}$ $(10^4 kg)$ | 7100 | 6800 | 6800 |
| $\tilde{\bar{D}}_2(k)$ $(10^6 kg)$ | $(37,\varphi(\omega),59)$ $\varphi(\omega)\sim\mathcal{N}(54,60)$ | $(39,\varphi(\omega),61)$ $\varphi(\omega)\sim\mathcal{N}(56,11)$ | $(40,\varphi(\omega),62)$ $\varphi(\omega)\sim\mathcal{N}(55,33)$ |
| $\tilde{\bar{c}}_2(k)$ | $(0.55,\varphi(\omega),0.90)$ $\varphi(\omega)\sim\mathcal{N}(0.80,14)$ | $(0.62,\varphi(\omega),0.85)$ $\varphi(\omega)\sim\mathcal{N}(0.76,19)$ | $(0.59,\varphi(\omega),0.84)$ $\varphi(\omega)\sim\mathcal{N}(0.78,27)$ |
| $Q_{3,k}^{min}$ $(10^4 kg)$ | 500 | 490 | 490 |
| $Q_{3,k}^{max}$ $(10^4 kg)$ | 1900 | 1800 | 2000 |
| $\tilde{\bar{D}}_3(k)$ $(10^6 kg)$ | $(8.2,\varphi(\omega),15.1)$ $\varphi(\omega)\sim\mathcal{N}(12.6,18)$ | $(7.8,\varphi(\omega),14.7)$ $\varphi(\omega)\sim\mathcal{N}(12.9,20)$ | $(9.4,\varphi(\omega),15.6)$ $\varphi(\omega)\sim\mathcal{N}(13.1,14)$ |
| $\tilde{\bar{c}}_3(k)$ | $(0.54,\varphi(\omega),0.87)$ $\varphi(\omega)\sim\mathcal{N}(0.72,5)$ | $(0.62,\varphi(\omega),0.79)$ $\varphi(\omega)\sim\mathcal{N}(0.81,9)$ | $(0.55,\varphi(\omega),0.93)$ $\varphi(\omega)\sim\mathcal{N}(0.78,13)$ |
| $Q_{4,k}^{min}$ $(10^4 kg)$ | 11000 | 13000 | 11000 |
| $Q_{4,k}^{max}$ $(10^4 kg)$ | 24000 | 22000 | 23000 |
| $\tilde{\bar{D}}_4(k)$ $(10^7 kg)$ | $(15.1,\varphi(\omega),18.2)$ $\varphi(\omega)\sim\mathcal{N}(16.1,18)$ | $(14.4,\varphi(\omega),17.6)$ $\varphi(\omega)\sim\mathcal{N}(15.9,22)$ | $(15.5,\varphi(\omega),18.4)$ $\varphi(\omega)\sim\mathcal{N}(16.9,14)$ |
| $\tilde{\bar{c}}_4(k)$ | $(0.58,\varphi(\omega),0.89)$ $\varphi(\omega)\sim\mathcal{N}(0.75,7)$ | $(0.54,\varphi(\omega),0.88)$ $\varphi(\omega)\sim\mathcal{N}(0.69,13)$ | $(0.57,\varphi(\omega),0.94)$ $\varphi(\omega)\sim\mathcal{N}(0.77,12)$ |

**Table 5.6** Results of goodness-of-fit test on the collected data of Item 1 and estimated parameters for their normal distributions

| Model parameters | Sample size $\sum_{j=1}^{c} O_j$ | Number of classes $c$ | Chi-square test $\chi^2$ | $\chi^2_{c-2-1,\alpha}$ $(\alpha = 0.05)$ | Normal distribution parameters $\zeta$ (mean) | $v^2$ (variance) |
|---|---|---|---|---|---|---|
| $\tilde{\bar{T}}_1^{sw}$ | 113 | 11 | 2.634 | 15.507 | 0.18 | 1.2 |
| $\tilde{\bar{T}}_1^{wc}$ | 97 | 10 | 5.298 | 14.067 | 0.04 | 2.6 |
| $\tilde{\bar{D}}_1(0)$ | 106 | 12 | 9.676 | 16.919 | 51 | 12 |
| $\tilde{\bar{D}}_1(1)$ | 112 | 10 | 10.843 | 14.067 | 50 | 14 |
| $\tilde{\bar{D}}_1(2)$ | 89 | 10 | 3.953 | 14.067 | 54 | 22 |
| $\tilde{\bar{D}}_1(3)$ | 124 | 12 | 4.829 | 16.919 | 53 | 24 |
| $\tilde{\bar{D}}_1(4)$ | 83 | 10 | 9.278 | 14.067 | 57 | 32 |
| $\tilde{\bar{D}}_1(5)$ | 107 | 11 | 7.279 | 15.507 | 52 | 42 |
| $\tilde{\bar{c}}_1(0)$ | 129 | 12 | 11.578 | 16.919 | 0.67 | 8 |
| $\tilde{\bar{c}}_1(1)$ | 93 | 10 | 2.612 | 14.067 | 0.72 | 13 |
| $\tilde{\bar{c}}_1(2)$ | 88 | 10 | 5.354 | 14.067 | 0.67 | 31 |
| $\tilde{\bar{c}}_1(3)$ | 121 | 12 | 6.789 | 16.919 | 0.74 | 15 |
| $\tilde{\bar{c}}_1(4)$ | 118 | 10 | 10.854 | 14.067 | 0.71 | 25 |
| $\tilde{\bar{c}}_1(5)$ | 91 | 10 | 3.754 | 14.067 | 0.77 | 19 |

**Table 5.7** Option premium and conversion coefficient of construction materials

| Item index $i$ | Item name | Variable name | Variable | Month index $k=0$ | $k=1$ | $k=2$ | $k=3$ | $k=4$ | $k=5$ |
|---|---|---|---|---|---|---|---|---|---|
| 1 | Reinforcing steel | Option premium $(10^4 yuan)$ | $\alpha_1(k)$ | 43.71 | 45.46 | 41.32 | 46.18 | 42.54 | 51.02 |
| | | Conversion coefficient $(10^{-5}kg^{-1})$ | $\beta_1(k)$ | -3.57 | -5.95 | -4.23 | -4.39 | -3.95 | -5.04 |
| 2 | Cement | Option premium $(10^4 yuan)$ | $\alpha_2(k)$ | 5.03 | 4.95 | 4.37 | 5.22 | 4.51 | 4.97 |
| | | Conversion coefficient $(10^{-5}kg^{-1})$ | $\beta_2(k)$ | -6.28 | -5.59 | -5.68 | -6.23 | -6.74 | -5.89 |
| 3 | Fly ash | Option premium $(10^4 yuan)$ | $\alpha_3(k)$ | 2.92 | 3.07 | 3.15 | 2.78 | 3.14 | 2.97 |
| | | Conversion coefficient $(10^{-4}kg^{-1})$ | $\beta_3(k)$ | -2.94 | -2.51 | -3.08 | -3.30 | -2.76 | -3.18 |
| 4 | Aggregate | Option premium $(10^4 yuan)$ | $\alpha_4(k)$ | 1.21 | 1.43 | 1.55 | 1.32 | 1.18 | 1.47 |
| | | Conversion coefficient $(10^{-5}kg^{-1})$ | $\beta_4(k)$ | -1.79 | -1.93 | -2.05 | -2.13 | -1.89 | -2.09 |

the population size, iteration number, acceleration constant, initial velocity, and inertia weight, the most reasonable parameters for the four construction materials were identified. Note that the population size (i.e., the number of particles) determines the evaluation runs, and thus impacts the optimization cost (Trelea 2003 [40]). After some testing of the solution algorithm, 50 particles were selected as the population size and 100 times as the iteration number. The inertia weight $w(\tau)$ was set to be varying with iteration as follows:

$$w(\tau) = w(T) + \frac{\tau - T}{1 - T}[w(1) - w(T)] \tag{5.70}$$

where $\tau$ = iteration index $= 1, 2, \cdots, T$; $T$ = iteration limit. Through further experiments, $w(1) = 0.9$ and $w(T) = 0.1$ were found to be the most suitable for controlling the impact of the previous velocities on the current velocity and influencing the trade-off between the global and local experiences. The other parameters were selected by comparing the results with observations of the dynamic search behavior of the swarm. Since various learning factors, i.e., $c_p$ and $c_g$, may lead to little difference in the PSO performance (Trelea 2003 [40]), they were set at values 2 and

**Table 5.8** Parameter selection for DP-based PSO

| | | Population size $L$ | Iteration number $T$ | Acceleration constant $c_p$ | $c_g$ | Inertia weight $w(1)$ | $w(T)$ |
|---|---|---|---|---|---|---|---|
| | | 50 | 100 | 2 | 3 | 0.9 | 0.1 |
| **Type index $i$** | **Item name** | **Initial velocity $v_{l((i-1)\times 6+k+1)}(0)$** $k=0$ | $k=1$ | $k=2$ | $k=3$ | $k=4$ | $k=5$ |
| 1 | Reinforcing steel | 1300 | 900 | 1300 | 900 | 1300 | 900 |
| 2 | Cement | 1000 | 700 | 1000 | 700 | 1000 | 700 |
| 3 | Fly ash | 600 | 300 | 600 | 300 | 600 | 300 |
| 4 | Aggregate | 8 | 5 | 8 | 5 | 8 | 5 |

3 respectively in this study. The values for initial velocity were selected based on the magnitude of the decision variables. Table 5.8 summarizes all parameter values selected for the DP-based PSO for the computational experiments.

### 5.5.4 *Results and Sensitivity Analysis*

Based on the optimization method developed in this chapter, the inventory management problem of the concrete construction of works contract Nos. *I&IIB* in TGP was solved using the DP-based PSO. The results are shown in Table 5.9. The results were obtained based on the following parameter values, $\lambda = 0.5$, $\sigma_1 = \sigma_2 = \sigma_3 = \sigma_4 = 0.2$, $r_1 = r_2 = r_3 = r_4 = 0.7$, $\omega_1 = 0.4$, $\omega_2 = 0.3$, $\omega_3 = 0.2$, $\omega_4 = 0.1$.

**Table 5.9** Detailed results for the IMP-LCP of concrete construction of works contract Nos. *I&IIB* in TGP

| Item index $i$ | Item name | Variable name | Variable | Month index $k=0$ | $k=1$ | $k=2$ | $k=3$ | $k=4$ | $k=5$ |
|---|---|---|---|---|---|---|---|---|---|
| 1 | Reinforcing steel | Purchase quantity ($10^4 kg$) | $Q_1(k)$ | 6142.3 | 3579.2 | 6426.8 | 3438.2 | 6327.3 | 3629.4 |
| | | Inventory level ($10^4 kg$) | $I_1(k)$ | 243.18 | 1164.36 | 198.91 | 1365.85 | 177.32 | 828.37 |
| 2 | Cement | Purchase quantity ($10^4 kg$) | $Q_2(k)$ | 6453.2 | 2928.7 | 6284.6 | 3108.5 | 6184.7 | 3372.9 |
| | | Inventory level ($10^4 kg$) | $I_2(k)$ | 196.54 | 1734.66 | 216.94 | 1845.47 | 208.58 | 1552.32 |
| 3 | Fly ash | Purchase quantity ($10^4 kg$) | $Q_3(k)$ | 1592.3 | 531.6 | 1466.2 | 706.8 | 1628.7 | 633.2 |
| | | Inventory level ($10^4 kg$) | $I_3(k)$ | 68.93 | 494.63 | 57.03 | 462.53 | 97.03 | 473.82 |
| 4 | Aggregate | Purchase quantity ($10^7 kg$) | $Q_4(k)$ | 21.64 | 13.02 | 22.68 | 11.86 | 21.74 | 10.05 |
| | | Inventory level ($10^7 kg$) | $I_4(k)$ | 1.63 | 4.32 | 1.25 | 5.04 | 1.03 | 3.57 |
| Fitness value ($10^6 yuan$): | | 469.6 | | | | | | | |

To gain insight into the selection principle for the parameters including the optimistic-pessimistic index (i.e., $\lambda$), probability levels (i.e., $\sigma_i$), and possibility levels (i.e., $r_i$), a sensitivity analysis was conducted against these parameters (see Table 5.10). The decision makers can then fine tune these parameters to obtain different solutions under different parameter levels. The solutions reflect the different optimistic-pessimistic attitudes for uncertainty and the different predictions for the probability and possibility levels. It should be noted that the probability and possibility coefficients of different items can be different. But due to the limited space here, all $\sigma_i$ were set at the same value $\sigma$, and all $r_i$ had the same value as for $r$, for

**Table 5.10** Sensitivity analysis on the total fitness value of the IMP-LCP

| $\sigma$ | $\lambda=0$ | | | $\lambda=0.25$ | | | $\lambda=0.75$ | | | $\lambda=1.0$ | | |
|---|---|---|---|---|---|---|---|---|---|---|---|---|
| | $r=0.7$ | $r=0.8$ | $r=0.9$ | $r=0.7$ | $r=0.8$ | $r=0.9$ | $r=0.7$ | $r=0.8$ | $r=0.9$ | $r=0.7$ | $r=0.8$ | $r=0.9$ |
| 0.2 | 409.6 | 398.9 | 385.8 | 429.5 | 420.7 | 403.7 | 503.9 | 519.1 | 530.7 | 524.9 | 537.9 | 549.6 |
| 0.3 | 416.1 | 404.8 | 396.5 | 436.9 | 429.9 | 416.9 | 492.6 | 506.9 | 513.6 | 518.2 | 529.7 | 535.9 |
| 0.4 | 423.7 | 413.8 | 402.5 | 443.6 | 432.7 | 427.5 | 486.9 | 495.8 | 507.9 | 509.7 | 517.6 | 528.4 |
| 0.5 | 432.8 | 420.4 | 410.9 | 452.9 | 443.9 | 432.6 | 480.5 | 487.2 | 495.5 | 502.6 | 510.8 | 513.9 |
| 0.6 | 439.9 | 427.7 | 418.8 | 460.5 | 451.5 | 440.9 | 471.9 | 478.3 | 488.7 | 494.9 | 499.9 | 508.7 |
| 0.7 | 445.2 | 437.2 | 425.4 | 468.8 | 460.9 | 444.9 | 462.8 | 472.5 | 481.4 | 489.8 | 493.6 | 502.5 |
| 0.8 | 450.8 | 442.5 | 436.6 | 479.6 | 465.6 | 457.5 | 453.6 | 466.4 | 472.6 | 483.5 | 488.3 | 496.9 |

$i=1,2,3,4$, and $\omega_1=0.4$, $\omega_2=0.3$, $\omega_3=0.2$, $\omega_4=0.1$. Table 5.10 summarizes the different total fitness values for the IMP-LCP with respect to different parameters $\lambda$, $\sigma$, and $r$.

As shown in Table 5.10, under the same probability and possibility level, if parameter $\lambda$ grows larger, the fitness value increases. It should be noted that the optimistic-pessimistic parameter should be interpreted according to the real world problem. For the IMP-LCP, since the objective is to minimize the total cost over the entire construction duration, $\lambda=1$ is actually a pessimistic extreme, and $\lambda=0$, is actually an optimistic extreme. Thus, the following conclusion can be drawn: under the same probability and possibility levels, the result is more pessimistic if the optimistic-pessimistic parameter goes up. It will be the decision maker who decides the optimal result based on their attitude.

For probability level $\sigma$, it can be seen that under the same optimistic-pessimistic parameter and possibility level, when $\lambda<0.5$, if $\sigma$ grows larger, then the fitness value increases; when $\lambda>0.5$, if $\sigma$ grows larger, the fitness value decreases. On the other hand, for the possibility level $r$, the results also indicate that under the same optimistic-pessimistic parameter and probability level, when $\lambda<0.5$, if $r$ grows larger, the fitness value decreases; when $\lambda>0.5$, if $r$ grows larger, the fitness value increases. These results are quite useful and may serve as a reference for decision-makers as it is their choice to identify an appropriate set of parameter values (i.e., $\lambda$, $\sigma$, and $r$) to optimize the decision making process.

### *5.5.5 Results Comparison with Standard PSO*

The optimization results for the IMP-LCP using a DP-based PSO were compared with the standard PSO. To carry out the comparison in a similar environment, the parameters stated in Table 5.8 were also adopted for the standard PSO; that is, population size = 50, iteration number = 100, the two acceleration constant $c_p$ and $c_g$ were set at 2 and 3 respectively, the inertia weight was the same as in Eq. (5.70), and $w(1)=0.9$, $w(T)=0.1$. The other parameters for the standard PSO were selected as follows: (1) initial velocities for the decision variables were the same as for the DP-based PSO; and (2) the initial velocities for the state variables were selected based on the magnitudes of the inventory level for every item.

Table 5.11 shows the comparison results, i.e., minimal fitness value, iterations to find optima and computation time. The convergence histories (i.e., the fitness value corresponding to the global best particle-represented sequence varies with iterations or generations) for the DP-based PSO and standard PSO for the practical application example were compared. The comparison results and convergence histories (Figure 5.13) were based on the average of the optimal results from 50 computational runs (for the DP-based PSO and the standard PSO respectively) that were able to obtain the minimal fitness value (i.e., 469.6), excluding the locally trapped particles.

**Table 5.11** Comparison results between DP-based PSO and standard PSO

| Approach | Minimal fitness value | Iterations to find optima | Computation time (s) |
|---|---|---|---|
| DP-based PSO | 469.6 | 49 | 683 |
| Standard PSO | 469.6 | 81 | 1456 |

This demonstrated that the DP-based PSO performed better than the standard PSO. The convergence histories in Figure 5.13 indicate that: (1) the DP-based PSO converges a little faster, that is, it requires fewer iterations than the standard PSO to find the optima; and (2) the DP-based PSO is more stable than the standard PSO in searching for optima. Because a dynamic programming model iterative mechanism is used to avoid redundant feasible check procedures for the state variables during both the initializing and updating process in each iteration, the DP-based PSO showed an improved search performance compared to the standard PSO under a similar environment.

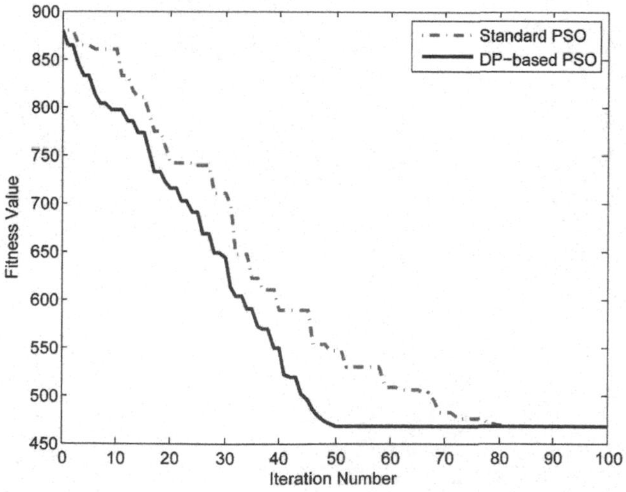


**Fig. 5.13** Convergence histories of DP-based PSO and standard PSO

# 6

# Fuzzy Random MOMSDM for Facilities Planning

Facilities planning problems (FPP) have two important parts; site layout planning and facility location. This chapter focuses on the current frontiers of these two problems, i.e., the dynamic construction site layout planning (DCSLP) and the dynamic temporary facilities location problem (DTFLP). Construction site layout planning is a fundamental task for any project undertaking. In practice, most construction site layout planning problems are dynamic, multi-objective and uncertain in nature. To enhance the general practice of dynamic construction site layout planning, this chapter looks at the development of a fuzzy random multi-objective decision making model. In this model, two objectives are considered: (1) the minimization of the total site layout cost; and (2) the maximization of the distance between 'high-risk' facilities and 'high-protection' facilities to reduce the possibility of safety or environmental accidents. After stating the problem and presenting the mathematical formulation, this chapter uses fuzzy random variables to propose an equivalent crisp model. From the characteristics of the proposed model, a multi-objective particle swarm optimization algorithm (MOPSO) with permutation-based representation is proposed to solve the problem. This approach is then applied to the Longtan hydropower construction project to illustrate the effectiveness of the proposed model and algorithm. This chapter comprehensively discusses multi-objective, dynamic and fuzzy random problems in construction site layout planning. Construction site layout planning in reality tends to be dynamic and uncertain in nature and conflicting objectives need to be dealt with. Therefore this chapter proposes a fuzzy random multi-objective decision making model for dynamic construction site layout planning problems. As pointed out by Mawdesley and Al-Jibouri (2003) [85], site layout planning research intends to concentrate on four main areas in the future: (1) Developing more suitable models; (2) Extending existing models to include a time element (dynamic layout); (3) Adding uncertainty; (4) Adding multiple criteria for evaluation and study into special cases for specific types of problems. This chapter offers an exploration in these four areas by investigating multi-objective dynamic construction site layout planning and the dynamic temporary facilities location problem under a fuzzy random environment using appropriate case studies.

J. Xu and Z. Zeng, *Fuzzy-Like Multiple Objective Multistage Decision Making*, Studies in Computational Intelligence 533,
DOI: 10.1007/978-3-319-03398-3_6, 

## 6.1 Statement of FPP

Facilities planning has taken on a whole new meaning in the past 10 years [151]. In the past, facilities planning was primarily considered to be a science. However, in today's competitive global marketplace, facilities planning is a strategy, as it is critical in many important industries, such as construction, manufacturing, mining, air and other transportation, public utilities, and communications. This section provides a comprehensive statement of the facilities planning problem (FPP).

### 6.1.1 Significance of FPP: DCSLP and DTFLP

Facilities planning can be divided into two; site layout planning and facilities location. The relationship structure for facilities planning is shown in Figure 6.1.

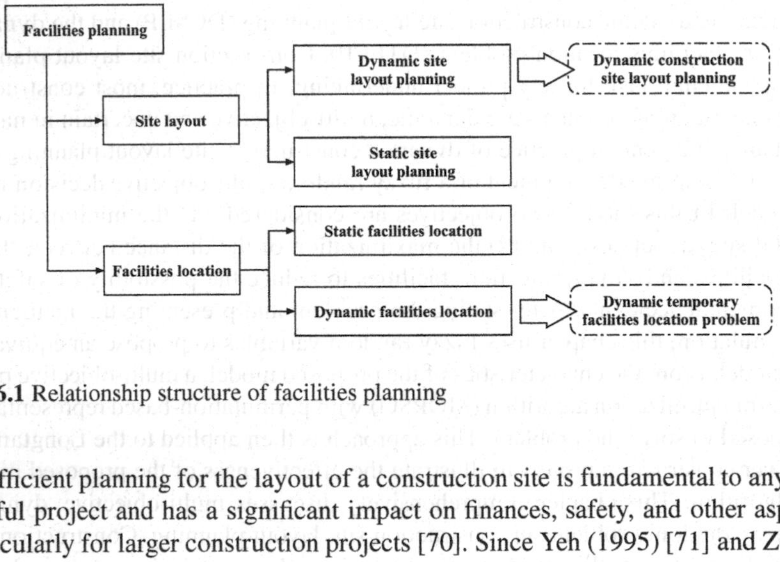


**Fig. 6.1** Relationship structure of facilities planning

Efficient planning for the layout of a construction site is fundamental to any successful project and has a significant impact on finances, safety, and other aspects, particularly for larger construction projects [70]. Since Yeh (1995) [71] and Zouein and Tommelein (1999) [72] identified the construction site layout problem, significant research has been conducted in this field. A review of this problem can be found in Meller and Gau (1996) [73]. This section presents a statement of a dynamic construction site layout planning (DCSLP) problem. Site layout problems can be formulated as a quadratic assignment problem (QAP) with the aim of achieving an optimal assignment of L facilities to L predetermined locations [74]. Previous research has mainly involved only a single objective function, e.g. [75, 76]. However, an optimal layout often demands the fulfillment of several competing and often conflicting design objectives, such as operational efficiency maximization, maintaining good employee morale, and minimizing travel time and distance for the movement of resources to decrease material handling time. Besides these concerns, legal obligations may also impose safety and permit constraints [77]. As stated by

Turskis et al. (2009) [78], the complex nature of decision-making in construction design and management requires practitioners to make decisions based on a wider variety of policy considerations in addition to cost benefit analysis and purely technical considerations. Previous construction site layout planning research has tended to concentrate on static problems, e.g. [71, 79]. In static construction site layout problems, the facilities serviced in the different construction phases in accordance with the requirements of the construction work during a construction project are assumed to be the same [80] and the possibility of site space reuse to accommodate different resources at different times, the relocation of resources, and the varying resource space needs over time are ignored [81]. An increasing number of studies focus on solving dynamic construction site layout planning problems. For instance, Baykasoglu et al. (2006) [82] made the first attempt to show how an Ant Colony Optimization Algorithm can be applied to dynamic construction site layout planning problems with the budget constraints. Ning et al. (2010) [81] used a continuous dynamic searching scheme to guide the max-min ant system algorithm towards a solution for a dynamic construction site layout planning problem, and further developed a computationally decision-making system to solve a dynamic, multi-objective and unequal-area construction site layout planning problem [84].

Facilities location is extremely important for construction projects. Take hydropower projects as an example. The Jinping-II Hydropower Station is symbolic of large-scale construction in the West-East Electric Transmission Project of China and has the largest hydraulic tunnel in the world. The characteristics of the Jinping-II Hydropower Station represent the characteristics of all large scale hydropower construction projects which are considered to be complex with uncertain environments, are of long duration and have high costs (Gang and Xu 2010 [137]; Xu et al. 2013b [586]). The facilities location planning management of a large scale hydropower construction is extensive in China, so effective facilities location planning is of utmost importance for the success of hydropower engineering construction projects. However, selecting the correct planning method is often difficult because of the influence of the many critical factors, such as technical factors, economic factors, safety factors and social factors. The Jinping-II Hydropower Station has adopted a lean management system to assist in making an optimal decision for facilities location planning to reduce cost and control project duration. Up to now, there have been few helpful applications for facilities location problems for large scale hydropower construction projects (Nicklow et al. 2010 [146]). This lack of applications is the stimulus for developing and presenting a facilities location problem case study at the Jinping-II Hydropower Station.

The environment of a large scale hydropower construction project is that a transportation condition is given and the available space for locations are discrete and limited. Construction projects are becoming increasingly more complex and dynamic in their nature. The duration of the project is often divided into several stages (Xu and Zeng 2011 [242]; Xu and Zhang 2012 [154]). According to the construction work requirements over the entire construction project, the location of temporary facilities changes in different stages. With this in mind then, it is necessary to consider discrete location planning (on networks) models (Wolfgang and

Gabriela 1997 [153]), dynamic models (Ning et al. 2010 [81]; Rosenblatt 1986 [148]) and quadratic assignment problems (Koopmans and Beckmann 1957 [139]). Thus, the facilities location planning and layout problem is discrete and dynamic on a given transportation network, which is called the Dynamic Temporary Facilities Location Problem (DTFLP).

### 6.1.2 Problem Description of DCSLP

Construction site layout deals with the assignment of appropriate site locations for temporary facilities such as warehouses, site offices, workshops, and batch plants. The scheduling of activities and space considerations for multiple sources is the basis of the site layout problem. Project managers, superintendents and subcontractors jointly agree on the facilities locations using past experience, trial and error, insight, preference, common sense and intuition [91]. Construction site layout can be delimited as a design problem for the arrangement of a set of predetermined facilities on the site, while satisfying a set of constraints and optimizing the layout objectives. The effective placement of facilities within the site is significantly influenced by the movement of resources or the interactions between the facilities [81].

In dynamic construction site layout planning problems, the facilities serviced are different in different construction phases in accordance with the construction work requirements. For example, in Figure 6.2, the construction project lasts for $n$ years, and according to the facilities requirements, the complete project can be divided into $t$ phases. In phase 1, six facilities are setup, located in six of eight locations; in phase 2, facilities 1, 2, 6 are still open, facility 4 is closed, whilst facility 8 changes. In the last phase, only four facilities remain open.

Four quantitative factors: material flows (MF), information flows (IF), personnel flows (PF) and equipment flows (EF) and two qualitative factors: safety/environmental concerns (SE) and users' preference (UP) are usually considered in a construction site layout planning problem. According to [81], these six factors are defined as:

(1) MF: the flow of parts, raw materials, works-in- process and finished products between departments. The MF can be measured as unit per time unit.

(2) IF: the communication (oral or reports) between facilities. IF can be measured using a personnel survey and can be expressed as the number of communications per time unit.

(3) PF: the number of employees from one or both facilities that perform tasks from one facility to another.

(4) EF: EF is defined as the equipment number needed for materials handling (trucks, mixers, etc.), which are used to transfer materials between facilities.

(5) SE represents the level of safety and environmental hazards, measured by the safety concerns, which may arise when two facilities are close to each other, and may affect site workers by an increase in the likelihood of accidents, excessive noise, uncomfortable temperature or pollution.

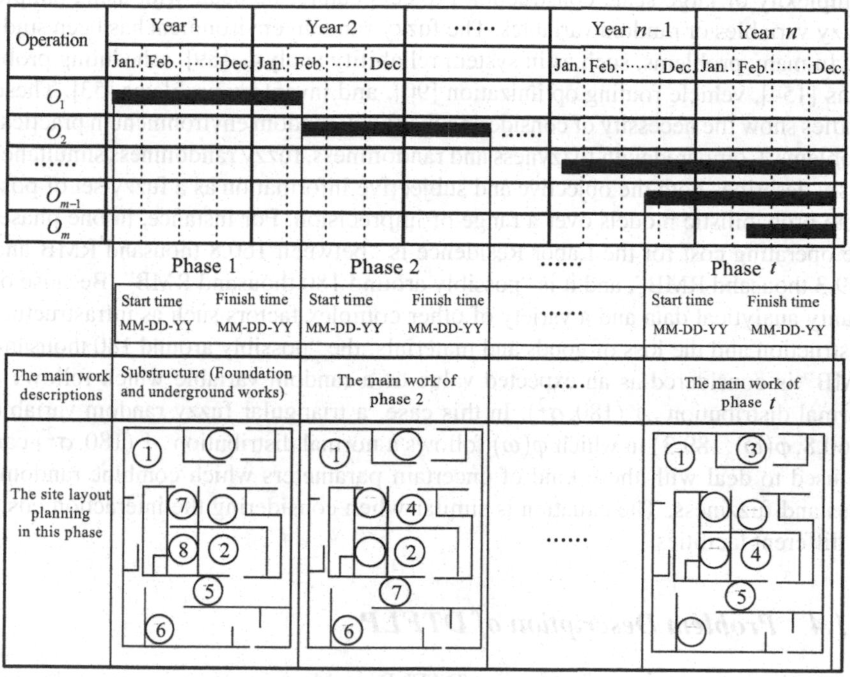


**Fig. 6.2** An example of dynamic construction site layout planning problem

(6) UP represents the project manager's desire to have the facilities close to or apart from each other.

This chapter considers the MF, IF, PF, EF and SE in the mathematical model, and considers the UP (users' preference) in the algorithm design.

### *6.1.3 Fuzzy Random Environment for DCSLP*

In practice, there are two types of uncertainties in layout planning problems [86]: the first uncertainty is subjective and internal, such as the decision makers' perception and dissension. The second uncertainty is objective and is affected by external forces, such as uncertainties in demand, product price, product mix, equipment breakdown, task times and queuing delays. In these two kinds of uncertainties, objective information can be dealt with using randomness and subjective information can be dealt using fuzziness. Current uncertainty research in construction site layout planning problems concentrates on fuzzy uncertainty, e.g. [87, 88]. In reality, because both objective and subjective information exist, both fuzziness and randomness should be considered.

In the real world, there is often a hybrid uncertain environment in construction projects, especially in large scale construction projects. The imprecision and complexity of large scale construction projects cannot be dealt with using simple fuzzy variables or random variables. The fuzzy random environment has been studied in many problems, such as in system reliability analysis [89], scheduling problems [154], vehicle routing optimization [90], and inventory problems [53]. These studies show the necessity of considering the fuzzy random environment in practical problems. Compared with fuzzyness and randomness, fuzzy randomness simultaneously describes both the objective and subjective information as a fuzzy set of possible probabilistic models over a range of imprecision. For instance, in one phase, the operating cost for the Labor Residence is "between 160.8 thousand RMB and 189.3 thousand RMB", and it is "possibly around 180 thousand RMB". Because of scanty analytical data and a variety of other complex factors such as infrastructure destruction and the loss of goods and materials, the "possibly around 180 thousand RMB" is considered as an expected value of a random variable which follows a normal distribution $\mathcal{N}(180, \sigma^2)$. In this case, a triangular fuzzy random variable $(160.8, \varphi(\omega), 189.3)$ in which $\varphi(\omega)$ follows a normal distribution $\mathcal{N}(180, \sigma^2)$ can be used to deal with these kind of uncertain parameters which combine randomness and fuzziness. The situation is similar when considering the interaction costs of different facilities.

### 6.1.4 Problem Description of DTFLP

The hydropower construction project DTFLP problem is a temporary production (i.e. the location plan and resources organization) system. The system is structured to deliver the product while maximizing value and minimizing waste. According to Glenn and Gregory (2003) [136], this can be called a "lean" project. Lean project management differs from traditional project management not only in the goals it pursues, but also in the structure of its phases, the relationship between the phases and the participants in each phase. These phases in the DTFLP problem not only refer to the different stages but also identify the different value streams, namely the material flows, information flows, personnel flows, equipment flows and safety/environmental concerns.

The difficulty in visualizing the value streams in a construction project hampers the management of the construction project. Large hydropower construction projects typically involve multiple discrete facilities which suffer from waste problems that arise in the relocation and unnecessary movement in the materials handling. Achieving minimal waste in the DTFLP problem requires not only appropriate construction planning, but also effective management. Lean thinking applied to construction has led to the development of planning and control systems and other practices that seek to improve this management (Rafael et al. 2010 [149]).

The research here deals with the selection of a more efficient location plan and an improvement in the ability to modify the movable facilities locations in various project stages in order to achieve an efficient and effective operation of the overall site layout. The temporary construction facilities in our DTFLP problem refer to the movable facilities that are able to be relocated at the start of any of the

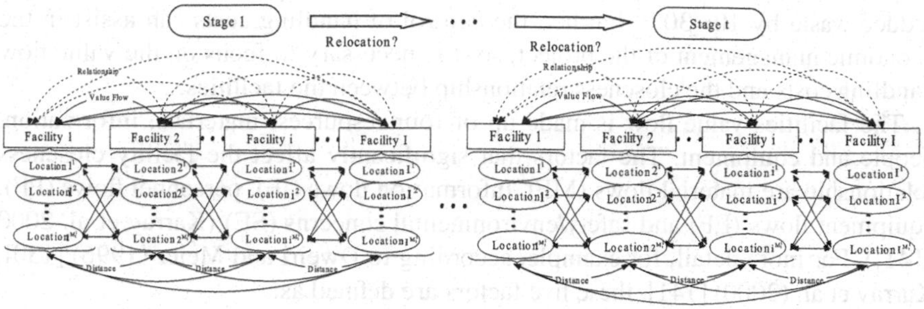


**Fig. 6.3** The DTFLP problem

identified project stages. For the large scale hydropower construction project DT-FLP problem, the duration is divided into $T$ stages. There are $I$ kind of temporary facilities and $M_i$ possible locations for each facility $i$ in a stage. The decision makers need to assign facilities to corresponding locations based on the requirements of the different stages. $x_{imk}(t) = \{0,1\}$ can be used for the decision variables which represent the moving of the temporary facility $i$ ($i = 1,2,\cdots,I$) from location $m$ to $k$ ($m,k = 1,2,\cdots,M$). Each of these facilities is associated with a corresponding value flow and relationship, see Figure 6.3. The closeness relationship is a comprehensive selection routine that is used by the decision makers to determine the layout of the facilities. The level of the closeness relationship is related to the value flow of the resources among the facilities. Empirically, for instance, the concrete system is closely linked to the cement/sand/aggregate storage yard, so the corresponding level of the closeness relationship is high. Similarly, the labor residence has close contact with the carpentry shop and the office. For operational management convenience, these two facility pairs need a proximity distance. On the contrary, if the safety/environmental concerns between the oil depot and the explosives storage are likely to cause safety and environmental accidents, they are identified as "very hazardous" facilities, so these facilities need to be located far from each other. In practice, however, for typical extensive hydropower station project management in China, the arbitrary closeness relationship numerical ratings are usually assigned through a relationship chart, which often leads to a significant waste of resources. Therefore, in this chapter an optimized location solution is sought.

Location optimization required an efficient arrangement of the facilities within the available area. The effective placement of facilities within the site is significantly influenced by the movement of resources, or basically the interactions, or in other words, the closeness relationship between the facilities which strongly affects effective and efficient project management. In the location planning process, many objectives need to be considered to effectively utilize resources, to reduce costs, and to ensure efficient project management. The movement of resources between the facilities during the construction project make up the value flow and handling costs in the DTFLP problem. Tompkins et al. (2010) [151] estimated that 20~50% of manufacturing costs are due to parts handling and so an optimal arrangement could reduce waste by 10∼30%. Further, the control of handling costs can assist in the economic management of the project, so it is necessary to focus on the value flow handling costs and the closeness relationship between the facilities.

The facilities value flow is made up of four resources; materials, information, people and equipment. The factors that significantly affect the facility closeness relationship are material flows (MF), information flows (IF), personnel flows (PF), equipment flows (EF) and safety/environmental concerns (SE) (Karray et al. 2000 [141]). For more detail, for example, according to Dweiri and Meier (1996) [130], Karray et al. (2000) [141], these five factors are defined as:

(1) Material Flows (MF): the flow of parts, raw materials, works-in-process and finished products between departments which can be measured by unit per time unit.
(2) Information flows (IF): the communication (oral or reports) between facilities. IF can be measured through surveys with involved personnel and can be expressed by the number of communications per time unit.
(3) Personnel flows (PF): the number of employees from one or both facilities that perform tasks from one facility to another.
(4) Equipment flows (EF): EF is defined by the material handling equipment numbers (trucks, mixers, etc.) used to transfer materials between facilities.
(5) Safety/environmental concerns (SE): represent the level of safety and environmental hazards, measured by the safety concerns which may arise when two facilities are close to each other, and which may affect site workers by an increase in the likelihood of accidents, noise, uncomfortable temperatures, and pollution.

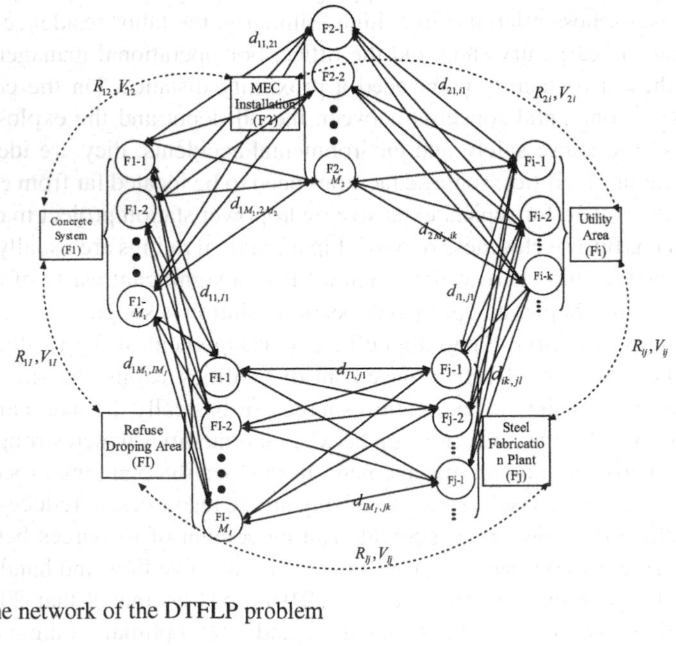


**Fig. 6.4** The network of the DTFLP problem

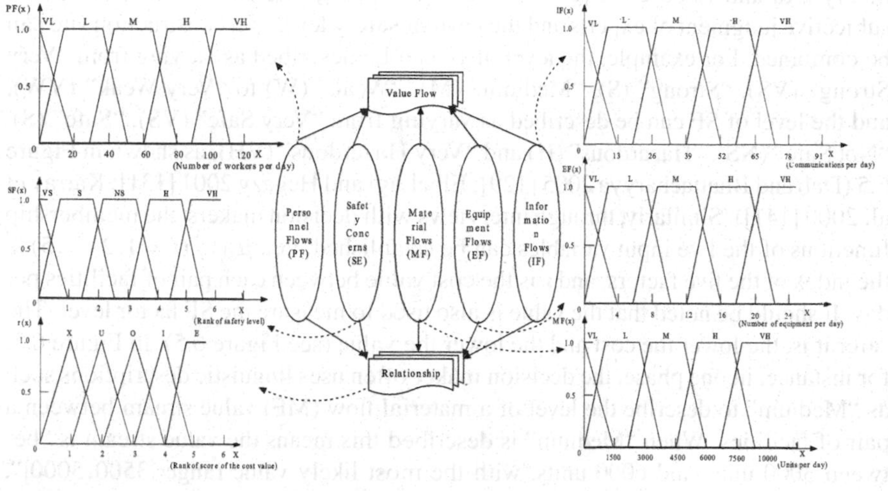

**Fig. 6.5** Variables fuzzification and the membership functions

In a large scale hydropower construction project network, there are many temporary facilities such as site offices, test laboratories, storage areas, fabrication areas, and rest areas. As the physical model shows in Figure 6.4, there are $M_1$, $M_2$, $M_i$, $M_j$ and $M_l$ possible locations for the Concrete System, MEC Installation, Utility Area, Steel Fabrication Plant and Refuse Dropping Area respectively. Each pair of facilities has three main factors: $d_{ik,jl}$ (distance between location $k$ of facility $i$ and location $l$ of facility $j$). Based on this network, the mathematical model can be built.

### 6.1.5 Fuzzy Environment for DTFLP

There is often inadequate data or imprecise information available and the nature of the construction ensures that uncertainties are inherent in every condition (Emre et al. 2009 [132]). Due to this complexity, construction managers often perform decision making tasks using previous experience, which can lead to ambiguity and even inefficiency (Michael and Saad 2003 [366]; Maravas and Pantouvakis 2012 [145]). Therefore, because these five factors are vague, fuzzy theory is needed (Xu and Zhou 2011 [281]; Zadeh 1965 [22]) as fuzzy measures can reduce vagueness (Xu and Zhou 2011 [281]; Zadeh 1965 [22]). These factors are calculated using the following process.

To define the fuzzy variables for these five factors, the most important consideration is the membership functions. The assignment of membership functions is subjective in nature as they are generally determined from human expertise. In real-world construction projects, therefore, project managers, resident engineers, and site superintendents can provide planners with the required data to establish the membership functions for the input variables (Karray et al. 2000 [141]). Using fuzzy theory (Xu and Zhou 2011 [281]; Zadeh 1965 [22]), historical accident data, the subjective judgment of experts and the current safety level of a construction site can be combined. For example, the level of IF can be described as varying from "Very Strong" (VS), "Strong" (S), "Medium" (M), "Weak" (W) to "Very Weak" (VW), and the level of SF can be described as varying from "Very Safe" (VS), "Safe" (S), "Not Safe" (NS), "Hazardous" (H) and "Very Hazardous" (VH), as shown in Figure 6.5 (Deb and Bhattacharyya 2005 [129]; Elbeltagi and Hegazy 2001 [131]; Karray et al. 2000 [141]). Similarly, through interviews with decision makers, the membership functions of the five input variables can be established, i.e., $\mu_s(v)$ $(s = 1, 2\cdots, 5)$ is the index of the five factors, and $v$ is the cost value between each pair of facilities per day. It should be noted that the value is also used to measure the SF factor level. The safer it is, the lower the cost and the lower the value (see Figure 6.5). In Figure 6.5, for instance, in one phase, the decision maker often uses linguistic descriptions such as "Medium" to describe the level of a material flow (MF) value stream between a pair of facilities. When "Medium" is described this means the value stream is "between 3000 units and 6000 units, with the most likely value range $[3500, 5000]$". This range is different for every decision maker because of their different preferences, experiences and the lack of detailed data. In this case, trapezoidal fuzzy numbers can be used $(1600, 3000, 4500, 6000)$ to deal with these kind of uncertain parameters. The level, namely "Medium" is considered a trapezoidal fuzzy number, because the range of the mean, left and right borders is available through surveys and analysis. The situation is similar with a "Low" and "High" level MF.

Since the level is considered a trapezoidal fuzzy number (i.e., $(l, a, b, m)$), it should be noted that the mean (i.e., $[a, b]$), the left border (i.e., $l$) and the right border (i.e., $r$) of the fuzzy number are obtained based on the experience of the engineers and an analysis of the collected data. To collect equipment failure rate data, interviews were conducted with a number of experienced engineers (i.e., $q = 1, 2, \cdots, E$, where $q$ is the index of engineers) who presented the linguistic terms based on observed data over time. Generally, the view of each engineer can be denoted as a range, (i.e., $[l_q, r_q]$) and a most likely range of values (i.e., $[a_q, b_q]$). Since different engineers have different views on equipment failure rate, the minimal value of all $l_q$ and the maximal value of all $r_q$ (for $q = 1, 2, \cdots, E$) are selected as the left border (i.e., $l$) and the right border (i.e., $r$), respectively. The minimal value of the expected range for all the $a_q$ (for $q = 1, 2, \cdots, E$) is regarded as the left border of the mean range (i.e., $a$). The maximal value of the expected range for all the $b_q$ (for $q = 1, 2, \cdots, E$) is regarded as the right border of the mean range (i.e., $b$). In this way, the membership function of the trapezoidal fuzzy number can be constructed, namely "Low", "Medium" and "High". For example, the membership function of the "Medium" level of MF (i.e. $s = 1$) is:

$$\mu(v)_s = \begin{cases} \frac{1}{1400}(v + 1600) & v \in [1600, 3000) \\ 1 & v \in [3000, 4500) \\ -\frac{1}{1500}(v - 6000) & v \in [4500, 6000) \end{cases} \tag{6.1}$$

Similarly, when it comes to the "Very Low" level, the value is "between $l$ and $r$, with the most likely value $l$". For example, the membership function of the "VL" level of MF (i.e. $s = 1$) can be stated as in the following equation:

$$\mu(v)_s = \{ -\frac{1}{500}v + 1 \; v \in [0, 1500) \tag{6.2}$$

As to the "Very High" level of MF, this can be described as "between $l$ and $+\infty$, with a most likely range of $[r, +\infty)$". The membership function of the "VH" level of MF (i.e. $s = 1$) can be stated as in the following equation:

$$\mu(v)_s = \begin{cases} \frac{1}{1000}(x - 6700) & v \in [6700, 7700) \\ 1 & v \in [7700, +\infty) \end{cases} \tag{6.3}$$

Therefore, all the membership functions for the fuzzy variable MF have been established. In the same way, the functions for the other four fuzzy variables (IF, PF, EF and SE) of the different facilities can be obtained.

The next process is to calculate the closeness relationship based on a five value flow. In practice, the closeness relationship is divided into five levels that range from "A", "E", "I", "O", "U". Based on the above five input variables, the if-then rule is often employed by many researchers to imitate the decision makers' decision processes regarding the membership functions for the fuzzy sets. However, in a large scale hydropower construction project, to ensure lean management, it is necessary to consider all five factors over the five levels, so the fuzzy decision system needs $5^5 = 3125$ rules. Because these kinds of decision rules are often very difficult to make, it is necessary to find an alternative for the determination of the relation between the five input factors and the closeness relationships. An intuitively developed strategy to find the rating for each move based on the value of their relationships can be summarized in the following process.

Each of these variables is assumed to have a weight factor that affects the closeness between each pair of facilities (Saaty 1980 [150]). In much of the previous research, the weight was set by the decision makers based on their experience and preferences. However, this can often lead to subject uncertainty. Here, the cost value weight $w_s$ is used as the coincidence indicator to balance these factors. The cost value weight is obtained using a modification of the defuzzification methodology (Gentile et al. 2003 [134]). $w_s$ is obtained as:

$$w_s = \frac{\int v\mu_s(v)dv}{\int \mu_s(v)dv} / \sum_{s=1}^{5} \frac{\int v\mu_s(v)dv}{\int \mu_s(v)dv}, \quad \sum_{s=1}^{5} w_s = 1, \tag{6.4}$$

where $s$ is the index of the five factors (namely, $s = 1, 2 \cdots, 5$ means MF, IF, PF, EF and SE respectively), and $v$ is the cost value between each pair of facilities per day. Thus, the function of the input variables and the output of the closeness relationship can be stated as:

$$\mu(r) = \sum_{s=1}^{5} w_s \mu_s(v), \tag{6.5}$$

where $r_{ij}$ is the total score of the cost value, $s$ is the index of the 5 factors (namely, $s = 1, 2 \cdots, 5$ means MF, IF, PF, EF and SE respectively), and $v$ is the cost value of interaction flow. For instance, the membership function for level "A" for the closeness relationship is as in the following:

$$\mu(r) = \begin{cases} r & r \in [0,1) \\ 1 & r = 1 \\ -r+2 & r \in (1,2) \end{cases} \tag{6.6}$$

Zétényi (1988) [158] pointed out that psychologists generally consider a good representation of a fuzzy set using its expected value. This method is used here to calculate the closeness relationship value of each pair of facilities, so the crisp output rating of the closeness relationship is equal to:

$$R = \frac{\int_{-\infty}^{+\infty} \mu(r) r \mathrm{d}r}{\int_{-\infty}^{+\infty} \mu(r) \mathrm{d}r}, \tag{6.7}$$

where $R$ is the final crisp rating, and $\mu(r)$ is the membership value of $r$, which can be obtained using Eq. (6.5). Using this method, these fuzzy input coefficients can be transformed to a determined value.

## 6.2 Modelling Process of DCSLP

The construction site layout problem is a type of QAP, which requires an equal number of facilities and locations. If the number $m$ of facilities is less than the number $n$ of locations, $n-m$ dummy facilities are created and a zero set-up and transportation cost assigned to them. If there are fewer locations than facilities, the problem is infeasible. The goals for the solution to dynamic construction site layout problems is to minimize the cost of both each facility and the interaction between the different facilities, as well as to ensure that the distance between some facilities is optimized to reduce the possibility of safety or environmental accidents. The facilities and locations are interrelated through the use of area constraints and logical constraints. For example, different facilities cannot be in the same location at the same time, and each facility can be located in only one location in one period. After presenting the mathematical formulation, fuzzy random variables are introduced to develop an equivalent crisp model. To model dynamic construction site layout planning problems under a fuzzy random environment, the following assumptions are made:

### *6.2.1 Assumptions*

To model the dynamic construction site layout planning problems under fuzzy random environment, the following assumptions are made:

(1) All possible locations for different facilities can be identified.

(2) $F$ facilities are to be positioned on a site, $L$ locations are available for each facility to position, $L \geq F$.

(3) The interaction cost and the facilities operating costs are regarded as fuzzy random variables.

(4) All the resource demands at each facility during every period can be satisfied.

(5) For different facility assignments, there are different set-up and removal costs.

(6) The possibility of safety or environmental accidents and the possible loss are proportional to the distance between 'high-risk' facilities and 'high-protection' facilities.

To formulate the model, the indices, variables, certain parameters, uncertain variables, and decision variables are as follows:

**Indices:**

$x,y$ : different types of site-level facilities index, $x,y \in \{1,2,\cdots,F\}$. Among those, 'high-risk' facilities which may cause safety or environmental accidents, are marked as $x^k$, 'high-protection' facilities which are vulnerable and are likely to suffer great losses, are marked as $y^r$,

$k$ : 'high-risk' facilities index, $k \in \{1,2,\cdots,K\}$,

$r$ : 'high-protection' facilities index, $r \in \{1,2,\cdots,R\}$,

$i,j$ : locations index, $i,j \in \{1,2,\cdots,L\}$,

$t$ : periods for the problem index, $t \in \{1,2,\cdots,T\}$.

**Variables:**

$\eta_{xit}$ : denotes the opening of facility $x$ at location $i$ during period $t$,

$\omega_{xit}$ : denotes the closure of facility $x$ at location $i$ during period $t$.

**Certain parameters:**

$\alpha_t$ : an appropriate discount rate,

$A_{ij}$ : the distance from location $i$ to location $j$,

$C^s_{xit}$ : the startup cost of facility $x$ at location $i$ during time period $t$,

$C^c_{xit}$ : the closure cost of facility $x$ at location $i$ during time period $t$,

$S_{xt}$ : the area required for facilities $x$ during time period $t$,

$D_i$ : the area of the location $i$,

$w_{kr}$ : the risk weight of 'high-risk' facility $x^k$ near 'high-protection' facility $y^r$.

**Uncertain variables:**

$\tilde{\bar{C}}^z_{xit}$ : the operating cost at location $i$ during time period $t$ for facility $x$,

$\tilde{\bar{C}}_{xyt}$ : the interaction cost per unit distance of facility $x$ and facility $y$ for unit -distance during time period $t$.

**Decision variables:**

$\delta_{xit}$ : denotes the existence of facility $x$ at location $i$ during period $t$, with the initial condition of $\delta_{xi0} = 0$, $\delta_{xj0} = 0$,

$\delta_{yit}$ : denotes the existence of facility $y$ at location $i$ during period $t$, with the initial condition of $\delta_{yi0} = 0$, $\delta_{yj0} = 0$.

Based on the assumptions and notations above, here a fuzzy random multi-objective decision making model for dynamic construction site layout planning problems is proposed.

### 6.2.2 *Objective Functions*

Various layout plans are evaluated to determine the optimum. From the six construction site layout planning factors, two objectives functions are derived.

**The Total Cost of Site Layout Planning**

The cost for each single facility is composed of three parts, the setup cost, the closure cost and the operating cost. Since $\eta_{xit}$ is the opening of facility $x$ at location $i$ during phase $t$, $C^s_{xit}$ is the startup cost. $\eta_{xit} = 1$ denotes facility $x$ is setup at location $i$ during phase $t$, and $C^s_{xit}$ is a certain variable forecast by the construction site layout managers, so that $C^s_{xit}\eta_{xit}$ denotes the opening cost. Similarly $C^c_{xit}\omega_{xit}$ denotes the closure cost, and $\bar{\tilde{C}}^z_{xit}\delta_{xit}$ denotes the facilities operating cost.

It is not enough to just minimize the cost for each single facility. There are imperative MF, IF, PF, EF between the different facilities. An optimum dynamic site layout plan minimizes the interaction costs between these facilities. In phase $t$, if facility $x$ is located in location $i$, facility $y$ is located in location $j$ and there are interaction activities between $x$ and $y$. Frequently, the future conditions are difficult to predict. Thus, as per the assumptions in the previous section, the interaction cost of $x$ and $y$ for unit-distance is a fuzzy random variable, marked as $\bar{\tilde{C}}_{xyt}$. Since $A_{ij}$ is the distance from $i$ to $j$, the interaction cost between $x$ and $y$ during phase $t$ is $\delta_{xit}\delta_{yjt}A_{ij}\bar{\tilde{C}}_{xyt}$.

In addition, for a large scale construction project, the time duration can be lengthy. Some construction projects last several years or even more than ten years, so it is necessary to consider fund time value. $a_t$ denotes an appropriate discount rate to determine the net present value.

Considering the fund time value, the total cost objective $C$ can be expressed as minimizing the following:

$$\sum_{t=1}^{T}\sum_{x,y=1}^{F}\sum_{i,j=1}^{L}\alpha_t\left(C^s_{xit}\eta_{xit} + C^c_{xit}\omega_{xit} + \bar{\tilde{C}}^z_{xit}\delta_{xit} + \delta_{xit}\delta_{yjt}A_{ij}\bar{\tilde{C}}_{xyt}\right). \tag{6.8}$$

**Safety and Environmental Objective**

The risk of a fatality in the construction industry is five times more than that in a manufacturing based industry, and the risk of a major injury is two and a half times higher [92]. Safety and environmental issues are therefore extremely important when designing construction site layout. A well-planned and well-run project should be both safe and efficient to save lives and money, and reduce injury and ill-health.

The 'high-risk' facilities which may cause safety or environmental accidents are marked as $x^k$, for example, oil depots, explosives storage and dangerous chemicals storage. The 'high-protection' facilities are vulnerable facilities which could suffer great losses if these accidents happen. From previous studies it is known that the nearer these two kinds of facilities, the greater the chance of serious accidents. Thus, it is better that these 'high-risk' facilities are located far from 'high-protection' facilities. $\delta_{x^k it}$ denotes facility $x^k$ is located in location $i$, and $\delta_{y^r jt}$ denotes facility $y^r$ is located in location $j$, so the distance between the 'high-risk' and 'high-protection' facility is denoted as: $\delta_{x^k it}\delta_{y^r jt}A_{ij}$. As $w_{kr}$ is the risk weight, and in accordance with the sixth assumption above, the safety and environmental objective $D$ can be expressed as minimizing the following:

$$\sum_{t=1}^{T}\sum_{i,j=1}^{L}\sum_{k=1}^{K}\sum_{r=1}^{R}(w_{kr}\delta_{x^k it}\delta_{y^r jt}A_{ij})^{-1}. \tag{6.9}$$

### 6.2.3 Constraints

There are two kinds of constraints in the DCSLP. The first is area constraints which describe the limitations for facility locations. The second is logical constraints which states that one facility can only be located in one location.

**Area Constraints**

The location area has to meet the facility requirement, therefore,

$$\delta_{xit}S_{xt} < D_i, \forall x \in \{1,2,\cdots,F\}, t \in \{1,2,\cdots,T\}, i \in \{1,2,\cdots,L\}. \tag{6.10}$$

Note that the use ratio of the area cannot reach 100%, so that $S_{xt} < D_i$ instead of $S_{xt} \le D_i$.

**Logical Constraints**

To obtain a feasible solution, there are some logical constraints. At most, one facility is located in one location, namely:

$$\delta_{xit} + \delta_{yit} \le 1, \delta_{xjt} + \delta_{yjt} \le 1, \tag{6.11}$$

yet a certain type of facility can be located in more than one location, namely:

$$\delta_{xit} + \delta_{xjt} \ge 0, \delta_{yit} + \delta_{yjt} \ge 0. \tag{6.12}$$

$\delta_{xit}, \delta_{xjt}, \delta_{yit}, \delta_{yjt}$ denotes the existence of the facility, with 1 denoting positive and 0 denoting negative.

$\eta_{xit}$ is the opening of facility $x$ at location $i$ in phase $t$, and $\omega_{xit}$ is the closure of facility $x$ at location $i$ or $j$ during phase $t$. Practically, if $\delta_{xi,t-1} = 0$, and $\delta_{xit} = 1$, namely $\delta_{xit} - \delta_{xi,t-1} = 1$, then facility $x$ is opened in location $i$ in phase $t$, so $\eta_{xit} = 1$. $\delta_{xit} - \delta_{xi,t-1} = 0$ denotes facility $x$ is neither open nor closed in phase $t$, and $\eta_{xit} = 0, \omega_{xit} = 0$. $\delta_{xit} - \delta_{xi,t-1} = -1$ denotes facility $x$ is closed at location $i$ during period $t$, so $\omega_{xit} = 1$, namely:

$$\eta_{xit} = \begin{cases} 1 \text{ if } \delta_{xit} - \delta_{xi,t-1} = 1 \\ 0 \text{ otherwise} \end{cases} \quad \text{and} \quad \omega_{xit} = \begin{cases} 1 \text{ if } \delta_{xit} - \delta_{xi,t-1} = -1 \\ 0 \text{ otherwise} \end{cases} \tag{6.13}$$

where $x,y \in \{1,2,\cdots,F\}, i,j \in \{1,2,\cdots,L\}, t \in \{1,2,\cdots,T\}$.

By integrating Eqs. (6.8) $\sim$ (6.13), a mathematical model for the dynamic construction site layout planning problem can be obtained. However, it is difficult to achieve optimal programming results when there are the dual uncertainties of randomness and fuzziness. To transform the fuzzy random variables into deterministic equivalents, the following method is used.

### 6.2.4 Dealing with Fuzzy Random Variables

To describe phenomenon containing random and fuzzy factors synchronously, Kwakernaak proposed the concept of fuzzy random variables for the first time in 1978 [94]. Subsequently, many scholars have defined fuzzy random variables from different perspectives [93, 105]. Here the fuzzy random variable is considered to be defined in a real number set, in which the above definitions are equivalent. This study utilizes the definition proposed by Puri (1986) [93]. In this definition, the fuzzy random variable is a measurable function from a probability space to a collection of fuzzy variables.

Without loss of generality, fuzzy random variables are denoted in this problem as $\tilde{\bar{\xi}} = ([\xi]_L, \varphi(\omega), [\xi]_R)$, where, $\varphi(\omega)$ is a random variable with a probability density function $p_\varphi(x)$. Since it has been supposed above that $\varphi(\omega)$ follows a Normal distribution $\mathcal{N}(\mu_0, \sigma_0^2)$. Let $\sigma$ be any given probability level of the random variable, $r$ be any given possibility level of the fuzzy variable. The parameters satisfy:

$$\sigma \in [0, \sup p_\varphi(x)] \text{ and } r \in \left[\frac{[\xi]_R - [\xi]_L}{[\xi]_R - [\xi]_L + \varphi_\sigma^R - \varphi_\sigma^L}, 1\right]. \tag{6.14}$$

$\sigma$ and $r$ reflect the decision-maker's degree of optimism, and are respectively called the probability level and the possibility level.

The method dealing with the fuzzy random variable is as follows:

(1) Use statistical methods to estimate the parameters $[\xi]_L, [\xi]_R, \mu_0$ and $\sigma_0$ from previous data and professional experience.

(2) Obtain the decision-maker's degree of optimism.

(3) Let $\varphi_\sigma$ be the $\sigma$-level sets (or $\sigma$-cut) of the random variable $\varphi(\omega)$, i.e. $\varphi_\sigma = [\varphi_\sigma^L, \varphi_\sigma^R] = \{x \in R | p_\varphi(x) \geq \sigma\}$, where,

$$\varphi_\sigma^L = \inf\{x \in R | p_\varphi(x) \geq \sigma\} = \inf p_\varphi^{-1}(\sigma) = \mu_0 - \sqrt{-2\sigma_0^2 \in \left(\sqrt{2\pi}\sigma_0\sigma\right)}, \tag{6.15}$$

$$\varphi_\sigma^R = \inf\{x \in R | p_\varphi(x) \geq \sigma\} = \sup p_\varphi^{-1}(\sigma) = \mu_0 + \sqrt{-2\sigma_0^2 \in \left(\sqrt{2\pi}\sigma_0\sigma\right)}. \tag{6.16}$$

The probability density function of the random variable $\varphi$ with $\varphi_\sigma^L$ and $\varphi_\sigma^R$ is shown in Figure 6.6.

(4) Transform the fuzzy random variable $\tilde{\bar{\xi}} = ([\xi]_L, \varphi(\omega), [\xi]_R)$ into the $(r,\sigma)$-level trapezoidal fuzzy variable by $\tilde{\bar{\xi}} \to \tilde{\xi}_{(r,\sigma)} = \left([\xi]_L, \underline{\xi}, \overline{\xi}, [\xi]_R\right)$, where

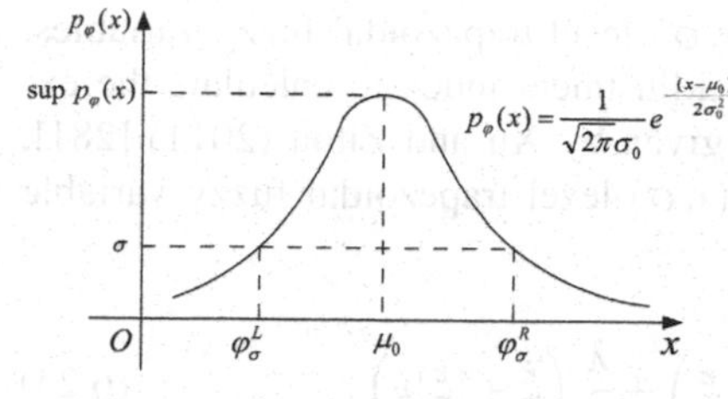


**Fig. 6.6** The probability density function of random variable $\varphi$

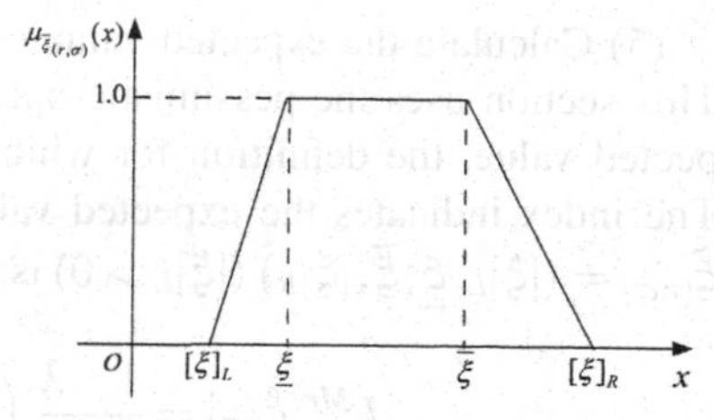


**Fig. 6.7** The $(r,\sigma)$-level trapezoidal fuzzy variable $\tilde{\xi}_{(r,\sigma)}$

$$\underline{\xi} = [\xi]_R - r([\xi]_R - \varphi_\sigma^L) = [\xi]_R - r\left([\xi]_R - \mu_0 + \sqrt{-2\sigma_0^2 \ln(\sqrt{2\pi}\sigma_0\sigma)}\right), \quad (6.17)$$

$$\overline{\xi} = [\xi]_L + r(\varphi_\sigma^R - [\xi]_L) = [\xi]_L + r\left(\mu_0 + \sqrt{-2\sigma_0^2 \ln(\sqrt{2\pi}\sigma_0\sigma)} - [\xi]_L\right). \quad (6.18)$$

The membership function of $\tilde{\xi}_{(r,\sigma)}$ is as below:

$$\mu_{\tilde{\xi}_{(r,\sigma)}}(x) = \begin{cases} 0 & \text{if } x < [\xi]_L, x > [\xi]_R, \\ \frac{x-[\xi]_L}{\underline{\xi}-[\xi]_L} & \text{if } -[\xi]_L \le x < \underline{\xi}, \\ 1 & \text{if } \underline{\xi} \le x \le \overline{\xi}, \\ \frac{[\xi]_R - x}{[\xi]_R - \overline{\xi}} & \text{if } \overline{\xi} < x \le [\xi]_R. \end{cases} \quad (6.19)$$

Therefore, the fuzzy random variable $\tilde{\bar{\xi}}$ is transformed into a $(r,\sigma)$-level trapezoidal fuzzy variable $\tilde{\xi}_{(r,\sigma)}$. The membership function of $\tilde{\xi}_{(r,\sigma)}$ is shown in Figure 6.7.

Based on the method above, the fuzzy random variables in this chapter are transformed into $(r,\sigma)$-level trapezoidal fuzzy variables as follows:

$$\tilde{\bar{C}}_{xyt} \to \tilde{C}_{xyt} = ([C_{xyt}]_L, \underline{C}_{xyt}, \overline{C}_{xyt}, [C_{xyt}]_R). \quad (6.20)$$

$$\mu_{\tilde{C}_{xyt}} = \begin{cases} 0 & \text{if } x < [C_{xyt}]_L, x > [C_{xyt}]_R, \\ \frac{x-[C_{xyt}]_L}{\underline{C}_{xyt}-[C_{xyt}]_L} & \text{if } -[C_{xyt}]_L \le x < \underline{C}_{xyt}, \\ 1 & \text{if } \underline{C}_{xyt} \le x \le \overline{C}_{xyt}, \\ \frac{[C_{xyt}]_R - x}{[C_{xyt}]_R - \overline{C}_{xyt}} & \text{if } \overline{C}_{xyt} < x \le [C_{xyt}]_R. \end{cases} \quad (6.21)$$

$$\tilde{\bar{C}}^z_{xit} \to \tilde{C}^z_{xit} = ([C^z_{xit}]_L, \underline{C}^z_{xit}, \overline{C}^z_{xit}, [C^z_{xit}]_R). \quad (6.22)$$

$$\mu_{\tilde{C}^z_{xit}} = \begin{cases} 0 & \text{if } x < [C^z_{xit}]_L, x > [C^z_{xit}]_R, \\ \frac{x-[C^z_{xit}]_L}{\underline{C}^z_{xit}-[C^z_{xit}]_L} & \text{if } -[C^z_{xit}]_L \le x < \underline{C}^z_{xit}, \\ 1 & \text{if } \underline{C}^z_{xit} \le x \le \overline{C}^z_{xit}, \\ \frac{[C^z_{xit}]_R - x}{[C^z_{xit}]_R - \overline{C}^z_{xit}} & \text{if } \overline{C}^z_{xit} < x \le [C^z_{xit}]_R. \end{cases} \quad (6.23)$$

(5) Calculate the expected value of the $(r,\sigma)$-level trapezoidal fuzzy variables. This section uses the pessimistic-optimistic adjustment index to calculate the expected value, the definition for which was given by Xu and Zhou (2011) [281]. The index indicates the expected value of $(r,\sigma)$-level trapezoidal fuzzy variable $\tilde{\xi}_{(r,\sigma)} = ([\xi]_L, \underline{\xi}, \overline{\xi}, [\xi]_R)$ $([\xi]_L > 0)$ is:

$$E^{Me}[\tilde{\xi}_{r,\sigma}] = \frac{1-\lambda}{2}\left([\xi]_L + \underline{\xi}\right) + \frac{\lambda}{2}\left(\overline{\xi} + [\xi]_R\right), \tag{6.24}$$

where $\lambda$ is the optimistic and pessimistic index to determine the combined attitude of decision makers. Note that the fuzzy measure, $Me$, is used to evaluate a confidence degree that a fuzzy variable takes values in an interval. $\lambda = 1$ denotes that the decision maker is optimistic and the decision maker deems that the best case of that event has the maximal chance of happening, while $\lambda = 0$ denotes the opposite.

Based on the objective function and the constraint formulas, the mathematical model for the dynamic construction site layout planning problem can be transformed into the following multi-objective expected value model:

$$\left\{\begin{array}{l}
\min C = \sum\limits_{t=1}^{T}\sum\limits_{x,y=1}^{F}\sum\limits_{i,j=1}^{L} \alpha_t\left(C^{s}_{xit}\eta_{xit} + C^{c}_{xit}\omega_{xit} + E^{Me}[C^{z}_{xit}]\delta_{xit} + \delta_{xit}\delta_{yjt}A_{ij}E^{Me}[C_{xyt}]\right)\\
\max D = \sum\limits_{t=1}^{T}\sum\limits_{i,j=1}^{L}\sum\limits_{k=1}^{K}\sum\limits_{r=1}^{R}\left(w_{kr}\delta_{x^k it}\delta_{y^r jt}A_{ij}\right)^{-1}\\
\text{s.t.}\left\{\begin{array}{l}
\delta_{xit}S_{xt} < D_i\\
\delta_{xit} + \delta_{yit} \le 1\\
\delta_{xjt} + \delta_{yjt} \le 1\\
\delta_{xit} + \delta_{xjt} \ge 0\\
\delta_{yit} + \delta_{yjt} \ge 0\\
S_{xt} \ge 0\\
D_i > 0\\
\delta_{xit}, \delta_{xjt}, \delta_{yit}, \delta_{yjt} = 0 \text{ or } 1\\
\eta_{xit} = \left\{\begin{array}{l} 1 \text{ if } \delta_{xit} - \delta_{xi,t-1} = 1\\ 0 \text{ otherwise}\end{array}\right.\\
\omega_{xit} = \left\{\begin{array}{l} 1 \text{ if } \delta_{xit} - \delta_{xi,t-1} = -1\\ 0 \text{ otherwise}\end{array}\right.\\
\alpha_t = \frac{1}{(1+\alpha)^{(t-1)}}\\
A_{ij} \ge 0\\
x \ne y, i \ne j
\end{array}\right.
\end{array}\right. \tag{6.25}$$

## 6.3 Modelling Process of DTFLP

In this section, based on the problem description for the DTFLP, the modelling process for the DTFLP is presented.

### 6.3.1 Assumptions

The assumptions for facility location planning models on networks (real-world distances, different costs at different potential locations, etc.) are generally more realistic. Using dynamic programming, a mathematical model is presented to tackle the facility location planning problem encountered by a large-scale construction project under a fuzzy environment. The assumptions for this model are as follows:

(1) The facilities locations are treated as dots and the areas and the orientation difference of the facilities are neglected.

(2) On the networks, the distances correspond to real road distances. The distances are measured between the location midpoints.

(3) The transportation costs are a main factor for the choice of locations. Transportation costs are usually assumed to be proportional to the distances and the material flow intensities. Transportation, resource and handling costs are assumed to be linear.

(4) Regardless of the unpredictable impact of the positioning of a facility, the subsequent decisions are made in the same and future stages.

(5) Planners specify the start of these stages to coincide with scheduled milestones, which represent the finish and start of major tasks and accordingly the release of and demand for significant facilities (which means $t = 0, 1, \cdots, T-1$).

Based upon the assumptions above, the relationship equation, initial conditions, constraint conditions, and objective functions of the DTFLP can be established.

### 6.3.2 State Equation

Modeling a complex problem using dynamic programming requires breaking it down into a set of simpler and easier sub-problems (decision epoch) that are solved sequentially to generate an optimal solution for the larger problem (Haidar 2009 [138]). The decision epoch is a minimal description of the system history at stage $k$ that is crucial for the computation of the possible reward or cost of the current decision. As each decision epoch can have a short-term and/or long-term effect on future decision epochs, the state equation should explain this effect between the different stages, but this effect is reflected in many aspects, such as cost, safety, environmental effect, and operations management, so it is very difficult to clearly state all of these aspects. Rosenblatt (1986) [148] used a combination of layouts with minimum total cost to formulate a DP model. Combining Rosenblatt's (1986) [148] model with the DTFLP problem presented here, a recursive formulation is developed, and the total cost for each of the layouts considered in the horizon is obtained. The state equation is established in the following forms:

$$C^*(t) = \min\{C^*(t-1) + \sum_{i=1}^{I} \sum_{m=1}^{M} \sum_{k=1}^{M} (a_{im} + c_i) x_{imk}(t)\} + F(t), \quad i \in \Phi, k \in \Upsilon, m \in \Omega, \tag{6.26}$$

where $\Phi = \{1, 2, \cdots, I\}$, $\Upsilon = \{1, 2, \cdots, K\}$, $\Omega = \{1, 2, \cdots, M\}$ and $t = \{0, \cdots, T\}$. $x_{imk}(t)$ is the decision variable of temporary facility $i$ in stage $t$ and $m, k$ is the location index. When $x_{imk}(t) = 1$, a change of locations takes place, namely, facility $i$ is moved from location $m$ to location $k$. $a_{im}$ is the variable cost when facility $i$ is built in location $m$. $c_i$ is the fixed location cost of facility $i$. $M$ is the number of the possible positions for each temporary facility.

The rearrangement costs are zero if no change in the layout is made from one period to another. Furthermore, it is assumed that the shifting costs are independent of the periods in which they are being made. $F(t)$ can be made a function of the period in which the total value of the interaction flow exists. $F(t)$ is obtained in the previous section. $C^*(t)$ is the minimum cost (shifting and handling) corresponding with the DTFLP problem for all periods up to $t$.

### 6.3.3 Initial and Terminal Conditions

At the beginning of the stage, when $t = 0$, the moveable facility can be moved because they have not yet been designed for any specific location. So, the decision variable is assumed to be

$$x_{imk}(0) = 0. \tag{6.27}$$

The following state equation represents the initial DTFLP cost condition:

$$C^*(0) = C_0. \tag{6.28}$$

where $C_0$ denotes the initial cost at the beginning of the whole period.

Similarly, at the end of the construction duration, $t = T$, the final security condition can be stated:

$$C^*(T) \leq C_e. \tag{6.29}$$

$C_e$ is the cost value allocated by the construction designer at the end of the construction site layout stage.

### 6.3.4 Constraint Conditions

As defined in the literature, the facility layout problem's aim is to minimize the objectives subject to two sets of constraints: area requirements and department location restrictions (Meller and Gau 1996 [73]). In a large scale hydropower construction project, the DTFLP is based on a specific and designed network with the facilities treated as discrete dots and the area requirements ignored. As assumed before, the real road distances between the boundaries are used to accord with the location restrictions.

$x_{imk}(t)$ is a binary variable which takes the value of 1 if the facility $i$ is assigned to location $k$ from location $m$, and 0 otherwise. If $m = k$, the facility at stage $t$ is not moved. The following constraints guarantee that one facility can be assigned to one location and one location can accommodate no more than one facility because of the environmental condition. The location of each facility is then assigned and located in accordance with these constraints.

$$\begin{cases} \sum_{m=1}^{M}\sum_{k=1}^{M} x_{imk}(t) = 1, & \forall i \in \Phi \\ \sum_{n=1}^{M}\sum_{l=1}^{M} x_{jnl}(t) = 1, & \forall j \in \Phi \\ \sum_{i=1}^{I} x_{imk}(t) \leq 1, & \forall k \in \Upsilon, m \in \Omega \\ \sum_{j=1}^{I} x_{jnl}(t) \leq 1, & l \in \Upsilon, n \in \Omega \end{cases} \tag{6.30}$$

where $x_{imk}(t), x_{jnl}(t) \in \{0,1\}$, $\Phi = \{1,2,\cdots,I\}$, $\Upsilon = \{1,2,\cdots,K\}$, $\Omega = \{1,2,\cdots,M\}$ and $t = \{1,\cdots,T\}$.

### 6.3.5 Objective Functions

The handling cost is commonly used in a DTFLP problem. The first objective of function $f_1$ is to minimize total site location costs and handling costs between the facilities in order to reduce construction site costs. The first part of $f_1$ is the location cost which is related with the variable cost $a_{im}$ and the fixed cost $c_i$ when the facility $i$ is built in location $m$. In addition, the handling cost refers to the costs resulting from the interaction flows between the facilities associated with the DTFLP over the entire project duration. Based on the function for material handing costs (Aiello et al. 2006 [123]) and the value of interaction flows in Ning et al.'s (2010) [81] research, the equation for the interaction flows in the DTFLP problem can be defined as in the following function:

$$F(t) = \sum_{i=1}^{I}\sum_{j=1}^{I}\sum_{m=1}^{M}\sum_{k=1}^{M}\sum_{n=1}^{M}\sum_{l=1}^{M} v_{ij}(t) d_{ik,jl} x_{imk}(t) x_{jnl}(t), \quad i,j \in \Phi, k,l \in \Upsilon, m,n \in \Omega, \tag{6.31}$$

where $\Phi = \{1,2,\cdots,I\}$, $\Upsilon = \{1,2,\cdots,K\}$, $\Omega = \{1,2,\cdots,M\}$ and $t = \{0,\cdots,T\}$. $d_{kl}$ is the distance between facilities $i$ and $ij$. $v_{ij}$ is the cost value of the interaction flow between facilities $i$ and $j$. $x_{imk}(t)$ is a binary variable which takes value 1 if facility $i$ is assigned to location $k$ from location $m$, and 0 otherwise.

As the state equation presented in Eq. (6.26) refers to the total cost from the initial stage to the $t^{th}$ stage, so the total cost over the entire DTFLP problem can be obtained in the last stage, i.e. when $t = T-1$:

$$C^*(T) = \min\{C^*(T-1) + \sum_{i=1}^{I}\sum_{m=1}^{M}\sum_{k=1}^{M}(a_{im}+c_i)x_{imk}(T-1)\} + F(T), \tag{6.32}$$

$i \in \Phi, k \in \Upsilon, m \in \Omega$, where $\Phi = \{1,2,\cdots,I\}$, $\Upsilon = \{1,2,\cdots,K\}$, $\Omega = \{1,2,\cdots,M\}$. Thus, the first objective function $f_1$ can be defined as the following function:

$$\min f_1 = C^*(T), \tag{6.33}$$

From a practical planning point of view, the closeness relationship level is neither achievable nor feasible (Soltani and Fernando 2004 [88]). Fortenberry and Cox

(1985) [133] used a distance-weighted flow to measure the cost terms of the closeness relationship. But such an approach had poor practicability due to the difficulty of normalizing the variables and quantifying the weights (Aiello et al. 2006 [123]). Here, the measurement of the second objective $f_2$ is modified with the aim of minimizing the closeness relationship effect. Thus, the second objective $f_2$ can be defined as:

$$\min f_2 = \sum_{i=1}^{I}\sum_{j=1}^{I}\sum_{m=1}^{M}\sum_{k=1}^{M}\sum_{n=1}^{M}\sum_{l=1}^{M} R_{ij} d_{ik,jl} x_{imk}(t) x_{jnl}(t), \quad i \in \Phi, k \in \Upsilon, m \in \Omega, \tag{6.34}$$

where $\Phi = \{1,2,\cdots,I\}$, $\Upsilon = \{1,2,\cdots,K\}$, $\Omega = \{1,2,\cdots,M\}$ and $t = \{1,\cdots,T\}$. As stated in the previous section, $R_{ij}$ is the relationship value, which can be calculated using Equation 6.7. $x_{imk}(t)$ is a binary variable which takes value 1 if facility $i$ is assigned to location $k$ from location $m$, and 0 otherwise. $d_{kl}$ is the distance between facilities $i$ and $ij$. $v_{ij}$ is the cost value of the interaction flow between facilities $i$ and $j$.

The two objectives are complementary rather than contradictory and guarantee that the project is carried out successfully and efficiently and all the facilities resources have been taken advantage of. The two objectives together show adequately and clearly the aim of optimization in the DTFLP problem as both the economic and resource aspects are taken into consideration, which ensures a comprehensive understanding of DTFLP problem optimization.

From the above statement, therefore, the DTFLP model can be proposed as:

$$\begin{cases} \min f_1 = C^*(T) \\ \min f_2 = \sum_{i=1}^{I}\sum_{j=1}^{I}\sum_{m=1}^{M}\sum_{k=1}^{M}\sum_{n=1}^{M}\sum_{l=1}^{M} R_{ij} d_{ik,jl} x_{imk}(t) x_{jnl}(t) \\ \text{s.t.} \begin{cases} C^*(t) = \min\{C^*(t-1) + \sum_{i=1}^{I}\sum_{m=1}^{M}\sum_{k=1}^{M}(a_{im}+c_i)x_{imk}(t)\} + F(t) \\ F(t) = \sum_{i=1}^{I}\sum_{j=1}^{I}\sum_{m=1}^{M}\sum_{k=1}^{M}\sum_{n=1}^{M}\sum_{l=1}^{M} v_{ij}(t) d_{ik,jl} x_{imk}(t) x_{jnl}(t) \\ C^*(0) = C_0, \quad C^*(T) \le C_e, \quad x_{imk}(0) = 0 \\ \sum_{m=1}^{M}\sum_{k=1}^{M} x_{imk}(t) = 1, \forall i \in \Phi \\ \sum_{n=1}^{M}\sum_{l=1}^{M} x_{jnl}(t) = 1, \forall j \in \Phi \\ \sum_{i=1}^{I} x_{imk}(t) \le 1, \forall k \in \Upsilon, m \in \Omega \\ \sum_{j=1}^{I} x_{jnl}(t) \le 1, l \in \Upsilon, n \in \Omega \end{cases} \end{cases} \tag{6.35}$$

where $x_{imk}(t), x_{jnl}(t) \in \{0,1\}$, $\Phi = \{1,2,\cdots,I\}$, $\Upsilon = \{1,2,\cdots,K\}$, $\Omega = \{1,2,\cdots,M\}$ and $t = \{1,\cdots,T\}$. $R_{ij}$ is the relationship value, which can be calculated using Equation 6.7. $x_{imk}(t)$ is a binary variable which takes value 1 if facility $i$ is assigned to location $k$ from location $m$, and 0 otherwise. $d_{kl}$ is the distance between facilities $i$ and $ij$. $v_{ij}$ is the cost value of the interaction flow between facilities

$i$ and $j$. $a_{im}$ is the variable cost when facility $i$ is built in location $m$. $c_i$ is the fixed part of the location cost of facility $i$. $M$ is the number of the possible positions for each temporary facility.

## 6.4 Modelling Analysis of Piecewise Linear Control

When there are piecewise linear functions in the model, the maximal theory becomes invalid when dealing with the facility location problem. This section focuses on a discussion of a piecewise linear control system encountered in the facility location problem, and presents a theoretical solution method for this problem.

### 6.4.1 *Introduction to Piecewise Linear Control*

Up to now, when using iterative dynamic programming to solve an optimal control problem, the given time interval was divided into $P$ time stages of equal length and at each time-stage the control was held constant. Such a piecewise constant control policy provides a reasonably good approximation for the continuous control policy, and the approximation can be improved by having a larger number of time stages. In many situations the optimal control policy is quite smooth, and therefore a very good approximation is achieved using piecewise linear, rather than constant, sections. A further constraint can be imposed so that the linear sections form a continuous curve. Piecewise linear continuous control was first used in iterative dynamic programming for several systems by Luus (1993a) [191]. It was found that the use of piecewise linear control does indeed give an excellent result with a relatively small number of time stages, and allows for an optimal control policy for very high dimensional systems to be determined accurately [192].

### 6.4.2 *Analysis for Piecewise Linear Control*

Consider the system described using the vector differential equation

$$\frac{d\mathbf{x}}{dt} = \mathbf{f}(\mathbf{x},\mathbf{u}) \tag{6.36}$$

with $\mathbf{x}(0)$ given, where $\mathbf{x}$ is an $(n\times 1)$ state vector and $\mathbf{u}$ is an $(m\times 1)$ control vector bounded by

$$\alpha_j \leq u_j \leq \beta_j, \quad j=1,2,\cdots,m. \tag{6.37}$$

Associated with the system is a performance index

$$I = \Phi[\mathbf{x}(t_f)] \tag{6.38}$$

where $\phi$ is a positive definite function of $\mathbf{x}$ at the given final time $t_f$. The optimal control problem is to find the control policy $\mathbf{u}(t)$ in the time interval $0 \leq t < t_f$ so that the performance index is minimized.

Divide the given time interval $(0,t_f)$ into $P$ subintervals $(0,t_1)$, $(t_1,t_2)$, $\cdots$, $(t_k,t_{k+1})$, $\cdots$, $(t_{P-1},t_P)$, each of length $L$, so that

$$L = \frac{t_f}{P}. \tag{6.39}$$

Piecewise linear continuous control policy is sought to allow for a minimization of the performance index. Therefore, the control policy in the time interval $(t_k,t_{k+1})$ is calculated using the expression

$$\mathbf{u}(t) = \mathbf{u}(k) + \frac{\mathbf{u}(k+1) - \mathbf{u}(k)}{L}(t - t_k) \tag{6.40}$$

where $\mathbf{u}(k)$ is the value of $\mathbf{u}$ at time $t_k$ and $\mathbf{u}(k+1)$ is the value of $\mathbf{u}$ at time $t_{k+1}$. The optimal control problem is now to find $\mathbf{u}(k)$, $k = 0,1,2,\cdots,P$, inside the range specified by Eq. (6.37), so that the performance index in Eq. (6.38) is minimized. In the last stage, both $\mathbf{u}(P)$ and $\mathbf{u}(P-1)$ need to be determined. This means that for the last stage it is necessary to determine $2m$ values for the $m$ control variables. For all other stages there are only $m$ values to be determined because of the continuity restriction.

### 6.4.3 *Algorithm for Piecewise Linear Control*

Here iterative dynamic programming with a single grid point is used, i.e., $N = 1$, in a multi-pass fashion, where the region size is contracted by an amount $\gamma$ after every iteration. After a specified number of iterations making up a pass, the region size is restored to a value $\eta$ times the region size at the beginning of the pass so the region size can be used in the subsequent pass. To illustrate this underlying logic in iterative dynamic programming, an algorithm is given to solve the optimal control problem as outlined in Eqs. (6.36)-(6.39), where it is required to minimize the performance index in Eq. (6.38) with the use of piecewise linear control, as specified in Eq. (6.40) over $P$ stages, each of the same length:

1. Divide the time interval $[0,t_f]$ into $P$ time stages, each of length $L$.
2. Choose the number of test values for $\mathbf{u}$ denoted by $R$, an initial control policy and the initial region size $\mathbf{r}_{in}$; also choose the region contraction factor $\gamma$ used after every iteration and the region restoration factor $\eta$ used after every pass.
3. Choose the number of iterations to be used in every pass and the number of passes to be used in the run.
4. Set the pass number index $q = 1$ and the iteration number index to $j = 1$.
5. Set the region size vector $\mathbf{r}^j = \eta^q \mathbf{r}_{in}$.
6. Using the best control policy (the initial control policy for the first iteration of the first pass), integrate Eq. (6.36) from $t = 0$ to $t_f$ to generate the $\mathbf{x}$-trajectory and store the value of $\mathbf{x}$ at the beginning of each time stage, so that $\mathbf{x}(k-1)$ corresponds to the value of $\mathbf{x}$ at the beginning of stage $k$.
7. Starting at stage $P$, corresponding to time $t_f - L$, integrate Eq. (6.36) from $t_f - L$ to $t_f$, using as the initial state the stored value $\mathbf{x}(P-1)$ from step 6, once with

each of the $R$ allowable values for the control vector calculated from Eq. (6.40), using

$$\mathbf{u}(P-1) = \mathbf{u}^{*j}(P-1) + \mathbf{D}_1\mathbf{r}^j \tag{6.41}$$

$$\mathbf{u}(P) = \mathbf{u}^{*j}(P) + \mathbf{D}_2\mathbf{r}^j \tag{6.42}$$

where $\mathbf{u}^{*j}(P-1)$ and $\mathbf{u}^{*j}(P)$ are the best values obtained in the previous iteration and $\mathbf{D}_1$ and $\mathbf{D}_2$ are diagonal matrices of different random numbers between $-1$ and 1. Clip suitably any of the control variables at $\alpha_j$ or $\beta_j$ to satisfy Eq. (6.37). Out of the $R$ values for the performance index, choose the control values that give the minimum value, and store these values as $\mathbf{u}^*(P-1)$ and $u^*(P)$. The best control for the last stage is now determined.

8. Step back to stage $P-1$, corresponding to time $t_f - 2L$. Choose $R$ values for $\mathbf{u}(P-2)$ as in the previous step, and by taking as the initial state $\mathbf{x}(P-2)$ integrate Eq. (6.36) over one stage length by using Eq. (6.40) for the control. Continue integration over the last time stage using the stored value of $\mathbf{u}^*(P-1)$ and $\mathbf{u}^*(P)$ from step 7. Compare the $R$ values of the performance index and store the best control as $\mathbf{u}^*(P-2)$ that gives the minimum value for the performance index.

9. Continue the procedure until stage 1, corresponding to the initial time $t = 0$ and the given initial state, is reached. As before, integrate Eq. (6.36) and compare the $R$ values of the performance index and store the control $\mathbf{u}(0)$ as $\mathbf{u}^*(0)$ that gives the minimum performance index. Store also the corresponding $\mathbf{x}$-trajectory.

10. Reduce the region size for allowable control

$$\mathbf{r}^{j+1} = \gamma\mathbf{r}^j \tag{6.43}$$

where $j$ is the iteration number index. Use the best control policy from step 9 as the midpoint for the allowable values for the control denoted by the superscript $*$.

11. Increment the iteration index $j$ by 1 and go to step 6; continue the procedure for the specified number of iterations to finish a pass.

12. Increment the pass index $q$ by 1 and go to step 5; continue the procedure for the specified number of passes, and interpret the results.

### 6.4.4 Numerical Example

To show the advantages of using piecewise linear control, several examples are considered. As a guide to the amount of computational effort required, the computation time on a Core *i*5 is reported.

*Example 6.1.* We now consider the system used by Thomopoulos and Papadakis (1991) [193], which showed difficulties in the convergence of several optimization procedures. The system, also considered by Luus (1993a) [191], is described by

$$\frac{dx_1}{dt} = x_2 \tag{6.44}$$

$$\frac{dx_2}{dt} = -x_1 - x_2 + 100[H(t-0.5) - H(t-0.6)] \tag{6.45}$$

$$\frac{dx_3}{dt} = 5x_1^2 + 2.5x_2^2 + 0.5u^2 \tag{6.46}$$

with the initial state

$$\mathbf{x}(0) = [0,0,0]^T. \tag{6.47}$$

In Eq. (6.45) there is a disturbance term in the form of a rectangular pulse of magnitude 100 from $t = 0.5$ until $t = 0.6$. The optimal control problem is to find control $u$ in the time interval $0 \le t < 2$ so that the performance index

$$I = x_3(t_f) \tag{6.48}$$

is minimized. Here the final time $t_f = 2$.

Using an exponential smoothing technique Thomopoulos and Papadakis (1991) [193] overcame the discontinuity problem and obtained a minimum value $I = 58.538$. With iterative dynamic programming the discontinuity produced by the rectangular pulse causes no difficulties provided that the integration step size is chosen so that the state equations can be properly integrated. Using iterative dynamic programming, Luus (1993a) [191] reported a minimum of $I = 58.20$ using 40 stages of piecewise constant control and $I = 58.18$ with the use of piecewise linear control over 40 time stages of equal length. Here this problem is considered with 20 stages of equal length.

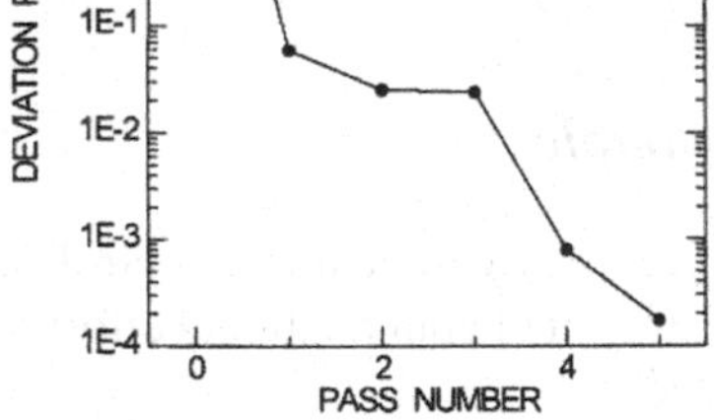

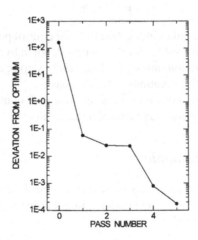


**Fig. 6.8** Convergence profile for the nondifferentiable system, showing the deviation of the performance index from 58.185 as a function of the pass number with $P = 20$

For the iterative dynamic programming $P = 20$ was chosen, $u^{(0)} = 0$, $r^{(0)} = 10$, $\gamma = 0.95$, $\eta = 0.7$, and there were 20 iterations per pass. As shown in Figure 6.8, convergence from the initial value of 220.395 to the minimum value of $I = 58.185$

was obtained very rapidly. To integrate the equations the Runge-Kutta method was used with an integration step-size of 0.025. The computation time for the 5 passes was 0.98*s*.

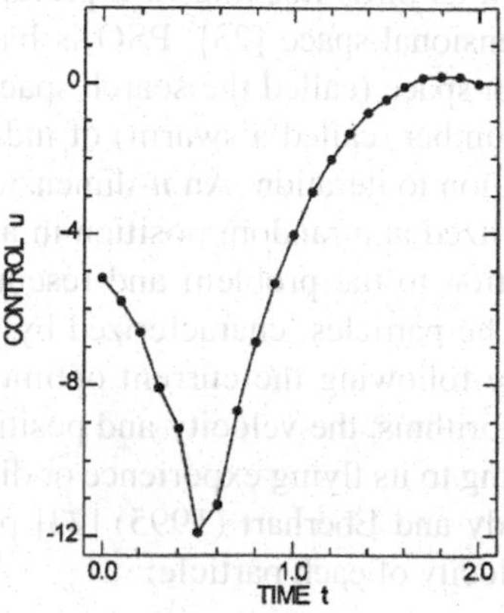


**Fig. 6.9** Piecewise linear continuous optimal control policy with $P = 20$ for the nondifferentiable system

The resulting piecewise linear optimal control policy is given in Figure 6.9 and the state trajectories are given in Figure 6.10. It can be observed that the rectangular pulse in the time interval from 0.5 to 0.6 caused a very rapid increase in the state variable $x_2$ during this time interval.

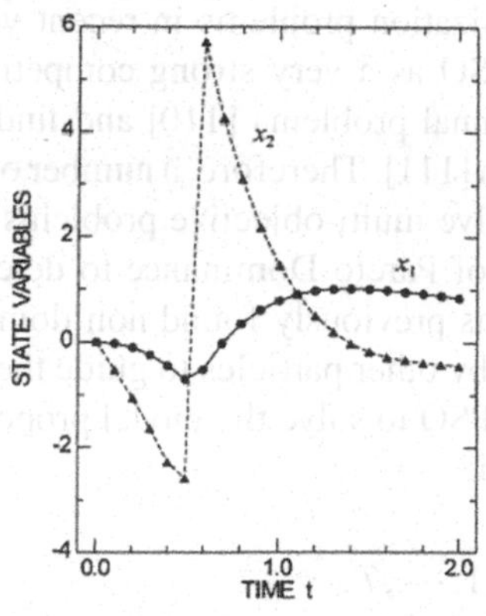


**Fig. 6.10** State trajectories for the nondifferentiable system

## 6.5 MOPSO with Permutation-Based Representation for DCSLP

PSO is a population-based self-adaptive search optimization technique which simulates social behavior, such as birds flocking to a promising position, for certain objectives in a multi-dimensional space [23]. PSO is based on a set of potential solutions defined in a given space (called the search space) and conducts searches using a fixed population number (called a swarm) of individuals (called particles) that are updated from iteration to iteration. An $n$-dimensional position of a particle (called the solution), initialized at a random position in a multidimensional search space, represents the solution to the problem and resembles a chromosome of a genetic algorithm [106]. The particles, characterized by position and velocity, fly through the problem space following the current optimum particles [23]. Unlike other population-based algorithms, the velocity and position of each particle is dynamically adjusted according to its flying experience or discoveries as well as those of its companions. Kennedy and Eberhart (1995) [23] proposed the following to update the position and velocity of each particle:

$$v_{ld}(\tau+1) = w(\tau)v_{ld}(\tau) + c_p r_1[p_{ld}^{\text{best}}(\tau) - p_{ld}(\tau)] + c_g r_2[p_{gd}^{\text{best}(\tau)} - p_{ld}(\tau)], \quad (6.49)$$

$$p_{ld}(\tau+1) = p_{ld}(\tau) + v_{ld}(\tau+1), \quad (6.50)$$

where $v_{ld}(\tau+1)$ is the velocity of $l^{th}$ particle at the $d^{th}$ dimension in the $\tau^{th}$ iteration, $w$ is an inertia weight, $p_{ld}(\tau)$ is the position of $l^{th}$ particle at the $d^{th}$ dimension, $r_1$ and $r_2$ are random numbers in the range $[0,1]$, $c_p$ and $c_g$ are personal and global best position acceleration constants respectively, while, $p_{ld}^{\text{best}}$ and $p_{gd}^{\text{best}}$ are the personal and global best position of $l^{th}$ particle at the $d^{th}$ dimension.

As PSO can be implemented easily and effectively, it has been widely applied in solving real-world optimization problems in recent years [107, 108, 109]. Researchers are also seeing PSO as a very strong competitor to other algorithms in solving multi-objective optimal problems [110] and find it especially suitable for multi-objective optimization [111]. Therefore, a number of proposals have been suggested to extend PSO to solve multi-objective problems [112, 107]. The MOPSO approach uses the concept of Pareto Dominance to determine the flight direction of a particle and it maintains previously found non-dominated vectors in a global repository that is later used by other particles to guide their own flight [110].

This section applies MOPSO to solve the model proposed in last section and the following notations are used:

- $t$ : iteration index, $t = 1, \cdots, T$,
- $i$ : particle index, $i = 1, \cdots, L$,
- $d$ : dimension index, $d = 1, \cdots, N \times \tau$,
- $j$ : phase index, $j = 1, \cdots, \tau$,
- $f$ : facility index, $f = 1, \cdots, F$,

| | | |
|---|---|---|
| $l$ | : | location index, $l = 1, \cdots, N$, |
| $P_i^{\max}(d)$ | : | maximum position value of particle $i$ at the d$^{th}$ dimension, |
| $P_i^{\min}(d)$ | : | minimum position value of particle $i$ at the d$^{th}$ dimension, |
| $r_1, r_2$ | : | uniform distributed random number within $[0,1]$, |
| $w(t)$ | : | inertia weight in the $t^{th}$ iteration, |
| $v_{id}^j(t)$ | : | velocity of the $j^{th}$ phase of the $i^{th}$ particle at the $d^{th}$ dimension in the $t^{th}$ iteration, |
| $x_{id}^j(t)$ | : | position of the $j^{th}$ phase of the $i^{th}$ particle at the $d^{th}$ dimension in the $t^{th}$ iteration, |
| $x_{id}^{\text{pbest}}(t)$ | : | personal best position of the $j^{th}$ phase of the $i^{th}$ particle at the $d^{th}$ dimension in the $t^{th}$ iteration, |
| $x_{id}^{\text{gbest}}(t)$ | : | global best position of the $j^{th}$ phase of the $i^{th}$ particle at the $d^{th}$ dimension in the $t^{th}$ iteration, |
| $c_p$ | : | personal best position acceleration constant, |
| $c_g$ | : | global best position acceleration constant, |
| $ARC$ | : | the positions of the particles that represent non-dominated vectors in the repository. |

### 6.5.1 Permutation-Based Representation

There have been various types of solution representations. For example Zhang and Wang (2008) [246] used priority-based representation for a static construction site layout planning problem. In this study, permutation-based representation is used in applying the MOPSO method where the candidate solution is represented by a multidimensional particle with ordinal numbers.

All dimensions of a multidimensional particle in the traditional PSO are independent of each other; thus, the updating of the velocity and particle based on Eqs. (6.49) and (6.50) is performed independently for each element. As a result, more than one element in an updated particle may have the same value, such as $\{3, 5, 7, 3, \cdots\}$. Particles with the same value at multiple elements represent an infeasible permutation, because two facilities cannot appear at the same location. Therefore, the MOPSO updating mechanism needs to be modified to eliminate conflicts [588, 589].

Various genetic crossover operators have been developed for permutation infeasibility, for example, the partially mapped crossover (PMX) [114], the cycle cross-over (CX) [115], and the order crossover (OX) [116]. For the PMX, at each of the arbitrary elements, the two values (i.e., location index) of the two parents are exchanged. Then, the repeated value of another element in one parent is replaced by the mapped value of the specified arbitrary element in the second parent, and vice versa. In other words, the PMX performs swaps for a series of arbitrary elements by exchanging each of them with an element where there is a mapped value for the arbitrary element in the other parent. Figure 6.11 illustrates the PMX, where one arbitrary element is specified. Zhang et al. (2006a) [27] successfully used permutation-based PSO with PMX for a resource-constrained project scheduling problem.

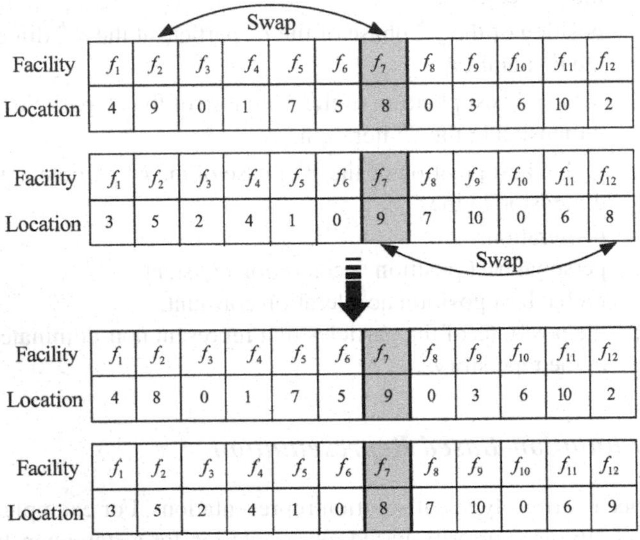


**Fig. 6.11** PMX for the permutation-based representation

### 6.5.2 *Hybrid Particle-Updating Mechanism*

Originally, the velocity computed by Eq. (6.49) is a distance measure used to decide on a new position for the updated particle. A larger velocity means a more distant position or area is being explored. When used for permutation-based particle representation, the $N$-dimension velocity, i.e., $V_i(t) = \{v_{i1}(t), v_{i2}(t), \cdots, v_{iN}(t)\}$ computed using Eq. (6.49) corresponds to the distance or gap between the current particle-represented sequence (i.e., the order of locations) and its own local best and the global best sequences (i.e., experiences) found so far. A larger gap indicates that such a particle-represented sequence is more likely to be updated. Therefore, the absolute value of the velocity is used here to represent the probability that the particle-represented sequence will change. Based on the PMX concept and the $N$-dimension probability, each dimension of a particle is randomly determined to see if the facility at such a location will be moved to another or if the element will be swapped with another. The "another element" has the value (i.e., facility-index number) the former element maps in the reference particle which is randomly selected from the current particle's local best (experience or sequence) and the swarm's global best. Here the particle to be updated and the reference particle respectively, resemble Parents 1 and 2 in the PMX.

Considering the permutation feasibility, the PMX concept is used in this chapter to develop the hybrid updating mechanism for the particle-represented sequence.

### 6.5.3 *Multi-objective Method*

The particle-represented solution is checked and adjusted for infeasibility caused by the violation of the resource constraints before transforming it to the layout planning. The taxonomy that Reyes-Sierra and Coello (2006) [117] proposed to classify the current MOPSOs are aggregating approaches, lexicographic ordering, sub-population approaches, Pareto-based approaches, combined approaches and other approaches. Previous multi-objective optimizations in construction site layout planning studies have tended to use the aggregating approach, such as in [118]. However, the weights in this approach can be arbitrary.

This section uses the Pareto Archived Evolutionary Strategy (PAES) [119], which is one of Pareto-based approaches used to update the best position. This approach uses leader selection techniques based on Pareto dominance. The basic idea is to select the particles that are non-dominated in the swarm as leaders. The PAES procedure is outlined below, $X_i^{\text{pbest}}$ is the best position vector of particle $i$, and is initially set equal to the initial position of particle $i$. $X_i^{\text{gbest}}$ and $X_i^{\text{lbest}}$ are similarly derived.

---

**Procedure.** The updating of the best position in the subsequent iterations

---

generate initial random solution $x_{id}^{\text{pbest}}(t)$ and add it to the archive;
update $x_{id}^{\text{pbest}}(t)$ to produce $x_{id}(t+1)$ and evaluate $x_{id}(t+1)$;
  **if** ($x_{id}^{\text{pbest}}(t)$ dominates $x_{id}(t+1)$) **then** discard $x_{id}(t+1)$;
  **else if** ($x_{id}(t+1)$ dominates $x_{id}^{\text{pbest}}(t)$) **then** replace $x_{id}^{\text{pbest}}(t)$ with $x_{id}(t+1)$, and add $x_{id}(t+1)$ to the archive $x_{id}^{\text{pbest}}(t+1) = x_{id}(t+1)$;
  **else if** ($x_{id}(t+1)$ is dominated by any member of the archive) **then** discard $x_{id}(t+1)$;
  **else** apply test ($x_{id}^{\text{pbest}}(t), x_{id}(t+1)$, archive) to determine which becomes the new current solution and whether to add $x_{id}(t+1)$ to the archive;
until a termination criterion has been reached, return to line 2.

---

### 6.5.4 *Framework of MOPSO*

The MOPSO framework with permutation-based representation to solve the dynamic construction layout planning problems is presented in this section.

*Step 1.* Initialize $L$ particles as a swarm. For $i = 1, \cdots, L$, $j = 1, \cdots, \tau$ generate the position of the $i^{th}$ particle with an integer random position:

**Procedure.** Test

**if** the archive is not full;
  add $x_{id}(t+1)$ to the archive;
  **if** ($x_{id}(t+1)$ is in a less crowded region of the archive than $x_{id}^{\text{pbest}}(t)$);
    accept $x_{id}^{\text{pbest}}(t+1) = x_{id}(t+1)$;
  **else** maintain $x_{id}^{\text{pbest}}(t+1) = x_{id}^{\text{pbest}}(t)$;
**else**
  **if** ($x_{id}(t+1)$ is in a less crowded region of the archive than $x_{id}^{\text{pbest}}(t)$ for some member $x_{id}^{\text{pbest}}(t)$ on the archive);
  add $x_{id}(t+1)$ to the archive, and remove a member of the archive from the most crowded region;
  **if** ($x_{id}(t+1)$ is in a less crowded region of the archive than $x_{id}^{\text{pbest}}(t)$);
    accept $x_{id}^{\text{pbest}}(t+1) = x_{id}(t+1)$;
  **else** maintain $x_{id}^{\text{pbest}}(t+1) = x_{id}^{\text{pbest}}(t)$;
 **else**
  **if** ($x_{id}(t+1)$ is in a less crowded region of the archive than $x_{id}^{\text{best}}(t)$);
    accept $x_{id}^{best}(t+1) = x_{id}(t+1)$;
  **else** maintain $x_{id}^{\text{pbest}}(t+1) = x_{id}^{\text{pbest}}(t)$.

$$X_{if}^{j}(t)=[x_{i1}^{1}(t),x_{i2}^{1}(t),\cdots,x_{iN}^{1}(t);x_{i1}^{2}(t),x_{i2}^{2}(t),\cdots,x_{iN}^{2}(t);\cdots;x_{i1}^{T}(t),x_{i2}^{T}(t),\cdots,x_{iN}^{T}(t)] \tag{6.51}$$

the value is $\{0,1,2,\cdots,N\}$, $P_i^{\max}(d)=N$, $P_i^{\min}(d)=0$. There can be more than one '0', and other values are unequal in one phase, otherwise, a different facility will be located in same location which is unacceptable.

As shown in Figure 6.12, there are $F$ facilities, $N$ locations and $N-F$ dummy facilities.

*Step 2.* Decoding particles into solutions. For $i^{th}$ particle, $X_f^j$, if $x_f^j = 0$ denotes that facility $f$ does not exist during phase $j$, while $x_f^j \neq 0$ denotes that facility $f$ is located in location $x_f^j$ during phase $j$.
*Step 3.* Check the feasibility of solutions. For $i = 1,\cdots,L$, if the feasibility criterion is met by all particles, i.e. $\delta_{xit}S_{xt} < D_i$, then continue. Otherwise, return to step 1.
*Step 4.* Initialization the speed of each particle and personal best position:. Let velocity $V_{if}(1) = [v_{i1}(1), v_{i2}(1),\cdots,v_{iN}(1)]$, and the value is between 0 and 1. The initial $x_{id}^{\text{pbest}}$ is current position.
*Step 5.* Evaluation particles' performance. For $i = 1,\cdots,L$, compute the performance measurement under every objective, and set the objective value as the fitness value of $X_i^j(t)$, represented by $Fitness_i(X_i^j(t))$. For each of the particles, $Fitness_i(X_i^j(t))$ is a $1\times 2$ matrix.
*Step 6.* Store the positions of the particles that represent non-dominated vectors in the archive denoted as $ARC$.

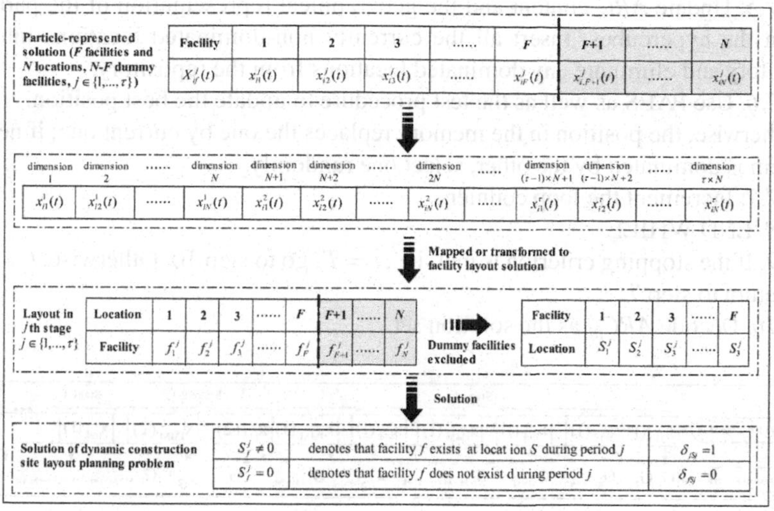


**Fig. 6.12** Decoding method and mapping between MOPSO particles and solutions to the problem

*Step 7.* WHILE the maximum number of cycles has not been reached DO

*Step 7.1.* Compute the speed of each particle using the following formula:

$$v^j_{id}(t+1) = w(t)v^j_{id}(t) + c_p r_1[pBest^j_{id} - x^j_{id}(t)] + c_g r_2[ARC_{fd}(t) - x^j_{id}(t)], \quad (6.52)$$

where $c_R$ is an acceleration constant, $w(t) = w(T) + \frac{t-T}{1-T}[w(1) - w(T)]$ [113], $ARC_{fd}(t)$ is a solution randomly selected from the repository in each iteration for facility $i$, the index $d$ is selected in the following way: hypercubes which contain more than one particle are assigned a fitness equal to the result of dividing any number into the number of contained particles. The assignment aims to decrease the fitness of those hypercubes that contain more particles and it can be seen as a form of fitness sharing [114]. Then, roulette-wheel selection method is applied using these fitness values to select the hypercube from which the corresponding particle is taken. Once the hypercube has been selected, a particle is selected randomly within such hypercube.

*Step 7.2.* Add the speed produced from the previous step, and compute the new positions of the particles using:

$$x^j_{id}(t+1) = x^j_{id}(t) + v^j_{id}(t+1). \quad (6.53)$$

*Step 7.3.* Maintain the particles within the search space to prevent them moving beyond their boundaries.

*Step 7.4.* Evaluate particles' performance.
*Step 7.5.* Update *ARC* content and the geographical representation of the particles within the hypercubes. Insert all the currently non-dominated locations into the repository and eliminate any dominated locations from the repository.
*Step 7.6.* Use PAES as well as the test procedure to update the best position.
Otherwise, the position in the memory replaces the one by current one; if neither of them is dominated by the other, select one randomly.
*Step 7.7.* Increment the loop counter.
*Step 8.* END WHILE
*Step 9.* If the stopping criterion is met, i.e., $t = T$, go to step 10. Otherwise, $t = t + 1$ and return to step 7.
*Step 10.* Decode $ARC_{id}$ as the solution set.

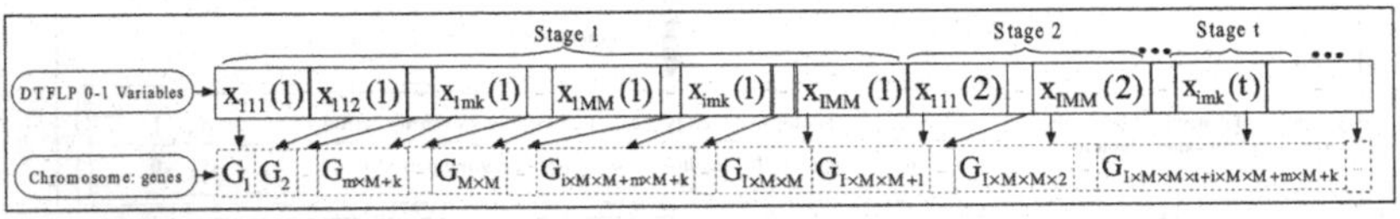


**Fig. 6.13** Representation of the DTFLP problem in SA-based MOGA

## 6.6 SA-Based MOGA for DTFLP

Due to the high combinatorial complexity of the DTFLP problem (McKendall and Hakobyan 2010 [144]), sound heuristic techniques are necessary. GA has become popular in engineering optimization problems (Gen and Cheng 2000 [436]; Xu and Li 2012 [585]) and in water resources planning (Nicklow et al. 2010 [146]) in the recent past but has been shown to have poor convergence near the global optimum and premature convergence to the suboptimum because restrictions on solution space are not carried out during the process (Ghoshal 2004 [135]; Zhou et al. 2009 [157]). Much of the research has realized it is very difficult to use a single technology such as GA to handle the DTFLP problem (Zhang et al. 2002 [156]). Osman et al. (2003) [147] proposed an optimization model to search for an optimal dynamic site layout by utilizing GA within a CAD (Computer Aided Design) environment. Erel et al. (2001) [83] also proposed several heuristics for a dynamic plant layout problem using dynamic programming and simulated annealing (SA). Ghoshal (2004) [135] improved the searching ability and dispensed with the GA local optimum by adopting a basic SA exponential acceptance probability as the criterion for testing, acceptance or rejection. Nevertheless, these methods do not employ a dynamic search scheme to lead the algorithm to layout different facilities involved in different time intervals during the dynamic search (Ning et al. 2010 [81]) and have not yet been comprehensively applied to a real case for a large scale hydropower construction project. In this chapter, the SA-based GA is used to solve the Jinping-II hydropower construction DTFLP problem. The presented methodology is relatively similar to what practitioners are using today but is considerably more effective and realistic in the modeling.

### 6.6.1 Representation of Chromosomes

The representation of the chromosomes should meet three principles (Goldberg 1989 [336]): completeness, soundness, and non-redundancy. The SA-based GA adopts the characteristics of the chromosome which is divided by level. As Figure 6.13 shows, the structure of each chromosome is like this: each chromosome can be divided into four levels. First, the chromosome is divided by stage, which can be divided into $T$ stages. Based on the first level, the second level is divided by the type of facilities, with $I$ parts. The third level is divided by the candidate locations of each facility. There are $M$ locations for each type of facility which is the same as the third level, with the last level also divided into $M$ parts. The genes are put in order according to the candidate facility location index. Thus, there are $M$ genes in each subpart of the last level. In total, there are $I \times M \times M \times T$ genes in one chromosome. As the variables of the DTFLP problem are 0 and 1, which are suitable for the GA gene, it is possible to omit the encoding and decoding procedure and initialize the population of $N$ chromosomes randomly. Each chromosome $C_n$ represents a sequence of genes $x_{imk}(t)$ for the DTFLP model (see Figure 6.13). In this way, the parallel structure of the chromosomes' representation in the SA-based MOGA is built. This kind of parallel structure can reduce the dimensionality of the solution so reducing computing difficulty. It should be noted that when initializing the first generation of the chromosomes randomly, the solution can be made feasible by combining the constraints with each level of chromosome separately (see Figure 6.14) which also saves the searching route and computing time.

### 6.6.2 Dealing with Multiple Objectives

To reduce the difficulty of measuring the different objectives and to avoid the subjective predilection process to obtain the weight, the $F$ fitness is adopted (Goldberg 1989 [336]) to deal with the multi-objective part of the SA-based GA algorithm. $F$ fitness (Goldberg 1989 [336]) uses a ranking matrix to aggregate the total fitness value for the GA algorithm. No previous research has combined the $F$ fitness (Goldberg 1989 [336]) with the SA-GA hybrid algorithm to solve multi-objectives, so this application can be seen to be original and inspired.

### 6.6.3 $T_f$ Control Method

To improve the SA-based GA algorithm proposed by Mohanta et al. (2007) [143] in the GA neighborhood search , $T_f$ is added at the end of annealing process to control the GA neighborhood search process, and then checked twice (see Figure 6.14). The $T_f$ control has a significant effect, which is explained in following section. To make it more suitable for the DTFLP problem and to avoid an unfeasible solution, it is checked three times; at initialization, at the SA neighborhood move, and when the new population after crossover and mutation of GA is derived for the algorithm.

These parts are different from pre-existing SA-GA hybrid algorithms.

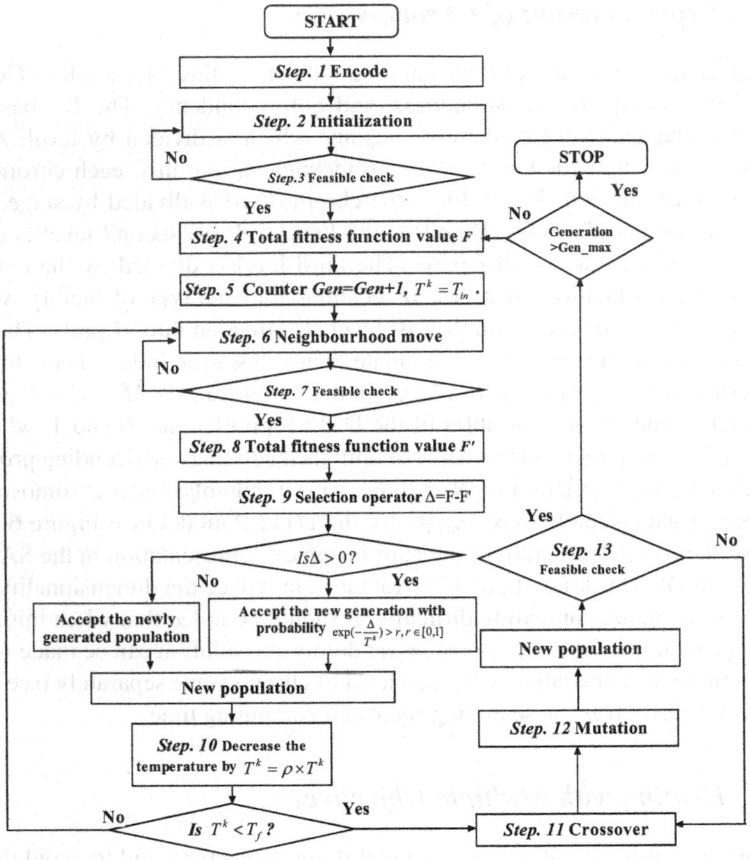


**Fig. 6.14** The SA-based MOGA proceeding

## 6.6.4 *Framework of SA-Based MOGA*

The proceeding of the SA-based MOGA algorithm is:

*Step 1.* Encode candidate solution population;
*Step 2.* Initialization: the size of population, maximum number of generations $G_{max}$, mutation probability, crossover rate, the population with each variable of length $I \times M \times M$ bits; Initial the SA parameters: the value of initial temperature, the value of acceptance probability ($P_c$), the final acceptance probability $P_f$ and cooling rate, $\rho$ (Bennage and Dhingra 1995 [125]) Eq.(6.54):

$$\rho = \frac{1}{el_{\max} - 1} \cdot \frac{\ln P_c}{\ln P_f}, \tag{6.54}$$

where *el* is the inner counter for every generation.

*Step 3.* Check: map the string for each variable to respective bounds, Equation (6.30), if the variable satisfy the constrains, go to Step 4. or return to Step 2;
*Step 4.* Calculate the objective functions of the new generation, $f_1$ Eq. (6.33) and $f_2$ Eq. (6.34) and evaluate the total fitness function Eqs. (6.55) and (6.56) to get $F$ (Goldberg 1989 [336]) as the fitness value to choose the best solution:

$$F_i(X_j) = \begin{cases} (N - Y_i(X_j))^2, & Y_i(X_j) > 1 \\ kN^2, & Y_i(X_j) = 1 \end{cases} \quad i = 1, 2, \cdots, n \tag{6.55}$$

$$F(X_j) = \sum_{i=1}^{n} F_i(X_j) \quad j = 1, 2, \cdots, n \tag{6.56}$$

where $n = 2$ is the number of the objectives. $N$ is the number of the chromosomes. $X_j$ is the chromosome. $Y_i$ is the rank number of chromosome $X_j$. $F_i(X_j)$ is the fitness value of $X_j$ in objective $i$. $F(X_j)$ is the total fitness value of all the objectives. $k \in (1,2)$ and $k$ is the factor to lay emphasis on the fitness of the best solution of one objective function.
*Step 5.* Counter: $Gen = Gen + 1$, and set $T^k = T_{in}$ according to Eq. (6.57) (Kirkpatrick et al. 1983 [140]). If $Gen > Gen_max$, go to Step 15. Or go to Step 6;

$$\begin{cases} T_{in} = \frac{F_{\min} - F_{\max}}{\ln P_c}, \\ T_f = \frac{F_{\min} - F_{\max}}{\ln P_f}, \end{cases} \tag{6.57}$$

where $F_{\min}$ and $F_{\max}$ are the lower and higher bound min and max of the fitness value.
*Step 6.* Neighborhood move: choose a position of every chromosome randomly and change its value;
*Step 7.* If the neighborhood solution is feasible, go to Step 8, or return to Step 6;
*Step 8.* Calculate the objective functions of the neighborhood solution, $f_1'$ Eq. (6.33) and $f_2'$ Eq. 6.34 and evaluate the total fitness function Eqs. (6.55) and (6.56) to get $F'$;
*Step 9.* Boltzmann selection operator (Nicklow et al. 2010 [146]): $\Delta = F - F'$, if $\Delta > 0$, Accept the new generation with probability $exp(-\frac{\Delta}{T^k}) > r$, see Eq. 6.58 (Ghoshal 2004 [135]); Or $\Delta < 0$, accept the newly generated population;

$$P(\Delta) = exp(-\Delta / K_b T), \tag{6.58}$$

$K_b$ is the Boltzman's constant (Ashtiani et al. 2007 [124]; Baykasoglu and Gindy 2001 [127]; Kirkpatrick et al. 1983 [140]; Lin et al. 1993 [142]).
*Step 10.* Decrease the temperature by $T^k = \rho T^k$. See Eq. 6.6.4, if $T^k < T_f$, go to Step 11; Or, return to Step 6.
*Step 11.* Adopt the uniform crossover technique (Back et al., 2000) to generate new offsprings from the selected parents, the crossover probability is $p_c$;
*Step 12.* Adopt the jump mutation to break the stagnation in improvement (Nicklow et al. 2010 [146]), the mutation probability is $P_m$;
*Step 13.* Check: map the new generation to respective bounds, Eq. 6.30, if the variable satisfy the constrains, go to Step 14. or return to Step 11;
*Step 14.* End: Choose the best individual in new population as results.

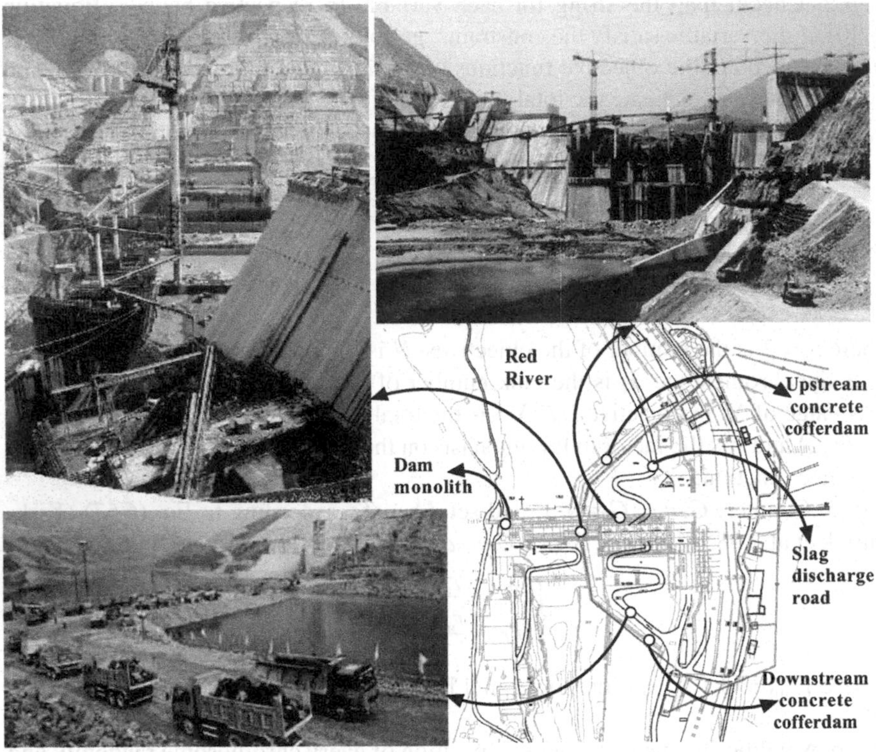


**Fig. 6.15** Longtan hydropower station

The complete process is repeated over several genetic iteration cycles until the required optimum is achieved. In this way, The SA-based MOGA can be used to solve DTFLP model.

It should be noted that the pre-existing GA package provided by the Matlab was not used as it can only solve simple problems and cannot be combined with the SA. The algorithm is implemented in Matlab 7.0 with a new program ,with a Core *i*5, 3.20 GHz with 4 GB memory.

## 6.7 Case Study for DCSLP

The problem considered in this section is from the Longtan hydropower station project. The following is the case representation, the model comparison and the algorithm evaluation.

### 6.7.1 Representation of Case Problem

The Longtan hydropower construction project located in Tian'e county of Guangxi Zhuang Autonomous Region is one of the ten major indicative projects for the national Great Western Development in China and the Power Transmission from West to East strategies. It is a control project for the cascade development of the Red River (Figure 6.15). Its layout is as follows: a roller compacted concrete gravity dam, a flood building with seven outlets, two bottom outlets on the river bed, a power stream systems capacity with nine installations on the left bank, and navigation structures on the right bank, including a 2-stage vertical ship lift which is used for navigation. The designed impoundment level of the Longtan hydropower station is 400m and the installed plant capacity is 5,400 MW. During the construction process, tunnel diversion is used to divert the river, with the two diversion openings on the left and right banks. The diversion standard is a ten-year flood with a corresponding flow of 14,700 $m^3$/s.

There are ten bid sections in the principal part of the Longtan large-scale water conservancy and hydropower construction project. Bid I is the excavation of the left bank slope, and the left bank diversion tunnel. Bid II is the excavation of the right bank slope, the diversion tunnel and the navigation structure. Bid I and Bid II can be done concurrently. Bid III1 is the river closure, the cofferdam and the excavation of the river bed. Bid III2 is the diversion and retaining dam concrete pouring on the left bank. Bid IV is plant construction, the water diversion and the tailrace system in the underground powerhouse. Bid V is the second excavation of the navigation structure and concrete pouring. Bid VI is steel tube processing. Bid VII, VIII and IX are the installation of the generating system's metal structures, the flood system and the ship lift system. Bid X is the electrical installation. Based on the plan for each bid section, the principal part of the Longtan hydropower construction project is divided into 8 phases. The facilities required in each phase are shown in Figure 6.16. There are 14 main temporary facilities involved in the principal part of the Longtan hydropower construction project. These facilities as well as their required area during each phase are listed in Table 6.1. As shown in Figure 6.17, 14 candidate locations were identified after an investigation of carrying capacity, slope stability and topographic, geological and traffic conditions. The location areas are denoted as $D_i$, $D_i = (6500, 6000, 5750, 7800, 7200, 25000, 26000, 6600, 7200, 4600, 7200, 7200, 7400, 4800)$ $(m^2)$. The distances between these 17 locations are shown in Table 6.2. The setup costs for the facilities at different locations are listed in Table 6.3. The facilities closure costs include dismantling costs, transport costs, material loss costs and function loss costs. In this section, the facilities closure costs at different locations are assumed to be 11.3% of startup costs.

Empirically, $F_8$ (Oil Depot) and $F_9$ (Explosive Storage) are the most likely to cause safety and environmental accidents. Thus, they are identified as 'high-risk' facilities, namely $\{x^1, x^2\} = \{F_8, F_9\}$. $F_2$ (Carpentry Shop), $F_{13}$ (Office) and $F_{14}$ (Labor Residence) are identified as 'high-protection' facilities, namely $\{y^1, y^2, y^3\} = \{F_2, F_{13}, F_{14}\}$. Decision makers can set the weight by preference. In Longtan's case, the weights were set at: $\{w_{11}, w_{12}, w_{13}\} = \{w_{21}, w_{22}, w_{23}\} = \{0.2, 0.3, 0.5\}$. The

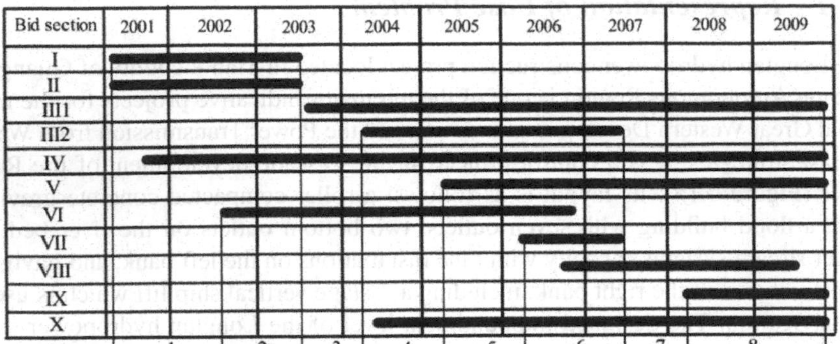


| Phase | 1 | 2 | 3 | 4 | 5 | 6 | 7 | 8 |
|---|---|---|---|---|---|---|---|---|
| Start time | 2001.07 | 2002.08 | 2003.09 | 2004.01 | 2005.01 | 2006.02 | 2007.06 | 2008.01 |
| Finish time | 2002.08 | 2003.09 | 2004.01 | 2005.01 | 2006.02 | 2007.06 | 2008.01 | 2009.12 |
| Layout | $F_1$ $F_4$ $F_5$ $F_6$ $F_8$ $F_9$ $F_{10}$ $F_{11}$ $F_{12}$ $F_{13}$ $F_{14}$ | $F_1$ $F_2$ $F_4$ $F_5$ $F_6$ $F_8$ $F_9$ $F_{10}$ $F_{11}$ $F_{12}$ $F_{13}$ $F_{14}$ | $F_1$ $F_2$ $F_4$ $F_5$ $F_6$ $F_8$ $F_9$ $F_{10}$ $F_{11}$ $F_{12}$ $F_{13}$ $F_{14}$ | $F_1$ $F_2$ $F_3$ $F_4$ $F_5$ $F_6$ $F_8$ $F_{10}$ $F_{11}$ $F_{12}$ $F_{13}$ $F_{14}$ | $F_1$ $F_2$ $F_3$ $F_4$ $F_5$ $F_6$ $F_8$ $F_9$ $F_{10}$ $F_{11}$ $F_{12}$ $F_{13}$ $F_{14}$ | $F_1$ $F_2$ $F_5$ $F_6$ $F_7$ $F_8$ $F_9$ $F_{10}$ $F_{11}$ $F_{12}$ $F_{13}$ $F_{14}$ | $F_1$ $F_2$ $F_5$ $F_6$ $F_7$ $F_8$ $F_{10}$ $F_{11}$ $F_{12}$ $F_{13}$ $F_{14}$ | $F_2$ $F_5$ $F_6$ $F_7$ $F_8$ $F_{11}$ $F_{12}$ $F_{13}$ $F_{14}$ |

**Fig. 6.16** The construction schedule and facilities required

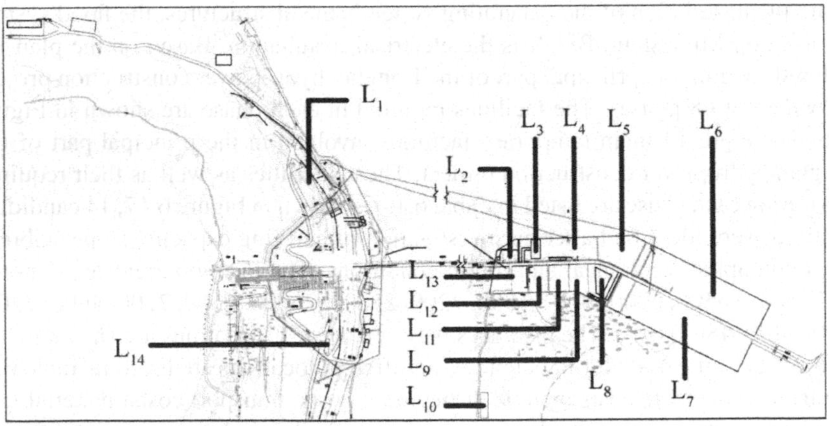


**Fig. 6.17** The distribution of candidate locations

**Table 6.1** Facilities to be distributed and their required area during each phase

| Index | Facility | Phase 1 | Phase 2 | Phase 3 | Phase 4 | Phase 5 | Phase 6 | Phase 7 | Phase 8 |
|---|---|---|---|---|---|---|---|---|---|
| $F_1$ | Reinforcing Steel Shop | 1800 | 1800 | 1800 | 2000 | 2000 | 1800 | 1000 | 0 |
| $F_2$ | Carpentry Shop | 0 | 2500 | 2500 | 3000 | 3000 | 2500 | 2000 | 2000 |
| $F_3$ | Concrete Precast Shop | 0 | 0 | 0 | 1360 | 1360 | 0 | 0 | 0 |
| $F_4$ | Drill Tools Repair Shop | 800 | 800 | 800 | 800 | 800 | 0 | 0 | 0 |
| $F_5$ | Equipment Repairing Workshop | 3500 | 3500 | 3500 | 3500 | 3500 | 3000 | 3000 | 3000 |
| $F_6$ | Truck Maintenance Shop | 4000 | 4000 | 4000 | 4000 | 4000 | 3500 | 3000 | 3000 |
| $F_7$ | Metal and Electrical Installing Workshop | 0 | 0 | 0 | 0 | 0 | 3000 | 3300 | 3300 |
| $F_8$ | Oil Depot | 1000 | 1000 | 1000 | 1000 | 1000 | 1000 | 1000 | 1000 |
| $F_9$ | Explosive Storage | 750 | 750 | 750 | 0 | 750 | 750 | 0 | 0 |
| $F_{10}$ | Rebar Storage | 2000 | 2000 | 2500 | 3510 | 3510 | 3000 | 2500 | 0 |
| $F_{11}$ | Steel Storage | 2000 | 2000 | 2000 | 2500 | 2500 | 2000 | 2000 | 2000 |
| $F_{12}$ | Integrated Warehouse | 800 | 1000 | 1000 | 1000 | 1000 | 800 | 1000 | 1000 |
| $F_{13}$ | Office | 4500 | 4500 | 4500 | 4500 | 4500 | 4500 | 4500 | 4500 |
| $F_{14}$ | Labor Residence | 8000 | 8000 | 8000 | 8000 | 10000 | 8000 | 8000 | 8000 |

**Table 6.2** The distances between locations ($m$)

| | $A_{ij}$ | | | | | | | | | | | | | |
|---|---|---|---|---|---|---|---|---|---|---|---|---|---|---|
| | $L_1$ | $L_2$ | $L_3$ | $L_4$ | $L_5$ | $L_6$ | $L_7$ | $L_8$ | $L_9$ | $L_{10}$ | $L_{11}$ | $L_{12}$ | $L_{13}$ | $L_{14}$ |
| $L_1$ | 0 | 1440 | 1584 | 1740 | 1908 | 2148 | 2256 | 1956 | 1800 | 2172 | 1680 | 1620 | 1476 | 1452 |
| $L_2$ | 1440 | 0 | 144 | 300 | 468 | 708 | 816 | 636 | 516 | 732 | 348 | 264 | 72 | 1140 |
| $L_3$ | 1584 | 144 | 0 | 156 | 324 | 564 | 672 | 492 | 372 | 1092 | 204 | 120 | 180 | 1284 |
| $L_4$ | 1740 | 300 | 156 | 0 | 168 | 408 | 480 | 336 | 216 | 1248 | 120 | 144 | 228 | 1440 |
| $L_5$ | 1908 | 468 | 324 | 168 | 0 | 240 | 348 | 108 | 132 | 1416 | 288 | 264 | 348 | 1608 |
| $L_6$ | 2148 | 708 | 564 | 408 | 240 | 0 | 108 | 240 | 312 | 1656 | 420 | 504 | 588 | 1848 |
| $L_7$ | 2256 | 816 | 672 | 480 | 348 | 108 | 0 | 264 | 360 | 1680 | 600 | 708 | 828 | 2028 |
| $L_8$ | 1956 | 636 | 492 | 336 | 108 | 240 | 264 | 0 | 120 | 1500 | 336 | 444 | 564 | 1656 |
| $L_9$ | 1800 | 516 | 372 | 216 | 132 | 312 | 360 | 120 | 0 | 1380 | 168 | 324 | 444 | 1776 |
| $L_{10}$ | 2172 | 732 | 1092 | 1248 | 1416 | 1656 | 1680 | 1500 | 1380 | 0 | 1212 | 1128 | 888 | 1872 |
| $L_{11}$ | 1680 | 348 | 204 | 120 | 288 | 420 | 600 | 336 | 168 | 1212 | 0 | 84 | 276 | 1488 |
| $L_{12}$ | 1620 | 264 | 120 | 144 | 264 | 504 | 708 | 444 | 324 | 1128 | 84 | 0 | 192 | 1404 |
| $L_{13}$ | 1476 | 72 | 180 | 228 | 348 | 588 | 828 | 564 | 444 | 888 | 276 | 192 | 0 | 1212 |
| $L_{14}$ | 1452 | 1140 | 1284 | 1440 | 1608 | 1848 | 2028 | 1656 | 1776 | 1872 | 1488 | 1404 | 1212 | 0 |

fuzzy random coefficients for the interaction costs per unit distance are listed in Table 6.4. The fuzzy random operating costs for each facility located at $L_2$, $L_3$, $L_4$, $L_9$, $L_{11}$, $L_{12}$ and $L_{13}$ are listed in Table 6.5. The facilities operating costs at $L_1$ and $L_{14}$ are 5.8% greater because of the distance from the work shop. Similarly, the facilities operating costs at $L_5$, $L_6$, $L_7$, $L_8$ and $L_{10}$ are each 7.5% greater. Using the method proposed in the previous section, the fuzzy random coefficients can be transformed to determined values.

**Table 6.3** The startup costs of facilities at different locations (1000 RMB)

| | $C^s_{xit}$ | | | | | | | | | | | | | |
|---|---|---|---|---|---|---|---|---|---|---|---|---|---|---|
| $x$ | $i$ | | | | | | | | | | | | | |
| | $L_1$ | $L_2$ | $L_3$ | $L_4$ | $L_5$ | $L_6$ | $L_7$ | $L_8$ | $L_9$ | $L_{10}$ | $L_{11}$ | $L_{12}$ | $L_{13}$ | $L_{14}$ |
| $F_1$ | 367.1 | 264.5 | 250.8 | 264.1 | 267.5 | 214.4 | 223.5 | 255.5 | 170.2 | 159.6 | 131.0 | 105.3 | 174.0 | 360.9 |
| $F_2$ | 157.3 | 126.4 | 130.4 | 84.7 | 85.5 | 150.0 | 138.6 | 127.7 | 106.3 | 180.1 | 97.3 | 117.4 | 98.8 | 174.9 |
| $F_3$ | 126.1 | 196.4 | 192.6 | 267.8 | 270.6 | 235.4 | 223.5 | 255.5 | 287.8 | 160.1 | 130.4 | 108.9 | 218.3 | 229.6 |
| $F_4$ | 105.9 | 116.1 | 120.0 | 127.6 | 128.3 | 182.2 | 95.2 | 159.6 | 148.8 | 138.1 | 131.4 | 179.1 | 163.7 | 131.2 |
| $F_5$ | 126.1 | 105.2 | 170.7 | 95.9 | 96.5 | 139.6 | 127.3 | 180.9 | 191.3 | 159.6 | 130.0 | 95.2 | 109.1 | 98.4 |
| $F_6$ | 126.4 | 100.6 | 80.2 | 74.1 | 74.6 | 85.7 | 63.1 | 95.8 | 127.5 | 148.4 | 140.7 | 95.2 | 131.8 | 109.3 |
| $F_7$ | 84.7 | 106.0 | 180.6 | 169.4 | 171.1 | 160.0 | 85.0 | 95.8 | 106.3 | 116.7 | 195.6 | 137.8 | 142.1 | 87.4 |
| $F_8$ | 96.0 | 107.6 | 100.3 | 105.8 | 106.4 | 192.6 | 191.4 | 202.8 | 148.8 | 138.1 | 151.1 | 190.9 | 87.4 | 98.4 |
| $F_9$ | 105.9 | 97.0 | 80.2 | 95.2 | 96.5 | 150.8 | 212.1 | 170.4 | 180.7 | 169.9 | 98.3 | 95.5 | 109.1 | 98.3 |
| $F_{10}$ | 84.7 | 95.0 | 90.3 | 105.8 | 106.4 | 128.6 | 116.9 | 106.6 | 85.0 | 95.2 | 98.0 | 63.0 | 81.9 | 109.7 |
| $F_{11}$ | 63.0 | 73.9 | 103.8 | 95.9 | 96.5 | 117.9 | 127.3 | 95.8 | 97.2 | 84.9 | 98.3 | 63.5 | 76.2 | 90.1 |
| $F_{12}$ | 63.5 | 74.9 | 90.0 | 63.5 | 64.6 | 96.4 | 106.6 | 95.8 | 95.6 | 74.7 | 86.9 | 74.0 | 98.8 | 89.1 |
| $F_{13}$ | 115.0 | 126.4 | 140.7 | 148.3 | 149.2 | 117.9 | 106.7 | 95.2 | 96.8 | 106.4 | 97.3 | 76.7 | 131.8 | 164.5 |
| $F_{14}$ | 167.3 | 73.3 | 60.2 | 74.0 | 74.6 | 107.2 | 116.9 | 117.1 | 85.0 | 95.6 | 97.3 | 63.0 | 76.5 | 89.6 |

After running the MATLAB program for the proposed MOPSO algorithm, the solutions were obtained and the efficiency of the proposed algorithm tested.

The algorithm parameters for the case problem were set as follows: swarm size popsize $= 500$; iteration max $T = 1000$; personal best position acceleration constant $c_p = 2$; global best position acceleration constant $c_g = 2$; local best position acceleration constant $c_l = 1$. The large number of particles and iterations were used to reduce the probability of becoming trapped in a stagnant state [122]. The computing environment was an inter Core $i2$, 3.20 GHz clock pulse with 2 GB memory.

The red dots in Figure 6.18 represent the pareto-optimal solutions, while the blue dots represent the best position of the particles at the current iteration. Decision makers are able to choose their preferred plan from these pareto-optimal solutions. For example, if decision makers believe that the safety and environmental objective is the most important, they may allow for an increased cost to ensure the safest site layout plan. Thus, they would choose the safest plan as shown in Table 6.6. On the contrary, if decision makers believe the cost objective is the most important, they may choose the minimum total cost plan as shown in Table 6.7.

### 6.7.2 Model Comparison

Here a comparison is given between a dynamic construction site layout model considering fuzzy random factors and other traditional models, including one without any uncertainty, and the other only considering fuzziness.

(1) Qualitatively, there are two kinds of uncertainty in construction site layout planning problems, i.e., objective uncertainty which can be dealt using randomness, and subjective uncertainty which can be dealt using fuzzyness. Only the fuzzy

**Table 6.4** The fuzzy random coefficients of the interaction costs per unit distance (RMB)

| | $\bar{\tilde{C}}_{1,3,t}$ | $\bar{\tilde{C}}_{1,10,t}$ | $\bar{\tilde{C}}_{2,3,t}$ |
|---|---|---|---|
| Phase 1 | 0 | $(3.6, \rho_{1,10,1}, 7.8)$ $\rho_{1,10,1} \sim \mathcal{N}(5.2, 5)$ | 0 |
| Phase 2 | 0 | $(3.8, \rho_{1,10,2}, 8)$ $\rho_{1,10,2} \sim \mathcal{N}(5.4, 6)$ | 0 |
| Phase 3 | 0 | $(4.1, \rho_{1,10,3}, 8)$ $\rho_{1,10,3} \sim \mathcal{N}(5.5, 4)$ | 0 |
| Phase 4 | $(3.6, \rho_{1,3,4}, 7.8)$ $\rho_{1,3,4} \sim \mathcal{N}(5, 5)$ | $(4.3, \rho_{1,10,4}, 8.4)$ $\rho_{1,10,4} \sim \mathcal{N}(5.5, 4)$ | $(5, \rho_{2,3,4}, 12.5)$ $\rho_{2,3,4} \sim \mathcal{N}(10, 4)$ |
| Phase 5 | $(4, \rho_{1,3,5}, 8.5)$ $\rho_{1,3,5} \sim \mathcal{N}(5.5, 6)$ | $(4.5, \rho_{1,10,5}, 9.5)$ $\rho_{1,10,5} \sim \mathcal{N}(5.5, 5)$ | $(6.7, \rho_{2,3,5}, 14)$ $\rho_{2,3,4} \sim \mathcal{N}(10.5, 5)$ |
| Phase 6 | 0 | $(4, \rho_{1,10,6}, 8.5)$ $\rho_{1,10,6} \sim \mathcal{N}(5, 4)$ | 0 |
| Phase 7 | 0 | $(4, \rho_{1,10,7}, 7.8)$ $\rho_{1,10,3} \sim \mathcal{N}(5.5, 4)$ | 0 |
| Phase 8 | 0 | 0 | 0 |
| | $\bar{\tilde{C}}_{5,8,t}$ | $\bar{\tilde{C}}_{6,8,t}$ | $\bar{\tilde{C}}_{7,11,t}$ |
| Phase 1 | $(8.4, \rho_{5,8,1}, 16)$ $\rho_{5,8,1} \sim \mathcal{N}(11.5, 8)$ | $(9, \rho_{6,8,1}, 18.5)$ $\rho_{6,8,1} \sim \mathcal{N}(12.4, 4)$ | 0 |
| Phase 2 | $(8.5, \rho_{5,8,2}, 15.8)$ $\rho_{5,8,2} \sim \mathcal{N}(11, 8)$ | $(8.5, \rho_{6,8,2}, 18.8)$ $\rho_{6,8,2} \sim \mathcal{N}(12.6, 8)$ | 0 |
| Phase 3 | $(8.5, \rho_{5,8,3}, 15.8)$ $\rho_{5,8,3} \sim \mathcal{N}(11, 10)$ | $(8.5, \rho_{6,8,3}, 18.8)$ $\rho_{6,8,3} \sim \mathcal{N}(12.6, 8)$ | 0 |
| Phase 4 | $(8.5, \rho_{5,8,4}, 15.8)$ $\rho_{5,8,4} \sim \mathcal{N}(11, 4)$ | $(8.5, \rho_{6,8,4}, 19)$ $\rho_{6,8,4} \sim \mathcal{N}(12.8, 8)$ | 0 |
| Phase 5 | $(10, \rho_{5,8,5}, 16.8)$ $\rho_{5,8,5} \sim \mathcal{N}(12, 6)$ | $(10, \rho_{6,8,5}, 20.8)$ $\rho_{6,8,5} \sim \mathcal{N}(12, 8)$ | 0 |
| Phase 6 | $(8, \rho_{5,8,6}, 15)$ $\rho_{5,8,6} \sim \mathcal{N}(11, 8)$ | $(8, \rho_{6,8,6}, 22)$ $\rho_{6,8,6} \sim \mathcal{N}(12, 6)$ | $(4, \rho_{7,11,6}, 6)$ $\rho_{7,11,6} \sim \mathcal{N}(5, 2)$ |
| Phase 7 | $(8.5, \rho_{5,8,7}, 15.8)$ $\rho_{5,8,7} \sim \mathcal{N}(11, 4)$ | $(8.5, \rho_{6,8,7}, 24)$ $\rho_{6,8,7} \sim \mathcal{N}(12, 8)$ | $(4, \rho_{7,11,7}, 7)$ $\rho_{7,11,7} \sim \mathcal{N}(5.4, 1)$ |
| Phase 8 | $(8.5, \rho_{5,8,8}, 15.8)$ $\rho_{5,8,8} \sim \mathcal{N}(11, 8)$ | $(8.5, \rho_{6,8,8}, 26)$ $\rho_{6,8,8} \sim \mathcal{N}(14, 8)$ | $(4, \rho_{7,11,8}, 8)$ $\rho_{7,11,8} \sim \mathcal{N}(6, 4)$ |

random programming approach explicitly considers the entire range of hybrid uncertain scenarios, which is more realistic.

(2) Quantitatively, for each type, the MOPSO programme was run 10 times with the same parameter values to study the performance. Considering that truly determinate data do not exist in practice, and most are obtained by ignoring the uncertainty, the determinate data are chosen randomly without any uncertainty considerations. For the fuzzy type, the statement "it is about 5.2 RMB" is understood as "it is 5.2 RMB" to ignore the ambiguous information with the other parameters being similar. From the results listed in Table 6.8, it is clear that the fuzzy random modeling has a much better performance than the others, not only in the value of the results, but also in stability.

**Table 6.5** The fuzzy random coefficients of the operating costs ($1 \times 10^3$ RMB)

| | Phase 1 | Phase 2 | Phase 3 |
|---|---|---|---|
| $F_1$ | $(80.3,\mathcal{N}(85.5,8),90)$ | $(76.1,\mathcal{N}(80.5,5),86)$ | $(97.1,\mathcal{N}(100.8,6),110)$ |
| $F_2$ | | $(57,\mathcal{N}(62,6),64.8)$ | $(55,\mathcal{N}(64,10),70.8)$ |
| $F_3$ | | | |
| $F_4$ | $(40.3,\mathcal{N}(42,4),44.3)$ | $(40.3,\mathcal{N}(42,4),44.3)$ | $(40.3,\mathcal{N}(42,4),44.3)$ |
| $F_5$ | $(53,\mathcal{N}(55.4,5),57)$ | $(54,\mathcal{N}(56.4,5),57,8)$ | $(54,\mathcal{N}(56.4,5),57,8)$ |
| $F_6$ | $(42,\mathcal{N}(44,2),45)$ | $(42,\mathcal{N}(44,2),45)$ | $(42,\mathcal{N}(44,2),45)$ |
| $F_7$ | | | |
| $F_8$ | $(35.1,\mathcal{N}(36.5,2),38)$ | $(35.1,\mathcal{N}(36.5,2),38)$ | $(36.3,\mathcal{N}(38.2,2),40)$ |
| $F_9$ | $(41,\mathcal{N}(43.6,4),45.6)$ | $(41,\mathcal{N}(43.6,4),45.6)$ | $(41,\mathcal{N}(43.6,4),45.6)$ |
| $F_{10}$ | $(29.3,\mathcal{N}(31,3),33)$ | $(29.3,\mathcal{N}(31,3),33)$ | $(29.3,\mathcal{N}(31,3),33)$ |
| $F_{11}$ | $(28.6,\mathcal{N}(30,3),33.2)$ | $(28.6,\mathcal{N}(30,3),33.2)$ | $(29.6,\mathcal{N}(31.4,4),34)$ |
| $F_{12}$ | $(33,\mathcal{N}(34.5,2),36)$ | $(33,\mathcal{N}(34.5,2),36)$ | $(33,\mathcal{N}(34.5,2),36)$ |
| $F_{13}$ | $(42.3,\mathcal{N}(43,3),44)$ | $(42.3,\mathcal{N}(43,3),44)$ | $(42.3,\mathcal{N}(43,3),44)$ |
| $F_{14}$ | $(52,\mathcal{N}(54.9,4),57)$ | $(52,\mathcal{N}(56.7,4),58)$ | $(58,\mathcal{N}(62.8,4),65)$ |
| | Phase 4 | Phase 5 | Phase 6 |
| $F_1$ | $(103,\mathcal{N}(110,10),118.9)$ | $(93.3,\mathcal{N}(95.5,4),97)$ | $(86.1,\mathcal{N}(88,4),89)$ |
| $F_2$ | $(68,\mathcal{N}(71.5,8),75.8)$ | $(76.4,\mathcal{N}(78,6),80)$ | $(77,\mathcal{N}(78,2),79.5)$ |
| $F_3$ | $(134,\mathcal{N}(138,8),144.5)$ | $(135,\mathcal{N}(138,10),146)$ | |
| $F_4$ | $(40.3,\mathcal{N}(42,4),44.3)$ | $(39,\mathcal{N}(40.5,2),41.8)$ | |
| $F_5$ | $(58,\mathcal{N}(59.4,4),62,5)$ | $(58,\mathcal{N}(59.4,4),62,5)$ | $(55,\mathcal{N}(56.4,3),57)$ |
| $F_6$ | $(44,\mathcal{N}(46.4,3),47)$ | $(44,\mathcal{N}(46.4,3),47)$ | $(42,\mathcal{N}(43.2,3),45)$ |
| $F_7$ | | | $(66.4,\mathcal{N}(67.5,4),68)$ |
| $F_8$ | $(36.3,\mathcal{N}(38.2,2),40)$ | $(36.3,\mathcal{N}(38.2,2),40)$ | $(36,\mathcal{N}(37.4,2),38.8)$ |
| $F_9$ | | $(41,\mathcal{N}(43.6,4),45.6)$ | $(41,\mathcal{N}(43.6,4),45.6)$ |
| $F_{10}$ | $(32.4,\mathcal{N}(34,3),36)$ | $(32.4,\mathcal{N}(34,3),36)$ | $(32.4,\mathcal{N}(34,3),36)$ |
| $F_{11}$ | $(34.5,\mathcal{N}(36,3),38)$ | $(34.5,\mathcal{N}(36,3),38)$ | $(33.8,\mathcal{N}(35,3),36.6)$ |
| $F_{12}$ | $(33,\mathcal{N}(34.5,2),36)$ | $(33,\mathcal{N}(34.5,2),36)$ | $(33,\mathcal{N}(34.5,2),36)$ |
| $F_{13}$ | $(42.3,\mathcal{N}(43,3),44)$ | $(42.3,\mathcal{N}(43,3),44)$ | $(42.3,\mathcal{N}(43,3),44)$ |
| $F_{14}$ | $(59,\mathcal{N}(64.8,4),67)$ | $(58,\mathcal{N}(62.8,4),65)$ | $(56,\mathcal{N}(58.5,2),63)$ |
| | Phase 7 | Phase 8 | |
| $F_1$ | $(72.1,\mathcal{N}(74.8,6),78)$ | | |
| $F_2$ | $(67,\mathcal{N}(71,2),73)$ | $(58,\mathcal{N}(60.5,3),63)$ | |
| $F_3$ | | | |
| $F_4$ | | | |
| $F_5$ | $(52,\mathcal{N}(53.4,4),55,3)$ | $(50,\mathcal{N}(52.1,4),53,6)$ | |
| $F_6$ | $(42,\mathcal{N}(43.2,3),45)$ | $(38,\mathcal{N}(40.4,5),43)$ | |
| $F_7$ | $(66.4,\mathcal{N}(67.5,4),68)$ | $(66.4,\mathcal{N}(67.5,4),68)$ | |
| $F_8$ | $(34.3,\mathcal{N}(35.2,2),36)$ | $(34.3,\mathcal{N}(35.2,2),36)$ | |
| $F_9$ | | | |
| $F_{10}$ | $(27.5,\mathcal{N}(29,4),32)$ | | |
| $F_{11}$ | $(31.2,\mathcal{N}(33,3),34.7)$ | $(28.9,\mathcal{N}(30.4,3),32.2)$ | |
| $F_{12}$ | $(33,\mathcal{N}(34.5,2),36)$ | $(33,\mathcal{N}(34.5,2),36)$ | |
| $F_{13}$ | $(42.3,\mathcal{N}(43,3),44)$ | $(42.3,\mathcal{N}(43,3),44)$ | |
| $F_{14}$ | $(53.2,\mathcal{N}(54.1,4),56)$ | $(51.2,\mathcal{N}(53.4,4),55.4)$ | |

### 6.7.3 Algorithm Analysis

In this part, the algorithm used in this section is analyzed from both qualitative and quantitative aspects.

(1) Qualitatively, this section compares the GA, which is mostly used in the construction site layout problems, with the algorithm used in this study. The merit of the GA is its strong evolutionary process to find an optimal solution using crossover operations, selection and mutation. However, the randomly generated initial generation at the algorithms' beginning affects the solution quality because of a bad gene inherited from the parent generation. Moreover, the searching capability is reduced as the GA does not rely on gradient or derivative information [120]. Further, most GA approaches adopt a sequence-based representation to encode the candidate solutions to the site layout problems, where each element in the sequence represents a facility, and the number in the element represents the facility location. Reproduction of the sequence-represented solutions based on the operators (e.g., crossover and mutation) could lead to infeasible solutions when several elements have the same number value leading to multiple facilities at one location. In the proposed algorithm, the permutation-based representation closely connects the PSO particles and the problem solutions. The hybrid particle-updating mechanism successfully avoids permutation infeasibility. At the same time, the multi-objective method is

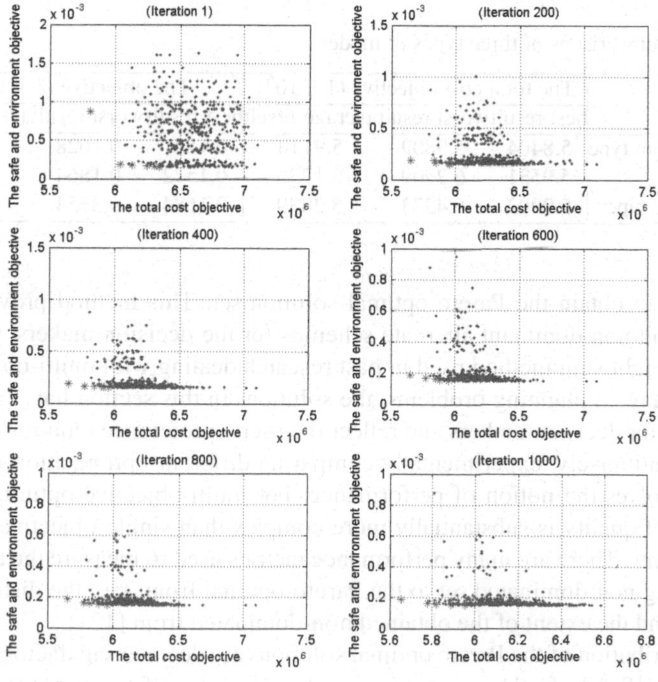


**Fig. 6.18** Pareto optimal solutions of the Longtan case

**Table 6.6** The safest dynamic site layout plan in the Longtan case

| Phase | $F_1$ | $F_2$ | $F_3$ | $F_4$ | $F_5$ | $F_6$ | $F_7$ | $F_8$ | $F_9$ | $F_{10}$ | $F_{11}$ | $F_{12}$ | $F_{13}$ | $F_{14}$ |
|---|---|---|---|---|---|---|---|---|---|---|---|---|---|---|
| 1 | 3 | - | - | 1 | 4 | 5 | - | 10 | 14 | 9 | 2 | 12 | 6 | 7 |
| 2 | 3 | 13 | - | 1 | 4 | 5 | - | 10 | 14 | 9 | 2 | 12 | 6 | 7 |
| 3 | 3 | 13 | - | 1 | 4 | 9 | - | 10 | 14 | 11 | 2 | 8 | 6 | 7 |
| 4 | 12 | 3 | 1 | 13 | 4 | 9 | - | 10 | - | 11 | 2 | 8 | 6 | 7 |
| 5 | 12 | 3 | 1 | 13 | 4 | 9 | - | 10 | 14 | 11 | 2 | 8 | 6 | 7 |
| 6 | 12 | 13 | - | - | 4 | 9 | 5 | 10 | 14 | 11 | 2 | 8 | 6 | 7 |
| 7 | 12 | 13 | - | - | 4 | 9 | 5 | 14 | - | 11 | 2 | 8 | 6 | 7 |
| 8 | - | 13 | - | - | 4 | 9 | 5 | 14 | - | - | 11 | 8 | 6 | 7 |

**Table 6.7** The most cost-effective dynamic site layout plan in the Longtan case

| Phase | $F_1$ | $F_2$ | $F_3$ | $F_4$ | $F_5$ | $F_6$ | $F_7$ | $F_8$ | $F_9$ | $F_{10}$ | $F_{11}$ | $F_{12}$ | $F_{13}$ | $F_{14}$ |
|---|---|---|---|---|---|---|---|---|---|---|---|---|---|---|
| 1 | 8 | - | - | 1 | 4 | 5 | - | 3 | 10 | 11 | 2 | 12 | 6 | 7 |
| 2 | 8 | 13 | - | 1 | 4 | 5 | - | 3 | 10 | 11 | 2 | 12 | 6 | 7 |
| 3 | 12 | 13 | - | 14 | 4 | 9 | - | 3 | 10 | 11 | 2 | 8 | 6 | 7 |
| 4 | 12 | 13 | 1 | 14 | 4 | 9 | - | 3 | - | 11 | 2 | 8 | 6 | 7 |
| 5 | 12 | 13 | 1 | 14 | 4 | 9 | - | 3 | 10 | 11 | 2 | 8 | 6 | 7 |
| 6 | 12 | 13 | - | - | 4 | 9 | 5 | 3 | 10 | 11 | 2 | 8 | 6 | 7 |
| 7 | 12 | 13 | - | - | 4 | 9 | 5 | 10 | - | 11 | 2 | 8 | 6 | 7 |
| 8 | - | 13 | - | - | 4 | 9 | 5 | 10 | - | - | 11 | 8 | 6 | 7 |

**Table 6.8** Comparisons of three types of model

| Type | The total cost objective ($1 \times 10^6$) | | | The objective ($1 \times 10^{-3}$) | | |
|---|---|---|---|---|---|---|
| | best result | worst result | average result | best result | worst result | average result |
| Fuzzy random type | 5.8404 | 5.9809 | 5.9114 | 0.1521 | 0.1628 | 0.1562 |
| Fuzzy type | 5.9531 | 6.2909 | 6.1736 | 0.1572 | 0.1864 | 0.1711 |
| Determinate type | 5.9902 | 6.4373 | 6.2889 | 0.1584 | 0.3754 | 0.1816 |

introduced to obtain the Pareto optimal solution set. This method provides more effective and non-dominant alternate schemes for the decision makers. Compared with the weight-sum method used in past research dealing with multi-objective dynamic site layout planning problems, the solutions in this section have more reference value for decision makers and reflect the users' preference requirements.

(2) Quantitatively, experimentally comparing different optimization techniques always involves the notion of performance. For multi-objective optimization, the definition of quality is substantially more complex than single-objective optimization problems. There are many performance metrics used to measure the distance of the resulting non-dominated set to the Pareto-optimal front, i.e., the distribution of solutions and the extent of the obtained non-dominated front [121].

The distribution of the Pareto optimal solutions are deemed satisfactory as shown in Figure 6.18. For further expression of the efficiency of the convergence, three performance metrics are studied. Table 6.9 shows the average distance to the Pareto optimal sets, the distribution in combination with the number of non-dominated solutions found and the extent of the Pareto optimal fronts proposed in [121]. The performance metrics for the Pareto optimal set shown in Table 6.9 provide a satisfactory result for convergence efficiency . Although there are some fluctuations in the three metrics, the final result was not affected.

**Table 6.9** Algorithm evaluation by performance metrics

| Iteration | The average distance metric | The distribution metric | The extent metric |
|---|---|---|---|
| 1 | 0.0936 | 0.1278 | 572.3830 |
| 200 | 0.0715 | 0.3860 | 864.0474 |
| 400 | 0.0552 | 0.1369 | 966.0624 |
| 600 | 0.0425 | 0.3328 | 678.6942 |
| 800 | 0.0363 | 0.5641 | 752.1259 |
| 1000 | 0.0336 | 0.6991 | 994.3038 |

## 6.8 Case Study for DTFLP

An actual DTFLP construction project is used here as a practical application example for the optimization method. The “Jingping Bend” is the main stream of the Yalong River, southwest of China, 28.12833°N 101.79083 °E, see Figure 6.19. The length of the “Jingping Bend” is 150 km, but the downstream part of the river on the opposite side is only separated by 16 km. Over that distance there is an elevation drop of 310 m, creating an excellent situation for hydroelectricity production. Two projects were planned for the bend, the Jinping I and Jinping II with a combined capacity of 8,400 MW. Located at 28°07’42”N 101°47’27”E, the Jinping-II Hydropower Station is symbolic of the large-scale constructions of the West-east Electric Transmission Project of China and has the largest hydraulic tunnel in the world. The total installed capacity of the hydropower construction is 4.8 million kw, and the generated energy is 242.3 billion kwh a year. Once completed, it will have a powerful regulating ability with a normal pool level of 1646 m and a reservoir capacity of 14.01 million m$^3$. The construction project began in July 2007 and the total construction duration is 8 years 3 months.

The main permanent structures are the Maomaotan Sluice Dam, the diversion system and the power house. The Maomaotan Sluice Dam on the west-side of the Jinping bend will divert water into a 16.6 km long headrace tunnel towards the Jinging-II Power Station. From the left sub-figure in Figure 6.19, it can be seen that the Maomaotan Sluice Dam and the power house are separated by mountains and there is no real traffic road between them except for the Jinping Highway along the Jinping Bend. There are 10 principal areas of the Jinping-II large-scale water conservancy and hydropower construction project. They are the Songlinping Area, the Dabenliu Stockpile Area, the Maomao Bank Sluice Dam, the Yinbazigou Area, the Jingfeng Bridge Stockpile Area, the Daping Area, the Santan Dropping Area, the Muofanggou Area, the Zhoujiaping Area and the Muos agou Area. The Songlinping Area is a work area on the west bank slope and the excavation treatment for

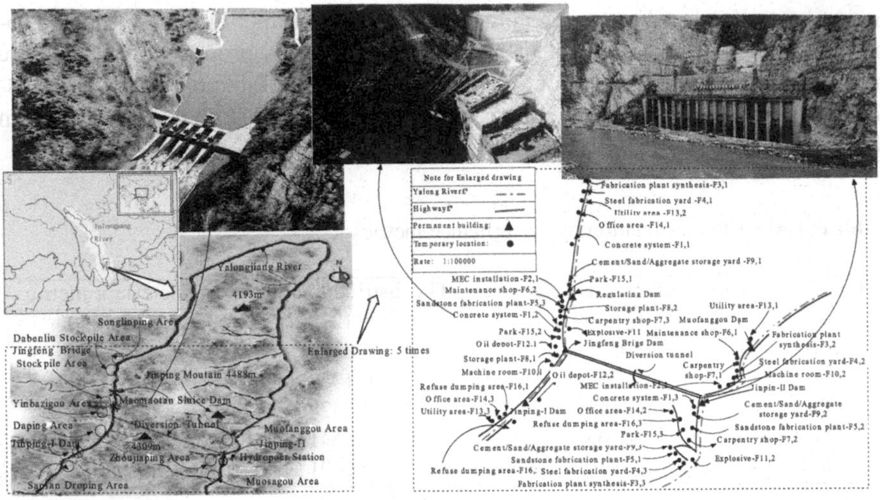


**Fig. 6.19** The map of Jinping II hydropower construction

west bank slope. The Dabenliu Stockpile Area and Jingfeng Bridge Stockpile Area are the reservoir area for the western area. The Maomao Bank Sluice Dam is the retaining dam. The Yinbazigou Area is the concrete pouring area for the west bank diversion tunnel. The Daping Area is the river closure, cofferdam and river bed excavation treatment. The Santan Dropping Area is the spoil dumping site for the western work areas. The Muofanggou Area is the excavation treatment area for the navigation structure and steel tube processing for the eastern areas. The Zhoujiaping Area is the concrete pouring area for the eastern area, the plant construction, and the installation of metal structures for the generating system and the electrical installation. The Muosagou Area is the water diversion area, the tailrace system in the underground powerhouse, the flood system and the ship lift system. When the left sub-figure of Figure 6.19 is enlarged 5 times, it is possible to see the detailed candidate locations, which were chosen because they lie along the Yalong River and the traffic lines, while other areas were in the mountains. The top left corner in Figure 6.19 is the blueprint for the Jinping-II station. The middle-upper of Figure 6.19 is the concrete pouring site for the Jinping-II station. The top right corner of Figure 6.19 is the entrance of the diversion tunnel on the west bank.

Based on the natural environment and the location of the main structures, the work was to provide a proper alternative construction location plan which included the warehouse, affiliated companies, and stationary facilities.

## *6.8.1 Data Collection*

In order to obtain robust results, the study collected data from multiple sources (Cao and Hoffman 2011 [128]). In the current study, data were obtained from three sources: the database of the Sanxia Lt. Company, the managerial practice observations from the Jinpin-II Power Station and interviews. As a further note, the investigation into the closeness relationship between the facilities in project management practice is able to ascertain the validity of the measurement techniques used in previous section. Interviews were conducted face to face with construction managers to provide an opportunity to improve data precision. The construction managers' project experience assisted in the comprehension of the specific nature of the facilities.

**Table 6.10** The list of temporary facilities and the corresponding variables

| No. | Facility | Number of locations | Fixed Cost | Variable Cost (in 10 thousand RMB) | | |
|---|---|---|---|---|---|---|
| | | $M$ | $c_i$ | $a_{i1}$ | $a_{i2}$ | $a_{i3}$ |
| F1 | Concrete system | 3 | 77.9546 | 11.74 | 15.63 | 43.89 |
| F2 | MEC installation | 2 | 129.91 | 19.56 | 27.24 | - |
| F3 | Fabrication plant synthesis | 3 | 3.6 | 0.78 | 1.16 | 0.53 |
| F4 | Steel fabrication yard | 3 | 1.775 | 1.12 | 0.89 | 1.34 |
| F5 | Sandstone fabrication plant | 3 | 39.74 | 4.76 | 10.33 | 6.36 |
| F6 | Maintenance shop | 2 | 2.43 | 0.32 | 0.73 | - |
| F7 | Carpentry shop | 3 | 1.449 | 0.19 | 0.89 | 0.16 |
| F8 | Storage plant | 2 | 3.26 | 0.28 | 1.89 | - |
| F9 | Cement/Sand/Aggregate storage yard | 3 | 4.38 | 0.57 | 0.26 | 1.03 |
| F10 | Machine room | 2 | 5.67 | 0.65 | 1.47 | - |
| F11 | Explosive depot | 2 | 1.674 | 0.12 | 0.54 | - |
| F12 | Oil depot | 2 | 1.93 | 0.65 | 0.28 | - |
| F13 | Utility area | 3 | 162.36 | 34.69 | 13.31 | 15.52 |
| F14 | Office area | 3 | 59.36 | 6.88 | 9.56 | 10.45 |
| F15 | Parking lot | 3 | 7.8 | 1.74 | 0.68 | 1.15 |
| F16 | Refuse dumping area | 3 | 11.18 | 1.36 | 1.52 | 3.13 |

As stated, the enlarged drawing of the Jinping-II Hydropower Station in Figure 6.15 includes some possible construction zones and available location dots for the temporary facilities. It can be seen that the main sub-projects of the Jinping-II construction and the sites of these sub-projects change over time. Thus, based on the plan for each construction section, the principal part of the Jinping-II hydropower construction project can be divided into 3 stages; excavation, concrete pouring and MEC installation. There are 16 kinds of temporary facilities involved in the principal process for the Jinping-II hydropower construction project. The facilities as well as their corresponding constant variables such as $a_{im}(t)$ and $c_i$ variables are listed in Table 6.10. It should be noted that, as in the last stage, i.e., $t = 3$, and some facility locations need not be relocated or will be closed, so the $a_{i3}$ of these facilities can be set at a very high level (i.e., for F2 MEC installation, F6 Maintenance shop, F8 Storage plant, F10 Machine room, F11 Explosive depot, F12 Oil depot, $a_{i3} = 1000$).

**Table 6.11** The closeness relationship between facilities

| No. | F1 | F2 | F3 | F4 | F5 | F6 | F7 | F8 | F9 | F10 | F11 | 12 | F13 | F14 | F15 | F16 |
|---|---|---|---|---|---|---|---|---|---|---|---|---|---|---|---|---|
| F1 | | | | | | | | | | | | | | | | |
| F2 | 3 | | | | | | | | | | | | | | | |
| F3 | 3 | 3 | | | | | | | | | | | | | | |
| F4 | 3 | 5 | 4 | | | | | | | | | | | | | |
| F5 | 4 | 2 | 4 | 3 | | | | | | | | | | | | |
| F6 | 2 | 3 | 5 | 3 | 3 | | | | | | | | | | | |
| F7 | 2 | 2 | 4 | 3 | 3 | 3 | | | | | | | | | | |
| F8 | 4 | 4 | 5 | 4 | 4 | 2 | 4 | | | | | | | | | |
| F9 | 5 | 2 | 2 | 5 | 5 | 2 | 2 | 3 | | | | | | | | |
| F10 | 4 | 5 | 5 | 4 | 4 | 5 | 4 | 3 | 3 | | | | | | | |
| F11 | 1 | 1 | 1 | 1 | 1 | 1 | 1 | 1 | 1 | 1 | | | | | | |
| F12 | 1 | 1 | 1 | 1 | 1 | 1 | 1 | 1 | 1 | 1 | 3 | | | | | |
| F13 | 1 | 1 | 1 | 1 | 1 | 1 | 1 | 3 | 2 | 2 | 1 | 1 | | | | |
| F14 | 2 | 2 | 2 | 2 | 2 | 3 | 2 | 2 | 2 | 2 | 1 | 1 | 5 | | | |
| F15 | 3 | 3 | 3 | 3 | 3 | 1 | 1 | 2 | 2 | 1 | 1 | 1 | 1 | 3 | | |
| F16 | 4 | 1 | 2 | 3 | 4 | 1 | 2 | 1 | 3 | 1 | 1 | 1 | 1 | 2 | 3 | |

By eliciting the preference information from the construction engineers, the preference-based membership functions for the value flow of these facilities at different stages and the closeness relationship were constructed, which are associated with the fuzzy factors in Elbeltagi and Hegazy (2001) [131] and Karray et al. (2000) [141]. Using the method proposed in the previous section, the closeness relationships between the facilities were calculated from the actual data and are shown in Table 6.11. The value flow of these facilities in different stages is listed in Table 6.12. The distances between the locations $d_{kl}$ were also known, (see Tables 6.13 and 6.14).

### 6.8.2 Results Discussion

The DTFLP of the Jinping-II hydropower station with sixteen facilities and three stages was solved optimally using the SA-based MOGA algorithm. However, when using genetic algorithms, different values for the variables are arrived at in different instances. Therefore, what method can the project manager use to select the values for the parameter variables from these different runs. The SA-based GA parameters are set from the results of preliminary experiments that were conducted to observe the behavior of the algorithm at different parameter settings. Many scholars such as Zouein et al. (2002) [159], Dimou and Koumousis (2003) [254] have given suggestions on the proper parameters. To reduce potential statistical errors, the convergence iteration number and computing time of the algorithm were also calculated. By comparing several sets of parameters, including population size (i.e., $N$), iteration number (i.e., $G_{max}$), crossover probability (i.e., $p_c$), and mutation probability (i.e., $P_m$), the most reasonable parameters were chosen. The adopted parameters in the

**Table 6.12** The value flow $v_{ij}(t)$ between facilities (in 10 thousand RMB)

| No. | F1 | F2 | F3 | F4 | F5 | F6 | F7 | F8 | F9 | F10 | F11 | 12 | F13 | F14 | F15 | F16 |
|---|---|---|---|---|---|---|---|---|---|---|---|---|---|---|---|---|
| | Stage $t = 1$ | | | | | | | | | | | | | | | |
| F1 | - | | | | | | | | | | | | | | | |
| F2 | 14.67 | - | | | | | | | | | | | | | | |
| F3 | 1.36 | 98.94 | - | | | | | | | | | | | | | |
| F4 | 1.9 | 297.73 | 947.96 | - | | | | | | | | | | | | |
| F5 | 6.09 | 0 | 711.76 | 0.8 | - | | | | | | | | | | | |
| F6 | 0.4 | 0.28 | 0.46 | 0.29 | 0.62 | - | | | | | | | | | | |
| F7 | 1.16 | 1.46 | 1.47 | 0.96 | 0 | 0.08 | - | | | | | | | | | |
| F8 | 6.68 | 0.35 | 8626.78 | 1002.95 | 232.89 | 0.35 | 2.81 | - | | | | | | | | |
| F9 | 1194.56 | 0 | 7605.75 | 3.24 | 8854.07 | 0 | 0.49 | 0 | - | | | | | | | |
| F10 | 7.8 | 6.21 | 6.21 | 3.58 | 3.75 | 6.21 | 0.41 | 8.17 | 280.73 | - | | | | | | |
| F11 | 0 | 0.1 | 0.26 | 0.01 | 0 | 0 | 0 | 0 | 0 | 0 | - | | | | | |
| F12 | 6.6 | 0.16 | 0.95 | 1.77 | 1.69 | 0.06 | 0.08 | 0 | 0 | 5.96 | 0 | - | | | | |
| F13 | 1.76 | 264.62 | 87.57 | 36.04 | 7.56 | 0.85 | 0.87 | 0.54 | 104.43 | 74.67 | 36.56 | 0.52 | - | | | |
| F14 | 2.64 | 19.85 | 32.34 | 24.32 | 5.15 | 0.57 | 0.69 | 0.38 | 88.68 | 49.78 | 28.69 | 0.34 | 538.93 | - | | |
| F15 | 49.6 | 6.7 | 454.49 | 1128.9 | 10456.26 | 3.41 | 4.17 | 236.59 | 13.6 | 49.72 | 0.31 | 6.67 | 557.08 | 371.39 | - | |
| F16 | 26.32 | 4.93 | 14.54 | 86.34 | 290.75 | 0.04 | 0.07 | 0 | 1.96 | 0 | 0.01 | 0 | 0 | 17.61 | 45.79 | - |
| | Stage $t = 2$ | | | | | | | | | | | | | | | |
| F1 | - | | | | | | | | | | | | | | | |
| F2 | 142.23 | - | | | | | | | | | | | | | | |
| F3 | 12.13 | 1.46 | - | | | | | | | | | | | | | |
| F4 | 16.78 | 1741.37 | 543.57 | - | | | | | | | | | | | | |
| F5 | 2643.79 | 0 | 417.56 | 0.67 | - | | | | | | | | | | | |
| F6 | 113.21 | 75.84 | 24.24 | 16.53 | 25.42 | - | | | | | | | | | | |
| F7 | 711.34 | 8.54 | 33.81 | 35.52 | 1.87 | 1.32 | - | | | | | | | | | |
| F8 | 1.36 | 1198.34 | 10675.29 | 1741.29 | 10014.27 | 122.15 | 107.15 | - | | | | | | | | |
| F9 | 3381.11 | 0.02 | 35.39 | 74.52 | 377626.89 | 0 | 18.13 | 0 | - | | | | | | | |
| F10 | 893.52 | 212.9 | 1460.98 | 118.47 | 576.78 | 260.82 | 16.54 | 13.06 | 864.19 | - | | | | | | |
| F11 | 0 | 0.18 | 0.26 | 0.34 | 0 | 0 | 0 | 0 | 0 | 0 | - | | | | | |
| F12 | 5.03 | 0.94 | 32.91 | 69.03 | 78.39 | 2.76 | 3.36 | 0 | 0 | 781.62 | 0 | - | | | | |
| F13 | 274.72 | 36.26 | 2152.53 | 400.35 | 209.96 | 17.43 | 39.15 | 5.67 | 115.95 | 475.96 | 72.47 | 18.72 | - | | | |
| F14 | 100.03 | 34.29 | 1891.51 | 280.14 | 139.48 | 1.22 | 14.94 | 3.76 | 28.96 | 266.54 | 61.65 | 11.23 | 1297.26 | - | | |
| F15 | 1163.92 | 1747.52 | 2752.89 | 2090.46 | 378628.16 | 4.32 | 140.31 | 25.23 | 1163.78 | 341.31 | 14.26 | 264.66 | 487.93 | 29.36 | - | |
| F16 | 372.76 | 45.44 | 1541.56 | 64.74 | 12094.41 | 6.86 | 8.57 | 61.75 | 11263.38 | 0 | 0.02 | 0 | 1979.63 | 322.58 | 1959.76 | - |
| | Stage $t = 3$ | | | | | | | | | | | | | | | |
| F1 | - | | | | | | | | | | | | | | | |
| F2 | 96.71 | - | | | | | | | | | | | | | | |
| F3 | 7.3 | 1.22 | - | | | | | | | | | | | | | |
| F4 | 11.45 | 1253.53 | 91.82 | - | | | | | | | | | | | | |
| F5 | 1770.81 | 0 | 284.74 | 0.26 | - | | | | | | | | | | | |
| F6 | 184.13 | 131.44 | 60.26 | 15.92 | 11.21 | - | | | | | | | | | | |
| F7 | 7.88 | 5.38 | 19.94 | 30.76 | 0.16 | 0.12 | - | | | | | | | | | |
| F8 | 170.08 | 1924.42 | 225.19 | 1212.37 | 40.73 | 68.57 | 46.77 | - | | | | | | | | |
| F9 | 3993.48 | 0.01 | 195.37 | 1497.54 | 1536.23 | 0 | 12.89 | 0 | - | | | | | | | |
| F10 | 842.74 | 244.65 | 198.89 | 149.04 | 53.34 | 63.44 | 18.25 | 8.71 | 562.25 | - | | | | | | |
| F11 | 0 | 0.17 | 0.24 | 0.32 | 0 | 0 | 0 | 0 | 0 | 0 | - | | | | | |
| F12 | 3.34 | 1.15 | 11.43 | 44.85 | 72.12 | 3.12 | 0.73 | 0 | 0 | 197.78 | 0 | - | | | | |
| F13 | 147.23 | 185.48 | 1048.87 | 521.34 | 8.98 | 17.97 | 16.84 | 2.56 | 94.42 | 31.73 | 71.58 | 13.48 | - | | | |
| F14 | 77.57 | 96.43 | 125.69 | 187.67 | 2.16 | 4.67 | 3.54 | 0.77 | 15.23 | 17.76 | 59.43 | 3.13 | 64.89 | - | | |
| F15 | 370.52 | 472.65 | 552.43 | 1385.56 | 1577.39 | 2.93 | 84.85 | 22.09 | 1261.25 | 573.54 | 4589.46 | 19.63 | 609.91 | 280.32 | - | |
| F16 | 443.4 | 12.71 | 16.16 | 87.76 | 57.77 | 3.84 | 8.74 | 66.84 | 260.47 | 0 | 0.23 | 0 | 109.38 | 22.86 | 472.22 | - |

SA-based GA were selected as follows: population size= 100, the crossover rate= 0.8, the mutation rate= 0.2, the acceptance probability $P_c = 0.95$ and $\rho = 0.95$. The DTFLP programme was run 30 times and the best result chosen as the solution. In this solution, the first objective was 72,422 million yuan and the second was 706654 with the total objective being 7574.8. The DTFLP program plan is shown in Figure 6.20. From Figure 6.20, it can be seen that some of the facilities were relocated at each stage, while some of them had no change over some of the stages. This difference was not only related to the cost balance for the DTFLP program model but also can be explained through practical experience. For example, the concrete system is located in the $2^{nd}$ candidate site in the first two stages while it is removed to the $3^{rd}$ candidate site in the last stage because in the first two stages the main concrete system work is concentrated in the western part of the construction project. It would waste a lot of time and money if the concrete system were located in the $3^{rd}$ candidate site which is in the eastern part. When it came to the last stage, the concrete work of the western part was almost completed and the main part of

**Table 6.13** a: The distance $d_{ik,jl}$ between different possible locations of the facilities

| No. | | F1 | | | F2 | | F3 | | | F4 | | | F5 | | | F6 | | F7 | | |
|---|---|---|---|---|---|---|---|---|---|---|---|---|---|---|---|---|---|---|---|---|
| | *m* | 1 | 2 | 3 | 1 | 2 | 1 | 2 | 3 | 1 | 2 | 3 | 1 | 2 | 3 | 1 | 2 | 1 | 2 | 3 |
| F2 | 1 | 3.3 | 1.6 | 18.5 | | | | | | | | | | | | | | | | |
| | 2 | 20.3 | 19 | 1.8 | | | | | | | | | | | | | | | | |
| F3 | 1 | 3.7 | 8.2 | 88.1 | 7 | 87.1 | | | | | | | | | | | | | | |
| | 2 | 88.7 | 93.2 | 3.3 | 92.2 | 1.4 | | | | | | | | | | | | | | |
| | 3 | 97.4 | 102 | | 100.3 | 6.8 | | | | | | | | | | | | | | |
| F4 | 1 | 3 | 8 | 88.6 | 5.8 | 87.7 | 0.6 | 86.3 | 93.2 | | | | | | | | | | | |
| | 2 | 90.3 | 94.9 | 2.8 | 92.6 | 1 | 85.4 | 1.2 | 7.8 | | | | | | | | | | | |
| | 3 | 96.6 | 100.9 | 4.2 | 100.2 | 6.5 | 93.2 | 7.9 | 0.4 | | | | | | | | | | | |
| F5 | 1 | 95 | 100.5 | 3.5 | 99.7 | 6.2 | 92.8 | 7.5 | 0.7 | 93.3 | 6.6 | 3.2 | | | | | | | | |
| | 2 | 92.4 | 97.8 | 1.2 | 96.9 | 3.1 | 90 | 4.9 | 3.8 | 90.5 | 3.8 | 0.9 | | | | | | | | |
| | 3 | 4 | 0.5 | 96.6 | 1.3 | 95.4 | 8.3 | 93.7 | 100.7 | 7.6 | 91.2 | 96.3 | | | | | | | | |
| F6 | 1 | 87.5 | 92.9 | 3.2 | 91.4 | 2 | 83.9 | 0.8 | 8.9 | 84.5 | 1.2 | 8.6 | 8.2 | 4.4 | 92.5 | | | | | |
| | 2 | 3.8 | 0.6 | 95.9 | 0.6 | 94.6 | 7.5 | 93 | 100.4 | 7 | 90.4 | 100.1 | 99.7 | 97.1 | 0.2 | | | | | |
| F7 | 1 | 90.1 | 94.9 | 3.5 | 91.6 | 1.7 | 84.2 | 0.4 | 8.6 | 84.8 | 0.9 | 8.3 | 7.9 | 2.3 | 94.5 | 0.3 | 92.2 | | | |
| | 2 | 26.5 | 93.4 | 2 | 97.7 | 3.8 | 90.8 | 5.5 | 2.7 | 90.8 | 4.6 | 2.4 | 2 | 0.8 | 93 | 6.2 | 98.3 | | | |
| | 3 | 4.2 | 0.9 | 96.4 | 0.9 | 95.1 | 7.8 | 93.3 | 100.4 | 7.3 | 90.7 | 100.1 | 99.7 | 97.6 | 0.5 | 92.6 | 0.3 | | | |
| F8 | 1 | 6.3 | 8 | 96 | 3 | 97.2 | 10.1 | 95.4 | 102.7 | 9.5 | 92.9 | 102.4 | 102 | 97.2 | 8.4 | 94.7 | 2.4 | 95 | 98 | 2.1 |
| | 2 | 3.1 | 1.5 | 98.8 | 0.3 | 94.4 | 7.3 | 92.7 | 100.3 | 6.7 | 9.2 | 100 | 99.6 | 100 | 0.9 | 92 | 0.3 | 92.3 | 100.8 | 0.6 |
| F9 | 1 | 2.3 | 2.2 | 94.6 | 1 | 93.1 | 6 | 91.4 | 98.4 | 5.7 | 91.9 | 98.1 | 97.7 | 95.8 | 1.8 | 91.7 | 1.6 | 91 | 96.6 | 2.9 |
| | 2 | 19.9 | 98.6 | 0.4 | 96.1 | 2.2 | 99.2 | 4 | 4.6 | 89.8 | 3 | 4.3 | 3.9 | 0.8 | 98.2 | 4.7 | 97.7 | 4.4 | 1.6 | 96 |
| | 3 | 95.2 | 100.8 | 3.7 | 100.1 | 5.7 | 92.3 | 7 | 1.2 | 92.8 | 6.5 | 0.9 | 0.5 | 2.7 | 100.4 | 7.7 | 101.7 | 7.3 | 2.2 | 102 |
| F10 | 1 | 7 | 2 | 99.3 | 3.2 | 97.3 | 10.6 | 96.1 | 103 | 10.1 | 95.6 | 102.7 | 102.3 | 100.5 | 2.4 | 95.4 | 2.6 | 95.7 | 101.1 | 2.3 |
| | 2 | 90.7 | 95.1 | 2.5 | 93.4 | 0.5 | 85.9 | 1.7 | 7.3 | 87 | 0.5 | 7 | 6.6 | 1.3 | 94.7 | 2.4 | 94 | 2.1 | 2.1 | 94.3 |
| F11 | 1 | 4.8 | 0.5 | 97.1 | 1.8 | 95.5 | 8.5 | 94.2 | 101.3 | 7.85 | 91.4 | 101 | 100.6 | 98.3 | 0.5 | 93.5 | 1.2 | 93.8 | 99.2 | 0.9 |
| | 2 | 95.5 | 99.6 | 2.6 | 98.6 | 5.2 | 91.7 | 6.7 | 0.5 | 94.1 | 8 | 0.2 | 1.5 | 1.4 | 99.2 | 7.5 | 99.2 | 7.2 | 0.6 | 99.5 |
| F12 | 1 | 6 | 1 | 98.2 | 2.8 | 96.7 | 9.6 | 94.9 | 102.2 | 9 | 92.4 | 101.9 | 101.5 | 99.4 | 1.4 | 94.2 | 1.2 | 94.5 | 100.2 | 0.9 |
| | 2 | 9 | 4 | 95.2 | 5.8 | 99.8 | 12.7 | 98.1 | 104 | 12.2 | 96 | 103.7 | 103.3 | 96.4 | 4.4 | 97.4 | 5.2 | 97.7 | 97.2 | 4.9 |
| F13 | 1 | 86.1 | 91.6 | 5 | 89.6 | 3.2 | 82.6 | 2 | 10 | 88.2 | 2.5 | 9.7 | 9.3 | 3.8 | 11.2 | 1.3 | 90.2 | 1.6 | 4.6 | 90.5 |
| | 2 | 2.5 | 7.5 | 89.8 | 5.8 | 88.2 | 1.1 | 86.6 | 93.7 | 0.5 | 87.1 | 93.4 | 93 | 91 | 7.1 | 85.9 | 6.4 | 86.2 | 91.8 | 6.7 |
| | 3 | 12 | 7 | 104.1 | 8.8 | 102.7 | 15.4 | 100.7 | 108.7 | 14.7 | 98.8 | 108.4 | 108 | 105.3 | 7.4 | 100 | 8.2 | 100.2 | 106.1 | 7.9 |
| F14 | 1 | 1.3 | 6.8 | 90.5 | 5.1 | 88.9 | 1.8 | 87.3 | 94.4 | 1.2 | 87.7 | 94.1 | 93.7 | 91.7 | 6.4 | 86.6 | 5.7 | 86.2 | 92.5 | 6 |
| | 2 | 95.1 | 84.6 | 5.2 | 101.2 | 7 | 93.8 | 8.7 | 4.1 | 94.2 | 7.6 | 3.8 | 3.4 | 4 | 84.2 | 8 | 101.8 | 7.7 | 3.2 | 102.1 |
| | 3 | 11.5 | 6 | 103.6 | 7.8 | 102.2 | 14.9 | 100.2 | 108.3 | 14.3 | 98.3 | 108 | 107.6 | 104.8 | 6.4 | 99.5 | 7.2 | 99.8 | 105.6 | 6.9 |
| F15 | 1 | 2.8 | 2.2 | 95.1 | 0.5 | 93.6 | 6.5 | 91.9 | 99.1 | 6 | 89.4 | 98.8 | 98.4 | 96.3 | 1.8 | 91.2 | 1.1 | 91.5 | 96.9 | 1 |
| | 2 | 5.9 | 0.4 | 97.6 | 2.1 | 96.1 | 9 | 94.4 | 101.7 | 8.2 | 91.9 | 101.4 | 101 | 99.8 | 0.8 | 93.7 | 1.5 | 94 | 100.6 | 1.2 |
| | 3 | 96.1 | 90.5 | 4 | 99.6 | 5.7 | 92.5 | 7.5 | 3.4 | 93 | 6.3 | 3.1 | 2.7 | 2.8 | 90.1 | 8.2 | 100.2 | 7.9 | 2 | 100.5 |
| F16 | 1 | 11 | 5.7 | 103.3 | 7.2 | 101.8 | 14.5 | 99.9 | 107.8 | 13.9 | 98 | 107.5 | 107.1 | 104.5 | 6.1 | 99.2 | 6.6 | 99.5 | 105.3 | 6.3 |
| | 2 | 96.7 | 91.4 | 4.5 | 100.2 | 6.3 | 93.1 | 8.1 | 4 | 93.7 | 7 | 3.7 | 3.3 | 3.3 | 91 | 8.8 | 100.8 | 8.5 | 2.5 | 101.1 |

**Table 6.14** b: The distance $d_{ik,jl}$ between different possible locations of the facilities

| No. | | F8 | | F9 | | | F10 | | F11 | | F12 | | F13 | | | F14 | | | F15 | | |
|---|---|---|---|---|---|---|---|---|---|---|---|---|---|---|---|---|---|---|---|---|---|
| | *m* | 1 | 2 | 1 | 2 | 3 | 1 | 2 | 1 | 2 | 1 | 2 | 1 | 2 | 3 | 1 | 2 | 3 | 1 | 2 | 3 |
| F9 | 1 | 3.8 | 1.3 | | | | | | | | | | | | | | | | | | |
| | 2 | 21.4 | 96.4 | | | | | | | | | | | | | | | | | | |
| | 3 | 96.4 | 100.4 | | | | | | | | | | | | | | | | | | |
| F10 | 1 | 0.6 | 2.9 | 4.7 | 99.7 | 102 | | | | | | | | | | | | | | | |
| | 2 | 92.2 | 93.7 | 92.4 | 2.9 | 6.1 | | | | | | | | | | | | | | | |
| F11 | 1 | 6.3 | 1.5 | 2.5 | 97.5 | 100.1 | 6.8 | 0.7 | | | | | | | | | | | | | |
| | 2 | 97 | 98.9 | 97.6 | 2.2 | 1.2 | 97.5 | 5.7 | | | | | | | | | | | | | |
| F12 | 1 | 0.5 | 2.5 | 3.1 | 98.6 | 101 | 1 | 96.2 | 3.40 | 92.90 | | | | | | | | | | | |
| | 2 | 10.5 | 5.5 | 6.8 | 95.6 | 102.8 | 10 | 99.3 | 6.40 | 96.50 | | | | | | | | | | | |
| F13 | 1 | 87.6 | 80.9 | 9.7 | 5.4 | 8.8 | 88.1 | 3.7 | 90.20 | 3.00 | 87.1 | 90.2 | | | | | | | | | |
| | 2 | 4 | 6.1 | 4.8 | 90.2 | 92.5 | 4.5 | 88.7 | 6.40 | 87.60 | 3.5 | 6.6 | | | | | | | | | |
| | 3 | 5.3 | 8.5 | 7.8 | 105.5 | 107.5 | 4.8 | 103.2 | 9.40 | 99.30 | 5.8 | 3 | | | | | | | | | |
| F14 | 1 | 2.8 | 5.4 | 4.1 | 90.9 | 93.2 | 3.3 | 89.4 | 5.70 | 88.20 | 2.3 | 5.4 | 84.4 | 0.8 | 8.1 | | | | | | |
| | 2 | 96 | 101.5 | 100.2 | 4.8 | 2.9 | 97.1 | 7.5 | 101.80 | 8.10 | 96.1 | 99.2 | 11.2 | 92.6 | 101.9 | | | | | | |
| | 3 | 4.8 | 7.5 | 9.2 | 104 | 107.1 | 4.3 | 102.7 | 8.40 | 98.80 | 5.3 | 2.5 | 97.5 | 14 | 0.5 | | | | | | |
| F15 | 1 | 4.3 | 0.8 | 0.5 | 95.5 | 97.9 | 4.8 | 94.1 | 1.10 | 89.90 | 3.8 | 6.9 | 89.1 | 5.3 | 9.6 | 4.5 | 100.1 | 9.1 | | | |
| | 2 | 1 | 1.7 | 3 | 98 | 100.5 | 1.5 | 96.6 | 2.70 | 92.40 | 0.5 | 3.6 | 91.6 | 8.4 | 6.3 | 7.6 | 108.8 | 5.8 | | | |
| | 3 | 97.6 | 99.9 | 98.6 | 3.6 | 2.2 | 98.1 | 6.2 | 100.20 | 6.80 | 97.1 | 100.2 | 9.9 | 93.6 | 102.9 | 94.4 | 5.2 | 102.4 | | | |
| F16 | 1 | 4.5 | 6.9 | 8.7 | 103.7 | 106.6 | 4 | 102.3 | 7.80 | 98.50 | 5 | 2.1 | 97.1 | 13.5 | 0.8 | 12.7 | 108.5 | 0.3 | 8.8 | 5.5 | 103.2 |
| | 2 | 98.2 | 100.5 | 99.2 | 4.1 | 2.8 | 98.7 | 6.8 | 100.80 | 7.50 | 97.7 | 100.9 | 10.5 | 94.2 | 103.5 | 95 | 5.7 | 103 | 102.8 | 91.8 | 0.6 |

the concrete work was in the eastern part of the project, so the concrete system needed to be moved to the east. In the first two stages, the 1$^{st}$ candidate site for the concrete system is closer to the 1$^{st}$ candidate site for the cement/sand/aggregate storage yard than the 2$^{nd}$ candidate site of the concrete system and these two pairs of facilities have a high closeness relationship (i.e. 5). At the same time, the 1$^{st}$ candidate site for the concrete system is also further from the 3$^{rd}$ candidate site for the sandstone fabrication plant than the 2nd candidate site for the concrete system and these two pairs of facilities also have a high closeness relationship (i.e. 4). There were other facilities where it was necessary to consider the relationship and the distance from the concrete system. Although it is difficult for the decision maker to solve, this typical pareto problem was successfully solved using the method in this section, which proves its practicality. Project management can select the value of the variables from these different runs by considering the real environment of the candidate solution; for example, personnel, transport conditions and safety. It should be noted that hydropower stations are often located in mountainous areas, so have difficult environments. Some candidates have an economic benefit advantage but may not be very safe because of such events as debris flow and flood. The selection is also influenced by local government policy. The decision makers need to comprehensively consider the conditions and make a final decision.

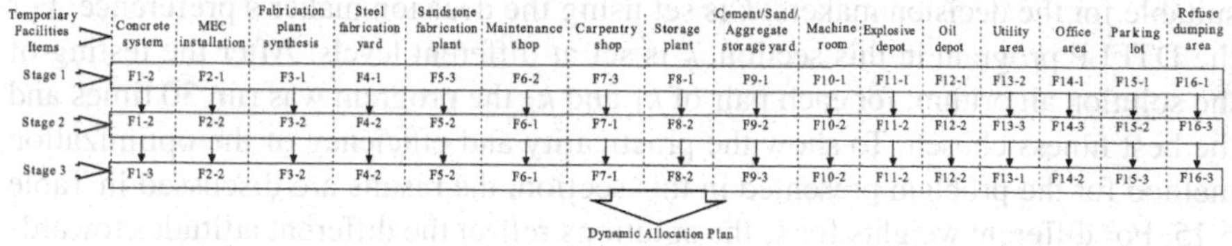


**Fig. 6.20** Dynamic allocation plan for the DTFLP problem of the Jinping-II hydropower station

To gain insight into the parameter selection principles including the population size (i.e., $N$), iteration number (i.e., $G_{max}$), crossover probability (i.e., $p_c$), mutation probability (i.e., $P_m$), the acceptance probability $P_c = 0.95$, cooling rate $\rho = 0.95$ and the weight of the best solution $k$ in each objective, a sensitivity analysis was conducted against these parameters as shown in Table 6.15. The decision makers can fine tune these parameters to obtain different solutions under different parameter levels. Under the same level for the other parameters, the average computing time for a population size $N = 50$ was about 1.65 times that of $N = 30$, so better results can be obtained when the population size is larger. In the SA-based GA algorithm, when the crossover probability (i.e., $p_c$) and mutation probability (i.e., $P_m$) increase respectively, the results are better, but the effect of the crossover probability is greater than that of the mutation probability. The cooling rate $\rho$ has little effect on the results, and the acceptance probability also had no significant effect on the results. It should be noted that, under the same parameters, no matter how many times the program code was run, the best solution was always the same. This proves the stability of the SA-based GA presented.

From Table 6.15, it can be seen that the parameters have some relationship with the results, which proves that the algorithm is not stable. Consequently, it is necessary to repeat the algorithm many times to find the most suitable solution. In addition, the weight $k$ in the evaluation value $F$ (see Eq. (6.55)) for finding the optimal feasible solution was chosen randomly, which has a strong relationship with the fitness value. This is a significant factor affecting the fluctuation in the results.

**Table 6.15** Sensitivity analysis for parameter selection of SA-based GA

| Parameter Combination | Population Size | Crossover Possibility | Mutation Possibility | Acceptance probability | Cooling rate | Fitness value | | | Convergence | Computing |
|---|---|---|---|---|---|---|---|---|---|---|
| | | | | $P_c$ | $\rho$ | $f_1$ (in million yuan) | $f_2$ | $f$ | Iteration Number | Time |
| 1 | 30 | 0.7 | 0.05 | 0.95 | 0.95 | 35,627 | 674270 | 2946.20 | 2 | 61.969 |
| 2 | 30 | 0.8 | 0.05 | 0.95 | 0.95 | 13,481 | 422060 | 6628.00 | 1 | 57.219 |
| 3 | 30 | 0.7 | 0.2 | 0.95 | 0.95 | 24,803 | 728700 | 3240.00 | 1 | 60.187 |
| 4 | 30 | 0.7 | 0.05 | 0.9 | 0.95 | 36,871 | 886170 | 2010.47 | 1 | 30.468 |
| 5 | 30 | 0.7 | 0.05 | 0.95 | 0.9 | 32,948 | 836030 | 3240.00 | 2 | 60.391 |
| 6 | 50 | 0.7 | 0.05 | 0.95 | 0.95 | 51,253 | 616670 | 5271.33 | 1 | 107.812 |
| 7 | 50 | 0.8 | 0.05 | 0.95 | 0.95 | 63,327 | 289080 | 7756.68 | 3 | 106.391 |
| 8 | 50 | 0.7 | 0.2 | 0.95 | 0.95 | 62,660 | 698950 | 9000.12 | 1 | 106.922 |
| 9 | 50 | 0.7 | 0.05 | 0.9 | 0.95 | 64,673 | 686150 | 9026.89 | 1 | 104.188 |
| 10 | 50 | 0.7 | 0.05 | 0.95 | 0.9 | 14,506 | 687120 | 5677.23 | 2 | 108.234 |

As a result, the objective value $F$ for each run is different. $k$ lays emphasis on the fitness of the best solution for each objective function. In practice, to make it more suitable for the decision makers, $k$ is set using the decision maker's preference. For the DTFLP program in this section, $k$ is set at different levels. After the testing of the solution algorithm, for each pair of $k_1$ and $k_2$ the program was run 30 times and the best fitness chosen. To show the practicality and efficiency of the optimization method for the problem presented in this section, the results are discussed in Table 6.15. For different weights for $k$, the solutions reflect the different attitudes towards the two objectives. The decision maker can choose a plan from these different runs using their preferences. For example, the decision maker may think that a safety and environmental objective is the most important objective, so they may sacrifice cost for the safest site layout plan. In that case, they will choose $k_1 < k_2$. On the contrary, if the decision maker thinks the cost objective is the most important, they may choose a minimum total cost for the dynamic site layout plan and sacrifice the safety and environmental objective, thus, they will choose $k_1 > k_2$.

These results are quite useful and may serve as references for decision-makers as it is their choice to identify an appropriate set of parameter values to optimize their decision making processes. Nevertheless, this proposed methodology provides a systematic approach to narrowing down the number of alternatives, and to facilitating decision making.

### 6.8.3 Model Comparison

In many existing research references, the DTFLP has been modeled with a surrogate function for flow distance or with simplified objectives that may be trapped into the local optimum, such as in Urban's (1998) [152] model, where the material flow was considered but the other factors were not. To determine how well each of the modeling methods contributed to the DTFLP, each objective was respectively considered. Rosenblatt's (1986) [148] model was used to solve the DTFLP problem for the Jinping-II hydropower station using the same given parameters. The objective was only $f_1$ this model and the result was 12,330 million yuan, (see Figure 6.21), 60,090 million yuan less than the 72,442 million yuan result of our multi-objectives model. While the second objective $f_2$ was 839,583 at the same time, 132,929 higher

(18.81%) than 706,654, the result of the multi-objectives model. Similarly, if $f_2$ is used as the only objective, the results are 46,122 (see Figure 6.22), 660,532 less than the model while the cost ($f_1$) was 79,189 million yuan, 6,767 million yuan (9.34%) higher. Thus, the integration method (Eqs. (6.55) and (6.56) (Goldberg 1989 [336])), for the multi-objectives used in the model here have a greater effect on the outcomes. Furthermore, this model is more suitable for construction as the project manager is able to consider both the cost and the relationships at the same time.

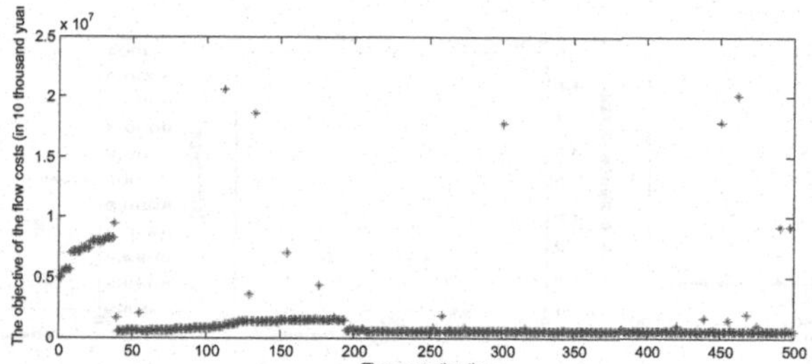


**Fig. 6.21** The searching process of $f_1$ in Rosenblatt's model

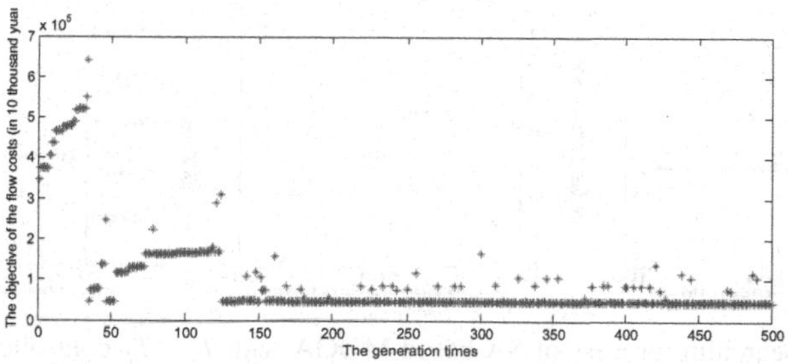


**Fig. 6.22** The searching process of $f_2$ in one objective model

Each layout design application is unique in nature as there are different attributes associated with different applications (Yang and Hung 2007 [155]). This model is not fit for a layout problem with a few facilities in a limited area, such as the Khaled's dynamic site layout planning problem (Haidar et al. 2009 [138]) because the orientation of the facilities or the geometric constraints are not considered.

### 6.8.4 Algorithm Evolution

The results showed some improvement but delayed the search time by about 20 times in the algorithm proposed than in the SA-based GA algorithm in Mohanta et al. (2007) [143], but this exceeded time was considered acceptable. The effect of the final time control on the SA is clearly reflected in the reduction of settling times and demerit figures, so the SA-based MOGA with $T < T_f$ controller is much slower in the problem solving. The SA-based MOGA with $T < T_f$ controller technique can also monitor its problem solving performance using diversification and intensification procedures for a number of the iterative loops. Thus, the SA-based MOGA modifies the solutions in different operators but demonstrates some important deficiencies, particularly in the need for improved solutions. To compare the performance of the algorithms, the DTFLP programme was also solved using the MOGA. For each type, the DTFLP program was run 10 times with the same parameter values, (see Figure 6.23 $\sim$ 6.26).

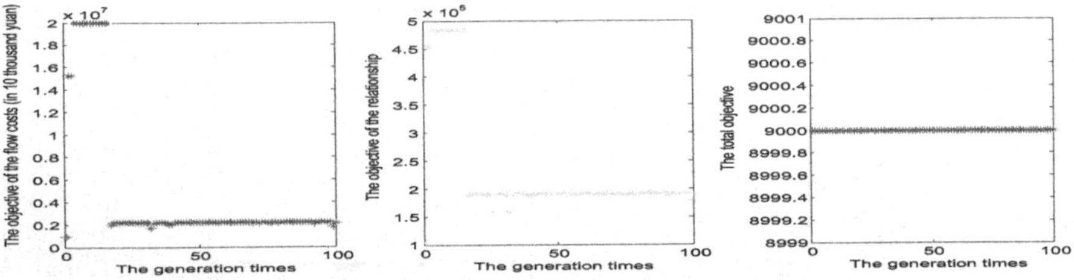


**Fig. 6.23** The searching process of SA-based MOGA with $T < T_f$ controller

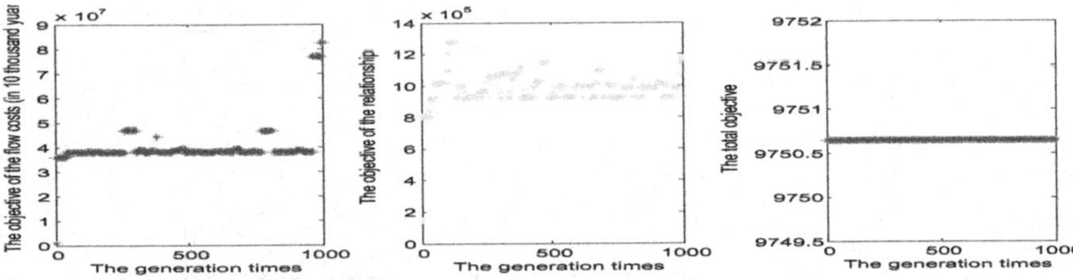


**Fig. 6.24** The searching process of SA-based MOGA with $T < T_f$ controller (generation= 1000)

When the generation is set = 1000, it was discovered that each of the objectives increased with the generation, while the total fitness value remains unchanged as it had reached the best solution in the SA-based MOGA algorithm (see Figure 6.24). However, every objective decreased with the generation in the MOGA algorithm.

The results and performance of the algorithm and the actual data for the project are listed in Table 6.16. By comparing the value of the objectives of different algorithms with the actual data (or the planning data) of the project, it was found that there were differences between them. The net increase of the first objective for the SA-based MOGA algorithm from Mohanta (2007) [143], the SA-based MOGA with the $T_f$ controller algorithm and the MOGA algorithm was 14,542 million yuan, 4,786 million yuan and 18,415 million yuan respectively. The rate of increase was 18.83%, 6.20% and 23.85%. The net rate of the increase for the second objective was 16.01%, 11.17% and 9.71%. Though the MOGA algorithm had a better searching ability for the first objective, the SA-based MOGA algorithm from Mohanta et al. (2007) [143] had a better ability in searching for both objectives. The total

**Table 6.16** The comparison of the results and the actual data (or Planning investment)

| Algorithm | Stage $t$ | Cost(in million yuan) $C$ | | | | Data=795470 $f_2$ | | | Fitness $f$ | Convergence | time |
|---|---|---|---|---|---|---|---|---|---|---|---|
| | | Result | Data | Increase | Rate | Result | Increase | Rate | $f$ | Iteration number | |
| SA-based MOGA[143] | 1 | 29,615 | 32,574 | 2,959 | 9.08% | 668120 | 127350 | 16.01% | 5754 | 2 | 7.5780 |
| | 2 | 55,899 | 66,556 | 10,657 | 16.01% | | | | | | |
| | 3 | 62,666 | 77,208 | 14,542 | 18.83% | | | | | | |
| SA-based MOGA with $T < T_f$ controller | 1 | 31,568 | 32,574 | 1,006 | 3.09% | 706654 | 88816 | 11.17% | 6400.6 | 1 | 132.875 |
| | 2 | 62,838 | 66,556 | 3,718 | 5.59% | | | | | | |
| | 3 | 72,422 | 77,208 | 4,786 | 6.20% | | | | | | |
| MOGA | 1 | 26,653 | 32,574 | 5,921 | 18.18% | 718261 | 77209 | 9.71% | 5041 | 5 | 7.047 |
| | 2 | 43,392 | 66,556 | 23,164 | 34.80% | | | | | | |
| | 3 | 58,793 | 77,208 | 18,415 | 23.85% | | | | | | |

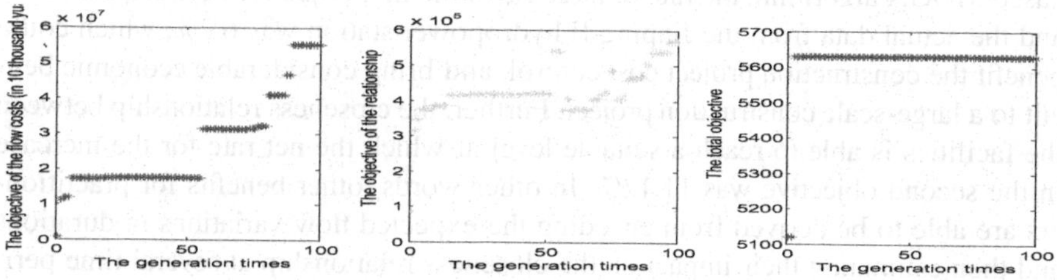


**Fig. 6.25** The searching process of SA-based MOGA without $T < T_f$ controller

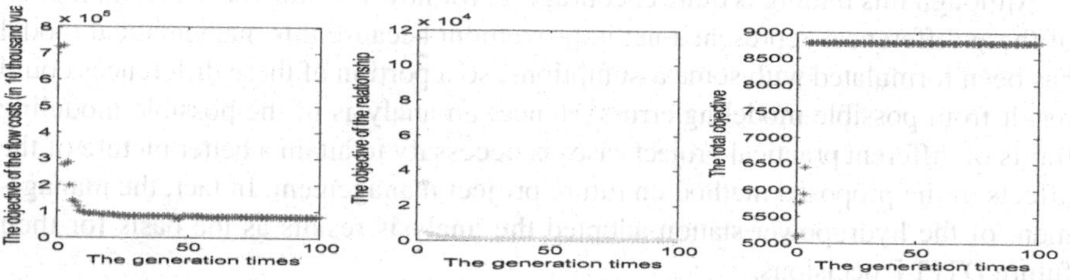


**Fig. 6.26** The searching process of MOGA

objectives for these three algorithms were 5,754, 6,400 and 5,041. The proposed algorithm had the best ability in aggregate for the different objectives. From Figures 6.25 and 6.26, it can be seen that the proposed SA-based MOGA shows a much better performance for the final objective as it is suited to the solution for discrete multi-objective problems just as for the 0-1 integer variables in the DTFLP problem. However, the convergence of the SA-based MOGA was too quick as the iteration number was around 2 because the number of variables in the proposed example case was small resulting in a the shorter solution search process. Thus, it showed very limited updating in the convergence but when the number of the variables increased, the SA-based GA had the greater advantage.

It is worth noting that the gap between the best and the worst and between the best and the average solutions for the SA-based MOGA with the proposed $T < T_f$ controller algorithm are narrower than the gap for the other two types. This shows that the proposed SA-based MOGA with the $T < T_f$ controller algorithm creates a more stable solution space. This suggests that the SA-based MOGA with the $T < T_f$ controller algorithm is an effective and relatively efficient approach for solving the DTFLP problem.

### 6.8.5 *Effectiveness Analysis*

There are significant differences between the proposed methodology and the actual outcomes. Project managers are interested in these differences because they are more likely to gain enlightenment for future practical cases. For the proposed SA-based MOGA algorithm, the rate of increase in the first objective between the result and the actual data from the Jinping-II hydropower station was 6.2%, which could benefit the construction project cost control, and bring considerable economic benefit to a large-scale construction project. Further, the closeness relationship between the facilities is able to reach a suitable level at which the net rate for the increase in the second objective was 11.17%. In other words, other benefits for practitioners are able to be derived from encoding the expected flow variations in durations and then estimating their impact on the closeness relationship at several time periods. Consequently, the safety relationship level in the construction project can be improved.

Although this finding is quite encouraging, for now it is not 100% certain that all of these differences represent a net improvement because this mathematical model has been formulated with some assumptions, so a portion of these differences could result from possible modeling errors. Hence, an analysis of the possible modeling limits of different practical project cases is necessary to attain a better picture of the effects of the proposed method on future project management. In fact, the management of the hydropower station adopted the analysis results as the basis for their future DTFLP decisions.

It can be concluded that the current practice is not optimal. In fact, in current practice, decision makers usually make their decisions based only on experience. Therefore, the results here are able to provide decision makers with a theoretical optimal dynamic temporary facilities location plan for guiding current practice. The goal of the proposed methodology to solve the DTFLP was not totally achievable due to the nature of a DTFLP, where other constraints form the limitations, and because of current management philosophy (Yang and Hung 2007 [155]).

# 7

# Fuzzy Random MOMSDM for Transportation Assignment

In large scale hydropower construction projects, the transportation system is crucial and because of the massive cost and construction duration, this comprehensive optimization allocation transportation problem must be dealt with urgently, especially when there is a complex uncertain environment. Therefore this chapter focuses on a multi-objective optimization for a two-stage based earth and rockfill dam construction transportation system under a fuzzy random environment with the aim of minimizing total operational costs, transportation duration and total waste. Dynamic transportation assignment problems (DTAP) are often encountered in many practical systems, such as urban planning traffic networks, closed traffic network systems, field service support systems, and container transportation and flow-shop-type production systems (Teodor et al. 2004 [591]; Wang et al. 2008 [592]; Ebben et al. 2004 [593]; Papadopoulos 1996 [233]; Pasquale et al. 2008 [594]; Celik et al. 2009 [595]; Soukhal et al. 2005 [596]; Hurink and Knust 2001 [597]). Linear programming, bi-level programming and multi-objective programming are usually used to solve such transportation problems, for example, Luathep et al. (2011) [598], Ali and Sik (2012) [599], Maher et al.(2001) [600], Cao et al. (2007) [601], Islam and Roy (2006) [602], EI-Wahed and Lee (2006) [603]. While these studies have made a significant contribution to the DTAP and solved many realistic problems, the transportation systems in these studies rarely discuss multi-stages, much less consider multi-objectives and multi-stages together from a dynamic view in large scale hydropower construction projects. In this chapter, because of quantity limitations and other constraints, decision-makers need to determine a suitable allocation by dynamically allocating traffic so that the total operational cost, transportation duration and total cost are minimized. Specifically, several kinds of transportation carriers and multi-stages are proposed, meaning that the two stage-based dynamic transportation assignment problem (TS-based DTAP) is a multi-objective dynamic programming (MODP) process for large scale hydropower projects. This chapter mainly focuses on a discussion of the multistage decision making for a TS-based DTAP under a fuzzy random environment.

J. Xu and Z. Zeng, *Fuzzy-Like Multiple Objective Multistage Decision Making*,
Studies in Computational Intelligence 533,
DOI: 10.1007/978-3-319-03398-3_7, 

## 7.1 Statement of TS-Based DTAP

The two stage-based dynamic transportation assignment problem (TS-based DTAP) is quite common in a large scale hydropower construction projects [590]. This section presents the TS-based DTAP which are encountered in an earth and rockfill dam construction. The motivation for employing fuzzy random variables in the TS-based DTAP is also explained.

### *7.1.1 Multistage Earth and Rockfill Dam Construction Project with Two-Stage Based Transportation System*

In a large scale hydropower construction project, earth and rockfill dam construction is crucial work. The earth and rockfill dam construction process can be divided into three stages depending on total construction duration, flood retention and discharge standards, and construction diversions, which in order are foundation filling, central dam filling and upper dam filling, as shown in Figure 7.1, with each stage needing different quantities of earth-rock depending on the required construction strength and other essential features. In the first stage, the main work is foundation filling, and the main work of the second stage is central dam filling and cofferdam filling, which is a key period for earth and rockfill dam construction and has a high construction intensity. The last task is the filling of the dam crest and the completion of the project's corresponding height, thickness and slope protection construction, all of which are dependent on variations in the water level in the upper dam area.

The earth and rockfill dam construction project transportation system can be divided into two stages according to the construction site's environmental features, and transportation route. The first stage is the transportation from the borrow areas to the stockpile areas using dump trucks, and the second stage is from the stockpile areas to their corresponding fill areas using belt conveyors, the details of which are in Figure 7.1. For each transportation route, the transportation distance, road condition, transportation intensity and other objective factors are quite dissimilar which means there are different and often uncertain parameters. The project manager prefers to allow each dump truck driver to work on fixed transportation routes between the borrow areas and the stockpile areas to improve operational efficiency and reduce traffic accidents. Therefore, the selection of transportation routes and the earth-rock allocation between the borrow areas, stockpile areas and fill areas needs to be appropriately planned to ensure the provision of adequate earth-rock to meet the various filling intensities at the corresponding fill areas and to optimize the objectives. The aim of the TS-based DTAP proposed in this chapter is to achieve the objectives of cost minimization, transportation duration minimization and waste minimization to assure optimization and maximum benefit, efficiency and resource utilization.

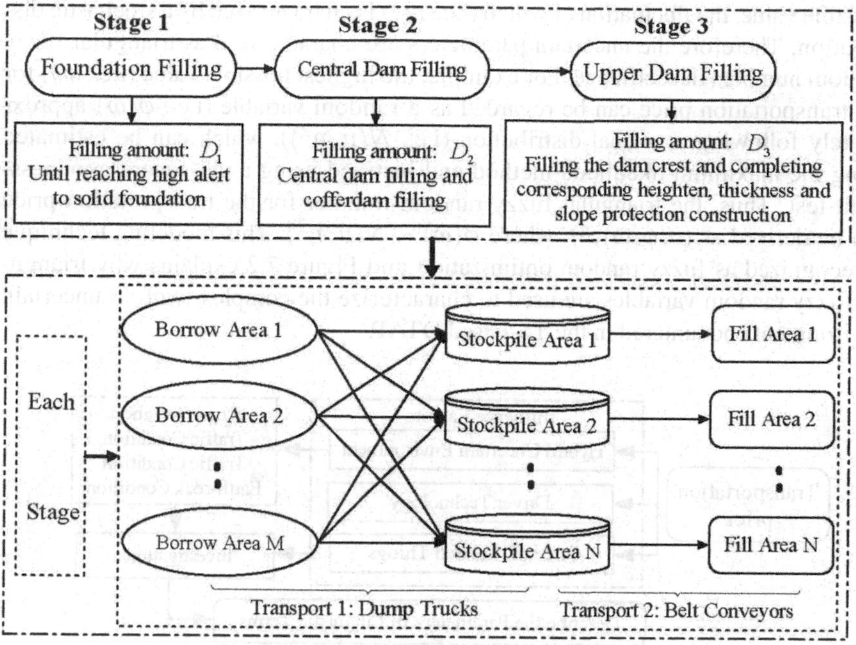


**Fig. 7.1** Construction stages and transportation system of earth and rockfill dam construction project

## 7.1.2 *Motivation for Employing Fuzzy Random Variables in TS-Based DTAP*

The need to address uncertainty in earth and rockfill dam construction projects is widely recognized. For example, the earth-rock excavation price at each stage may be different because of the uncertainty of excavation difficulty, the cost of mechanical production and construction management, and indirect shop labor expense. In fact, over the years, advancements in engineering technology have allowed for more construction operation alternatives. Subsequently, decision makers are faced with more imprecise information than ever before. Therefore, fuzzy random variables that can take account of both fuzziness and randomness are favored by decision makers to describe the uncertainty and vague information encountered in reality.

In practice, to collect the data, investigations and surveys were done with related project engineers. First, the above parameters were determined for each day, which is an interval (i.e., $[l_d, r_d]$) with a highest possible value (i.e., $m_d$). For each stage, the minimum value of all $l_d$ and the maximal value of all $r_d$ for each parameter in the survey data was determined, and the left border (i.e., $a$) and right border (i.e., $b$) of the fuzzy random number were selected. For more accurate and reasonable calculations, a new method for determining the highest possible value for each parameter at each stage is proposed, as in each stage there are many possibilities for the highest possible value, the fluctuation of which (i.e., $m_d$) is characterized by a stochastic distribution. Therefore the uncertain parameters are characterized as triangular fuzzy random numbers in each stage. For example, the highest possible value (i.e., $m_d$) for the transportation price can be regarded as a random variable (i.e., $\varphi(\omega)$ approximately following a normal distribution (i.e., $N(\mu,\eta^2)$), which can be estimated using the maximum likelihood method and justified using a chi-square goodness-of-fit test. Thus, the triangular fuzzy random number for the transportation price can be derived as $(a,\varphi(\omega),b)$, where $\varphi(\omega) \sim N(\mu,\eta^2)$. This modeling technique is recognized as fuzzy random optimization and Figure 7.2 explains why triangular fuzzy random variables are used to characterize the complex twofold uncertain environment encountered in the TS-based DTAP.

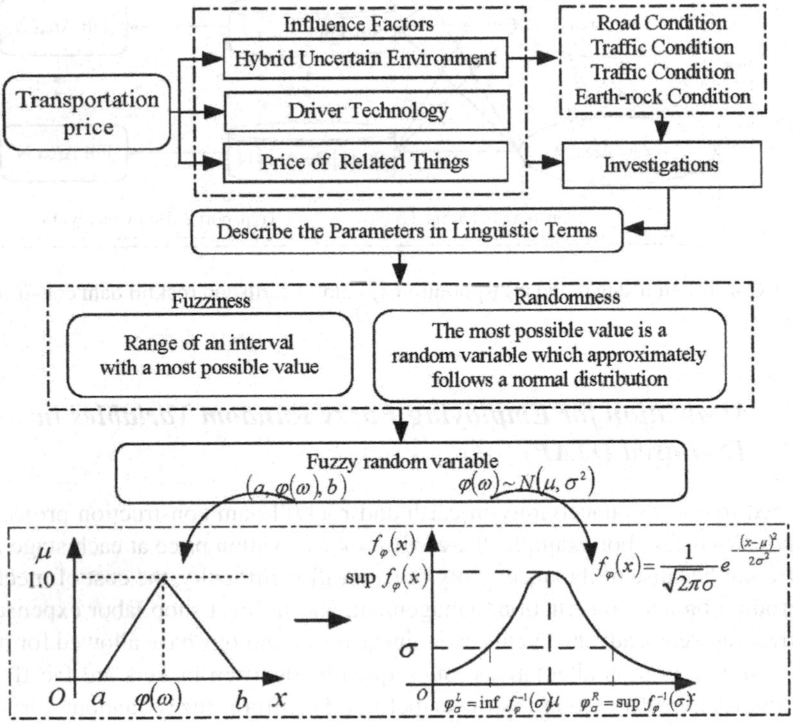


**Fig. 7.2** Characteristic of the complex twofold uncertain environment using triangular fuzzy random variables

## 7.2 Modelling Process of TS-Based DTAP

To model the multiple objective optimization for a multistage earth and rockfill dam construction transportation system under a fuzzy random environment in this section, the state equations, initial conditions, constraint conditions, and objective functions are presented. As mentioned in the problem statement above, in each stage, the earth-rock can be transported from any borrow area to any stockpile area, but each stockpile area is in charge of its corresponding fill area, so we can use the same set $S$ to express the stockpile areas and fill areas. In order to make the model easier to understand, the following notations are introduced:

*Sets and Subscripts*

$B$ : set of all borrow areas, $i \in B$

$S$ : set of all stockpile areas or set of all fill areas, $j \in S$

$K$ : set of all stages, $k \in K$

$A_j(k)$ : set of all borrow areas from which there is earth-rock transported to stockpile area $j$ at stage $k$

*Parameters*

$\tilde{\bar{c}}_i(k)$ : excavation price of earth-rock in borrow area $i$ at stage $k$, $i \in B, k \in K$

$\tilde{\bar{c}}_j^1(k)$ : storage price of earth-rock in stockpile area $j$ during stage $k$, $j \in S, k \in K$

$\tilde{\bar{c}}_j^2(k)$ : storage price of earth-rock in stockpile area $j$ between stage $k$ and stage $k+1$, $j \in S, k \in K$

$\tilde{\bar{c}}_{ij}(k)$ : transportation price from borrow area $i$ to stockpile area $j$ at stage $k$, $i \in B$, $j \in S, k \in K$

$\tilde{\bar{c}}_{jj}(k)$ : transportation price from stockpile area $j$ to fill area $j$ at stage $k$, $j \in S$, $k \in K$

$\tilde{\bar{t}}_{ij}(k)$ : transportation time of the first dump truck from borrow area $i$ to stockpile area $j$ at stage $k$, $i \in B, j \in S, k \in K$

$\tilde{\bar{t}}_{jj}(k)$ : transportation time per unit earth-rock from stockpile area $j$ to fill area $j$ at stage $k$, $j \in S, k \in K$

$\tilde{\bar{w}}_{ij}(k)$ : waste per unit of earth-rock from borrow area $i$ to stockpile area $j$ at stage $k$, $i \in B, j \in S, k \in K$

$\tilde{\bar{w}}_{jj}(k)$ : waste per unit of earth-rock from stockpile area $j$ to fill area $j$ at stage $k$, $j \in S, k \in K$

$\tilde{d}_j(k)$ : total demand of fill area $j$ at stage $k$, $j \in S, k \in K$

*Decision Variables*

$$\delta_{ij}(k) = \begin{cases} 1, & \mathrm{i} \in \mathrm{A_j(k)} \\ 0, & \text{otherwise} \end{cases}$$

$x_{ij}(k)$ : amount of earth-rock transported from borrow area $i$ to stockpile area $j$ at stage $k$, $i \in A_j(k), j \in S, k \in K$

$y_{jj}(k)$ : amount of earth-rock transported from stockpile area $j$ to fill area $j$ at stage $k$, $j \in S, k \in K$

## 7.2.1 Dealing with Fuzzy Random Parameters

The introduction of the fuzzy random coefficient is of great applicability but makes the process much more difficult to solve. One strategy is to employ a transformation method to convert the fuzzy random parameters into real numbers. In this section, the hybrid crisp approach put forward in Xu et al. (2013a) [584] is used to integrate the decision maker's optimistic-pessimistic attitudes in adopting real world practice. As described before, $\tilde{\bar{c}}_i(k)$, $\tilde{\bar{c}}_j^1(k)$, $\tilde{\bar{c}}_j^2(k)$, $\tilde{\bar{c}}_{ij}(k)$, $\tilde{\bar{c}}_{jj}(k)$, $\tilde{\bar{t}}_{ij}(k)$, $\tilde{\bar{t}}_{jj}(k)$, $\tilde{\bar{w}}_{ij}(k)$, $\tilde{\bar{w}}_{jj}(k)$ can be characterized as triangular fuzzy random numbers, generally, let $\tilde{\bar{\xi}} = ([\xi]_L, \varphi(\omega), [\xi]_R)$ be any of them, where $\varphi(\omega)$ is a random variable with a probability density function $f_\varphi(x)$ and Normal distribution $N(\mu, \eta^2)$, $\gamma$ and $\sigma$ are any given possibility level of the fuzzy variable and any given probability level of the random variable respectively.

Transform the fuzzy random variable $\tilde{\bar{\xi}} = ([\xi]_L, \varphi(\omega), [\xi]_R)$ into the $(\gamma,\sigma)$-level trapezoidal fuzzy variable $\tilde{\xi}_{(\gamma,\sigma)}$ using the equation: $\tilde{\bar{\xi}} \to \tilde{\xi}_{(\gamma,\sigma)} = ([\xi]_L, \underline{\xi}, \overline{\xi}, [\xi]_R)$, wherein $\underline{\xi} = [\xi]_R - \gamma([\xi]_R - \varphi_\sigma^L)$ and $\overline{\xi} = [\xi]_L + \gamma(\varphi_\sigma^R - [\xi]_L)$, here $\varphi_\sigma^L = \inf\{x \in R | f_\varphi(x) \geq \sigma\} = \mu - \sqrt{-2\eta^2 \ln(\sqrt{2\pi}\sigma\eta)}$ and $\varphi_\sigma^R = \inf\{x \in R | f_\varphi(x) \geq \sigma\} = \mu + \sqrt{-2\eta^2 \ln(\sqrt{2\pi}\sigma\eta)}$.

Then the $(\gamma,\sigma)$-level trapezoidal fuzzy variables are subsequently defuzzified using an expected value operator with an optimistic-pessimistic index as follows:

$$E^{Me}[\tilde{\xi}_{(\gamma,\sigma)}] = \frac{(1-\lambda)}{2}([\xi]_L + \underline{\xi}) + \frac{\lambda}{2}(\overline{\xi} + [\xi]_R), \tag{7.1}$$

where $\lambda$ is the optimistic-pessimistic index to determine the combined attitude of the decision makers. Xu and Zhou (2011) [281] proposed that the fuzzy measure Me was a convex combination of Pos and Nec. Based on the above method, for example, the fuzzy random variables $\tilde{\bar{c}}_{ij}(k)$ in this chapter can be transformed into the $(\sigma_{ij}, \gamma_{ij})$-level trapezoidal fuzzy variables using the formula $\tilde{\bar{c}}_{ij}(k) \to \tilde{c}_{ij}(k)_{(\sigma_{ij},\gamma_{ij})} = ([c_{ij}(k)]_L, \underline{c}_{ij}(k), \overline{c}_{ij}(k), [c_{ij}(k)]_R)$. It should be noted that the probability and possibility coefficients for the different fuzzy random variables may be different, but in this discussion, all probability levels $\sigma_{ij}$, $\sigma_{jj}$, $\sigma_i$, $\sigma_j$, $\sigma_j^1$ and $\sigma_j^2$ are set to the same value $\sigma$, and all possibility levels $\gamma_{ij}$, $\gamma_{jj}$, $\gamma_i$, $\gamma_j$, $\gamma_j^1$ and $\gamma_j^2$ are set to the same value $\gamma$.

### 7.2.2 State Equations

From the decision variable notations, $x_{ij}(k)\delta_{ij}(k)$ is the amount of earth-rock transported from borrow area $i$ to stockpile area $j$ at stage $k$. It is assumed that there are $M$ borrow areas, $N$ stockpile areas and $N$ fill areas. Let $B_i(k)$ and $S_j(k)$ be the quantity of available earth-rock in borrow area $i$ ($i = 1,2,\cdots,M$) and stockpile area $j$ ($j = 1,2,\cdots,N$) respectively at the end of stage $k$, Therefore Eq. (7.2) and (7.3) can be used to describe the relationship between the state variables (i.e., $B_i(k)$ and $S_j(k)$) and the decision variables (i.e., $\delta_{ij}(k)$, $x_{ij}(k)$ and $y_{jj}(k)$) as follows,

$$B_i(k) = B_i(k-1) - \sum_{j=1}^{N} x_{ij}(k)\delta_{ij}(k), \quad \forall i \in B, k \in K, \tag{7.2}$$

$$S_j(k) = S_j(k-1) + \sum_{i=1}^{M} x_{ij}(k)\delta_{ij}(k) \cdot (1 - E^{Me}[\tilde{w}_{ij}(k)_{(\gamma,\sigma)}]) - y_{jj}(k), \quad \forall j \in S, k \in K. \tag{7.3}$$

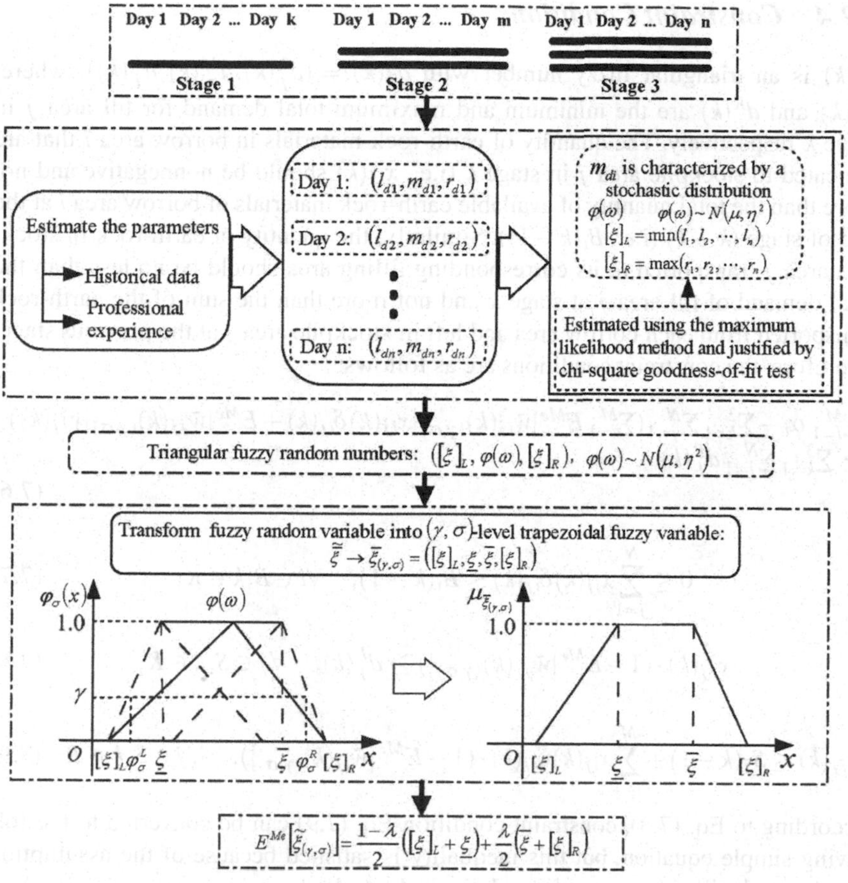


**Fig. 7.3** Transformation method of the fuzzy random parameters

## 7.2.3 Initial Conditions

It should be noted that the quantity of earth-rock materials in every borrow area is fixed at the beginning of the construction project, so let $\alpha_i$ be the total amount of earth-rock materials in borrow area $i$. The quantity of available earth-rock materials at stockpile area $j$ ($j = 1, 2, \cdots, N$) at the beginning of the construction project is zero. The total demand of fill area $h$ at stage $k$ may be different in different stages, which is a decision that would be made by the project manager according to the construction requirements at that stage, thus the state variables $B_i(k)$ and $S_j(k)$ can be initialized as follows,

$$B_i(0) = \alpha_i, \quad \forall i \in B, \tag{7.4}$$

$$S_j(0) = 0, \quad \forall j \in J. \tag{7.5}$$

### 7.2.4 Constraint Conditions

$\tilde{d}_j(k)$ is an triangular fuzzy number with $\tilde{d}_j(k) = (d_j^L(k), d_j^M(k), d_j^R(k))$, where, $d_j^L(k)$ and $d_j^R(k)$ are the minimum and maximum total demand for fill area $j$ in stage $k$ respectively. The quantity of earth-rock materials in borrow area $i$ that are allocated to stockpile area $j$ in stage $k$ (i.e., $x_{ij}(k)$ should be nonnegative and not more than the total quantity of available earth-rock materials of borrow area $i$ at the end of stage $(k-1)$ (i.e., $B_i(k-1)$). Similarly, the quantity of earth-rock in stockpile area $j$ transported to its corresponding filling area should be no less than the total demand of fill area $j$ at stage $k$ and not more than the sum of the earth-rock transported from each borrow area and left in stockpile area $j$ at the previous stage. Therefore, the constraint conditions are as follows:

$$\sum_{i=1}^{M} \alpha_i - \sum_{k=1}^{3} \sum_{j=1}^{N} \left(\sum_{i=1}^{M} E^{Me}[\tilde{w}_{ij}(k)_{(\gamma,\sigma)}] x_{ij}(k)\delta_{ij}(k) + E^{Me}[\tilde{w}_{jj}(k)_{(\gamma,\sigma)}] y_{jj}(k)\right) \geq \sum_{k=1}^{3} \sum_{j=1}^{N} d_j^R(k), \tag{7.6}$$

$$0 \leq \sum_{j=1}^{N} x_{ij}(k)\delta_{ij}(k) \leq B_i(k-1), \quad \forall i \in B, k \in K, \tag{7.7}$$

$$y_{jj}(k) \cdot (1 - E^{Me}[\tilde{w}_{jj}(k)_{(\gamma,\sigma)}]) \geq d_j^L(k), \quad \forall j \in S, k \in K, \tag{7.8}$$

$$y_{jj}(k) \leq S_j(k-1) + \sum_{i=1}^{M} x_{ij}(k)\delta_{ij}(k) \cdot (1 - E^{Me}[\tilde{w}_{ij}(k)_{(\gamma,\sigma)}]), \quad \forall j \in S, k \in K. \tag{7.9}$$

According to Eq. (7.3), constraint condition Eq. (7.9) can be converted to the following simple equation, but this inequality is satisfied because of the assumption that demand will never exceed supply in each stockpile area.

$$S_j(k) \geq 0, \quad \forall j \in S, k \in K. \tag{7.10}$$

On the other hand, it is assumed that each stockpile area has its own maximum capacity at the end of each stages, so let $\beta_j$ be the maximum capacity of stockpile area $j$. The quantity of available earth-rock in stockpile $j$ $(j = 1, 2, \cdots, N)$ $S_j(k)$ at the end of stage $k$ should be no more than the maximum capacity of stockpile area $j$. Thus Eq. (7.11) is also considered a constraint condition.

$$S_j(k) \leq \beta_j, \quad \forall j \in S, k \in K. \tag{7.11}$$

To simplify the problem, it is assumed that the quantity of earth-rock transported from borrow area $i$ to stockpile area $j$ at each stage is at least one dump truck maximum transported when $\delta_{ij}(k) = 1$. Let C be the heaped capacity of the dump truck. Then the following constraint condition is derived.

$$x_{ij}(k) \geq C \quad \forall i \in B, j \in S, k \in K. \tag{7.12}$$

### 7.2.5 Objective Functions

The objective of the decision maker is to minimize the total cost, the transportation duration and the total waste over the whole transportation system. This leads to the need for a multiple objectives decision making problem. The details of the three objective functions are presented in this part.

#### Total Cost

The objective function defines the total cost over the complete construction duration. The aim of the contractor is to minimize total cost, which is comprised of excavation costs, transportation costs and storage costs. The details for each component are given below.

***Excavation Cost***: In a construction project, as in the problem statement description, the excavation price for the earth-rock in each borrow area at each stage may be different because of the uncertainty related to excavation difficulty, the cost of mechanical production and construction management, and indirect shop labor expenses, so the fuzzy random variables $\tilde{\bar{c}}_i(k)$ are used to express these characteristics. Let $C_e$ be the total earth-rock excavation cost which can be expressed as follows:

$$C_e = \sum_{k=1}^{3}\sum_{i=1}^{M}\sum_{j=1}^{N} x_{ij}(k)\delta_{ij}(k)\cdot E^{Me}[\tilde{\bar{c}}_i(k)_{(\gamma,\sigma)}]$$

***Transportation Cost***: The transportation system in the TS-based DTAP discussed here contains two subsystems. One transportation subsystem is the transportation from the borrow areas to the stockpile areas using dump trucks; the other is from the stockpile areas to their corresponding fill areas using a belt conveyor. The transportation costs for the different transportation routes between the different borrow areas, stockpile areas and fill areas in each stage are quite dissimilar and uncertain because of differences in distance, road conditions and transport intensity. $\tilde{\bar{c}}_{ij}(k)$ and $\tilde{\bar{c}}_{jj}(k)$ can express these respectively; let $C_t$ be the total transportation cost which can be expressed as follows:

$$C_t = \sum_{k=1}^{3}\sum_{j=1}^{N}\sum_{i=1}^{M} E^{Me}[\tilde{\bar{c}}_{ij}(k)_{(\gamma,\sigma)}]\cdot x_{ij}(k)\delta_{ij}(k) + \sum_{k=1}^{3}\sum_{j=1}^{N} E^{Me}[\tilde{\bar{c}}_{jj}(k)_{(\gamma,\sigma)}]\cdot y_{jj}(k)$$

***Storage Cost***: The storage cost is the cost of storage at each stage and between stages. Let $h_j^k(t)$ express the earth-rock in stockpile area $j$ at time $t$ during stage $k$. It can be seen that $h_j^k(t)$ changes over time which leads to changes in the storage cost. Let $T_j(k)$ be the transportation duration at stockpile area $j$ at stage $k$. The maximum value of $h_j^k(t)$ is denoted as $h_j^k(\bar{t})$. Then, the storage cost in stockpile area $j$ during stage $k$ is $\int_0^{T_j(k)} \tilde{\bar{c}}_j^1(k)\cdot h_j^k(t)dt$, which is difficult to calculate. In this section, a conversion factor is introduced to balance the difference between $h_j^k(\bar{t})$ and the actual storage cost of stockpile area $j$ during stage $k$. Let $\tilde{\bar{c}}_j(k)$ be the conversion factor for stockpile area $j$ at stage $k$, then $\tilde{\bar{c}}_j(k)$ is defined using the following equation:

$$\int_0^{T_j(k)} E^{Me}[\tilde{c}_j^1(k)_{(\gamma,\sigma)}]\cdot h_j^k(t)dt = E^{Me}[\tilde{c}_j(k)_{(\gamma,\sigma)}]\cdot E^{Me}[\tilde{c}_j^1(k)_{(\gamma,\sigma)}]\cdot h_j^k(\bar{t}) \quad \forall j\in, k\in K.$$

Where $\tilde{c}_j(k)$ is a fuzzy random number and its expected value $E^{Me}[\tilde{c}_j(k)_{(\gamma,\sigma)}]$ is between 0 and 1. The storage cost in stockpile area $j$ between stage $k$ and stage $(k+1)$ is $E^{Me}[\tilde{c}_j^2(k)_{(\gamma,\sigma)}]\cdot S_j(k)$.

$$C_s = \sum_{k=1}^{2}\sum_{j=1}^{N} E^{Me}[\tilde{c}_j^2(k)_{(\gamma,\sigma)}]\cdot S_j(k) + \sum_{k=1}^{3}\sum_{j=1}^{N} E^{Me}[\tilde{c}_j(k)_{(\gamma,\sigma)}]\cdot E^{Me}[\tilde{c}_j^1(k)_{(\gamma,\sigma)}]\cdot h_j^k(\bar{t})$$

Therefore the total cost of the whole system is,

$$f_c(x_{ij}(k), \delta_{ij}(k), y_{jj}(k)) = C_e + C_t + C_s - C_i,$$

i.e.,

$$\begin{aligned} f_c(x_{ij}(k), \delta_{ij}(k), y_{jj}(k)) = & \sum_{k=1}^{3}\sum_{i=1}^{M}\sum_{j=1}^{N} x_{ij}(k)\delta_{ij}(k)\cdot E^{Me}[\tilde{c}_i(k)_{(\gamma,\sigma)}] + \\ & \sum_{k=1}^{3}\sum_{j=1}^{N}\sum_{i=1}^{M} E^{Me}[\tilde{c}_{ij}(k)_{(\gamma,\sigma)}]\cdot x_{ij}(k)\delta_{ij}(k) + \\ & \sum_{k=1}^{3}\sum_{j=1}^{N} E^{Me}[\tilde{c}_{jj}(k)_{(\gamma,\sigma)}]\cdot y_{jj}(k) + \\ & \sum_{k=1}^{2}\sum_{j=1}^{N} E^{Me}[\tilde{c}_j^2(k)_{(\gamma,\sigma)}]\cdot S_j(k) + \\ & \sum_{k=1}^{3}\sum_{j=1}^{N} E^{Me}[\tilde{c}_j(k)_{(\gamma,\sigma)}]\cdot E^{Me}[\tilde{c}_j^1(k)_{(\gamma,\sigma)}]\cdot h_j^k(\bar{t}). \end{aligned} \tag{7.13}$$

**Transportation Duration**

In a transportation system, because of the distance, road condition, intensity and some other important factors, transportation times between the different borrow areas, stockpile areas and fill areas are quite dissimilar and uncertain. At each construction stage, earth-rock is transported using dump truck from the borrow areas to the stockpile areas, then, using belt conveyors, from the stockpile areas to the fill areas. When the first dump truck arrives at stockpile area $j$ from the different borrow areas, earth-rock from stockpile area $j$ is then transported to the corresponding fill area. The dump trucks and belt conveyors are moving the earth-rock at the same time, with the belt conveyor being the last to finish. Therefore the transportation duration for filling area $j$ at each stage includes the transportation time of the first dump truck arriving at stockpile area $j$ of all dump trucks from the borrow areas and all the earth-rock transported to its corresponding fill area $j$. The waiting time is ignored at the stockpile areas. Let $T_j(k)$ and $T(k)$ be the transportation duration of fill area $j$ at stage $k$ and the transportation duration at stage $k$ respectively. Then $T_j(k)$ and $T(k)$ can be expressed as follows,

$$T_j(k) = \min_i\{E^{Me}[\tilde{t}_{ij}(k)_{(\gamma,\sigma)}] | \delta_{ij}(k) = 1\} + E^{Me}[\tilde{t}_{jj}(k)_{(\gamma,\sigma)}] \cdot y_{jj}(k),$$

where $i \in A_j(k)$, $j \in S$, $k \in K$,

$$T(k) = \max_j T_j(k).$$

Assume that $\tilde{\bar{t}}_{ij}(k)$ is infinite when $\delta_{ij}(k) = 0$, so let $\{\tilde{\bar{t}}_{ij}(k)\delta_{ij}(k) | \tilde{\bar{t}}_{ij}(k)\delta_{ij}(k) > 0\}$ express the transportation time of the first dump truck from borrow area $i$ to stockpile area $j$ at stage $k$. Therefore,the total transportation duration is,

$$\begin{aligned} &f_d(x_{ij}(k), \delta_{ij}(k), y_{jj}(k)) \\ &= \Sigma_{k=1}^3 T(k) \\ &= \Sigma_{k=1}^3 \max_j(\min_i\{E^{Me}[\tilde{t}_{ij}(k)_{(\gamma,\sigma)}] | \delta_{ij}(k) = 1\} + E^{Me}[\tilde{t}_{jj}(k)_{(\gamma,\sigma)}] y_{jj}(k)). \end{aligned} \tag{7.14}$$

**Total Waste**

There is inevitably some waste because of the road conditions, driving technology and some other factors. Similarly, because of the differences in the road condition on different routes and on different sections of the same route, the waste per unit of earth-rock is uncertain under a fuzzy random environment. At the end of each stage, there may be some earth-rock unused at fill areas which is discarded. Therefore the total construction project waste is,

$$f_w(x_{ij}(k), \delta_{ij}(k), y_{jj}(k)) = \sum_{k=1}^{3}\sum_{j=1}^{N}\sum_{i=1}^{M} x_{ij}(k)\delta_{ij}(k) - \sum_{k=1}^{3}\sum_{j=1}^{N} E^{Me}[\tilde{d}_j(k)]. \tag{7.15}$$

### 7.2.6 Expected Value Model

From the background of the earth-rock transportation system at the Pubugou Hydropower Project, appropriate planning for transportation route selection and earth-rock allocation between the borrow areas, stockpile areas and the fill areas is proposed in this section, the aim of which is to achieve the objectives (i.e. cost minimization, transportation duration minimization and waste minimization). For the state equations, constraints, and objective functions outlined above, a multistage multi-objective optimal control model is established to solve the TS-based DTAP under a fuzzy random environment. The fuzzy random coefficients are estimated using the maximum likelihood method and justified using a chi-square goodness-of-fit test. The expected value of the $(\gamma, \sigma)$-level trapezoidal fuzzy variables can be calculated using Eq.(7.1). Therefore, by integrating Eqs.(7.2)-(7.15), the research problem TS-based DTAP can be formulated in the following Eq.(7.16) multi-objective expected value model:

$$
\left\{
\begin{array}{l}
\min f_c(x_{ij}(k),\delta_{ij}(k),y_{jj}(k)) = \sum\limits_{k=1}^{3}\sum\limits_{i=1}^{M}\sum\limits_{j=1}^{N} x_{ij}(k)\delta_{ij}(k)E^{Me}[\tilde{c}_i(k)_{(\gamma,\sigma)}]+ \\
\qquad \sum\limits_{k=1}^{3}\sum\limits_{j=1}^{N}\sum\limits_{i=1}^{M} E^{Me}[\tilde{c}_{ij}(k)_{(\gamma,\sigma)}]x_{ij}(k)\delta_{ij}(k)+ \\
\qquad \sum\limits_{k=1}^{3}\sum\limits_{j=1}^{N} E^{Me}[\tilde{c}_{jj}(k)_{(\gamma,\sigma)}]y_{jj}(k)+ \\
\qquad \sum\limits_{k=1}^{2}\sum\limits_{j=1}^{N} E^{Me}[\tilde{c}_j^2(k)_{(\gamma,\sigma)}]\cdot S_j(k)+ \\
\qquad \sum\limits_{k=1}^{3}\sum\limits_{j=1}^{N} E^{Me}[\tilde{c}_j(k)_{(\gamma,\sigma)}]E^{Me}[\tilde{c}_j^1(k)_{(\gamma,\sigma)}]h_j^k(\bar{t}), \\
\min f_d(x_{ij}(k),\delta_{ij}(k),y_{jj}(k)) = \sum\limits_{k=1}^{3} \max_j(\min_i\{E^{Me}[\tilde{t}_{ij}(k)_{(\gamma,\sigma)}]|\delta_{ij}(k)=1\}+ \\
\qquad E^{Me}[\tilde{t}_{jj}(k)_{(\gamma,\sigma)}]y_{jj}(k)), \\
\min f_w(x_{ij}(k),\delta_{ij}(k),y_{jj}(k)) = \sum\limits_{k=1}^{3}\sum\limits_{j=1}^{N}\sum\limits_{i=1}^{M} x_{ij}(k)\delta_{ij}(k) - \sum\limits_{k=1}^{3}\sum\limits_{j=1}^{N} E^{Me}[\tilde{d}_j(k)] \\
\text{s.t.}\left\{
\begin{array}{l}
B_i(k) = B_i(k-1) - \sum\limits_{j=1}^{N} x_{ij}(k)\delta_{ij}(k), \quad \forall i\in B, k\in K, \\
S_j(k) = S_j(k-1) + \sum\limits_{i=1}^{M} x_{ij}(k)\delta_{ij}(k)\cdot(1-E^{Me}[\tilde{w}_{ij}(k)_{(\gamma,\sigma)}]) - y_{jj}(k), \\
\forall j\in S, k\in K, \\
B_i(0) = \alpha_i, \quad \forall i\in B, \\
S_j(0) = 0, \quad \forall j\in S, \\
0 \le E^{Me}[\tilde{c}_j(k)_{(\gamma,\sigma)}] \le 1, \quad \forall j\in S, k\in K, \\
0 \le S_j(k) \le \beta_j, \quad \forall j\in S, k\in K, \\
0 \le \sum\limits_{j=1}^{N} x_{ij}(k)\delta_{ij}(k) \le B_i(k-1), \quad \forall i\in B, k\in K, \\
y_{jj}(k)\cdot(1-E^{Me}[\tilde{w}_{jj}(k)_{(\gamma,\sigma)}]) \ge d_j^L(k), \quad \forall j\in S, k\in K, \\
\sum\limits_{i=1}^{M}\alpha_i - \sum\limits_{k=1}^{3}\sum\limits_{j=1}^{N}(\sum\limits_{i=1}^{M} E^{Me}[\tilde{w}_{ij}(k)_{(\gamma,\sigma)}]\cdot x_{ij}(k)\delta_{ij}(k) + E^{Me}[\tilde{w}_{jj}(k)_{(\gamma,\sigma)}]\cdot y_{jj}(k)) \\
\ge \sum\limits_{k=1}^{3}\sum\limits_{j=1}^{N} d_j^R(k) \\
x_{ij}(k) \ge C, y_{jj}(k) > 0 \quad \forall i\in B, j\in S, k\in K, \\
\delta_{ij}(k) = \left\{\begin{array}{ll} 1, & if \quad i\in A_j(k), j\in S, k\in K \\ 0, & otherwise \end{array}\right.
\end{array}
\right.
\end{array}
\right.
\tag{7.16}
$$

## 7.3 Dynamic Programming-Based Contraction Particle Swarm Optimization

Particle swarm optimization (PSO) is an optimization technique based on swarm intelligence, and was first proposed by Kennedy and Eberhart (1995) [23]. It simulates the social behavior, such as birds flocking to a promising position, for certain objectives in a multidimensional space (Eberhart and Shi 2001 [24]; Clerc and Kennedy 2002 [25]). There have been many variants (Anghinolfi and Paolucci 2009 [211]; Xu and Zeng 2012 [198]; Xu et al. 2013a [584]; Yapicioglu et al. 2007 [244];

Xu et al. 2012 [194]; Ali and Kachitvichyanukul 2009 [604]; Xu and Zeng 2011 [242]) of the PSO algorithm since its debut.

***Notations***

| | | |
|---|---|---|
| $\tau$ | : | iteration index; $\tau = 1, \cdots T$, $T$ is the iteration limit |
| $l$ | : | particle index, $l = 1, \cdots L$, $L$ is the population size |
| $h$ | : | dimension index, $h = 1, \cdots H$, $H$ is the problem dimension |
| $r_p, r_g$ | : | uniform random number in the interval $[0, 1]$ |
| $w(\tau)$ | : | inertia weight at the $\tau^{th}$ iteration |
| $v_{lh}(\tau)$ | : | velocity of the $l^{th}$ particle at the $h^{th}$ dimension in the $\tau^{th}$ iteration |
| $p_{lh}(\tau)$ | : | position of the $l^{th}$ particle at the $h^{th}$ dimension in the $\tau^{th}$ iteration |
| $\psi_{lh}$ | : | personal best position (pbest) of the $l^{th}$ particle at the $h^{th}$ dimension |
| $\psi_{gh}$: | : | global best position (gbest) at the $h^{th}$ dimension |
| $c_p, c_g$ | : | personal best and global best position acceleration constants respectively |
| $p^{max}, p^{min}$ | : | maximum and minimum position values respectively |
| $P_l$ | : | vector position of the $l^{th}$ particle, $P_l = (p_{l1}, p_{l2}, \cdots, p_{lH})$ |
| $V_l$ | : | vector velocity of the $l^{th}$ particle, $V_l = (v_{l1}, v_{l2}, \cdots, v_{lH})$ |
| $\vartheta_{lh}$ | : | initial velocity at the $h^{th}$ dimension, $V_l(1) = (\vartheta_{l1}, \vartheta_{l2}, \cdots, \vartheta_{lH})$ |
| $\Psi_l$ | : | vector personal best position of the $l^{th}$ particle, $\Psi_l = (\psi_{l1}, \psi_{l2}, \cdots, \psi_{lH})$ |
| $\Psi_g$ | : | vector global best position, $\Psi_g = (\psi_{g1}, \psi_{g2}, \cdots, \psi_{gH})$ |
| $R_l$ | : | the $l^{th}$ set of solution |
| $Z(P_l)$ | : | fitness value of $P_l$ |

The basic PSO formula is shown below:

$$V_l(\tau+1) = w(\tau)V_l(\tau) + c_p r_p(\Psi_l - P_l(\tau)) + c_g r_g(\Psi_g - P_l(\tau)), \tag{7.17}$$

$$P_l(\tau+1) = V_l(\tau+1) + P_l(\tau). \tag{7.18}$$

$$w(\tau) = w(T) + \frac{\tau - T}{1 - T}[w(1) - w(T)] \tag{7.19}$$

Eq. (7.17) is used to calculate the particle's new velocity according to its previous velocity and the distance of its current position from its own best experience and the group's best experience. Eq. (7.18) is traditionally used to update the particle that flies toward a new position (Shi and Eberhart 1998 [113]). The adaptive inertia weights(Shi and Eberhart 1998 [34]) provide good convergent behavior in this study, which is in accordance with the results provided by Eberhart and Shi (2001) [24]. The adaptive inertia weights are set to be varying with iterations as shown in Eq. (7.19).

Unlike the standard PSO (Bratton and Kennedy 2007 [99]), the DP-based CPSO in this section can reduce the dimensions of a solution representation using a state equation. Each particle is represented by its position in a H-dimensional space for the TS-based DTAP, and the problem dimensions should include decision variables (i.e., $x_{ij}(k)$,$\delta_{ij}(k)$ and $y_{jj}(k)$) and state variables (i.e., $B_i(k)$ and $S_j(k)$).

### 7.3.1 DP-Based CPSO Solution Representation

In this study, the DP-based CPSO dimensions are $(6MN + 3N)$ (i.e., $H = 6MN + 3N$), smaller than the size $(6MN + 6N + 3M)$ required by the standard PSO. As noted earlier, based on the state equation (i.e., Eq. (7.2)) and (i.e., Eq. (7.3)), the state variables can be determined. Figure 7.4 shows the structure of the particle and solution representation for the DP-based CPSO to generate solutions for the proposed multistage and multi-objective model.

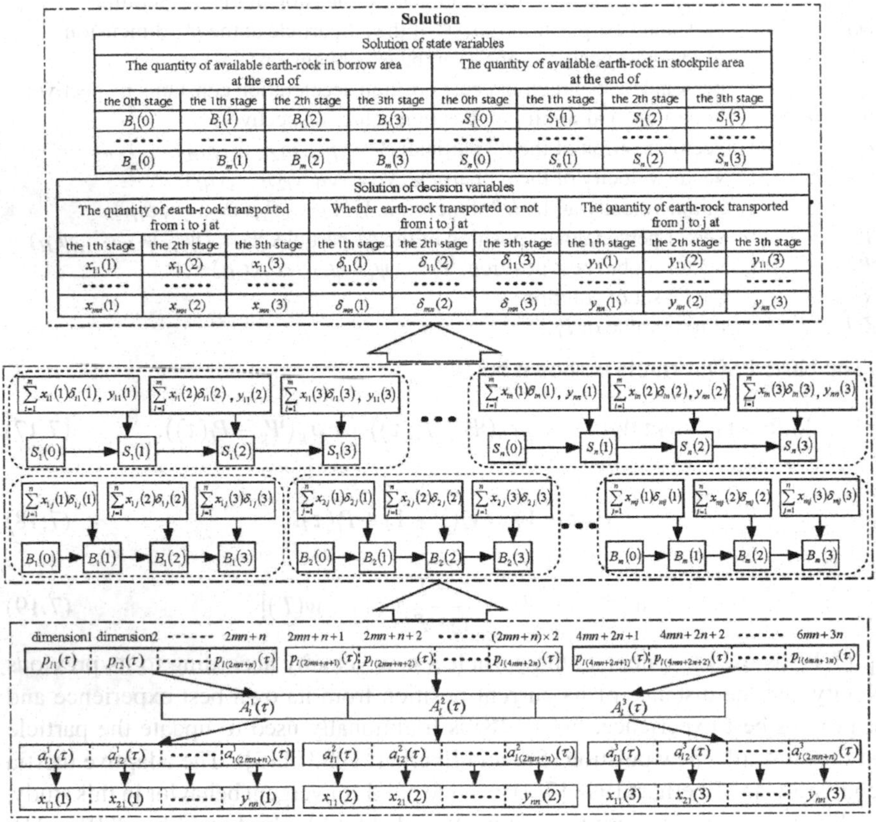


**Fig. 7.4** Solution representation of DP-Based CPSO

### 7.3.2 Weight-Sum Procedure for Dealing with the Multi-objective Factor

There are many methods for dealing with multiple objectives, such as the aggregating approach, lexicographic ordering, the sub-population approach, the Pareto-based approach, and the combined approach. In this section, the weight-sum procedure is adopted to deal with the three objectives above as the aggregated objective in the

form of a weighted-sum makes it possible to find the optimal Pareto solutions only when the solution set is convex (Gen et al. 2008 [605]). As the convexity of the three objectives and the constraints in the model above can be proved, the TS-based DTAP mathematical model uses convex programming and its solution set is also convex. Before the weight-sum procedure, the estimated maximal value is used to divide the dimensions and unify the orders of magnitude (Xu et al. 2013e [69]) to make sure the conformity in the three different objectives is effective. The basic procedure for aggregating the three objectives is as follows:
(1) Estimate maximal value $f_c^{max}$, $f_d^{max}$ and $f_w^{max}$ of $f_c$, $f_d$ and $f_w$ respectively;
(2) Calculate and standardize the $f_c'$, $f_d'$ and $f_w'$ as follows:

$$\begin{cases} f_c' = \frac{f_c}{f_c^{max}} \\ f_d' = \frac{f_d}{f_d^{max}} \\ f_w' = \frac{f_w}{f_w^{max}} \end{cases} \tag{7.20}$$

(3) The weighted-sum objective function $f$ is given by the following equation:

$$f = \min(\eta_c f_c' + \eta_d f_d' + \eta_w f_w') \tag{7.21}$$

Where, the weights $\eta_c$, $\eta_d$ and $\eta_w$ are proposed for the total project cost, total transportation duration and the total project waste during the transportation and construction system respectively, which are given by the decision makers and reflect the importance of each objective from the view of decision makers and satisfy $\eta_c + \eta_d + \eta_w = 1$. For a given individual, the fitness value function is expressed as follows:

$$Fitness(P_l(\tau)) = Z(P_l) = \eta_c f_c' + \eta_d f_d' + \eta_w f_w' \tag{7.22}$$

### *7.3.3 DP-Based CPSO Initialization*

The element in the multidimensional particle position can be initialized as follows to avoid an infeasible position:

*Step 1*. Initialize $L$ particles as a swarm: Set iteration $\tau = 1$. For $l = 1, 2, \cdots, L$, generate the position of particle $l$ with a random real number position $P_l$ in the range $[p^{min}, p^{max}]$: $P_l(\tau) = [p_{l1}(\tau), p_{l2}(\tau), \cdots, p_{lH}(\tau)]$, the value is $x_{ij}(k)$,$\delta_{ij}(k)$ and $y_{jj}(k)$. From the state transition equations as shown in Eq. (7.2) and Eq. (7.3), the state variables $B_i(k)$ and $S_j(k)$ are decided.

*Step 2*. Check the feasibility: If all particles satisfy the constraints in the expected value Model (7.16), then continue. Otherwise, return to Step 1.

*Step 3*. Initialize the speed of each particle: For $l = 1, 2, \cdots, L$, velocity $V_l(1) = (\vartheta_{l1}, \vartheta_{l2}, \cdots, \vartheta_{lH})$.

*Step 4*. Evaluate each of the particles: For $l = 1, \cdots L$, compute the performance measurement for $R_l$, and set this as the fitness value for $P_l$ , represented by $Z(P_l)$.

*Step 5*. Initialize the memory of each particle, the initial $\Psi_l$ is the current position $P_l(1)$, i.e, update the personal best position $\Psi_l = P_l(1)$.

### 7.3.4 DP-Based CPSO Updating and Schematic Procedure

After initialization, the updating procedure is as follows
Step 1. WHILE the maximum number of cycles has not been reached
DO
(1) Calculate the fitness value of the objective functions $f_c$ and $f_w$ respectively, then calculate the fitness value $Z(P_l)$ of the particle based on Eq. (21) and Eq. (22).
(2) Update pbest: For $l = 1,\cdots,L$, update $\Psi_l = P_l$, if $Z(P_l) < Z(\Psi_l)$.
(3) Update gbest: For $l = 1,\cdots,L$, update $\Psi_g = \Psi_l$, if $Z(\Psi_l) < Z(\Psi_g)$.
(4) Update the velocity and the position of each $l^{th}$ particle based on Eq. (17) Eq. (18) and Eq. (19).
(5) If $p_{lh}(\tau+1) > p^{max}$, then

$$p_{lh}(\tau+1) = p^{max}, v_{lh}(\tau+1) = 0 \tag{7.23}$$

If $p_{lh}(\tau+1) < p^{min}$, then

$$p_{lh}(\tau+1) = p^{min}, v_{lh}(\tau+1) = 0 \tag{7.24}$$

(6) Check the feasibility and make adjustments: After the updating, the decision variables (i.e., $\delta_{ij}(k)$, $x_{ij}(k)$ and $y_{jj}(k)$) are determined in accordance with the properties and characteristics of the constraint conditions. If some constraint conditions are not satisfied, the adjustment is made as follows:

1) Through the updating, if $\delta_{ij}(k)$ is not a 0-1 variable, it is adjusted as follows:

$$\hat{\delta}_{ij}(k) = \begin{cases} 1, & \delta_{ij}(k) \geq 0.8 \\ 0, & \delta_{ij}(k) < 0.8 \end{cases}$$

Then, assume that the 0-1 variables $\hat{\delta}_{ij}(k)$ do not change whatever constraint conditions can not be satisfied;

2) Adjust $y_{jj}(k)$ according to the unsatisfied constraint conditions which do not contain $x_{ij}(k)$ such as in Eq. (7.8). As the $\hat{\delta}_{ij}(k)$ has been decided, if $y_{jj}(k)$ cannot satisfy Eq. (7.8), adjustments are made as shown in Item (5) in Step 1;

3) After the adjustment above, $\hat{\delta}_{ij}(k)$ and $\hat{y}_{jj}(k)$ have been decided, then $x_{ij}(k)$ is adjusted according to the unsatisfied constraint conditions. Let $H$ and $H_1$ be the number of $\delta_{ij}(k)$ and $\delta_{ij}(k) = 1$ in the unsatisfied constraint condition respectively. If the unsatisfied constraint condition in the rest of the constraint conditions is in the form of $\sum_{h=1}^{H} (x\delta)_h \leq A$, then every $\{x_{ij}(k)|\delta_{ij}(k) = 1\}$ should reduce $\frac{A}{H_1}$. In the model above, the rest of the constraint conditions are shown in Eq.(7.6), Eq.(7.7) and Eq.(7.11)(Converted into $S_j(k-1) + \sum_{i=1}^{M} x_{ij}(k)\delta_{ij}(k) \cdot (1 - E^{Me}[\tilde{w}_{ij}(k)_{(\gamma_{ij},\sigma_{ij})}]) - y_{jj}(k) \leq \beta_j$ based on Eq. (7.3)) and they all have the same form $\sum_{h=1}^{H} (x\delta)_h \leq A$. Further, if there is more than one constraint condition not satisfied, then in the whole model every $\{x_{ij}(k)|\delta_{ij}(k) = 1\}$ should reduce $\max_r\{(\frac{A}{H_1})_r\}$, where $r$ is the number

of unsatisfied constraint conditions in the rest of the constraint conditions, so the adjustment for $x_{ij}(k)$ is $\hat{x}_{ij}(k) = x_{ij}(k) - \max_r\{(\frac{A}{H_1})_r\}$;

4) After the adjustment above, if there is still no feasible solution $\hat{y}_{jj}(k)$ and $\hat{x}_{ij}(k)$, then adjust $\hat{\delta}_{ij}(k)$ again, find $\{E^{Me}[\tilde{t}_{ij}(k)_{(\gamma_{ij},\sigma_{ij})}]|\delta_{ij}(k) = 0\}$, where $(E^{Me}[\tilde{t}_{ij}(k)_{(\gamma_{ij},\sigma_{ij})}]|\delta_{ij}(k) = 0) \leq (E^{Me}[\tilde{t}_{ij}(k)_{(\gamma_{ij},\sigma_{ij})}]|\delta_{ij}(k) > 0)$, and choose one or more $\delta_{ij}(k)$ properly and let $\delta_{ij}(k) = 1$. Continuously adjust using the above methods until all equations are satisfied;

5) After the adjustment for $\delta_{ij}(k)$, $x_{ij}(k)$ and $y_{jj}(k)$, make a generation into the state equation and determine the state variables.

Step 2. END WHILE.

Step 3. If the stopping criterion is met, i.e. $\tau = T$, go to Step 4. Otherwise, $\tau = \tau + 1$, and return to Step 1.

Step 4. Decode the global best position $\Psi_g$ as the solution set.

Figure 7.5 shows the schematic procedure for the DP-based CPSO to generate solutions for the proposed multistage multi-objective model.

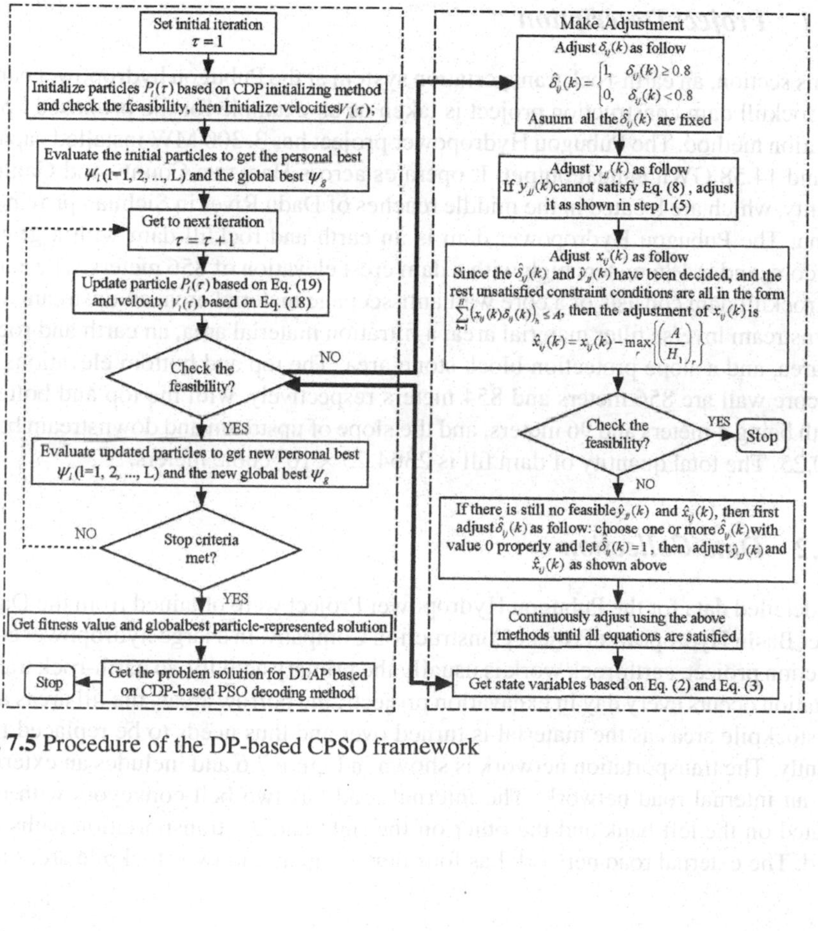


**Fig. 7.5** Procedure of the DP-based CPSO framework

## 7.4 Case Study: Earth-Rock Transportation System

Rapid economic growth and social advancement have created more pressing demands on energy all over the world. New and renewable sources of energy have become more important, of which the hydropower resources are recognized as one of the best. Hydropower plays an important role in China, especially in Sichuan Province. The Chinese government has emphasized renewable energy development particularly in areas of water conservancy and hydropower. The Ertan Hydropower Development Company, Ltd. (EHDC) was appointed to supply clean and renewable energy to support the economic and social development of the Sichuan-Chongqing region through the development of the hydropower resources on the Yalong River in Sichuan. The Jinping-I Hydropower Project is one of the EHDC's projects under construction. This section focuses on vehicle and facility scheduling in the queueing network optimization for the concrete transportation systems at the Jinping-I Hydropower Project with the aim of minimizing total operating costs and construction duration.

### *7.4.1 Project Description*

In this section, an earth-rock transportation system at the Pubugou hydropower earth and rockfill dam construction project is taken as an example for the proposed optimization method. The Pubugou Hydropower project has 3,300 MW installed capacity and 14.58 GWh annual output. It operates across Hanyuan County and Ganluo County, which are located in the middle reaches of Dadu River in Sichuan province, China. The Pubugou Hydropower dam is an earth and rockfill dam with a gravel soil core, and is 186 meters high with a dam crest elevation of 856 meters. The earth and rockfill dam consists of a core wall anti-seepage material area, an upstream and downstream inverse filter material area, a filtration material area, an earth and rockfill area, and a slope protection block stone area. The top and bottom elevations of the core wall are 856 meters and 854 meters respectively, with the top and bottom width being 4 meters and 96 meters, and the slope of upstream and downstream both $1:0.25$. The total quantity of dam fill is $2364.25 \times 10^4$ cubic meters.

### *7.4.2 Data Collection*

All detailed data for the Pubugou Hydropower Project were obtained from the Dadu River Basin Hydropower Project Construction Company. In a large hydropower construction project, earth-rock work is usually the primary activity, so earth-rock transportation occurs every day in excavation projects, the borrow areas, the fill areas and the stockpile areas as the material is turned over and thus needs to be replaced frequently. The transportation network is shown in Figure 7.6 and includes an external and an internal road network. The internal road has two belt conveyors with one located on the left bank and the other on the right and the transportation paths are fixed. The external road network has four borrow areas and two stockpile areas and

**Fig. 7.6** Pubugou Hydropower Construction Project

**Table 7.1** Information of carriers in Pubugou Hydropower Project

| Carriers | Kind index | Type |
|---|---|---|
| Dump truck | 1 | Terex TA 28(24$m^3$; 35 t) |
| Belt conveyer | 2 | BC(1000 mm; 4 km) |
| Belt conveyer | 3 | BC(1200 mm; 4.5 km) |

therefore it is possible that the earth-rock could be transported from any of the borrow areas to any of the stockpile areas.

Three types of transportation equipment (dump truck and two types of belt conveyor) are used in the construction project. All necessary data for each carrier type is calculated as shown in Table 7.1. Table 7.2 shows the details for each borrow area and stockpile area.

To apply the proposed methods conveniently, adjacent roads of the same type have been combined and road shapes have been ignored. An abstracted transportation network is illustrated in Figure 7.7. For each link in the transportation network, there is a free flow L-R fuzzy random travel time $\tilde{\bar{t}}_{ij}(k)$ and $\tilde{\bar{t}}_{jh}(k)$, price $\tilde{\bar{c}}_{ij}(k)$ and $\tilde{\bar{c}}_{jh}(k)$ and unit waste $\tilde{\bar{w}}_{ij}(k)$ and $\tilde{\bar{w}}_{jh}(k)$ during the transportation. The corresponding data are stated in Tables 7.3, 7.4 and 7.5. Similarly, the earth-rock excavation

**Table 7.2** Information of borrow areas and stockpile areas in Pubugou Hydropower Project

| Areas | Area index | maximum capacity |
|---|---|---|
| Borrow area | 1 | $732.4 \times 10^4 m^3$ |
| Borrow area | 2 | $581.7 \times 10^4 m^3$ |
| Borrow area | 3 | $528.3 \times 10^4 m^3$ |
| Borrow area | 4 | $790.2 \times 10^4 m^3$ |
| Stockpile area | 1 | $15 \times 10^4 m^3$ |
| Stockpile area | 2 | $15 \times 10^4 m^3$ |

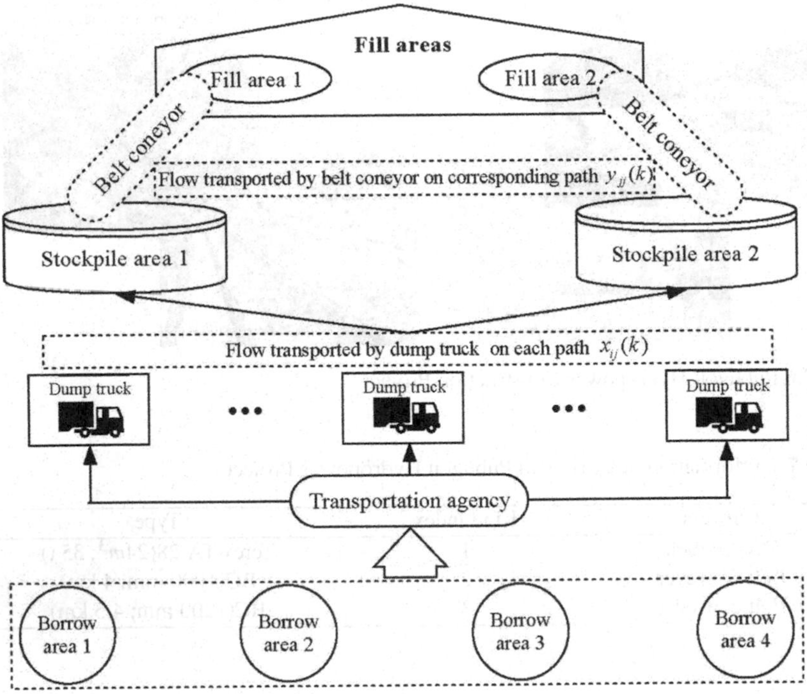


**Fig. 7.7** Location and detailed information of borrow areas, stockpile areas and fill areas in Pubugou Hydropower Project

price, storage cost and total demand for each fill area at each stage are also L-R fuzzy random parameters which are stated in Table 7.6. To collect the price, time and waste data, investigations and surveys were conducted to obtain historical data from both the financial department and experienced Dadu River Basin Hydropower Project Construction Company construction team engineers.

**Table 7.3** The data information of transportation price of each route at each stage

| Model parameters | | Stage index $k=1$ | $k=2$ | $k=3$ | Measurement unit |
|---|---|---|---|---|---|
| $\tilde{\tilde{c}}_{ij}(k)$ | $\tilde{\tilde{c}}_{11}(k)$ | $(4.82,\varphi(\omega),5.56)$<br>$\varphi(\omega)\sim N(5.20,3.1)$ | $(5.26,\varphi(\omega),6.44)$<br>$\varphi(\omega)\sim N(6.00,4.2)$ | $(5.28,\varphi(\omega),6.18)$<br>$\varphi(\omega)\sim N(5.82,3.8)$ | $CNY/unit$ |
| | $\tilde{\tilde{c}}_{21}(k)$ | $(3.02,\varphi(\omega),3.57)$<br>$\varphi(\omega)\sim N(3.25,2.1)$ | $(3.23,\varphi(\omega),4.09)$<br>$\varphi(\omega)\sim N(3.66,2.2)$ | $(3.18,\varphi(\omega),4.25)$<br>$\varphi(\omega)\sim N(3.72,1.8)$ | $CNY/unit$ |
| | $\tilde{\tilde{c}}_{31}(k)$ | $(3.89,\varphi(\omega),4.67)$<br>$\varphi(\omega)\sim N(4.23,1.7)$ | $(3.97,\varphi(\omega),4.55)$<br>$\varphi(\omega)\sim N(4.34,2.1)$ | $(4.16,\varphi(\omega),4.83)$<br>$\varphi(\omega)\sim N(4.59,2.4)$ | $CNY/unit$ |
| | $\tilde{\tilde{c}}_{41}(k)$ | $(5.78,\varphi(\omega),6.47)$<br>$\varphi(\omega)\sim N(6.12,3.8)$ | $(6.04,\varphi(\omega),6.58)$<br>$\varphi(\omega)\sim N(6.44,4.2)$ | $(6.06,\varphi(\omega),6.75)$<br>$\varphi(\omega)\sim N(6.40,4.0)$ | $CNY/unit$ |
| | $\tilde{\tilde{c}}_{12}(k)$ | $(5.22,\varphi(\omega),5.91)$<br>$\varphi(\omega)\sim N(5.57,2.8)$ | $(5.04,\varphi(\omega),5.65)$<br>$\varphi(\omega)\sim N(5.30,3.2)$ | $(5.30,\varphi(\omega),6.09)$<br>$\varphi(\omega)\sim N(5.84,3.4)$ | $CNY/unit$ |
| | $\tilde{\tilde{c}}_{22}(k)$ | $(5.97,\varphi(\omega),6.66)$<br>$\varphi(\omega)\sim N(6.21,4.1)$ | $(6.18,\varphi(\omega),6.47)$<br>$\varphi(\omega)\sim N(6.33,4.2)$ | $(6.06,\varphi(\omega),6.75)$<br>$\varphi(\omega)\sim N(6.50,3.8)$ | $CNY/unit$ |
| | $\tilde{\tilde{c}}_{32}(k)$ | $(5.36,\varphi(\omega),5.94)$<br>$\varphi(\omega)\sim N(5.60,3.1)$ | $(5.27,\varphi(\omega),5.65)$<br>$\varphi(\omega)\sim N(5.41,3.0)$ | $(5.32,\varphi(\omega),6.01)$<br>$\varphi(\omega)\sim N(5.77,4.2)$ | $CNY/unit$ |
| | $\tilde{\tilde{c}}_{42}(k)$ | $(3.38,\varphi(\omega),3.87)$<br>$\varphi(\omega)\sim N(3.63,2.1)$ | $(3.56,\varphi(\omega),4.09)$<br>$\varphi(\omega)\sim N(3.84,2.2)$ | $(3.82,\varphi(\omega),4.35)$<br>$\varphi(\omega)\sim N(4.00,2.0)$ | $CNY/unit$ |
| $\tilde{\tilde{c}}_{jj}(k)$ | $\tilde{\tilde{c}}_{11}(k)$ | $(3.08,\varphi(\omega),3.40)$<br>$\varphi(\omega)\sim N(3.34,2.2)$ | $(3.30,\varphi(\omega),3.84)$<br>$\varphi(\omega)\sim N(3.47,2.1)$ | $(3.36,\varphi(\omega),3.78)$<br>$\varphi(\omega)\sim N(3.62,2.6)$ | $CNY/unit$ |
| | $\tilde{\tilde{c}}_{22}(k)$ | $(3.07,\varphi(\omega),3.46)$<br>$\varphi(\omega)\sim N(3.21,2.0)$ | $(3.08,\varphi(\omega),3.50)$<br>$\varphi(\omega)\sim N(3.27,2.2)$ | $(3.16,\varphi(\omega),3.65)$<br>$\varphi(\omega)\sim N(3.40,1.8)$ | $CNY/unit$ |

**Table 7.4** The data information of transportation time of each route at each stage

| Model parameters | | Stage index $k=1$ | $k=2$ | $k=3$ | Measurement unit |
|---|---|---|---|---|---|
| $\tilde{\tilde{t}}_{ij}(k)$ | $\tilde{\tilde{t}}_{11}(k)$ | $(0.28,\varphi(\omega),0.40)$<br>$\varphi(\omega)\sim N(0.34,0.21)$ | $(0.30,\varphi(\omega),0.44)$<br>$\varphi(\omega)\sim N(0.37,0.22)$ | $(0.26,\varphi(\omega),0.38)$<br>$\varphi(\omega)\sim N(0.32,0.18)$ | $hour$ |
| | $\tilde{\tilde{t}}_{21}(k)$ | $(0.20,\varphi(\omega),0.27)$<br>$\varphi(\omega)\sim N(0.25,0.1)$ | $(0.23,\varphi(\omega),0.29)$<br>$\varphi(\omega)\sim N(0.26,0.15)$ | $(0.18,\varphi(\omega),0.25)$<br>$\varphi(\omega)\sim N(0.22,0.18)$ | $hour$ |
| | $\tilde{\tilde{t}}_{31}(k)$ | $(0.19,\varphi(\omega),0.27)$<br>$\varphi(\omega)\sim N(0.23,0.12)$ | $(0.17,\varphi(\omega),0.25)$<br>$\varphi(\omega)\sim N(0.21,0.1)$ | $(0.16,\varphi(\omega),0.23)$<br>$\varphi(\omega)\sim N(0.19,0.13)$ | $hour$ |
| | $\tilde{\tilde{t}}_{41}(k)$ | $(0.38,\varphi(\omega),0.47)$<br>$\varphi(\omega)\sim N(0.42,0.21)$ | $(0.40,\varphi(\omega),0.48)$<br>$\varphi(\omega)\sim N(0.44,0.32)$ | $(0.36,\varphi(\omega),0.45)$<br>$\varphi(\omega)\sim N(0.40,0.28)$ | $hour$ |
| | $\tilde{\tilde{t}}_{12}(k)$ | $(0.22,\varphi(\omega),0.31)$<br>$\varphi(\omega)\sim N(0.27,0.11)$ | $(0.24,\varphi(\omega),0.35)$<br>$\varphi(\omega)\sim N(0.30,0.22)$ | $(0.20,\varphi(\omega),0.29)$<br>$\varphi(\omega)\sim N(0.24,0.12)$ | $hour$ |
| | $\tilde{\tilde{t}}_{22}(k)$ | $(0.17,\varphi(\omega),0.26)$<br>$\varphi(\omega)\sim N(0.21,0.08)$ | $(0.18,\varphi(\omega),0.27)$<br>$\varphi(\omega)\sim N(0.23,0.12)$ | $(0.16,\varphi(\omega),0.25)$<br>$\varphi(\omega)\sim N(0.20,0.12)$ | $hour$ |
| | $\tilde{\tilde{t}}_{32}(k)$ | $(0.36,\varphi(\omega),0.44)$<br>$\varphi(\omega)\sim N(0.40,0.23)$ | $(0.37,\varphi(\omega),0.45)$<br>$\varphi(\omega)\sim N(0.41,0.22)$ | $(0.32,\varphi(\omega),0.41)$<br>$\varphi(\omega)\sim N(0.37,0.24)$ | $hour$ |
| | $\tilde{\tilde{t}}_{42}(k)$ | $(0.28,\varphi(\omega),0.37)$<br>$\varphi(\omega)\sim N(0.33,0.20)$ | $(0.26,\varphi(\omega),0.39)$<br>$\varphi(\omega)\sim N(0.34,0.22)$ | $(0.22,\varphi(\omega),0.35)$<br>$\varphi(\omega)\sim N(0.29,0.16)$ | $hour$ |
| $\tilde{\tilde{t}}_{jj}(k)$ | $\tilde{\tilde{t}}_{11}(k)$ | $(1.7,\varphi(\omega),2.2)$<br>$\varphi(\omega)\sim N(2.0,1.4)$ | $(1.9,\varphi(\omega),2.3)$<br>$\varphi(\omega)\sim N(2.1,1.2)$ | $(1.7,\varphi(\omega),2.2)$<br>$\varphi(\omega)\sim N(1.9,1.0)$ | $hour/10^4m^3$ |
| | $\tilde{\tilde{t}}_{22}(k)$ | $(1.6,\varphi(\omega),2.1)$<br>$\varphi(\omega)\sim N(1.8,1.1)$ | $(1.8,\varphi(\omega),2.1)$<br>$\varphi(\omega)\sim N(2.0,1.2)$ | $(1.3,\varphi(\omega),2.0)$<br>$\varphi(\omega)\sim N(1.7,1.1)$ | $hour/10^4m^3$ |

**Table 7.5** The data information of transportation waste of each route at each stage

| Model parameters | | Stage index $k=1$ | Stage index $k=2$ | Stage index $k=3$ | Measurement unit |
|---|---|---|---|---|---|
| $\tilde{\bar{w}}_{ij}(k)$ | $\tilde{\bar{w}}_{11}(k)$ | $(0.028,\varphi(\omega),0.040)$ $\varphi(\omega)\sim N(0.034,0.016)$ | $(0.032,\varphi(\omega),0.042)$ $\varphi(\omega)\sim N(0.037,0.02)$ | $(0.028,\varphi(\omega),0.038)$ $\varphi(\omega)\sim N(0.032,0.018)$ | $m^3/unit$ |
| | $\tilde{\bar{w}}_{21}(k)$ | $(0.020,\varphi(\omega),0.029)$ $\varphi(\omega)\sim N(0.025,0.018)$ | $(0.023,\varphi(\omega),0.029)$ $\varphi(\omega)\sim N(0.027,0.02)$ | $(0.019,\varphi(\omega),0.026)$ $\varphi(\omega)\sim N(0.022,0.010)$ | $m^3/unit$ |
| | $\tilde{\bar{w}}_{31}(k)$ | $(0.020,\varphi(\omega),0.026)$ $\varphi(\omega)\sim N(0.023,0.015)$ | $(0.018,\varphi(\omega),0.026)$ $\varphi(\omega)\sim N(0.021,0.016)$ | $(0.016,\varphi(\omega),0.023)$ $\varphi(\omega)\sim N(0.020,0.010)$ | $m^3/unit$ |
| | $\tilde{\bar{w}}_{41}(k)$ | $(0.036,\varphi(\omega),0.047)$ $\varphi(\omega)\sim N(0.041,0.023)$ | $(0.042,\varphi(\omega),0.047)$ $\varphi(\omega)\sim N(0.044,0.022)$ | $(0.036,\varphi(\omega),0.045)$ $\varphi(\omega)\sim N(0.040,0.030)$ | $m^3/unit$ |
| | $\tilde{\bar{w}}_{12}(k)$ | $(0.022,\varphi(\omega),0.031)$ $\varphi(\omega)\sim N(0.028,0.018)$ | $(0.025,\varphi(\omega),0.035)$ $\varphi(\omega)\sim N(0.030,0.016)$ | $(0.019,\varphi(\omega),0.025)$ $\varphi(\omega)\sim N(0.023,0.012)$ | $m^3/unit$ |
| | $\tilde{\bar{w}}_{22}(k)$ | $(0.019,\varphi(\omega),0.026)$ $\varphi(\omega)\sim N(0.022,0.014)$ | $(0.019,\varphi(\omega),0.029)$ $\varphi(\omega)\sim N(0.025,0.012)$ | $(0.016,\varphi(\omega),0.025)$ $\varphi(\omega)\sim N(0.020,0.011)$ | $m^3/unit$ |
| | $\tilde{\bar{w}}_{32}(k)$ | $(0.038,\varphi(\omega),0.043)$ $\varphi(\omega)\sim N(0.041,0.023)$ | $(0.037,\varphi(\omega),0.047)$ $\varphi(\omega)\sim N(0.043,0.022)$ | $(0.032,\varphi(\omega),0.041)$ $\varphi(\omega)\sim N(0.038,0.025)$ | $m^3/unit$ |
| | $\tilde{\bar{w}}_{42}(k)$ | $(0.027,\varphi(\omega),0.036)$ $\varphi(\omega)\sim N(0.032,0.018)$ | $(0.027,\varphi(\omega),0.039)$ $\varphi(\omega)\sim N(0.035,0.023)$ | $(0.022,\varphi(\omega),0.035)$ $\varphi(\omega)\sim N(0.029,0.022)$ | $m^3/unit$ |
| $\tilde{\bar{w}}_{jj}(k)$ | $\tilde{\bar{w}}_{11}(k)$ | $(0.028,\varphi(\omega),0.037)$ $\varphi(\omega)\sim N(0.032,0.015)$ | $(0.031,\varphi(\omega),0.043)$ $\varphi(\omega)\sim N(0.037,0.021)$ | $(0.026,\varphi(\omega),0.038)$ $\varphi(\omega)\sim N(0.032,0.020)$ | $m^3/unit$ |
| | $\tilde{\bar{w}}_{22}(k)$ | $(0.016,\varphi(\omega),0.026)$ $\varphi(\omega)\sim N(0.021,0.012)$ | $(0.018,\varphi(\omega),0.028)$ $\varphi(\omega)\sim N(0.024,0.014)$ | $(0.016,\varphi(\omega),0.025)$ $\varphi(\omega)\sim N(0.020,0.011)$ | $m^3/unit$ |

### 7.4.3 Parameters Selection for DP-Based CPSO

From the results of some preliminary experiments carried out to observe the behavior of the algorithm at different parameter settings, and through a comparison of several sets of parameters, including population size, iteration number, acceleration constant, initial velocity, and inertia weight, the most reasonable parameters were identified, Table 7.7 summarizes all the parameter values selected for the DP-based CPSO in the computational experiments. Note that the population size determines the evaluation runs which impact the optimization cost, and the learning factors $c_p$ and $c_g$ may lead to small differences in the PSO performance (Trelea 2003 [40]). The inertia weight $w(\tau)$ is set to be varying with iterations as follows:

$$w(\tau) = w(T) + \frac{\tau - T}{1 - T}[w(1) - w(T)] \tag{7.25}$$

where $\tau$ = iteration index = $1, 2, \cdots, T$. $w(1) = 0.9$ and $w(T) = 0.1$ were found to be the most suitable controlling the impact of the previous velocities on the current velocity and the influence of the trade-off between the global and local experiences.

### 7.4.4 Results and Analysis

In the model discussed above, compared with actual data (total costs, transportation duration and wastes) from the Pubugou Hydropower Project, the practicality and efficiency of the optimization model under a fuzzy random environment

**Table 7.6** The data information of earth-rock excavation price, storable price and total demand of each fill area at each stage

| Model parameters | | Stage index $k=1$ | $k=2$ | $k=3$ | Measurement unit |
|---|---|---|---|---|---|
| $\tilde{\bar{c}}_i(k)$ | $\tilde{\bar{c}}_1(k)$ | $(14.82,\varphi(\omega),15.56)$ | $(15.26,\varphi(\omega),16.44)$ | $(16.28,\varphi(\omega),17.18)$ | $CNY/unit$ |
| | | $\varphi(\omega)\sim N(15.20,12.2)$ | $\varphi(\omega)\sim N(16.00,13.0)$ | $\varphi(\omega)\sim N(16.82,13.8)$ | |
| | $\tilde{\bar{c}}_2(k)$ | $(14.02,\varphi(\omega),14.97)$ | $(13.23,\varphi(\omega),14.09)$ | $(14.18,\varphi(\omega),15.25)$ | $CNY/unit$ |
| | | $\varphi(\omega)\sim N(14.55,12.0)$ | $\varphi(\omega)\sim N(13.66,12.1)$ | $\varphi(\omega)\sim N(14.72,12.8)$ | |
| | $\tilde{\bar{c}}_3(k)$ | $(14.89,\varphi(\omega),15.67)$ | $(13.97,\varphi(\omega),14.55)$ | $(15.16,\varphi(\omega),15.83)$ | $CNY/unit$ |
| | | $\varphi(\omega)\sim N(15.23,13.2)$ | $\varphi(\omega)\sim N(14.34,11.0)$ | $\varphi(\omega)\sim N(15.59,11.9)$ | |
| | $\tilde{\bar{c}}_4(k)$ | $(15.78,\varphi(\omega),16.47)$ | $(16.04,\varphi(\omega),16.58)$ | $(16.46,\varphi(\omega),17.45)$ | $CNY/unit$ |
| | | $\varphi(\omega)\sim N(16.12,13.5)$ | $\varphi(\omega)\sim N(16.44,13.2)$ | $\varphi(\omega)\sim N(16.80,14.0)$ | |
| $\tilde{\bar{c}}_j^1(k)$ | $\tilde{\bar{c}}_1^1(k)$ | $(1.00,\varphi(\omega),1.32)$ | $(1.10,\varphi(\omega),1.44)$ | – | $CNY/unit$ |
| | | $\varphi(\omega)\sim N(1.14,1)$ | $\varphi(\omega)\sim N(1.27,1.1)$ | – | |
| | $\tilde{\bar{c}}_2^1(k)$ | $(0.91,\varphi(\omega),1.57)$ | $(1.23,\varphi(\omega),1.69)$ | – | $CNY/unit$ |
| | | $\varphi(\omega)\sim N(1.25,1.1)$ | $\varphi(\omega)\sim N(1.36,1.2)$ | – | |
| $\tilde{\bar{c}}_j^2(k)$ | $\tilde{\bar{c}}_1^2(k)$ | $(0.40,\varphi(\omega),0.60)$ | $(0.52,\varphi(\omega),0.68)$ | $(0.48,\varphi(\omega),0.62)$ | $CNY/unit$ |
| | | $\varphi(\omega)\sim N(0.50,0.40)$ | $\varphi(\omega)\sim N(0.60,0.49)$ | $\varphi(\omega)\sim N(0.56,0.36)$ | |
| | $\tilde{\bar{c}}_2^2(k)$ | $(0.36,0.40,0.45)$ | $(0.50,0.56,0.60)$ | $(0.40,0.0.42,0.50)$ | $CNY/unit$ |
| | | $\varphi(\omega)\sim N(0.40,0.25)$ | $\varphi(\omega)\sim N(0.56,0.36)$ | $\varphi(\omega)\sim N(0.42,0.49)$ | |
| $\tilde{d}_j(k)$ | $\tilde{d}_1(k)$ | $(364.6,368.2,371.1)$ | $(424.2,426.6,428.9)$ | $(373.1,374.5,376.7)$ | $10^4m^3$ |
| | $\tilde{d}_2(k)$ | $(375.6,377.1,379.8)$ | $(447.3,449.4,452.1)$ | $(370.0,372.3,373.8)$ | $10^4m^3$ |
| $\tilde{\bar{c}}_j(k)$ | $\tilde{\bar{c}}_1(k)$ | $(0.42,0.50,0.55)$ | $(0.50,0.60,0.65)$ | $(0.48,0.0.54,0.60)$ | $CNY/unit$ |
| | | $\varphi(\omega)\sim N(0.50,0.36)$ | $\varphi(\omega)\sim N(0.60,0.49)$ | $\varphi(\omega)\sim N(0.54,0.40)$ | |
| | $\tilde{\bar{c}}_2(k)$ | $(0.36,0.40,0.46)$ | $(0.48,0.52,0.60)$ | $(0.40,0.0.48,0.52)$ | $CNY/unit$ |
| | | $\varphi(\omega)\sim N(0.40,0.25)$ | $\varphi(\omega)\sim N(0.52,0.38)$ | $\varphi(\omega)\sim N(0.48,0.30)$ | |
| $h_j^k(\bar{t})$ | $h_1^k(\bar{t})$ | 25 | 23 | 22 | $10^4m^3$ |
| | $h_2^k(\bar{t})$ | 24 | 22 | 23 | $10^4m^3$ |

**Table 7.7** Parameter selection for DP-based CPSO

| Population size | Iteration number | Acceleration constant | | Inertia weight | |
|---|---|---|---|---|---|
| $L$ | $T$ | $c_p$ | $c_g$ | $w(1)$ | $w(T)$ |
| 100 | 100 | 2 | 2 | 0.9 | 0.1 |

presented in this section were verified by implementing the DP-based CPSO algorithm to solve the flow assignment between the transportation paths described previously. After running the proposed DP-based CPSO using MATLAB 7.0, the computational results were obtained based on an optimistic-pessimistic index $\lambda = 0.5$, weights $\eta_c = 0.5$, $\eta_d = 0.3$ and $\eta_w = 0.2$, probability $\sigma = 0.2$ and possibility $\gamma = 0.8$, where the maximal value of $f_c$ and $f_w$ were estimated and $f_c^{max} = 628.47 \times 10^6$, $f_d^{max} = 285.32$ and $f_w^{max} = 196.61 \times 10^4$. The results and performance of the algorithm, as compared to the actual data from the project, are listed in Table 7.8. Figure 7.8 shows the results of the DP-based CPSO Algorithm for the transportation allocation decision at the Pubugou Hydropower Project.

In this section, since there are some undetermined parameters such as the optimistic-pessimistic index $\lambda$, weights $\eta_c$, $\eta_d$ and $\eta_w$, probability levels $\sigma$ and possibility levels $\gamma$ which were given by the decision makers, therefore, based on the proposed model, further research needs to be done to analyze the sensitivity and the advantages compared with other models and algorithms.

**Table 7.8** Results of DP-based CPSO Algorithm for TS-based DTAP ($\lambda = 0.5$, $\eta_c = 0.5$, $\eta_d = 0.3$, $\eta_w = 0.2$, $\sigma = 0.2$ , $\gamma = 0.8$)

| Objective | Total Construction Cost, Duration and Waste | | | | Convergence | Computing |
|---|---|---|---|---|---|---|
| | Fitness value | Actual data | Net decrease | Rate of decrease | Iteration number | Time(s) |
| $f$ | 0.8773 | 0.9220 | 0.0379 | 4.47% | 46 | 17.34 |
| $f_c(10^6)$ | 582.50 | 594.74 | 12.24 | 2.06% | 46 | 17.34 |
| $f_d(10^1)$ | 244.47 | 262.36 | 17.89 | 6.82% | 46 | 17.34 |
| $f_w(10^4)$ | 154.16 | 170.06 | 15.90 | 9.35% | 46 | 17.34 |

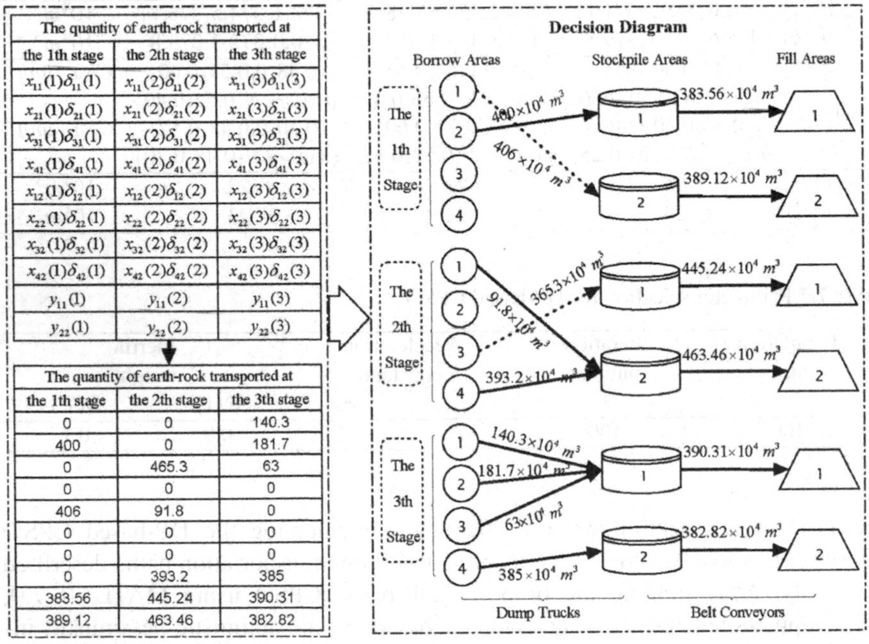


**Fig. 7.8** Results of DP-based CPSO Algorithm for transportation allocation decision in Pubugou Hydropower Project

**Sensitivity Analysis for the Optimistic-Pessimistic Index**

It should be noted that the results were obtained based on $\lambda = 0.5$, $\eta_c = 0.5$, $\eta_d = 0.3$, $\eta_w = 0.2$, $\sigma = 0.2$ and $\gamma = 0.8$. To gain further insight into the optimistic-pessimistic index (i.e.,$\lambda$) selection principle, a sensitivity analysis was conducted using the same weights and probability and possibility levels. Decision makers can fine tune this parameter to obtain different solutions under different levels. The solutions in Table 7.9 reflect the different optimistic-pessimistic attitudes for uncertainty, where $\lambda = 1$ and $\lambda = 0$ are the pessimistic extreme and optimistic extreme respectively. Since the objective is minimization and based on Eq.(7.1), it can be seen that if the optimistic-pessimistic $\lambda$ goes up, the result is more pessimistic, which means that the total cost, duration and total waste gradually increases. The optimistic-pessimistic index is given by the decision makers and reflects their attitude.

**Table 7.9** Sensitivity analysis on the optimistic-pessimistic index $\lambda$ of decision makers ($\eta_c = 0.5$, $\eta_d = 0.3$ $\eta_w = 0.2$, $\sigma = 0.2$ , $\gamma = 0.8$)

| Objective | Optimistic-pessimistic index | | | | | | | | | | |
|---|---|---|---|---|---|---|---|---|---|---|---|
| | $\lambda = 0$ | $\lambda = 0.1$ | $\lambda = 0.2$ | $\lambda = 0.3$ | $\lambda = 0.4$ | $\lambda = 0.5$ | $\lambda = 0.6$ | $\lambda = 0.7$ | $\lambda = 0.8$ | $\lambda = 0.9$ | $\lambda = 1.0$ |
| $f$ | 0.8173 | 0.8294 | 0.8402 | 0.8543 | 0.8664 | 0.8773 | 0.8887 | 0.9008 | 0.9132 | 0.9250 | 0.9371 |
| $f_c(10^6)$ | 566.93 | 570.08 | 573.17 | 576.29 | 579.41 | 582.50 | 585.61 | 588.73 | 591.86 | 594.98 | 598.12 |
| $f_d(10^1)$ | 216.30 | 221.78 | 226.36 | 234.12 | 239.45 | 244.47 | 249.37 | 254.52 | 260.28 | 265.85 | 271.31 |
| $f_w(10^4)$ | 136.49 | 140.25 | 143.76 | 147.12 | 151.03 | 154.16 | 157.87 | 162.02 | 165.81 | 169.22 | 173.01 |

**Sensitivity Analysis for the Probability and Possibility Levels**

As shown above, all probability levels were set at the same value $\sigma$ and all possibility levels were set to the same value $\gamma$. Reviewing this method for dealing with the fuzzy random variables, it can be seen that under different probability and possibility levels the objective function values are different. Thus, Table 7.10 summarizes the different total fitness values with respect to the different parameters $\lambda$, $\sigma$ and $\gamma$. The results can be used as a reference for decision-makers when choosing an appropriate parameter set (i.e., $\lambda$, $\sigma$ and $\gamma$) to optimize the decision making process. This is summarized as follows:

(1) At probability level $\sigma$, parameters $\lambda$, $\eta_c$, $\eta_w$ and the possibility level $\gamma$ are all fixed, when $\lambda < 0.5$, the larger the $\sigma$, the larger the objective function values; when $\lambda > 0.5$, the larger the $\sigma$, the smaller the objective function values.

(2) At possibility level $\gamma$, parameters $\lambda$, $\eta_c$, $\eta_w$ and the probability level $\sigma$ are all fixed, when $\lambda < 0.5$, the larger the $\gamma$, the smaller the objective function values; when $\lambda > 0.5$, the larger the $\gamma$, the larger the objective function values.

**Table 7.10** Sensitivity analysis on the probability level $\sigma$ and possibility level $\gamma$ of decision makers ($\eta_c = 0.5$, $\eta_d = 0.3$, $\eta_w = 0.2$)

| $\sigma$ | Objective | $\lambda=0$ | | | $\lambda=0.25$ | | | $\lambda=0.75$ | | | $\lambda=1.0$ | | |
|---|---|---|---|---|---|---|---|---|---|---|---|---|---|
| | | $\gamma=0.7$ | $\gamma=0.8$ | $\gamma=0.9$ | $\gamma=0.7$ | $\gamma=0.8$ | $\gamma=0.9$ | $\gamma=0.7$ | $\gamma=0.8$ | $\gamma=0.9$ | $\gamma=0.7$ | $\gamma=0.8$ | $\gamma=0.9$ |
| 0.2 | $f$ | 0.8285 | 0.8173 | 0.8047 | 0.8586 | 0.8473 | 0.8367 | 0.8958 | 0.9073 | 0.9143 | 0.9257 | 0.9371 | 0.9487 |
| | $f_c(10^6)$ | 569.89 | 566.93 | 562.51 | 578.34 | 574.73 | 571.86 | 586.71 | 590.13 | 593.14 | 594.91 | 598.12 | 601.52 |
| | $f_d(10^1)$ | 221.13 | 216.30 | 211.80 | 234.92 | 230.26 | 225.53 | 253.15 | 257.82 | 262.73 | 266.16 | 271.31 | 276.25 |
| | $f_w(10^4)$ | 140.14 | 136.49 | 132.23 | 148.95 | 145.45 | 142.18 | 160.07 | 163.91 | 163.35 | 169.62 | 173.01 | 176.61 |
| 0.4 | $f$ | 0.8383 | 0.8282 | 0.8171 | 0.8697 | 0.8582 | 0.8474 | 0.8858 | 0.8963 | 0.9084 | 0.9155 | 0.9267 | 0.9378 |
| | $f_c(10^6)$ | 572.51 | 569.82 | 566.12 | 581.75 | 577.85 | 574.21 | 583.37 | 587.24 | 590.92 | 591.58 | 594.86 | 598.23 |
| | $f_d(10^1)$ | 225.40 | 220.82 | 216.49 | 239.67 | 234.74 | 230.18 | 249.00 | 253.10 | 257.96 | 262.20 | 266.73 | 271.46 |
| | $f_w(10^4)$ | 143.34 | 140.24 | 136.68 | 152.27 | 149.13 | 146.05 | 157.14 | 160.22 | 164.21 | 166.29 | 170.08 | 173.46 |
| 0.6 | $f$ | 0.8497 | 0.8393 | 0.8284 | 0.8797 | 0.8683 | 0.8582 | 0.8749 | 0.8864 | 0.8972 | 0.9043 | 0.9151 | 0.9264 |
| | $f_c(10^6)$ | 575.89 | 572.64 | 569.35 | 584.20 | 580.95 | 577.37 | 580.04 | 584.65 | 587.69 | 588.25 | 591.51 | 594.75 |
| | $f_d(10^1)$ | 230.05 | 225.51 | 221.28 | 244.23 | 238.90 | 234.46 | 244.15 | 248.65 | 253.42 | 257.14 | 261.82 | 266.53 |
| | $f_w(10^4)$ | 147.06 | 144.15 | 140.33 | 155.46 | 152.30 | 149.74 | 154.06 | 157.13 | 160.40 | 163.15 | 166.37 | 170.07 |
| 0.8 | $f$ | 0.8601 | 0.8500 | 0.8396 | 0.8904 | 0.8795 | 0.8698 | 0.8643 | 0.8748 | 0.8869 | 0.8926 | 0.9040 | 0.9155 |
| | $f_c(10^6)$ | 578.22 | 575.90 | 572.53 | 587.46 | 584.26 | 581.50 | 577.70 | 580.00 | 584.36 | 584.94 | 588.18 | 591.57 |
| | $f_d(10^1)$ | 234.90 | 230.20 | 225.86 | 248.94 | 243.85 | 239.36 | 239.25 | 243.89 | 248.36 | 251.68 | 256.74 | 261.70 |
| | $f_w(10^4)$ | 150.50 | 147.28 | 144.17 | 158.58 | 155.62 | 152.83 | 150.51 | 154.22 | 158.14 | 159.88 | 163.25 | 166.85 |

### Weights of Objective Functions Sensitivity Analysis

From the discussion above, it can be seen that a difference in weights leads to a difference in objective function values. The results of this analysis are shown in Table 7.11 with respect to different weights, in which the optimistic-pessimistic index $\lambda = 0.5$, probability level $\sigma = 0.2$, possibility level $\gamma = 0.8$. These comparative results show that the difference in the solutions with different weights is not great as the weights reflect the importance of each objective from the view of the decision makers. Therefore, the results become progressively worse with an increase in the importance of the objective function $f_w$. However, in a real situation, decision makers would control the weights within a reasonable range according to the project.

**Table 7.11** Sensitivity analysis on the weights $\eta_c$ and $\eta_w$ of decision makers ($\lambda = 0.5$, $\sigma = 0.2$, $\gamma = 0.8$)

| Combinations | Weights | | | Objective Values | | | |
|---|---|---|---|---|---|---|---|
| | $\eta_c$ | $\eta_d$ | $\eta_w$ | $f$ | $f_c(10^6 yuan)$ | $f_d(10^1 hour)$ | $f_w(10^4 m^3)$ |
| Combination 1 | 0.6 | 0.2 | 0.2 | 0.8814 | 574.92 | 247.30 | 156.45 |
| Combination 2 | 0.5 | 0.3 | 0.2 | 0.8773 | 582.50 | 244.47 | 154.16 |
| Combination 3 | 0.5 | 0.2 | 0.3 | 0.8707 | 584.35 | 249.63 | 151.18 |
| Combination 4 | 0.4 | 0.4 | 0.2 | 0.8647 | 590.26 | 237.20 | 153.85 |
| Combination 5 | 0.4 | 0.3 | 0.3 | 0.8598 | 591.70 | 241.53 | 150.24 |
| Combination 6 | 0.4 | 0.2 | 0.4 | 0.8500 | 590.50 | 248.25 | 147.53 |

### Algorithm Evaluation

To carry out comparisons under a similar environment, the parameters stated in Table 7.7 and the initial velocities for the decision variables in the DP-based CPSO are also adopted for the standard PSO. However, the initial velocities for the state variables are selected based on the inventory level of the borrow areas and stockpile areas at each stage. Table 7.12 shows the comparison results and the convergence histories of the DP-based CPSO and standard PSO for the practical application example respectively based on an optimistic-pessimistic index $\lambda = 0.5$, weights $\eta_c = 0.5$, $\eta_D = 0.3$ and $\eta_w = 0.2$, probability level $\sigma = 0.2$ and possibility level $\gamma = 0.8$. The DP-based CPSO proposed in this chapter has shown to have an obvious advantage compared with the standard PSO when solving TS-based DTAP. The first advantage is that the DP-based CPSO is more stable than a standard PSO in searching for the optima. Another is that the proposed DP-based CPSO is faster when determining the optima than the standard PSO and converges faster than the standard PSO, that is, the DP-based CPSO needs less iterations to find the optimal solutions compared with standard PSO. Thus the DP-based CPSO shows its improved search performance compared with the standard PSO under a similar environment.

**Table 7.12** Comparison results between DP-based CPSO and Standard PSO

| Approach | Objective functions | Minimal fitness value($10^6$) | Iterations to find optima | Computation time (s) |
|---|---|---|---|---|
| DP-based CPSO | $f$ | 0.8773 | 46 | 17.34 |
| | $f_c(10^6)$ | 582.50 | 46 | 17.34 |
| | $f_d(10^1)$ | 244.47 | 46 | 17.34 |
| | $f_w(10^4)$ | 154.16 | 46 | 17.34 |
| Standard PSO | $f$ | 0.8773 | 74 | 32.20 |
| | $f_c(10^6)$ | 582.50 | 74 | 32.20 |
| | $f_d(10^1)$ | 244.47 | 74 | 32.20 |
| | $f_w(10^4)$ | 154.16 | 74 | 32.20 |

# References


1. Sheu, D., Lin, J.: Equipment management strategy under machine capacity loss. International Journal of Production Economics 103, 308–315 (2006)
2. Segelod, E.: Resource allocation in a few industries: Determinants and trends. International Journal of Production Economics 77, 63–77 (2002)
3. Khattab, M., Søyland, K.: Limited-resource allocation in construction projects. Computers and Industrial Engineering 31, 229–232 (1996)
4. Jaturonnatee, J., Murthy, D.N.P., Boondiskulchok, R.: Optimal preventive maintenance of leased equipment with corrective minimal repairs. European Journal of Operational Research 174(1), 201–215 (2006)
5. Pongpech, J., Murthy, D.N.P.: Optimal periodic preventive maintenance policy for leased equipment. Reliability Engineering and System Safety 91(7), 772–777 (2006)
6. Panagiotidou, S., Tagaras, G.: Optimal preventive maintenance for equipment with two quality states and general failure time distributions. European Journal of Operational Research 180(1), 329–353 (2007)
7. de León Hijes, F.G., Cartagena, J.J.R.: Maintenance strategy based on a multicriterion classification of equipments. Reliability Engineering and System Safety 91(4), 444–451 (2006)
8. Akselfod, L.M.: Refractory castables of the new generation in the production of cast iron and steel. Refractories and Industrial Ceramics 40(7-8), 363–370 (1999)
9. Varty, A., Boyle, R., Pritchard, E., Gill, R.: Construction of concrete face rockfill dams. Concrete Face Rockfill Dams-Design, Construction, and Performance 40(7-8), 435–458 (1985)
10. Dogramaci, A., Fraiman, N.: Replacement decisions with maintenance under uncertainty: an imbedded optimal control model. Operations Research 52, 785–794 (2004)
11. Das, K., Lashkari, R., Sengupta, S.: Machine reliability and preventive maintenance planning for cellular manufacturing systems. European Journal of Operational Research 183, 162–180 (2007)
12. McCool, J.: Testing for dependency of failure times in life testing. Technometrics 48, 41–48 (2006)
13. Seifried, A.: About statistics in fatigue strength. Materialwissenschaft Und Werkstofftechnik 35, 93–111 (2004)
14. Patankar, J., Mitra, A.: Effects of warranty execution on warranty reserve costs. Management Science 41, 395–400 (1995)
15. Friebelova, J., Friebel, L.: Using the Weibull distribution for simulation of machine lifetime. In: Proc. the 24th Int. Conf. Mathematical Methods in Economics, pp. 183–186 (2006)

16. Samanta, B., Sarkar, B., Mukherjee, S.: Reliability analysis of shovel machines used in an open cast coal mine. Mineral Resources Engineering 10, 219–231 (2001)
17. Erlichson, H.: Johann Bernoulli's brachistochrone solution using Fermat's principle of least time. European Journal of Physics 20, 299–304 (1999)
18. Hegab, M., Smith, G.: Delay time analysis in microtunneling projects. J. Constr. Eng. Manage. 133, 191–195 (2007)
19. Conover, W.J.: Practical non-parametric statistics, 2nd edn. Wiley, New York (1980)
20. Bendell, A.: Analysis methodologies. In: Cannon, A.G., Bendell, A. (eds.) Reliability Data Banks. Galliard Ltd., London (1991)
21. Chua, D.K.H., Goh, Y.M.: Poisson model of con-struction incident occurrence. Journal of Construction Engineering and Management 131, 715–722 (1995)
22. Zadeh, L.A.: Fuzzy sets. Information and Control 8, 338–353 (1965)
23. Kennedy, J., Eberhart, R.C.: Particle swarm optimization. In: Proc. IEEE Conf. on Neural Networks, pp. 1942–1948. Institute of Electrical and Electronics Engineers, New York (1995)
24. Eberhart, R.C., Shi, Y.: Tracking and optimizing dynamic systems with particle swarms. In: Proc. IEEE Congress on Evolutionary Computation, pp. 94–97. Institute of Electrical and Electronics Engineers, New York (2001)
25. Clerc, M., Kennedy, J.: The particle swarm-explosion, stability, and convergence in a multidimensional complex space. IEEE Transactions on Evolutionary Computation 6(1), 58–73 (2002)
26. Dai, H.C., Cao, G.J., Su, H.Z.: Management and construction of the Three Gorges Project. Journal of Construction Engineering and Management 132(6), 615–619 (2006)
27. Zhang, H., Li, H., Tam, C.M.: Permutation-based particle swarm optimization for resource-constrained project scheduling. J. Comput. Civ. Eng. 20(2), 141–149 (2006a)
28. Zhang, H., Li, H., Tam, C.M.: Particle swarm optimization for preemptive scheduling under break and resource-constraints. J. Constr. Eng. Manage. 132(3), 259–267 (2006b)
29. Zimmermann, H.J.: Fuzzy programming and linear programming with several objective functions. Fuzzy Sets and Systems 1, 45–55 (1978)
30. Zhang, H., Tam, C.M., Li, H., Shi, J.J.: Particle swarm optimization-supported simulation for construction operations. J. Constr. Eng. Manage. 132(12), 1267–1274 (2006c)
31. Zhang, H., Wang, J.: Particle swarm optimization for construction site unequal-are layout. J. Constr. Eng. Manage. 134(9), 739–748 (2008)
32. Langdon, W.B., Poli, R.: Evolving problems to learn about particle swarm optimizers and other search algorithms. IEEE Trans. Evol. Comput. 11(5), 561–578 (2007)
33. Feng, J., Liu, L., Wan, Y.: Order-based backorders in multi-item inventory systems. Operations Research Letters 38(1), 27–32 (2010)
34. Shi, Y., Eberhart, R.C.: Parameter selection in particle swarm optimization. In: Porto, V.W., Waagen, D. (eds.) EP 1998. LNCS, vol. 1447, pp. 591–600. Springer, Heidelberg (1998a)
35. Liu, F., Yu, Q., Zhang, S., Hou, Q.: Geological and geotechnical characteristics of Xiaolangdi dam, Yellow River, China. Bulletin of Engineering Geology and the Environment 65(3), 289–295 (2006)
36. Roxy, P.: Introduction to Statistic Data Analysis. Thomson Learning, Duxbury (2001)
37. Montgomery, D.C., Runger, G.C.: Applied statistics and probability for engineers, 2nd edn. Wiley, Singapore (1999)
38. Johnsonbaugh, R., Pfaffenberger, W.E.: Foundations of mathematical analysis. Dover Publications, UK (2002)
39. Skitmore, R.M., Pettitt, A.N., McVinish, R.: Gates' bidding model. J. Constr. Eng. Manage. 133(11), 855–863 (2007)

40. Trelea, I.C.: The particle swarm optimization algorithm: Convergence analysis and parameter selection. Information Processing Letters 85(6), 317–325 (2003)
41. Livingston, E.H.: Who was student and why do we care so much about his t-test? Journal of Surgical Research 118, 58–65 (2004)
42. Lomax, R.G.: Statistical Concepts: A Second Course. Routledge, New York (2007)
43. Naddor, E.: Inventory Systems, 1st edn. John Wiley, New York (1966)
44. Kiesmüller, G.P.: Multi-item inventory control with full truckloads: A comparison of aggregate and individual order triggering. European Journal of Operational Research 200(1), 54–62 (2010)
45. Suo, M.Q., Li, Y.P., Huang, G.H.: An inventory-theory-based interval-parameter two-stage stochastic programming model for water resources management. Engineering Optimization 43(9), 999–1018 (2011)
46. Rezaei, J., Davoodi, M.: A deterministic, multi-item inventory model with supplier selection and imperfect quality. Applied Mathematical Modelling 32(10), 2106–2116 (2008)
47. Shah, J., Avittathur, B.: The retailer multi-item inventory problem with demand cannibalization and substitution. International Journal of Production Economics 106(1), 104–114 (2007)
48. Bhattacharya, D.K.: On multi-item inventory. European Journal of Operational Research 162(3), 786–791 (2005)
49. Çetinkaya, S., Parlar, M.: Optimal myopic policy for a stochastic inventory problem with fixed and proportional backorder costs. European Journal of Operational Research 110(1), 20–41 (1998)
50. Rahim, M.A., Ohta, H.: An integrated economic model for inventory and quality control problems. Engineering Optimization 37(1), 65–81 (2005)
51. Rahim, M.A.: Inventory systems with random arrival of shipments. International Journal of Advanced Manufacturing Technology 29(1-2), 197–201 (2006)
52. Chen, X., Simchi-Levi, D.: Coordinating inventory control and pricing strategies with random demand and fixed ordering cost: The infinite horizon case. Mathematics of Operations Research 29(3), 698–723 (2004)
53. Xu, J.P., Liu, Y.G.: Multi-objective decision making model under fuzzy random environment and its application to inventory problems. Information Sciences 178(14), 2899–2914 (2008)
54. Xu, J.P., Yao, L.M.: Random-Like Multiple Objective Decision Making. Springer, Heidelberg (2011)
55. Bertsekas, D.P.: Dynamic Programming and Optimal Control, vol. I. Athena Scientific, Belmont (2005)
56. Missbauer, H., Hauber, W.: Bid calculation for construction projects: Regulations and incentive effects of unit price contracts. European Journal of Operational Research 171(3), 1005–1019 (2006)
57. Baker, E., Mellors, B., Chalmers, S., Lavers, A.: FIDIC Contracts: Law and Practice, 1st edn. Informa Publishing, UK (2009)
58. Tommelein, I.D.: Pull-driven scheduling for pipe-spool installation: Simulation of lean construction technique. Journal of Construction Engineering and Management 124(4), 279–288 (1998)
59. Zhang, H., Tam, C.M., Shi, J.J.: Application of fuzzy logic to simulation for construction operations. Journal of Computing in Civil Engineering 17(1), 38–45 (2003)
60. Tajbakhsh, M.M., Lee, C.G., Zolfaghari, S.: An inventory model with random discount offerings. Omega-International Journal of Management Science 39(6), 710–718 (2011)

61. Arifoglu, K., Ozekici, S.: Inventory management with random supply and imperfect information: A hidden Markov model. International Journal of Production Economics 134(1), 123–137 (2011)
62. Widyadana, G.A., Wee, H.M.: Optimal deteriorating items production inventory models with random machine breakdown and stochastic repair time. Applied Mathematical Modelling 35(7), 3495–3508 (2011)
63. Taleizadeh, A.A., Niaki, S.T., Aryanezhad, M.B.: Multi-product multi-constraint inventory control systems with stochastic replenishment and discount under fuzzy purchasing price and holding costs. American Journal of Applied Sciences 6(1), 1–12 (2009)
64. Taleizadeh, A.A., Niaki, S.T., Aryanezhad, M.B.: Optimising multi-product multi-chance-constraint inventory control system with stochastic period lengths and total discount under fuzzy purchasing price and holding costs. International Journal of Systems Science 41(10), 1187–1200 (2010)
65. Wu, Z.B., Xu, J.P.: A consistency and consensus based decision support model for group decision making with multiplicative preference relations. Decision Support Systems 52(3), 757–767 (2012)
66. Herrera, F., Herrera-Viedma, E., Verdegay, J.L.: A model of consensus in group decision making under linguistic assessments. Fuzzy Sets and Systems 78(1), 73–87 (1996)
67. Long, L.D., Ohsato, A.: Fuzzy critical chain method for project scheduling under resource constraints and uncertainty. International Journal of Project Management 26(6), 688–698 (2008)
68. Marler, R.T., Arora, J.S.: The weighted sum method for multi-objective optimization: New insights. Structural and Multidisciplinary Optimization 41(6), 853–862 (2010)
69. Xu, J.P., Zheng, H., Zeng, Z.Q., Wu, S.Y., Shen, M.B.: Discrete time-cost-environment trade-off problem for large-scale construction systems with multiple modes under fuzzy uncertainty and its application to Jinping-II Hydroelectric Project. International Journal of Project Management 30, 950–966 (2013e)
70. Hamiani, A., Popescu, G.: CONSITE: a knowledge-based expert system for site layout. Journal of Computing in Civil Engineering, 248–256 (1988)
71. Yeh, I.C.: Construction-site layout using annealed neural network. Journal of Computing in Civil Engineering 9(3), 201–208 (1995)
72. Zouein, P.P., Tommelein, I.D.: Dynamic layout planning using a hybrid incremental solution method. Journal of Construction Engineering and Management 125(6), 400–408 (1999)
73. Meller, R.D., Gau, K.Y.: The facility layout problem: Recent and emerging trends and perspectives. Journal of Manufacturing Systems 15(5), 351–366 (1996)
74. Domschke, W., Krispin, G.: Location and layout planning. OR Spectrum 19(3), 181–194 (1997)
75. Elbeltagi, E., Hagazy, T., Hosny, A.H., Eldosouky, A.: Schedule-dependent evolution of site layout planning. Construction Management and Economics 19(7), 689–697 (2001)
76. Elbeltagi, E., Hegazy, T., Eldosouky, A.: Dynamic layout of construction temporary facilities considering safety. Journal of Construction Engineering and Management 130(4), 534–541 (2004)
77. Khalafallah, A., El-Rayes, K.: Automated multi-objective optimization system for airport site layouts. Automation in Construction 20, 313–320 (2011)
78. Turskis, Z., Zavadskas, E.K., Peldschus, F.: Multi-criteria optimization system for decision making in construction design and management. Inzinerine Ekonomika (Engineering Economics) 61(1), 7–17 (2009)
79. Cheung, S.O., Tong, T.K.L., Tam, C.M.: Site pre-cast yard layout arrangement through genetic algorithms. Automation in Construction 11(1), 35–46 (2002)

80. Wang, M.J., Hu, M.H., Ku, M.Y.: A solution to the unequal area facilities layout problem by genetic algorithm. Computers in Industry 56(2), 207–220 (2005)
81. Ning, X., Lam, K.C., Lam, M.C.K.: Dynamic construction site layout planning using max-min ant system. Automation in Construction 19, 55–65 (2010)
82. Baykasoglu, A., Dereli, T., Sabuncu, I.: An ant colony algorithm for solving budget constrained and unconstrained dynamic facility layout problems. Omega 34, 385–396 (2006)
83. Erel, E., Sabuncuoglu, I., Aksu, B.A.: Balancing of U-type assembly systems using simulated annealing. International Journal of Production Research 39(13), 3003–3015 (2001)
84. Ning, X., Lamb, K.C., Lam, M.C.K.: A decision-making system for construction site layout planning. Automation in Construction 20, 459–473 (2011)
85. Mawdesley, M.J., Al-Jibouri, S.H.: Proposed genetic algorithms for construction site layout. Engineering Applications of Artificial Intelligence 16(5-6), 501–509 (2003)
86. Sethi, A.K., Sethi, S.P.: Flexibility in manufacturing: a survey. International Journal of Flexible Manufacturing Systems 2(4), 289–328 (1990)
87. Ning, X., Wang, L.G.: Construction site layout evaluation by intuitionistic fuzzy TOPSIS model. Applied Mechanics and Materials 71, 583–588 (2011)
88. Soltani, A.R., Fernando, T.: A fuzzy based multi-objective path planning of construction sites. Automation in Construction 13, 717–734 (2004)
89. Wang, Z., Huang, H.Z., Li, Y., Pang, Y., Xiao, N.C.: An approach to system reliability analysis with fuzzy random variables. Mechanism and Machine Theory 52, 35–46 (2012)
90. Xu, J.P., Yan, F., Li, S.: Vehicle routing optimization with soft time windows in a fuzzy random environment. Transportation Research Part E: Logistics and Transportation Review 47, 1075–1091 (2011)
91. Lam, K.C., Ning, X., Ng, T.: The application of the ant colony optimization algorithm to the construction site layout planning problem. Construction Management and Economics 25(4), 359–374 (2007)
92. Gangolells, M., Casals, M., Forcada, N., Roca, X., Fuertes, A.: Mitigating construction safety risks using prevention through design. Journal of Safety Research 41, 107–122 (2010)
93. Puri, M.L., Ralescu, D.A.: Fuzzy random variables. Journal of Mathematical Analysis and Applications 114, 409–422 (1986)
94. Kwakernaak, H.: Fuzzy random variables-I, definitions and theorems. Information Sciences 15, 1–29 (1978)
95. Kwakernaak, H.: Fuzzy random variables-II, algorithms and examples for the discrete case. Information Sciences 17, 253–278 (1979)
96. Lia, S.T., Cheng, Y.C.: Deterministic fuzzy time series model for forecasting enrollments. Computers and Mathematics with Applications 53(12), 1904–1920 (2007)
97. Pardo, M.J., de la Fuente, D.: Optimizing a priority-discipline queueing model using fuzzy set theory. Computers and Mathematics with Applications 54(2), 267–281 (2007)
98. Bojadziev, G., Bojadziev, M.: Fuzzy logic for business, finance, and management, 1st edn. World Scientific, Singapore (1997)
99. Bratton, J., Kennedy, D.: Defining a standard for particle swarm optimization. In: Proceedings of the IEEE Swarm Intelligence Symposium, pp. 120–127. IEEE Press, Washington DC (2007)
100. Cagnina, L.C., Esquivel, S.C., Coello, C.A.C.: A fast particle swarm algorithm for solving smooth and non-smooth economic dispatch problems. Engineering Optimization 43(5), 485–505 (1997)

101. Su, R.H., Yang, D.Y., Pearn, W.L.: Decision-making in a single-period inventory environment with fuzzy demand. Expert Systems with Applications 38(3), 1909–1916 (2011)
102. Handfield, R., Warsing, D., Wu, X.M.: $(Q, r)$-inventory policies in a fuzzy uncertain supply chain environment. European Journal of Operational Research 197(2), 609–619 (2009)
103. Shah, N.H., Soni, H.: Continuous review inventory model for fuzzy price dependent demand. International Journal of Modelling in Operations Management 1(3), 209–222 (2011)
104. Schmitt, A.J., Snyder, L.V., Shen, Z.J.M.: Inventory systems with stochastic demand and supply: Properties and approximations. European Journal of Operational Research 206(2), 313–328 (2010)
105. Gil, M., Lopez-Diaz, M., Ralescu, D.: Overview on the development of fuzzy random variables. Fuzzy Sets and Systems 157, 2546–2557 (2006)
106. Robinson, J., Sinton, S., Rahmat-Samii, Y.: Particle swarm, genetic algorithm, and their hybrids: Optimization of a profiled corrugated horn antenna. In: IEEE Antennas and Propagation Society International Symposium and URSI National Radio Science Meeting, San Antonio, pp. 168–175 (2002)
107. Zhang, H., Xing, F.: Fuzzy-multi-objective particle swarm optimization for time-cost-quality tradeoff in construction. Automation in Construction 19(8), 1067–1075 (2010)
108. Jiang, A.N., Wang, S.Y., Tang, S.L.: Feedback analysis of tunnel construction using a hybrid arithmetic based on Support Vector Machine and Particle Swarm Optimisation. Automation in Construction 20(4), 482–489 (2011)
109. Yang, I., Hsieh, Y.: Reliability-based design optimization with discrete design variables and non-smooth performance functions: AB-PSO algorithm. Automation in Construction 20(5), 610–619 (2011)
110. Coello, C.C., Lechuga, M.S.: MOPSO: A proposal for multiple objective particle swarm optimization. In: Proceedings of Congress of Evolutionary Computation, pp. 1051–1056 (2002)
111. Coello, C.C., Pulido, G., Lechuga, M.S.: Handling multiple objectives with particle swarm optimization. IEEE Transactions on Evolutionary Computation 8(3), 256–279 (2004)
112. Huang, V.L., Suganthan, P.N., Liang, J.J.: Comprehensive learning particle swarm optimizer for solving multiobjective optimization problems. International Journal of Intelligent Systems 21, 209–226 (2006)
113. Shi, Y., Eberhart, R.C.: A modified particle swarm optimizer. In: Proceedings of the IEEE Congress on Evolutionary Computation, Piscataway, pp. 69–73 (1998b)
114. Deb, K., Goldberg, D.E.: An investigation of niche and species formation in genetic function optimization. In: Proceedings of the 3rd International Conference on Genetic Algorithms, pp. 42–50 (1989)
115. Oliver, I.M., Smith, D.J., Holland, J.R.C.: A study of permutation crossover operators on the traveling salesman problems. In: Proceedings of the 2nd International Conference on Genetic Algorithms and Their Application, pp. 227–230. Erlbaum, Hillsdale (1987)
116. Davis, L.: Applying adaptive algorithms to epistatic domains. In: Proceedings of the 9th International Joint Conference on Artificial Intelligence, pp. 162–164. Kaufmann, San Francisco (1985)
117. Reyes-Sierra, M., Coello, C.A.C.: Multi-objective particle swarm optimizers: a survey of the state-of-the-art. International Journal of Computational Intelligence Research 2(3), 287–308 (2006)

118. Chen, C.W., Sha, D.Y.: Heuristic approach for solving the multi-objective facility layout problem. International Journal of Production Research 43(21), 4493–4507 (2005)
119. Knowles, J.D., Corne, D.W.: Approximating the nondominated front using the pareto archived evolution strategy. Evolution Computation 8, 149–172 (2000)
120. Easa, S.M., Hossain, K.M.A.: New mathematical optimization model for construction site layout. Journal of Construction Engineering and Management 134(8), 653–662 (2008)
121. Zitzler, E., Deb, K., Thiele, L.: Comparison of multiobjective evolutionary algorithms: empirical results. Evolutionary Computation 8(2), 195 (2000)
122. Van den Bergh, F., Engelbrecht, A.P.: A Convergence Proof for the Particle Swarm Optimiser. Fundamenta Informaticae 105(4), 341–374 (2010)
123. Aiello, G., Enea, M., Galante, G.: A multi-objective approach to facility layout problem by genetic search algorithm and Electre method. Robotics and Computer-Integrated Manufacturing 22, 447–455 (2006)
124. Ashtiani, B., Aryanezhad, M.B., Moghaddam, B.F.: Multi-start simulated annealing for dynamic plant layout problem. Journal of Industrial Engineering International 3(4), 44–50 (2007)
125. Bennage, W.A., Dhingra, A.K.: Single and multiobjective structural optimization in discrete-continuous variables using simulated annealing. International Journal for Numerical Methods in Engineering 38, 2753 (1995)
126. Back, T., Fogel, D., Michalewicz, Z.: Handbook of evolutionary computation. IOP Publishing Ltd. and Oxford University Press, Bristol (2000)
127. Baykasoglu, A., Gindy, N.N.: A simulated annealing algorithm for dynamic layout problem. Computers and Operations Research 28, 1403–1426 (2001)
128. Cao, Q., Hoffman, J.: A case study approach for developing a project performance evaluation system. International Journal of Project Management 29, 155–164 (2011)
129. Deb, S.K., Bhattacharyya, B.: Fuzzy decision support system for manufacturing facilities layout planning. Decision Support Systems 40, 305–314 (2005)
130. Dweiri, F., Meier, F.A.: Application of fuzzy decision-making in facilities layout planning. International Journal of Production Research 34(11), 3207–3225 (1996)
131. Elbeltagi, E., Hegazy, T.: A Hybrid AI-based system for site layout planning in construction site. Computer-Aided Civil and Infrastructure Engineering 16(2), 79–93 (2001)
132. Emre, G., Gürcanli, G., Muüngen, U.: An occupational safety risk analysis method at construction sites using fuzzy sets. International Journal of Industrial Ergonomics 39, 371–387 (2009)
133. Fortenberry, J., Cox, J.: Multiple criteria approach to the facilities layout problem. International Journal of Production Research 23(4), 773–782 (1985)
134. Gentile, M., Rogers, W.J., Mannan, M.S.: Development of an inherent safety index based on fuzzy logic. AIChE Journal 49(4), 959–968 (2003)
135. Ghoshal, S.P.: Application of GA/GA-SA based fuzzy automatic generation control of a multi-area thermal generating system. Electric Power Systems Research 70, 115–127 (2004)
136. Glenn, B., Gregory, H.: Lean project management. Building Research & Information 31(2), 119–133 (2003)
137. Gang, J., Xu, J.: The resource-constraint project scheduling with multi-mode under fuzzy random environment in the drainage engineering of LT hydropower. International Journal of Logistics and Transportation 4(2), 53–80 (2010)
138. Haidar, A., Naoum, S., Howes, R., Tah, J.: Genetic algorithms application and testing for equipment selection. Journal of Construction Engineering and Management 125(1), 32–38 (2009)

139. Koopmans, T.C., Beckmann, M.: Assignment problems and the location of economic activities. Econometrica 25(1), 53–76 (1957)
140. Kirkpatrick, S., Gelatt, J.C.D., Vecchi, M.P.: Optimisation by simulated annealing. Science 220, 671–680 (1983)
141. Karray, F., Zaneldin, E., Hegazy, T., Shabeeb, A.H.M., Elbeltagi, E.: Tools of soft computing as applied to the problem of facilities layout planning. IEEE Transactions on Fuzzy Systems 8(4), 367–379 (2000)
142. Lin, K.T., Kao, C.Y., Hsu, C.C.: Applying the genetic approach to simulated annealing in solving some NP-hard problems. IEEE Transactions on Systems Man and Cybernetics 23(6), 1752–1767 (1993)
143. Mohanta, D.K., Sadhu, P.K., Chakrabarti, R.: Deterministic and stochastic approach for safety and reliability optimization of captive power plant maintenance scheduling using GA/SA-based hybrid techniques: A comparison of results. Reliability Engineering and System Safety 92, 187–199 (2007)
144. McKendall, A.R., Hakobyan, J.A.: Heuristics for the dynamic facility layout problem with unequal-area departments. European Journal of Operational Research 201, 171–182 (2010)
145. Maravas, A., Pantouvakis, J.: Project cash flow analysis in the presence of uncertainty in activity duration and cost. International Journal of Project Management 30, 374–384 (2012)
146. Nicklow, J., Reed, P., Savic, D., Dessalegne, T., Harrell, L., Chan-Hilton, A., Karamouz, M., Minsker, B., Ostfeld, A., Singh, A.: State of the Art for Genetic Algorithms and Beyond in Water Resources Planning and Management. Journal of Water Resources Planning and Management 136(4), 412–432 (2010)
147. Osman, H.M., Georgy, G.E., Ibrahim, M.E.: A hybrid CAD-based construction site layout planning system using genetic algorithms. Automat. Constr. 12, 749–764 (2003)
148. Rosenblatt, M.: The dynamics of plant layout. Management Science 32(1), 76–86 (1986)
149. Rafael, S., Milan, R., Ronen, B.: Requirements for building information modeling based lean production management systems for construction. Automation in Construction 19, 641–655 (2010)
150. Saaty, T.L.: The Analytical Hierarchy Process. McGraw-Hill, New York (1980)
151. Tompkins, J.A., White, J.A., Bozer, Y.A., Frazelle, E.H., Tanchoco, J.M., Trevino, J.: Facilities planning, 4th edn. Wiley, New York (2010)
152. Urban, T.L.: Solution procedures for the dynamic facility layout problem. Annals of Operations Research 76, 323–342 (1998)
153. Wolfgang, D., Gabriela, K.: Location and layout planning—A survey. OR Spektrum 19, 181–194 (1997)
154. Xu, J., Zhang, Z.: A Fuzzy Random Resource-constrained Scheduling Model with Multiple Projects and Its Application to a Working Procedure in a Large-Scale Water Conservancy and Hydropower Construction Project. Journal of Scheduling 15(2), 253–272 (2012)
155. Yang, T., Hung, C.C.: Multiple-attribute decision making methods for plant layout design problem. Robotics and Computer-Integrated Manufacturing 23, 126–137 (2007)
156. Zhang, J.P., Liu, L.H., Coble, R.J.: Hybrid intelligence utilization for construction site layout. Automation in Construction 20, 44–55 (2002)
157. Zhou, F., AbouRizk, S.M., AL-Battaineh, H.: Optimisation of construction site layout using a hybrid simulation-based system. Simulation Modelling Practice and Theory 17, 348–363 (2009)
158. Zétényi, T.: Fuzzy sets in Psychology. Elsevier, North Holland (1988)

159. Zouein, P.P., Harmanani, H., Hajar, A.: Genetic algorithm for solving site layout problem with unequal-size 19 and constrained facilities. J. Comput. Civ. Eng. 16(2), 143–151 (2002)
160. Wu, H.C.: Optimality conditions for linear programming problems with fuzzy coefficients. Computers and Mathematics with Applications 55(12), 2807–2822 (2008)
161. Mula, J., Poler, R., Garcia-Sabater, J.P.: Material requirement planning with fuzzy constraints and fuzzy coefficients. Fuzzy Sets and Systems 158(7), 783–793 (2007)
162. Kacprzyk, J.: Multistage fuzzy control: A model-based approach to fuzzy control and decision making, 1st edn. Wiley, New York (1997)
163. Figueira, J., Greco, S., Ehrgott, M.: Multiple criteria decision analysis: state of the art surveys. Springer, Berlin (2005)
164. Bellman, R.E.: Dynamic programming. Princeton University Press, Princeton (1957)
165. Bellman, R.E., Dreyfus, S.E.: Applied dynamic programming. Princeton University Press, Princeton (1962)
166. Rao, S.S.: Engineering optimization: theory and practice, 4th edn. Wiley, New York (2009)
167. Xu, J.P., Wei, P.: Production-distribution planning of construction supply chain management under fuzzy random environment for large-scale construction project. Journal of Industrial and Management Optimization 9(1), 31–56 (2013)
168. Nandalal, K.D.W., Bogardi, J.J.: Dynamic programming based operation of reservoirs: applicability and limits. Cambridge University Press, New York (2007)
169. Kacprzyk, J., Safteruk, K., Staniewski, P.: On the control of stochastic systems in a fuzzy environment over infinite horizon. Systems Science 7, 121–131 (1981)
170. Kacprzyk, J., Staniewski, P.: Control of a deterministic system in a fuzzy environment over infinite horizon. Fuzzy Sets and Systems 10, 291–298 (1983)
171. Fung, L.W., Fu, K.S.: Characterization of a class of fuzzy optimal control problems. In: Gupta, M.M., Saridis, G.N., Gaines, B.R. (eds.) Fuzzy Automata and Decision Processes, pp. 209–219. North-Holland, New York (1977)
172. Mousavi, S.J., Mahdizadeh, K., Afshar, A.: A stochastic dynamic programming model with fuzzy storage states for reservoir operations. Advances in Water Resources 27(11), 1105–1110 (2004)
173. Oh, S.H., Luus, R.: Optimal feedback control of time-delay systems. AIChE J 22, 140–147 (1976)
174. Luus, R., Zhang, X., Hartig, F., Keil, F.J.: Use of piecewise linear control continuous optimal control for time-delay systems. Ind. Eng. Chem. Res. 34, 4136–4139 (1995)
175. Hairer, E., Norsett, S.P., Wanner, G.: Solving ordinary differential equations 1. Springer Series in Computational Mathematics, vol. 8. Springer, Berlin (1987)
176. Chan, H.C., Perkins, W.R.: Optimization of time delay systems using parameter imbedding. Automatica 9, 257–261 (1973)
177. Palanisamy, K.R., Balachandran, K., Ramasamy, R.S.: Optimal control of linear time-varying delay systems via single-term Walsh series. In: Proc. IEE 135 pt. D, vol. 332 (1988)
178. Dadebo, S., Luus, R.: Optimal control of time-delay systems by dynamic programming. Optimal Contr. Applic. Methods 13, 29–41 (1992)
179. Ko, D.Y.C., Stevens, W.F.: Study of singular solutions in dynamic optimization. AIChE J. 17, 160–166 (1971)
180. Reddy, K.V., Husain, A.: Computation of optimal control policy with singular subarc. Can. J. Chem. Eng. 59, 557–559 (1981)
181. Luus, R.: Optimal control of batch reactors by iterative dynamic programming. J. Process Control 4, 217–226 (1994)

182. Luus, R.: Determination of the region sizes for LJ optimization procedure. Hung. J. Ind. Chem. 26, 281–286 (1998b)
183. Luus, R., Hennessy, D.: Optimization of fed-batch reactors by the Luus-Jaakola optimization procedure. Ind. Eng. Chem. Res. 38, 1948–1955 (1999)
184. Bojkov, B., Luus, R.: Optimal control of nonlinear systems with unspecified final times. Chem. Eng. Sci. 51, 905–919 (1996)
185. Luus, R.: Handling difficult equality constraints in direct search optimization. Hung. J. Ind. Chem. 24, 285–290 (1996)
186. Luus, R.: Use of iterative dynamic programming with variable stage lengths and fixed final time. Hung. J. Ind. Chem. 24, 279–284 (1996)
187. Luus, R., Storey, C.: Optimal control of final state constrained systems. In: Proc. IASTED International Conference on Modelling, Simulation and Control, Singapore, August 11-14, pp. 245–249 (1997)
188. Goh, C.J., Teo, L.K.: Control parametrization: a unified approach to optimal control problems with generalized constraints. Automatica 24, 3–18 (1988)
189. Luus, R.: Application of iterative dynamic programming to state constrained optimal control problems. Hung. J. Ind. Chem. 19, 245–254 (1991)
190. Dadebo, S.A., McAuley, K.B.: Dynamic optimization of constrained chemical engineering problems using dynamic programming. Comput. Chem. Eng. 19, 513–525 (1995)
191. Luus, R.: Piecewise linear continuous optimal control by iterative dynamic programming. Ind. Eng. Chem. Res. 32, 859–865 (1993a)
192. Luus, R.: Application of iterative dynamic programming to very high-dimensional systems. Hung. J. Ind. Chem. 21, 243–250 (1993b)
193. Thomopoulos, S.C.A., Papadakis, I.N.M.: A single shot method for optimal step computation in gradient algorithms. In: Proceedings of the 1991 American Control Conference, Boston, MA, American Control Council, pp. 2419–2422. IEEE Service Center, Piscataway (1991)
194. Xu, J.P., Zeng, Z.Q., Wu, S.Y., Shen, M.B.: Multistage optimization for closed multiclass queueing networks in concrete transportation systems under uncertainty: case study of Jinping-I Hydropower Project in China. Technical Report (2012)
195. Abo-Sinna, M.A.: Multiple objective (fuzzy) dynamic programming problems: a survey and some applications. Applied Mathematics and Computation 157, 861–888 (2004)
196. Mine, H., Fukushima, M.: Decomposition of multiple criteria mathematical programming by dynamic programming. International Journal of Systems Sciences 10(5), 557–566 (1979)
197. Eykhoff, P.: System identification-parameter and state estimation. John Wiley and Sons, New York (1974)
198. Xu, J.P., Zeng, Z.Q.: A discrete time optimal control model with uncertainty for dynamic machine allocation problem and its application to manufacturing and construction industries. Applied Mathematical Modelling 36(8), 3513–3544 (2012)
199. Dombi, J.: A fuzzy, heuristic, interactive approach to the optimal network problem, pp. 253–275 (1983)
200. Smets, P., Magrez, P.: The measure of the degree of truth and of the grade of membership. Fuzzy Sets and Systems 25, 67–72 (1988)
201. Zadeh, L.A.: Probability measure of fuzzy events. Journal of Mathematical Analysis and Applications 23, 421–427 (1968a)
202. Zadeh, L.A.: A fuzzy-set-theoretic interpretation of linguistic hedges. Journal of Cybernetics 2(3), 4–34 (1972)

203. Giles, R.: The concept of grade of membership. Fuzzy Sets and Systems 25, 297–323 (1988)
204. Bezdek, J.C., Hathaway, R.J.: Clustering with relational *c*-means partitions from pairwise distance data. Mathematical Modelling 9, 435–439 (1987)
205. Pedrycz, W.: Fuzzy models and relational equations. Mathematical Modelling 9, 427–434 (1987)
206. Bandemer, H.: From fuzzy data to functional relationships. Mathematical Modelling 9, 419–426 (1987)
207. Rapoport, A., Wallsten, T.S., Cox, J.A.: Direct and indirect scaling of membership functions of probability phrase. Mathematical Modelling 9, 397–417 (1987)
208. Lew, A., Mauch, H.: Dynamic Programming: A Computational Tool. Springer, New York (2007)
209. Abo-Sinna, M.A.: Multiple objective (fuzzy) dynamic programming problems: a survey and some applications. Applied Mathematics and Computation 157, 861–888 (2004)
210. Abramov, V.M.: Some results for large closed queueing networks with and without bottleneck: Up-and down-crossings approach. Queueing Systems 38(2), 149–184 (2001)
211. Anghinolfi, D., Paolucci, M.: A new discrete particle swarm optimization approach for the single-machine total weighted tardiness scheduling problem with sequence-dependent setup times. European Journal of Operational Research 193, 73–85 (2009)
212. Balsamo, S., Persone, V.D., Inverardi, P.: A review on queueing network models with finite capacity queues for software architectures performance prediction. Performance Evaluation 51(2-4), 269–288 (2003)
213. Berger, A., Bregman, L., Kogan, Y.: Bottleneck analysis in multiclass closed queueing networks and its application. Queueing Systems 31(3-4), 217–237 (1999)
214. Casale, G.: A generalized method of moments for closed queueing networks. Performance Evaluation 68(2), 180–200 (2011)
215. Chen, S.P.: Solving fuzzy queueing decision problems via a parametric mixed integer nonlinear programming method. European Journal of Operational Research 177(1), 445–457 (2007)
216. Daduna, H., Szekli, R.: Conditional job-observer property for multitype closed queueing networks. Journal of Applied Probability 39(4), 865–881 (2002)
217. De Vuyst, S., Wittevrongel, S., Bruneel, H.: Mean value and tail distribution of the message delay in statistical multiplexers with correlated train arrivals. Performance Evaluation 48(1-4), 103–129 (2002)
218. Shuttleworth, M.: Definition of Research. Experiment Resources. Experiment-Research.com (2008)
219. Creswell, J.W.: Educational Research: Planning, conducting, and evaluating quantitative and qualitative research. Pearson, Upper Saddle River (2008)
220. Unattributed. Research in Dictionary tab. Merriam Webster (m-w.com). Encyclopaedia Britannica (2011)
221. Dubois, D., Prade, H.: Possibility Theory: An Approach to computerized Processing of uncertainty. Plenum, New York (1988)
222. Huang, G.H., Loucks, D.P.: An inexact two-stage stochastic programming model for water resources management under uncertainty. Civ. Eng. Environ. Syst. 17(2), 95–118 (2000)
223. George, D.K., Xia, C.H.: Fleet-sizing and service availability for a vehicle rental system via closed queueing networks. European Journal of Operational Research 211(1), 198–207 (2011)
224. Gerasimov, A.I.: Synthesis of closed queueing networks with several classes of messages on the basis of their optimization and balancing. Cybernetics and Systems Analysis 36(6), 906–915 (2000)

225. Govil, M.K., Fu, M.C.: Queueing theory in manufacturing: A survey. Journal of Manufacturing Systems 18(3), 214–240 (1999)
226. Lee, D.M., Pietrucha, M.T., Donnell, E.T.: Incorporation of transportation experts' opinions of median safety using a hierarchical fuzzy inference system. In: Proceeding of 85 Annual Meeting of the Transportation Research Board. Transportation Research Board, Washington DC (2006)
227. Lee, D.M., Donnell, E.T.: Analysis of nighttime driver behavior and pavement marking effects using fuzzy inference system. Journal of Computing in Civil Engineering 21(3), 200–210 (2006)
228. Harrison, P.G., Coury, S.: On the asymptotic behaviour of closed multiclass queueing networks. Performance Evaluation 47(2-3), 131–138 (2002)
229. Gross, D., Harris, C.M.: Fundamentals of Queueing Theory, 3rd edn. Wiley, New York (1998)
230. Hillier, F.S., Lieberman, G.J.: Introduction to Operations Research, 7th edn. McGraw-Hill, Singapore (2001)
231. Ivnitski, V.A.: Network of single-server queues with dependent service times. Queueing Systems 37(4), 363–377 (2001)
232. Papadopoulos, H.T., Heavey, C.: Queueing theory in manufacturing systems analysis and design: A classification of models for production and transfer lines. European Journal of Operational Research 92(1), 1–27 (1996)
233. Papadopoulos, H.T.: A field service support system using a queueing network model and the priority MVA algorithm. Omega 24(2), 195–203 (1996)
234. Kim, S.: The toll plaza optimization problem: Design, operations, and strategies. Transportation Research Part E-Logistics and Transportation Review 45(1), 125–137 (2009)
235. Kogan, Y.: Asymptotic expansions for large closed and loss queueing networks. Mathematical Problems in Engineering 8(4-5), 323–348 (2002)
236. Kunigahalli, R., Russell, J.S.: Sequencing for concrete placement using RAG-CAD data structure. Journal of Computing in Civil Engineering 9(3), 216–225 (1995)
237. Lazowska, E.D., Zahorjan, J., Graham, G.S., Sevcik, K.C.: Quantitative system performance: Computer system analysis using queueing network models. Prentice-Hall, Englewood Cliffs (1984)
238. Nahmias, S.: Fuzzy variables. Fuzzy Sets and Systems 1, 97–110 (1978)
239. Puterman, M.L.: Markov Decision Processes: Discrete Stochastic Dynamic Programming. John Wiley & Sons, Inc., Hoboken (2005)
240. Taha, H.A.: Operations Research: An Introduction, 7th edn. New Jersey. Prentice-Hall, New Jersey (2003)
241. Van, W.T., Vandaele, N.: Modeling traffic flows with queueing models: A review. ASIA-Pacific Journal of Operational Research 24(4), 435–461 (2007)
242. Xu, J., Zeng, Z.: Applying optimal control model to dynamic equipment allocation problem: case study of concrete-faced rockfill dam construction project. Journal of Construction Engineering and Management 137(7), 536–550 (2011)
243. Xu, J., Zhou, X.: Approximation based fuzzy multi-objective models with expected objectives and chance constraints: application to earth-rock work allocation. Information Sciences 238, 75–95 (2013)
244. Yapicioglu, H., Smith, A.E., Dozier, G.: Solving the semi-desirable facility location problem using bi-objective particle swarm. European Journal of Operational Research 177, 733–749 (2007)
245. Zhang, H., Li, H., Tam, C.M.: Permutation-based particle swarm optimization for resource-constrained project scheduling. J. Comput. Civ. Eng. 20(2), 141–149 (2006)

246. Zhang, H., Wang, J.Y.: Particle Swarm Optimization for Construction Site Unequal-Area Layout. Journal of Construction Engineering and Management 134, 739 (2008)
247. Andjel, E., López, F.J., Sanz, G.: Ergodicity of one-dimensional resource sharing systems. Stochastic Processes and their Applications 98(1), 1–22 (2002)
248. Bashiri, M., Badri, H.: A group decision making procedure for fuzzy interactive linear assignment programming. Expert Systems with Applications 38, 5561–5568 (2011)
249. Bellman, R., Zadeh, L.: Decision-making in a fuzzy environment. Management Science 17(4), 141–164 (1970)
250. Bendoly, E., Perry-Smith, J.E., Bachrach, D.G.: The perception of difficulty in project-work planning and its impact on resource sharing. Journal of Operations Management 28(5), 385–397 (2010)
251. Bogomolnaia, A., Moulin, H.: A new solution to the random assignment problem. Journal of Economic Theory 100(2), 295–328 (2001)
252. Burkard, D.: Selected topics on assignment problems. Discrete Applied Mathematics 123, 257–302 (2002)
253. da Silva, G.C., Bahiense, L., Ochi, L.S., Boaventura-Netto, P.O.: The dynamic space allocation problem: Applying hybrid GRASP and Tabu search metaheuristics. Computers & Operations Research 39(3), 671–677 (2012)
254. Dimou, C.K., Koumousis, V.K.: Genetic algorithms in competitive environments. Journal of Computing in Civil Engineering 17(3), 142–149 (2003)
255. Zadeh, L.A.: Fuzzy algorithm. Information and Control 12(2), 94–102 (1968b)
256. Zadeh, L.A.: Outline of a new approach to the analysis of complex systems and decision processes. IEEE Transactions on Systems Man and Cybernetics 3(1), 28–44 (1973)
257. Dubios, D., Prade, H.: Fuzzy Sets and Systems: Theory and Application. Academic Press, New York (1980)
258. Dubios, D., Prade, H.: Gradualness, uncertainty and bipolarity: Making sense of fuzzy sets. Fuzzy Sets and Systems 192(1), 3–24 (2012)
259. Erdoğan, G., Tansel, B.: A branch-and-cut algorithm for quadratic assignment problems based on linearizations. Computers & Operations Research 34, 1085–1106 (2007)
260. Feng, C.W., Liu, L.A., Burns, S.A.: Using genetic algorithms to solve construction time-cost trade-off problems. Journal of Computing in Civil Engineering-ASCE 11(3) (1997)
261. Gomar, J., Haas, C., Morton, D.: Assignment and allocation optimization of partially multiskilled workforce. Journal of Construction Engineering and Management-ASCE 128(2), 103–109 (2002)
262. Grefenstette, J.: Optimization of control parameters for genetic algorithms. IEEE Transaction on Systems Man and Cybernetics 16(1), 122–128 (1986)
263. Ho, S.Y., Lin, H.S., Liauh, W.H., Ho, S.J.: OPSO: Orthogonal particle swarm optimization and its application to task assignment problems. IEEE Transactions on Systems Man and Cybernetics Part A-Systems and Humans 38(2), 288–298 (2008)
264. Hoogendoorn, S., Bovy, P.: Dynamic user-optimal assignment in continuous time and space. Transportation Research 38, 571–592 (2004)
265. Kim, J.L., Ellis Jr., R.D.: Permutation-Based Elitist Genetic Algorithm for Optimization of Large-Sized Resource-Constrained Project Scheduling. Journal of Construction Engineering and Management-ASCE 134(11), 904–913 (2008)
266. Krokhmal, P.A., Pardalos, P.M.: Random assignment problems. European Journal of Operational Research 194(1), 1–17 (2009)
267. Lew, A., Mauch, H. (eds.): Dynamic Programming, A Computational Tool. SCI, vol. 38. Springer, Heidelberg (2007)

268. Li, R., Lee, E.: Fuzzy approaches to multicriteria de novo programs. Mathematical Analysis and Applications 153, 97–111 (1990)
269. Liu, L., Li, Y.: The fuzzy quadratic assignment problem with penalty: New models and genetic algorithm. Applied Mathematics and Computation 174, 1229–1244 (2006)
270. Liu, L.Z., Gao, X.: Fuzzy weighted equilibrium multi-job assighment problem and genetic algorithm. Applied Mathematical Modeling 33(10), 3926–3935 (2009)
271. Lo, H., Szeto, W.: A cell-based variational inequality formulation of the dynamic user optimal assignment problem. Transportation Research 36, 421–443 (2002)
272. Majundar, J., Bhunia, A.K.: Elitist genetic algorithm for assignment problem with imprecise goal. European Journal of Operational Research 177(2), 68–692 (2007)
273. Munkres, J.: Algorithms for the Assignment and Transportation Problems. Journal of the Society for Industrial and Applied Mathematics 5(1), 32–38 (1957)
274. Park, K., Kim, W.: A systolic parallel simulation system for dynamic traffic assignment: SPSS-DTA. Expert System with Applications 21, 217–227 (2001)
275. Samaddar, S., Kadiyala, S.S.: An analysis of interorganizational resource sharing decisions in collaborative knowledge creation. European Journal of Operational Research 170(1), 192–210 (2006)
276. Shirazi, A., Amindavar, H.: Channel assignment for cellular radio using extended dynamic programming. Internation Journal of Electronics and Communications 59, 401–409 (2005)
277. Srinivas, M., Patnaik, L.: Adaptive probabilities of crossover and mutation in genetic algorithms. IEEE Transaction on Systems, Man and Cybernatics 69(24), 408–422 (1994)
278. Vaziri, K., Carr, P., Nozick, L.: Project planning for construction under uncertainty with limited resources. Journal of Construction Engineering and Management-ASCE 133(4), 268–276 (2007)
279. Wilkins, D.: The Bathtub Curve and Product Failure Rate. The e-Magazine for the Reliability Professional 21 and 22 (2002)
280. Xu, J.P., Zhou, X.Y.: A class of multi-objective expected value decision-making model with birandom coefficients and its application to flow shop scheduling problem. Information Sciences 179, 2997–3017 (2009)
281. Xu, J., Zhou, X.: Fuzzy-Like Multiple Objective Decision Making. STUDFUZZ, vol. 263. Springer, Heidelberg (2011)
282. Xu, J.P., Yao, L.M.: Random-Like Multiple Objective Decision Making. Springer, Heidelberg (2011)
283. Markusen, A.: Fuzzy concepts, scanty evidence, policy distance: The case for rigour and policy relevance in critical regional studies. Regional Studies 37(6-7), 701–717 (2003)
284. Kacprzyk, J.: Multistage decision-making under fuzziness. Verlag TÜV Rheinland, Cologne (1983a)
285. Bellman, R.E., Zadeh, L.A.: Decision-making in a Fuzzy Environment. Management Science 17, 141–164 (1970)
286. Kacprzyk, J., Iwański, C.: A generalization of discounted multistage decision making and control via fuzzy linguistic quantifiers. International Journal of Control 45, 1909–1930 (1987)
287. Kacprzyk, J., Staniewski, P.: A new approach to the control of stochastic systems in a fuzzy environment. Archiwum Automatyki i Telemechaniki XXV, 443–444 (1980)
288. Kacprzyk, J., Staniewski, P.: Long-term inventory policy-making through fuzzy decision-making models. Fuzzy Sets and Systems 8, 117–132 (1982)

289. Kacprzyk, J.: Yager's probability of a fuzzy event in stochastic control under fuzziness. In: Sanchez, E., Gupta, M.M. (eds.) Proc. IFAC Symp. on Fuzzy Inf. Proc. of the Knowledge Represent. and Dec. Anal., Marseille, France, pp. 379–384. Pergamon Press, Oxford (1983b)
290. Zadeh, L.A.: The concept of a linguistic variable and its application to approximate reasoning-II. Information Sci. 8, 301–357 (1975b)
291. Kacprzyk, J., Staniewski, P.: Control of a deterministic system in a fuzzy environment over infinite horizon. Fuzzy Sets and Systems 10, 291–298 (1983)
292. Kacprzyk, J.: Yager's probability of a fuzzy event in stochastic control under fuzziness. In: Sanchez, E., Gupta, M.M. (eds.) Proc. IFAC Symp. on Fuzzy Inf. Proc. of the Knowledge Represent. and Dec. Anal., Marseille, France, pp. 379–384. Pergamon Press, Oxford (1983)
293. Zadeh, L.A.: Similarity relations and fuzzy orderings. Information Sci. 3, 177–200 (1971)
294. Zadeh, L.A.: Fuzzy sets as a basis for a theory of possibility. Fuzzy Sets and Systems 1, 3–28 (1978)
295. Dubios, D., Prade, H.: Fuzzy Sets and Systerms: Theory and Application. Academic Press, New York (1980)
296. Haack, S.: Deviant logic, fuzzy logic: beyond the formalism. University of Chicago Press, Chicago (1996)
297. Bellman, R.E., Zadeh, L.A.: Local and fuzzy logics, ERL Memo. M-584. University of California, Berkeley (1976); Epstein, D.: Modern Uses of Multiple-Valued Logics, D. Reidel, Dordrecht (IY77)
298. Lai, Y.J., Hwang, C.L.: Interactive fuzzy linearing programming. Fuzzy Sets and Systerms 45, 169–183 (1992)
299. Verdegay, J.L.: Fuzzy mathematical programming, pp. 231–256 (1982)
300. Verdegay, J.L.: S dual approach to solve the fuzzy linear programming problem. Fuzzy Sets and Systerms 14, 131–141 (1984)
301. Tanaka, H., Okuda, T., Asai, K.: On fuzzy mathematical programming. Journal of Cybernetics 3, 37–46 (1974)
302. Orlovski, S.A.: On programming with fuzzy constraints set. Kebernetes 6, 197–202 (1977)
303. Dubois, D., Prade, H.: The mean value of a fuzzy number. Fuzzy Sets and Systems 24, 279–300 (1987)
304. Campos, L., Verdegay, J.L.: Linear programming problems and ranking of fuzzy numbers. Fuzzy Sets and Systems 32, 1–11 (1989)
305. González, A.: A study of the ranking function approach through mean values. Fuzzy Sets and Systems 35, 29–41 (1990)
306. Yager, R.R.: A procedure for ordering fuzzy subsets of the unit interval. Information Sciences 24, 143–161 (1981)
307. Yager, R.R.: On the evaluation of uncertain courses of action. Fuzzy Optimization and Decision Making 1, 13–41 (2002)
308. Lu, M.: On crisp equivalents and solutions of fuzzy programming with different chance measures. Technical Report (2001)
309. DeFinetti, B.: Probability Theory. Wiley, New York (1974)
310. Fine, T.: Theories ol Probability. Academic Press, New York (1973)
311. Dempster, A.: Upper and lower probabilities induced by multi-valued mapping. Ann. Math. Statist. 38, 325–339 (1967)
312. Sharer, G.: A Mathematical Theory or Evidence. Princeton University Press, Princeton (1976)

313. Shorflifle, E.H.: A model ofinexact reasoning in medicine. Math. Biosciences 23, 351–379 (1975)
314. Duda, R.O., Hart, P.F., Nilsson, N.J.: Subjective Bayesian methods for rule-based inference systems. Stanford Research Institute Tech. Note 124, Stanford, CA (1976)
315. Bellman, R.E., Zadeh, L.A.: Local and fuzzy logics, ERL Memo. M-584. University of California, Berkeley (1976); Epstein, D.: Modern Uses of Multiple-Valued Logics. D. Reidel, Dordrecht (IY77)
316. Zimmermann, H.-J.: Using fuzzy sets in operational research. European Journal of Operational Research 13, 244–260 (1983)
317. Zimmermann, H.-J.: Fuzzy programming and linear programming with several objective functions. Fuzzy Sets and Systerms 1, 45–55 (1978)
318. Markowitz, H.: Portfolio selection. Journal of Finance 7, 77–91 (1952)
319. Markowitz, H.: Portfolio Selection: Efficient Diversification of Investments. Wiley, New York (1959)
320. Dubios, D., Prade, H.: Possibility theory. Plenum Press, New York (1988)
321. Tanaka, H., Guo, P., Turksen, I.B.: Portfolio selection based on fuzzy probabilities and possibility distributions. Fuzzy Sets and Systems 111, 387–397 (2000)
322. Inuiguchi, M., Tanino, T.: Portfolio selection under independent possibilistic information. Fuzzy Sets and Systems 115, 83–92 (2000)
323. Parra, M.A., Terol, A.B., Uría, M.V.R.: A Fuzzy Goal Programming Approach to Portfolio Selection. European Journal of Operational Research 133, 287–297 (2001)
324. Zhang, W.G., Nie, Z.K.: On admissible efficient portfolio selection problem. Applied Mathematics and Computation 159, 357–371 (2004)
325. Ramaswamy, S.: Portfolio selection using fuzzy decision theory. Working Paper of Bank for International Settlements (59) (1998)
326. Fang, Y., Lai, K.K., Wang, S.Y.: Portfolio rebalancing model with transaction costs based on fuzzy decision theory. European Journal of Operational Research 175, 879–893 (2006)
327. Huang, X.X.: Fuzzy chance-constrained portfolio selection. Applied Mathematics and Computation 177, 500–507 (2006)
328. Perez Gladish, B., Jones, D.F., Tamiz, M., Bilbao Terol, A.: An interactive three-stage model for mutual funds portfolio selection. Omega 35, 75–88 (2007)
329. Sakawa, M., Yano, H.: An interactive fuzzy satisfying method for generalized multiobjective linear programming problems with fuzzy parameters. Fuzzy Sets and Systems 35, 125–142 (1990)
330. Wang, J., Hwang, W.L.: A fuzzy set approach for R&D portfolio selection using a real options valuation model. Omega 35, 247–257 (2007)
331. Wang, S.Y., Xia, Y.S.: Portfolio Selection and Asset Pricing. Springer, Berlin (2002)
332. Pareto, V.: Manuale di Economica Polittica. Societa Editrice Libraia, Milan (1906)
333. Xu, J., Liu, Q., Wang, R.: A class of multi-objective supply chain networks optimal model under random fuzzy enviroment and its application to the industry of Chinese liquor. Information Sciences 178, 2022–2043 (2008)
334. Holland, H.: Adaption in Natural and Artifical Systems. University of Michigan, Ann. Arbor (1975)
335. Ishibuchi, H., Murata, T.: A multiobjective genetic local search algorithm and its application to flowshop scheduling. IEEE Transactions on Systems, Man and Cybernetics 28(3), 3403–3922 (1969)
336. Goldberg, D.E.: Genetic Algorithms in Search, Optimization and Machine Learning. Addison-Wesley, New York (1989)

337. Dubois, D., Prade, H.: Operations on fuzzy numbers. International Journal of System Sciences 9, 613–626 (1978)
338. Liu, B.: Random fuzzy dependent-chance programming and its hybrid intelligent algorithm. Information Sciences 141, 259–271 (2002)
339. Liu, B.: Theory and practice of uncertain programming. Physica (2002)
340. Wei, Q., Yan, H.: Generalized Optimization Theory and models. Science Publishing Company, Beijing (2003)
341. Luhandjula, M.K.: Optimisation under hybrid uncertainty. Fuzzy Sets and Systems 146, 187–203 (2004)
342. Stock, J.R.: Reverse logistics (Council of Logistics Management) (1992)
343. Stock, J.R.: Development and Implementation of Reverse Logistics Programs (Council of Logistics Management) (1998)
344. Spengler, T., Puchert, H., Penkuhn, T., Rentz, O.: Environmental integrated introduction and recycling management. European Journal of Operation Research 2, 308–326 (1997)
345. Barros, A.I., Dekekr, R., Scholten, V.: A two-level network for recycling sand: a case study. European Journal of Operation Research 110, 199–215 (1998)
346. Marin, A., Pelegrin, B.: The return plant location problem: modeling and resolution. European Journal of Operational Research 104,375–392 (1998)
347. Krikkle, H.R., Van Harten, A., Schuur, P.C.: Business case one: reverse logistics redesign for copiers. OR Spectrum 3, 381–409 (1999)
348. Jayaraman, V., Patterson, R.A., Rolland, E.: The design of reverse distribution networks: Models and solution procedures. European Journal of Operational Research 150, 128–149 (2003)
349. Ovidiu Liste, A.: generic stochastic model for supply-and-return network design. Computers & Operations Research 34, 417–442 (2007)
350. Salema, M.I.G., Barbosa-Povoa, A.P., Novais, A.Q.: An optimization model for the design of a capacitated multi-product reverse logistics network with uncertainty. European Journal of Operational Research 179, 1063–1077 (2007)
351. Felischmann, M., Bloemhof-Ruwaard, J.M., Dekker, R., van der Laan, E., van Nunen, J.A.E.E., Van Wassenhove, L.N.: Quantitative models for reverse logistics: A review. European Journal of Operational Research 103, 1–17 (1997)
352. Ko, H.J., Evans, G.W.: A genetic algorithm-based heuristic for the dynamic integrated forward/reverse logistics network for 3PLs. Computers & Operations Research 34, 346–366 (2007)
353. Xu, J., Li, J.: Multiple Objective Decision Making Theory and Methods. Tsinghua University Press, Beijin (2005) (in Chinese)
354. Xu, J., Yao, L.: A class of multiobjective linear programming model with random rough coeffients. Mathematical and Computer Moddelling 49, 189–206 (2009)
355. Chen, M.L., Lu, Y.Z.: A novel elitist multiobjective optimization algorithm: Multiobjective extremal optimization. European Journal of Operational Research 188, 637–651 (2008)
356. Kim, K., Gen, M., Kim, M.: Adaptive genetic algorithms for multi-recource constrained project scheduling problem with multiple modes. International Journal of Innovative Computing, Information and Control 2(1), 41–49 (2006)
357. Mendesa, J., Goncalvesb, J., Resendec, M.: A random key based genetic algorithm for the resource constrained project scheduling problem. Computers & Operations Research 36, 92–109 (2007)
358. Ke, H., Liu, B.: Project scheduling problem with mixed uncertainty of randomness and fuzziness. European Journal of Operational Research 183, 135–147 (2007)
359. Holland, J.: Adaptation in Natural and Artificial. University of Michigan Press (1975)

360. Kumar, A., Pathak, R., et al.: A genetic algorithm for distributed system topology design. Computers and Industrial Engineering 28, 659–670 (1995)
361. Blazewicz, J., Lenstra, J., Rinnooy, K.: Scheduiling subject to resource constrains: Classification & complexity. Discrete Applied Mathematics 5, 11–24 (1983)
362. Jozefowska, J., Mika, M., et al.: Solving the discrete-containuous project scheduling problem via its discretization. Mathematical Methods of Operations Research 52, 489–499 (2000)
363. Yun, Y., Gen, M.: Advanced scheduling problem using constrained programming techniques in scm environment. Computer & Industrial Engineering 43, 213–229 (2002)
364. Michalewicz, Z.: Genetic Algorithm + Data Structure = Evolution Programs, 3rd edn. Springer, New York (1996)
365. Mak, K., Wong, Y., Wang, X.: An adaptive genetic algorithm for manufacturing cell formation. International Journal of Manufacturing Technology 16, 491–497 (2000)
366. Michael, J.M., Saad, H.A.: Proposed genetic algorithms for construction site layout. Engineering Applications of Artificial Intelligence 16, 501–509 (2003)
367. Li, S.L., Nair, K.P.K.: Fuzzy models for single-period inventory problem. Fuzzy Sets and Systems 132, 273–289 (2002)
368. Zimmermann, H.: Fuzzy linear programming with several objective functions. Fuzzy Sets and System 1, 46–55 (1978)
369. Chen, E.J., Lee, L.H.: A multi-objective selection procedure of determining a Pareto set. Computers & Operations Research 36, 1872–1879 (2009)
370. Youness, Y.E.: European Journal of Operational Research 81, 440–443 (1995)
371. Hu, Y.: Practical Multiobjective Optimization. Shanghai Science and Technology Press, Shanghai (1990) (in Chinese)
372. Lin, C., Dong, J.: Methods and Theort for Multiobjective Optimization. Jilin Education Press, Changchun (1992) (in Chinese)
373. Nakayama, H., Sawaragi, Y., Tanino, T.: Theory of Multiobjective Optimization. Academic Press, Inc. (1985)
374. Luc, D.T.: On the domination property in vector optimization. Optim. Theory Appl. 43, 327–330 (1984)
375. Ijiri, Y.: Management Goals and Accounting for Control. Rand McNally, Chicago (1965)
376. Szidarovszky, F., Gershon, M.E., Duckstein, L.: Techniques for Multiobjective Decision Making in Systems Management. Elsevier, Amsterdam (1986)
377. Werners, B.: Interactive multiple objective programming subject to flexible constraints. European Journal of Operational Research 31, 342–349 (1987)
378. Leberling, H.: On finding compromise solution in multicriteria problems using the fuzzy min-operator. Fuzzy Sets and Systems 6, 105–118 (1981)
379. Zimmermann, H.J.: Fuzzy programming and linear programming with several objective functions. Fuzzy Sets and Systems 1, 45–55 (1978)
380. Wierzbicki, A.: Multiple Creteria Decision Making Theory and Application. Springer, Berlin (1980)
381. Shimizu, K.: Theory of Multiobjective and Conflict. Kyoritsu Syuppan (1982) (in Japanese)
382. Choo, E.V., Atkins, D.R.: An interactive algorithm for multicriteria programming. Computers and Operations Research 7, 81–88 (1980)
383. Teargny, J., Benayoun, R., Montgolfier, J., Larichev, O.: Linear programming with multiobjective functions: Step Method (STEM). Math. Programming. 1(2), 366–375 (1971)
384. Lee, S.M.: Goal Programming for Decision Analysis. Auerbach Publishers, Philadelphia (1972)

385. Kendall, K.E., Lee, S.M.: Formulating blood rotation policies with multiple objectives. Management Sciences 26(11), 1145–1157 (1980)
386. Sakawa, M.: Interactive fuzzy goal programming for multiobjective nonlinear programming problems and its applications to water quality management. Control and Cybernetics 13, 217–228 (1984)
387. Sakawa, M., Yumine, T., Yano, Y.: An interactive fuzzy satisfying method for multiobjective linear programming problems and its applications. IEEE Transaction on Systems, Man, and Cybernetics 17, 654–661 (1987)
388. Sakawa, M., Yano, Y.: Interactive fuzzy decision making for multiobjective nonlinear programming using augmented minimax problems. Fuzzy Sets and Systems 20, 31–43 (1986)
389. Sakawa, M., Seo, F.: Multiple Criteria Decision Analysis in Regional Planning-Concepts, Methods and Applications. D. Reidel, Dordrecht (1988)
390. Bazaraa, M.S., Shetty, C.M.: Nonlinear Programming: Theory and Algorithms. Wiley, New York (1979)
391. Li, J., Xu, J.P., Gen, M.: A class of fuzzy random multiobjective programming problem. Mathematical and Computer Modelling 44, 1097–1113 (2006)
392. Liu, B.: Uncertainty Theory: An Introduction to its Axiomatic Foundations. Springer, Berlin (2004)
393. Yoshida, Y.: The valuation of European options in uncertain environment. European Journal of Operational Research 145, 221–229 (2003)
394. Wu, H.C.: Pricing European options based on the fuzzy pattern of Black-Scholes formula. Computers & Operations Research 31, 1069–1081 (2004)
395. Zmeskal, Z.: Application of the fuzzy-stochastic methodology to appraising the firm value as a european call option. European Journal of Operational Research 135, 303–310 (2001)
396. Zmeskal, Z.: Value at risk methodology under soft conditions approach (fuzzy-stochastic approach). European Journal of Operational Research 161, 337–347 (2005)
397. Luhandjula, M.K., Gupta, M.M.: On fuzzy stochastic optimization. Fuzzy Sets and Systems 81, 41–55 (1996)
398. Feng, Y.H., Hu, L.J., Shu, H.S.: The variance and convariance of fuzzy random variables and their applications. Fuzzy Sets and Systems 120, 487–497 (2001)
399. Korner, R.: On the variance of fuzzy random variable. Fuzzy Sets and System 92, 83–93 (1997)
400. de Campos, L.M., González, A.: A subjective approach for ranking fuzzy numbers. Fuzzy Sets and Systems 29, 145–153 (1989)
401. López-Díaz, M., Gil, M.A.: The $\lambda$-average value and the fuzzy expectation of a fuzzy random variable. Fuzzy Sets and Systems 99, 347–352 (1998)
402. Liu, B.: Theory and Practice of Uncertain Programming. Physica-Verlag, Heidelberg (2002)
403. Liu, B.: Uncertainty Theory: An Introduction to its Axiomatic Foundations. Springer, Berlin (2004)
404. Keown, A.J., Martin, J.D.: A chance constrained goal programming model for working capital management. The Engineering Economists 22, 153–174 (1977)
405. Charnes, A., Cooper, W.W.: Chance-constrained programming. Management Science 6, 73–79 (1959)
406. Xu, J., Li, J.: A novel portfolio selection model in hybrid uncertain environment. Omega (2008)
407. Ammar, E., Khalifa, H.A.: Fuzzy portfolio optimization a quadratic programming approach. Chaos, Solitons and Fractals 18, 1045–1054 (2003)

408. Antczak, T.: A modified objective function method for solving nonlinear multiobjective fractional programming problems. J. Math. Anal. Appl. 322, 971–989 (2006)
409. Ben-Israel, A., Robers, P.D.: A Decomposition method for interval linear programming. Operations Research 21, 1154–1157 (1973)
410. Bitran, G.R.: Linear multiple objective problems with interval coefficient. Management Science 26, 694–706 (1980)
411. Black, F., Scholes, M.: The pricing of options and corporate liabilities. Journal of Political Economy 81, 637–654 (1973)
412. Buckley, J.J.: Stochastic versus possibilistic programming. Fuzzy Sets and Systems 34, 173–177 (1990)
413. Carlsson, C., Fuller, R.: On possibilistic mean value and variance of fuzzy numbers. Fuzzy Sets and Systems 122, 315–326 (2001)
414. Campos, L.M., Gonzalez, A.: A subjective approach for ranking fuzzy numbers. Fuzzy Sets and Systems 29, 143–153 (1989)
415. Chakraborty, M., Gupta, S.: Fuzzy mathematical programming for multi-objective linear fractional programming problem. Fuzzy Sets and Systems 125, 335–342 (2002)
416. Chanas, S., Kuchta, D.: Multiple objectiveprogramming in optimization of the interval objective function—a generalized approach. European Journal of Operational Research 94, 594–598 (1996)
417. Chang, T.J., Meade, N., Beasley, J.B., Sharaiha, Y.: Heuristic for cardinality constrained portfolio optimization. Computers and Operations Research 27, 1271–1302 (2000)
418. Chankong, V., Haimes, Y.Y.: Multiobjective Decision Making: Theory and Methodology. North-Holland, New York (1983)
419. Charnes, A., Cooper, W.W.: Chance-constrained programming. Management Science 6(1), 73–79 (1959)
420. Charnes, A., Cooper, W.W.: Management Models and Application of Linear Programming. Wiley, New York (1961)
421. Charnes, A., Granot, F., Philips, F.: An algorithm for solving interval linear programming probelms. Operations Research 25, 688–695 (1977)
422. Chen, M.S., Yao, J.S., Lu, H.F.: A fuzzy stochastic single-period model for cash management. European Journal of Operational Research 170, 72–90 (2006)
423. Chen, Y.J., Liu, Y.K.: Portfolio selection in fuzzy environment. In: Proceeedings of the Fourth International Conference on Machine Learning and Cybernetics, Guangzhou, pp. 2694–2699 (2005)
424. Chiodi, L., Mansini, R., Speranza, M.G.: Semi-absolute deviation rule for mutual funds portfolio selection. Annals of Operations Research 124, 245–265 (2003)
425. Colubi, A., Domonguez-Menchero, J.S., Lopez-Diaz, M., Ralescu, D.A.: On the formalization of fuzzy random variables. Information Sciences 133, 3–6 (2001)
426. Costa, J.P.: Computing non-dominated solutions in MOLFP. European Journal of Operational Research 181(3), 1464–1475 (2007)
427. Deb, K.: Multi-Objective Optimization Using Evolutionary Algorithms. John Wiley, New York (2001)
428. Doerner, K., Gutjahr, W.J., Hartl, R.F., Strauss, C., Stummer, C.: Pareto ant colony optimization: a metaheuristic approach to multiobjective portfolio Selection. Annals of Operations Research 131, 79–99 (2004)
429. Elton, E.J., Gruber, M.J.: The multi-period consumption investment problem and single period analysis. Oxford Economics Papers 9, 289–301 (1974)
430. Elton, E.J., Gruber, M.J.: On the optimality of some multiperiod portfolio selection criteria. Journal of Business 7, 231–243 (1974)

431. Feng, Y., Hu, L., Shu, H.: The variance and covariance of fuzzy random variables and their appliactions. Fuzzy Sets and Systems 120, 487–497 (2001)
432. Fonseca, C., Fleming, P.: An overview of evolutionary algorithms in multiobjective optimization. Evolutionary Computation 3(1), 1–16 (1995)
433. Fuller, R., Majlender, P.: On weighted possibilistic mean and variance of fuzzy numbers. Fuzzy Sets and Systems 136, 363–374 (2003)
434. Gao, J., Liu, B.: New primitive chance measures of fuzzy random event. International Journal of Fuzzy Systems 3(4), 527–531(2001)
435. Gao, Y., Zhang, G.Q., Lu, J., Wee, H.M.: Particle swarm optimization for bi-level pricing problems in supply chains. Journal of Global Optimization 51(2), 245–254 (2011)
436. Gen, M., Cheng, R.: Genetic Algorithms & Engineering Optimization. John Wiley & Sons, New York (2000)
437. Gil, M.A., Lopez-Diaz, M.: The $\lambda$-average value and the fuzzy expectation of a fuzzy random variable. Fuzzy Sets and Systems 99, 347–352 (1998)
438. Gil, M.A., Lopez-Diaz, M., Ralescu, D.A.: Overview on the development of fuzzy random variables. Fuzzy Sets and Systems 157, 2546–2557 (2006)
439. Giove, S., Funari, S., Nardelli, C.: An interval portfolio selection problem based on regret function. European Journal of Operational Research 170, 253–264 (2006)
440. Gladish, B.P., Parra, M.A., Terol, A.B., Uria, M.V.R.: Solving a multiobjective possibilistic problem through compromise programming. European Journal of Operational Research 164, 748–759 (2005)
441. Hamza, F., Janssen, J.: The mean-semivariances approach to realistic portfolio optimization subject to transaction costs. Applied Stochastic Models and Data Analysis 14, 275–283 (1998)
442. Horn, J.: Multicriterion Decision Making. Oxford University Press, New York (1997)
443. Huang, J.J., Tzeng, G.H., Ong, C.S.: A novel algorithm for uncertain portfolio selection. Applied Mathematics and Computation 173, 350–359 (2006)
444. Huang, X.: Fuzzy chance-constrained portfolio selection. Applied Mathematics and Computation 177, 500–507 (2006)
445. Alefeld, G., Herzberger, J.: Introduction to Interval Computations. Academic Press, New York (1983)
446. Hansen, E.: Global Optimization Using Interval Analysis. Marcel Dekker, New York (1992)
447. Halmos, P.R.: Measure theory. Van Nostrand, Princeton (1950) (republished by Spinger in 1974)
448. Ichihashi, H., Inuiguchi, M., Kume, Y.: Modality constrained programming models: a unified approach to fuzzy mathematical programming problems in the setting of possibilisty theory. Information Sciences 67, 93–126 (1993)
449. Ida, M.: Portfolio selelction problem with interval coefficients. Applied Mathematics Letters 16, 709–713 (2003)
450. Ida, M.: Solutions for the portfolio selection problem with interval and fuzzy coefficients. Reliable Computing 10, 389–400 (2004)
451. Ignizio, J.P.: Linear Programming in Single & Multi-Objective Systems. Prentice-Hall, Englewood Cliffs (1982)
452. Inuiguchi, M., Kume, Y.: Goal programming problems with interval coefficients and target intervals. European Journal of Operational Reseach 52, 345–360 (1991)
453. Inuiguchi, M., Ramik, J.: Possibilistic linear programming: A brief review of fuzzy mathematical programming and a comparison with stochastic programming in portfolio selection problem. Fuzzy Sets and Systems 111, 3–28 (2000)

454. Inguiguchi, M., Ramik, J., Tanino, T., Vlach, M.: Satisficing solutions and duality in interval and fuzzy linear programming. Fuzzy Sets and Systems 135, 151–177 (2003)
455. Inuiguchi, M., Sakawa, M.: Minimax regret solution to linear programming problems with an interval objective function. European Journal of Operational Research 86, 526–536 (1995)
456. Inuiguchi, M., Tanino, T.: Portfolio selection under independent possibilistic information. Fuzzy Sets and Systems 115, 83–92 (2000)
457. Ishibuchi, H., Tanaka, H.: Formulation and analysis of linear programming problems with interval coefficients. J. Jpn. Ind. Manage. Assoc. 40, 320–329 (1989)
458. Ishibuchi, H., Tanaka, H.: Multiobjective programming in optimization of the interval objective function. European Journal of Operational Research 48, 219–225 (1990)
459. Itoh, T., Ishii, H.: One machine scheduling problem with fuzzy random due-dates. Fuzzy Optimization and Decision Making 4, 71–78 (2005)
460. Feng, G.: Analysis and Synthesis of Fuzzy Control Systems: A Model-Based Approach. CRC Press, Taylor & Francis Group, New York (2010)
461. Chih, H.H.: Optimization of fuzzy production inventory models. Information Sciences 146, 29–40 (2002)
462. Roy, T.K., Maiti, M.: Multi-objective inventory models of deterioration items with some constraints in a fuzzy envirinment. Computers and Operations Researchs, 1085–1095 (1998)
463. Luhandjula, M.K., Gupta, M.M.: On fuzzy stochastic optimization. Fuzzy Sets and Systems 81, 41–55 (1996)
464. Feng, Y.H., Hu, L.J., Shu, H.S.: The variance and convariance of fuzzy random variables and their applications. Fuzzy Sets and Systems 120, 487–497 (2001)
465. López-Díaz, M., Gil, M.A.: The $\lambda$-average value and the fuzzy expectation of a fuzzy random variable. Fuzzy Sets and Systems 99, 347–352 (1998)
466. Korner, R.: On the variance of fuzzy random variable. Fuzzy Sets and System 92, 83–93 (1997)
467. de Campos, L.M., González, A.: A subjective approach for ranking fuzzy numbers. Fuzzy Sets and Systems 29, 145–153 (1989)
468. Katagiri, H., Ishii, H.: Fuzzy portfolio selection problem. In: IEEE SMC 1999 Conference Proceedings, vol. 3, pp. 973–978 (1999)
469. Katagiri, H., Ishii, H.: Linear programming problem with fuzzy random constraint. Mathematica Japonica 52, 123–129 (2000)
470. Katagiri, H., Sakawa, M., Ishii, H.: Fuzzy random bottleneck spanning tree problems using possibility and necessary measures. European Journal of Operational Research 152, 88–95 (2004)
471. Katagiri, H., Sakawa, M., Kato, K., Nishizaki, I.: A fuzzy random multiobjective 0-1 programming based on the expectation optimization model using possibility and necessary measures. Mathematical and Computer Modeling 40, 411–421 (2004)
472. Kataoka, S.: A stochastic programming model. Econometrica 31, 181–196 (1963)
473. Kato, K., Katagiri, H., Sakawa, M., Ohsaki, S.: An interactive fuzzy satisficing method based on the fractile optimization model using possibility and necessity measures for a fuzzy random multiobjective linear programming problem. Electronics and Communications in Japan 88(5), 20–28 (2005)
474. Feng, Y.H., Hu, L.J., Shu, H.S.: The variance and covariance offuzzy random variables and their applications. Fuzzy Sets and Systems 120, 487–497 (2001)
475. Kaufmann, A.: Introduction to the Theory of Fuzzy Subsets, vol. I. Academci Press, New York (1975)

476. Kita, H., Tamaki, H., Kobayashi, S.: Multiobjective optimization by genetic algorithms: A review. In: Fogel, D. (ed.) Proceedings of the IEEE International Conference on Evolutionary Computation, pp. 517–522. IEEE Press, Piscataway (1996)
477. Konno, K., Yamazika, H.: Mean absolute deviation portfolio optimization model and its application to Tokyo stock market. Mangement Science 37, 519–531 (1991)
478. Korner, R.: On the variance of fuzzy random variables. Fuzzy Sets and Systems 92, 83–93 (1997)
479. Kruse, R., Meyer, K.D.: Statistics with Vague Data. Reidel Publishing Company, Dordrecht (1987)
480. Lai, K.K., Wang, S.Y., Xu, J.P., Fang, Y.: A class of linear interval programming problems and its applications to portfolio selection. IEEE Transaction on Fuzzy Systems 10, 698–704 (2002)
481. Li, R.J., Lee, E.S.: Ranking fuzzy numbers-A comparison. In: Proceedings of North American Fuzzy Information Processing Society Workshop, West Lafayette, IL, pp. 169–204 (1987)
482. Liu, B.: Fuzzy random chance-constraint programming. IEEE Transactions on Fuzzy Systems 9(5), 713–720 (2001)
483. Liu, B.: Fuzzy random dependent-chance programming. IEEE Transactions on Fuzzy Systems 9(5), 721–726 (2001)
484. Liu, B.: Theory and Practice of Uncertain Programming. Physica Verlag, Heidelberg (2002)
485. Liu, B., Iwamura, K.: Chance constrained programming with fuzzy parameters. Fuzzy Sets and Systems 94, 227–237 (1998)
486. Liu, B., Iwamura, K.: A note on chance constrained programming with fuzzy coefficients. Fuzzy Sets and Systems 100, 229–233 (1998)
487. Liu, W.A., Zhang, W.G., Wang, Y.L.: On admissible efficient portfolio selection: Models and algorithms. Applied Mathematics and Computation 176, 208–218 (2006)
488. Liu, Y.K.: Convergent results about the use of fuzzy simulation in fuzzy optimization problems. IEEE Transactions on Fuzzy Systems 14(2), 295–304 (2006)
489. Liu, Y.K., Liu, B.: A class of fuzzy random optimization: Expected value models. Information Sciences 155, 89–102 (2003)
490. Liu, Y.K., Liu, B.: Fuzzy random variables: A scalar expected value operator. Fuzzy Optimization and Decision Making 2, 143–160 (2003)
491. Liu, Y.K., Liu, B.: On minimum-risk problems in fuzzy random decision systems. Computers & Operations Research 32, 257–283 (2005)
492. Lopez-Diaz, M., Angeles Gil, M.: Constructive definitions of fuzzy random variables. Stochastics & Probability Letters 36, 135–143 (1997)
493. Hiai, F., Umegaki, H.: Integrals, conditional expectations and martingales of multivalued functions. J. Multivar. Anal. 7, 149–182 (1977)
494. Feng, Y.: Mean-square integral and differential of fuzzy stochastic processes. Fuzzy Sets and Systems 102, 271–280 (1999)
495. Himmelberg, C.J.: Measurable relations. Fund. Math. 87, 53–72 (1975)
496. Luhandjula, M.K.: Linear programming under randomness and fuzziness. Fuzzy Sets and Systems 10, 45–55 (1983)
497. Luhandjula, M.K.: Fuzziness and randomness in an optimization framework. Fuzzy Sets and Systems 77, 291–297 (1996)
498. Luhandjula, M.K., Gupta, M.M.: On fuzzy stochastic optimization. Fuzzy Sets and Systems 81, 47–55 (1996)
499. Luhandjula, M.K.: Optimisation under hybrid uncertainty. Fuzzy Sets and Systems 146, 187–203 (2005)

500. Luhandjula, M.K.: Fuzzy stochastic linear programming: Survey and future research directions. European Journal of Operational Research 174, 1353–1367 (2006)
501. Markowitz, H.: Portfolio selection. Journal of Finance 7, 77–91 (1952)
502. Markowitz, H.: Portfolio Selection: Efficient Diversification of Investments. John Wiley & Sons, New York (1959)
503. Matlab. Optimization Toolbox for Use with Matlab, Version 7.0.1 (R14). Mathworks, Inc., Natick (2004)
504. Merton, R.C.: Lifetime portfolio selection under uncertainty: The continuous time case. Review of Economics and Statistics 51, 247–257 (1969)
505. Mausser, H.E., Laguna, M.: A heuristic to minimax absolute regret for linear programs with interval objective function coefficients. European Journal of Operational Research 117, 157–174 (1999)
506. Merton, R.C.: The theory of rational option pricing. Bell Journal of Economics and Management Science 4, 141–183 (1973)
507. Metev, B.: Use of reference points for solving MONLP problems. European Journal of Operational Research 80, 193–203 (1995)
508. Metev, B., Gueorguieva, D.: A simple method for obtaining weakly efficient points in multiobjective linear fractional programming problems. European Journal of Operational Research 126, 386–390 (2000)
509. Michalewicz, Z.: Genetic Algorithms + Data Structures = Evolution Programs. Springer, New York (1994)
510. Mohan, C., Nguyen, H.T.: An interactive satisfying method for solving mixed fuzzy-stochastic programming problems. Fuzzy Sets and Systems 117, 67–79 (2001)
511. Moore, R.E.: Method and Applications of Interval Analysis. SIAM, Philadelphia (1979)
512. Negoita, C.V., Ralescu, D.: Simulation, Knowledge-Based Computing, and Fuzzy Statistics. Van Nostrand Reinhold, New York (1987)
513. Nguyen, V.H.: Fuzzy stochastic goal programming problems. European Journal of Operational Research 176, 77–86 (2007)
514. Ogryczak, W.: Multiple criteria linear programming model for portfolio selection. Annals of Operations Research 97, 143–162 (2000)
515. Perold, A.F.: Large-scale portfolio optimization. Management Science 31(10), 1143–1159 (1984)
516. Pratap, A., Deb, K., Agrawal, S., Meyarivan, T.: A fast and elitist multi-objective genetic algorithm: NSGA-II. IEEE Transactions on Evolutionary Computation 6(2), 182–197 (2002)
517. Qi, Y., Hirschberger, M., Steuer, R.E.: Tri-criterion quadratic-linear programming. Technical report, Working paper, Department of Banking and Finance, University of Georgia, Athens (2006)
518. Qi, Y., Steuer, R.E., Hirschberger, M.: Suitable-portfolio investors, nondominated frontier sensitivity, and the effect of multiple objectives on standard portfolio selection. Annals of Operations Research 152(1), 297–317 (2007)
519. Qiao, Z., Wang, G.: On solutions and distributions problems of the linear programming with fuzzy random variable coefficients. Fuzzy Sets and Systems 58, 155–170 (1993)
520. Qiao, Z., Zhang, Y., Wang, G.: On fuzzy random linear programming. Fuzzy Sets and Systems 65, 31–49 (1994)
521. Ramaswamy, S.: Portfolio selection using fuzzy decision theory. Working Paper of Bank for International Settlements (59) (1998)
522. Ross, S.A.: The arbitrage theory of capital asset pricing. Journal of Economic Theory 13, 341–360 (1976)

523. Roubens, M., Teghem, J.: Comparisons of methodologies for fuzzy and stochastic multiobjective programming problems. Fuzzy Sets and Systems 42, 119–132 (1991)
524. Roy, A.D.: Safety-first and the holding of assets. Econometrics 20, 431–449 (1952)
525. Sakawa, K.: Fuzzy Sets and Interactive Multiobjective Optimization. Plenum Press, New York (1993)
526. Sawaragi, Y., Nakayama, H., Tanino, T.: Theory of Multiobjective Optimization. Academic Press, New York (1985)
527. Schaerf, A.: Local search techniques for constrained portfolio selection problems. Computational Economics 20, 177–190 (2002)
528. Schrage, L.: LINGO User's Guide. Lindo Publishing, Chicago (2004)
529. Sengupta, A., Pal, T.K.: On comparing interval numbers. European Journal of Operational Research 127, 28–43 (2000)
530. Sengupta, A., Pal, T.K., Chakraborty, D.: Interpretation of inequality constraints involving interval coefficients and a solution to interval linear programming. Fuzzy Sets and Systems 119, 129–138 (2001)
531. Sharpe, W.: Capital asset prices: a theory of market equilibrium under conditions of risk. Journal of Finance 19, 425–442 (1964)
532. Shih, C.J., Wangsawidjaja, R.A.S.: Mixed fuzzy-probabilistic programming approach for multiobjective engineering optimization with random variables. Computers and Structures 59(2), 283–290 (1996)
533. Shing, C., Nagasawa, H.: Interactive decision system in stochastic multiobjective portfolio selection. Int. J. Production Economics 60-61, 187–193 (1999)
534. Shukla, P.K., Deb, K.: On finding multiple pareto-optimal solutions using classical and evolutionary generating methods. European Journal of Operational Research 181(3), 1630–1652 (2007)
535. Siddharha, S.S.: A dual ascent method for the portfolio selection problem with multiple constraints and linked proposals. European Journal of Operational Research 108, 196–207 (1998)
536. Slowinski, R., Teghem, J.: Stochastic versus Fuzzy Approaches to Multiobjective Mathematical Programming under Uncertainty. Kluwer Academci Publishers (1990)
537. Stancu-Minasian, I.M.: Stochastic Programming with Mutiple Objective Functions. Rediel, Dordrecht (1984)
538. Stancu-Minasiana, I.M., Pop, B.: On a fuzzy set approach to solving multiple objective linear fractional programming problem. Fuzzy Sets and Systems 134, 397–405 (2003)
539. Steuer, R.E.: Mulitple objective linear programming with interval criterion weights. Management Science 23, 305–316 (1976)
540. Steuer, R.E., Qi, Y., Hirschberger, M.: Developments in Multi-Attribute Portfolio Selection. Working Paper, Department of Banking and Finance. University of Georgia, Athens (2006)
541. Streichert, F., Ulmer, H., Zell, A.: Evolutionary algorithms and the cardinality constrained portfolio selection problem. In: Ahr, D., Fahrion, R., Oswald, M., Reinelt, G. (eds.) Operations Research Proceedings 2003, Selected Papers of the International Conference on Operations Research (OR 2003), pp. 253–260. Springer, Berlin (2003)
542. Sturm, J.: Using SeDuMi 1.02, a MATLAB Toolbox for Optimization Over Symmetric Cones. Optimization Methods and Software 11, 625–653 (1999)
543. Tanaka, H., Okuda, T., Asai, K.: On fuzzy mathematical programming. Journal of Cybernetics 3, 37–46 (1974)
544. Tanaka, H., Guo, P., Turksen, I.B.: Portfolio selection based on fuzzy probabilities and possibility distributions. Fuzzy Sets and Systems 111, 387–397 (2000)
545. Telser, L.G.: Safety first and hedging. Review of Economic Studies 23, 1–16 (1955)

546. Tiryaki, F., Ahlatcioglu, M.: Fuzzy stock selection using a new fuzzy ranking and weighting algorithm. Applied Mathematics and Computation 170, 144–157 (2005)
547. Ulmer, H., Streichert, F., Zell, A.: Evaluating a hybrid encoding and three crossover operations on the constrained portfolio selection problem. In: Congress of Evolutionary Computation (CEC 2004), pp. 932–939. IEEE Press, Portland (2004)
548. Vandenberghe, L., Boyd, S.: Semidefinite Programming. SIAM Review 38, 49–95 (1996)
549. Wang, S.Y., Zhu, S.S.: On Fuzzy Portfolio Selection Problem. Fuzzy Optimization and Decision Making 1, 361–377 (2002)
550. Wang, G., Qiao, Z.: Linear programming with fuzzy random variable coefficients. Fuzzy Sets and Systems 57, 3295–3311 (1993)
551. Wierzbicki, A.: A mathematical basis for satisficing decision making. In: Morse, J.N. (ed.) Organizations: Multiple Agents with Multiple Criteria, Proceedings, pp. 465–485. Springer, Berlin (1981)
552. Wierzbicki, A.: On the completeness and constructiveness of parametric characterization to vector optimization problems. OR Spektrum 8, 73–87 (1986)
553. Yazenin, A.V.: Fuzzy and stochastic programming. Fuzzy Sets and Systems 22, 171–188 (1987)
554. Williams, J.O.: Maximizing the probability of achieving investment goals. Journal of Portfolio Management 24, 77–81 (1997)
555. Yoshimoto, A.: The mean-variance approach to portfolio optimization subject to transaction costs. Journal of Operations Research Society of Japan 39, 99–117 (1996)
556. Zadeh, L.A.: The concept of a linguistic variable and its application to approximate reasoning-I. Information Science 8, 199–249 (1975a)
557. Zadeh, L.A.: Fuzzy sets as a basis for a theory of possibility. Fuzzy Sets and Systems 1, 3–28 (1978)
558. Zeleny, M.: Compromise programming in multiple criteria decision making. In: Cochrane, J.L., Zeleny, M. (eds.) University of South Carolina Press, Columbia (1973)
559. Kumar, A., Pathak, R., et al.: A genetic algorithm for distributed system topology design. Computers and Industrial Engineering 28, 659–670 (1995)
560. Mendesa, J., Goncalvesb, J., Resendec, M.: A random key based genetic algorithm for the resource constrained project scheduling problem. Computers & Operations Research 36, 92–109 (2009)
561. Ke, H., Liu, B.: Project scheduling problem with mixed uncertainty of randomness and fuzziness. European Journal of Operational Research 183, 135–147 (2007)
562. Blazewicz, J., Lenstra, J., Rinnooy, K.: Scheduiling subject to resource constrains: Classification & complexity. Discrete Applied Mathematics 5, 11–24 (1983)
563. Jozefowska, J., Mika, M., et al.: Solving the discrete-containuous project scheduling problem via its discretization. Mathematical Methods of Operations Research 52, 489–499 (2000)
564. Kim, K., Gen, M., Kim, M.: Adaptive genetic algorithms for multi-recource constrained project scheduling problem with multiple modes. International Journal of Innovative Computing, Information and Control 2(1), 41–49 (2006)
565. Yun, Y., Gen, M.: Advanced scheduling problem using constrained programming techniques in scm environment. Computer & Industrial Engineering 43, 213–229 (2002)
566. Jain, R.: Decision making in the presence of fuzzy variables. IEEE Trans. on Systens, Man and Cybernetics 6, 698–703 (1976)
567. Mizumoto, M., Tanaka, K.: Algebraic properties of fuzzy numbers. In: IEEE International Conference on Cybernatics and Society, pp. 559–563 (1976)

568. Mizumoto, M., Tanaka, K.: Algebraic properties of fuzzy numbers. In: Gupta, M.M. (ed.) Advances in Fuzzy Set Theory and Applications, pp. 153–164. North-Holland, New York (1979)
569. Baas, S.M., Kwakernaak, H.: Rating and ranking of multiple aspect alternative using fuzzy sets. Automatica 13, 47–58 (1977)
570. Lee, C.C.: Fuzzy logic in control systems: Fuzzy logic controller-Part I. IEEE Transactions on Systems Man and Cybernetics 92(2), 404–418 (1990)
571. Dubois, D., Prade, H.: Fuzzy Sets and System: Theory and Applications. Academic Press, New York (1980)
572. Cheng, C.M.: Multi-model fuzzy control for nonlinear systems. Ph.D. thesis, University of New South Wales, Australia (1998)
573. Zimmermann, H.J.: Fuzzy Set Theory and Its Application, 2nd edn. Kluwer Academic, Boston (1991)
574. Dijkman, J., Van Haeringen, H., Delange, S.J.: Fuzzy numbers. Journal of Mathematical Analysis and Applications 92(2), 302–341 (1983)
575. Gupta, M.M.: Fuzzy information in decision making. In: International Conference on Advances in Information Sciences and Thechnology, Golden Jubilee Conference at the Indian Conference at the Indian Statistical Institute, Calcutta, Indian (1982)
576. Kaufmann, A., Gupta, M.M.: Introduction to Fuzzy Arithmetic. Van Nostrand, New York (1985)
577. Laarhoven, P.J.M., Pedrycz, W.: A fuzzy extension of Satty's priority theory. Fuzzy Sets and Systems 11(3), 229–241 (1983)
578. Dubois, D., Prade, H.: The mean-value of a fuzzy number. Fuzzy Sets and Systems 24(3), 279–300 (1987)
579. Buckley, J.J.: The multiple-judge, multiple-criteria ranking problem: A fuzzy-set approach. Fuzzy Sets and Systems 13(1), 139–147 (1984)
580. Buckley, J.J.: Generalized and extended fuzzys sets with applications. Fuzzy Sets and Systems 25(2), 159–174 (1988)
581. Bonissone, P.P.: A fuzzy set based linguistic approach: Theory and applications. In: Proceedings of the 1980 Winter Simulation Conference, Orlando, Florida, pp. 99–111 (1980)
582. Bonissone, P.P.: A fuzzy set based linguistic approach: Theory and applications. In: Gupta, M.M., Sanchez, E. (eds.) Approximate Reasoning in Decision Making, pp. 329–339. North-Holland (1982)
583. Xu, J., Meng, J., Zeng, Z.Q., Wu, S.Y., Shen, M.B.: Resource sharing-based multiobjective multistage construction equipment allocation under fuzzy environment. Journal of Construction Engineering and Management 139(2), 161–173 (2013d)
584. Xu, J., Zeng, Z.Q., Han, B., Lei, X.: A dynamic programming-based particle swarm optimization algorithm for an inventory management problem under uncertainty. Engineering Optimization 45(7), 851–880 (2013a)
585. Xu, J., Li, Z.: Multi-objective dynamic construction site layout planning in fuzzy random environment. Automation in Construction 27, 155–169 (2012)
586. Xu, J., Liu, Q., Lei, X.: A fuzzy multi-objective model and application for the discrete dynamic temporary facilities location planning problem. Journal of Civil Engineering and Management (accepted, 2013b)
587. Xu, J., Tu, Y., Zeng, Z.: Bilevel optimization of regional water resources allocation problem under fuzzy random environment. Journal of Water Resources Planning and Management 139(3), 246–264 (2013c)
588. Xu, J., Tu, Y., Zeng, Z.: A nonlinear multiobjective bilevel model for minimum cost network flow problem in a large-scale construction project. Mathematical Problems in Engineering Article ID 463976, 40 (2012)

589. Xu, J., Tao, Z.: Rough Multiple Objective Decision Making. Taylor & Francis Publishers (2011)
590. Xu, J., Feng, C.: Two-stage based dynamic earth-rock transportation assignment problem under fuzzy random environment to earth-rockfill dam construction. Journal of Civil Engineering and Management (accepted, 2013)
591. Teodor, G.C., Nicoletta, R., Giovanni, S.: Advanced freight transportation systems for congested urbanareas. Transportation Research Part C: Emerging Technologies 12(2), 119–137 (2004)
592. Wang, J.F., Lu, H.P., Peng, H.: System dynamics model of urban transportation system and its application. Journal of Transportation Systems Engineering and Information Technology 8(3), 83–89 (2008)
593. Ebben, M., van der Zee, D.J., van der Heijdena, M.: Dynamic one-way traffic control in automated transportation systems. Transportation Research Part B: Methodological 38(5), 441–458 (2004)
594. Pasquale, D.M., Giovanni, Q., Domenico, U.: A decision supportsystem for designing new services tailored to citizen profiles in a complex and distributed e-government scenario. Data Knowledge Engineering 67(1), 161–184 (2008)
595. Celik, M., Cebi, S., Kahraman, C., Er, I.D.: Application of axiomatic design and TOPSIS methodologies under fuzzy environment for proposing competitive strategies on Turkish container ports in maritime transportation network. Expert Systems with Applications 36(3), 4541–4557 (2009)
596. Soukhal, A., Oulamara, A., Martineau, P.: Complexity of flowshop scheduling problems with transportation constraints. European Journal of Operational Research 161(1), 32–41 (2005)
597. Hurink, J., Knust, S.: Makespan minimization for flow-shop problems with transportation times and a single robot. Discrete Applied Mathematics 112(1-3), 199–216 (2001)
598. Luathep, P., Sumalee, A., Lam, W.H.K., Li, Z.C., Lo, H.K.: Global optimization method for mixed transportation network design problem: A mixed-integer linear programming approach. Transportation Research Part B: Methodological 45(5), 808–827 (2011)
599. Ali, M.A.M., Sik, Y.H.: Transportation problem: A special case for linear programing problems in mining engineering. International Journal of Mining Science and Technology 22(3), 371–377 (2012)
600. Maher, M.J., Zhang, X.Y., Vliet, D.V.: A bi-level programming approach for trip matrix estimation and traffic control problems with stochastic user equilibrium link flows. Transportation Research Part B: Methodological 35(1), 23–40 (2001)
601. Cao, S.H., Yuan, Z.Z., Li, Y.H., Wu, X.Y.: Model for road network stochastic user equilibrium based on bi-level programming under the action of the traffic flow guidance system. Journal of Transportation Systems Engineering and Information Technology 7(4), 36–42 (2007)
602. Islam, S., Roy, T.K.: A new fuzzy multi-objective programming: Entropy based geometric programming and its application of transportation problems. European Journal of Operational Research 173(2), 387–404 (2006)
603. EI-Wahed, W.F.A., Lee, S.M.: Interactive fuzzy goal programming for multi-objective transportation problems. Omega 34(2), 158–166 (2006)
604. Ai, T.J., Kachitvichyanukul, V.: A particle swarm optimization for vehicle routing problem with time windows. International Journal of Operational Research 6(4), 519–537 (2009)
605. Gen, M., Cheng, R., Lin, L.: Network Models and Optimization: Multiobjective Genetic Algorithm Approach. Springer, Berlin (2008)

# Index

Printed in the United States
By Bookmasters